AF352869

New Science Based Concepts for Increased Efficiency in Battery Recycling 2020

New Science Based Concepts for Increased Efficiency in Battery Recycling 2020

Editor

Bernd Friedrich

MDPI • Basel • Beijing • Wuhan • Barcelona • Belgrade • Manchester • Tokyo • Cluj • Tianjin

Editor
Bernd Friedrich
IME Process Metallurgy and
Metal Recycling, RWTH
Aachen University
Aachen, Germany

Editorial Office
MDPI
St. Alban-Anlage 66
4052 Basel, Switzerland

This is a reprint of articles from the Special Issue published online in the open access journal *Metals* (ISSN 2075-4701) (available at: https://www.mdpi.com/journal/metals/special_issues/battery_recycling).

For citation purposes, cite each article independently as indicated on the article page online and as indicated below:

LastName, A.A.; LastName, B.B.; LastName, C.C. Article Title. *Journal Name* **Year**, *Volume Number*, Page Range.

ISBN 978-3-0365-5925-4 (Hbk)
ISBN 978-3-0365-5926-1 (PDF)

Cover image courtesy of IME Process Metallurgy and Metal Recycling Department, RWTH Aachen University

Contents

About the Editor

Bernd Friedrich

Prof. Dr. Bernd Friedrich is a full professor of process metallurgy and metal recycling. He has been the head of the corresponding IME institute at RWTH Aachen University since 1 July 1999, and he received an honorary doctorate from the Donetsk National Technical University in 2011.

His fields of research include: the eco-friendly recycling of metals; circular economy contributions; pyro- and hydrometallurgy; mineral waste conditioning and utilization; applied electrochemistry and molten salt electrolysis; vacuum metallurgy; the synthesis of multi-metal alloys; metal purification; and (nano)-powder synthesis.

His activities in working parties and expert committees include:

- GDMB—leader of special metals committee (2004–2010);
- Dean and vice-dean at the Faculty of Georesources and Materials Engineering and senator at RWTH Aachen (2004–2012);
- Advisory board member ACCESS e. V. Aachen (since 2010);
- Founding member and spokesman of aec e.V. and AMAP GmbH (open innovation cluster advanced Metals and Processes);
- Founder and spokesman of CCE (Center of Circular Economy) at RWTH Aachen University.

metals

Editorial

New Science Based Concepts for Increased Efficiency in Battery Recycling

Bernd Friedrich * and Lilian Schwich

IME, Institute for Process Metallurgy and Metal Recycling, RWTH Aachen University, 52056 Aachen, Germany; lschwich@ime-aachen.de

* Correspondence: bfriedrich@metallurgie.rwth-aachen.de

Citation: Friedrich, B.; Schwich, L. New Science Based Concepts for Increased Efficiency in Battery Recycling. *Metals* **2021**, *11*, 533. https://doi.org/10.3390/met11040533

Received: 19 March 2021
Accepted: 23 March 2021
Published: 25 March 2021

Publisher's Note: MDPI stays neutral with regard to jurisdictional claims in published maps and institutional affiliations.

It is a common understanding worldwide that electromobility will have a significant share in passenger transport and that there will be a very dynamic increase in the return volumes of discarded batteries in the future. Whilst, currently, recycling is in the hands of a few and mostly small companies, large companies are increasingly preparing for the circular economy scenario of electric vehicles. This requires robust, safe and efficient processes, on which many research centers are currently working in the international environment. At present, there is no preferred concept for the processing of battery scrap, not to mention any standardization or norming. Politically, the EU specification of a 50% weight-based recovery rate (recycling efficiency, RE) based on cell level is causing widespread discussion. On one hand, such a requirement seems low, but taking into account that 15% and 20% weight shares of electrolytes and graphite, respectively, have not been recoverable to date, this RE alone already poses challenges for many companies. The overall question therefore is whether the holistic weight-based RE is the right way to go, or if element-based quotas for Ni, Co, Cu, Li, etc. are clearly the more target-oriented way [1]. This request has been partly implemented recently by the EU in the new draft of the Battery Recycling Directive [2], where, exemplarily, Li is addressed, with a target of 70% by 2030.

Regardless of this discussion, the recovering of technology elements from Li-based batteries requires mechanical and metallurgical processes in combination. Many options for treating discarded batteries are currently being discussed and investigated. Three exemplary recycling process pathways, A, B, C, are shown as a modular scheme, as the following figure (Figure 1) simplifies. These three process options are already realized at an industrial or at least at a pilot scale and comprise different approaches regarding elements recovered and modules selected.

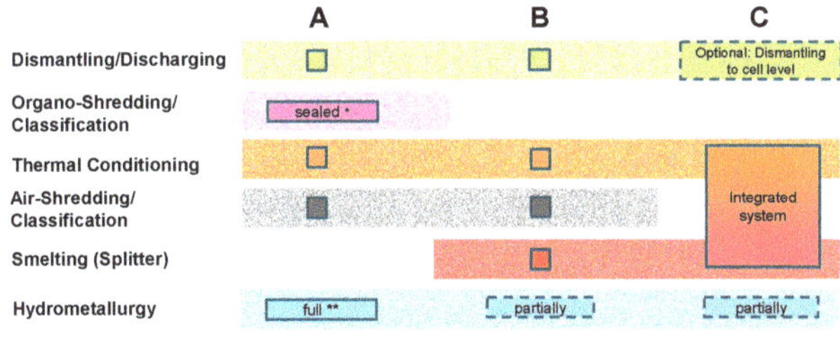

Figure 1. Options and flexibility of battery recycling routes indicating three already industrially applied process paths: **A** (inert shredding and separation before dedicated chemical processing), **B** (thermal conditioning and mechanical separation prior to large-scale production), **C** (direct smelting without pre-conditioning), based on [1].

It can be seen that different process modules are in place, which can be incorporated into a recycling process. The resulting process A combines inert shredding and mechanical comminution and separation. Thermal conditioning is performed afterwards, whereas process B starts with a thermal treatment and a mechanical comminution and classification is then performed in atmosphere before entering large-scale units. Process C can be inserted into a pyrometallurgical unit with or without dismantling and discharging. Hence, the selection of such modules starting from dismantling/discharging up to hydrometallurgy forms recycling routes A, B, C. Each process path entails specific benefits and drawbacks, for example, based on energy input, eco-footprint or recoverable elements and components. In this scheme, the focus is not set on the economic viability of the main points, but on providing options or scenarios with the focus on the respective technological strengths and weaknesses. Depending on the path taken, the components of a cell can be converted into commodity products, and even raw materials for a closed circle to batteries, or environmentally sensitive substances; Figure 2 is helpful to illustrate the options for obtainable recycling products.

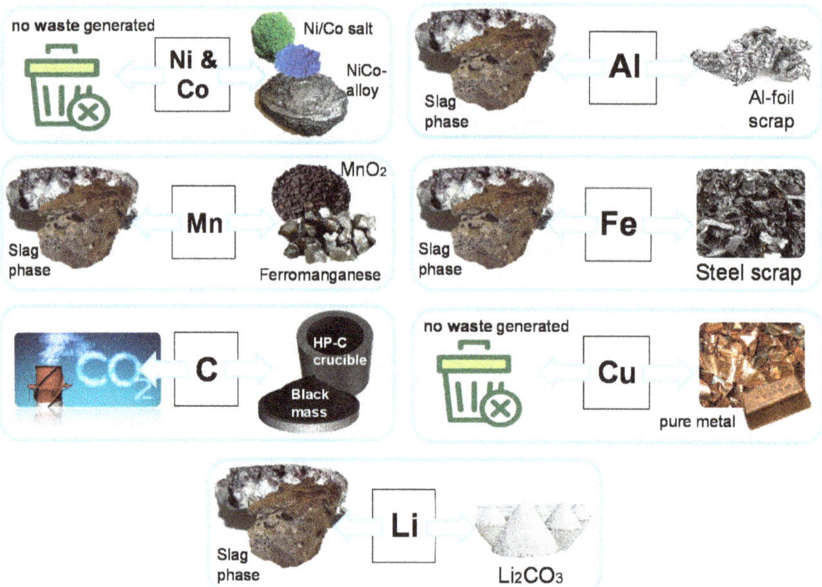

Figure 2. Alternatives of obtained recycling chain products depending on the process paths selected, based on [3]. Here, HP-C refers to high-purity graphite materials used for smelting crucibles.

Noble metals, such as copper (Cu), nickel (Ni) and cobalt (Co), are recovered generally as a marketable and profitable product, independent from the process modules selected. These metals comprise the highest value within a battery, which is why their recovery is a crucial goal in all recycling paths. However, losses in by-products can hardly be avoided since a yield of 100% is practically unrealistic, nevertheless, research always focuses on minimizing process-related losses. Moreover, iron (Fe), represented as steel casings or LFP-cathode systems, and manganese (Mn), represented as an alloying element in steel casings or NMC-cathode systems, are comparatively ignoble metals. Aiming for near zero-waste recycling, Fe and Mn are to be considered as potential products, as well. The mineral phase resulting as slag from pyrometallurgical operations can be transferred to the construction sector, but an elemental recovery of Mn and Fe is not realized in most scenarios. Aluminum (Al) from casing and foils can, similarly to Fe and Mn, be transferred to a slag in a direct smelting route or recovered to a high degree by advanced mechanical processes, like dismantling, shredding and classification. The graphite in the battery's anodes is a critical element according to the European Union [4], hence, its recovery must play an important role in terms of resource efficiency [5]. Using it as a reducing agent in pyrometallurgical operations is technically possible [6,7] but questionable, since a carbothermic smelting process is under critical view, when aiming at a carbon-free industry approach. Hence, dedicated recovery steps are required. Lithium (Li) represents the mainstay of proven and all innovative battery systems, and its recovery always requires hydrometallurgical operations. Lithium carbonate, but also lithium hydroxide, are marketable products, whose production in recycling ensures the principle of a circular economy.

The obtainable products correspond to a wide range of processes alternatives, which has already been outlined by the different process combinations A, B, C. The established large-scale operations for comparable commodity materials (outer ring) are certainly the most economic ones and, thus, it should be a battery recycling process target to reach an entry point into these production levels as fast as possible. As a matter of fact, the available options are highly diversified, leading to more recycling process alternatives. Figure 3 visualizes this by showing most of the battery recycling options as a modular wheel.

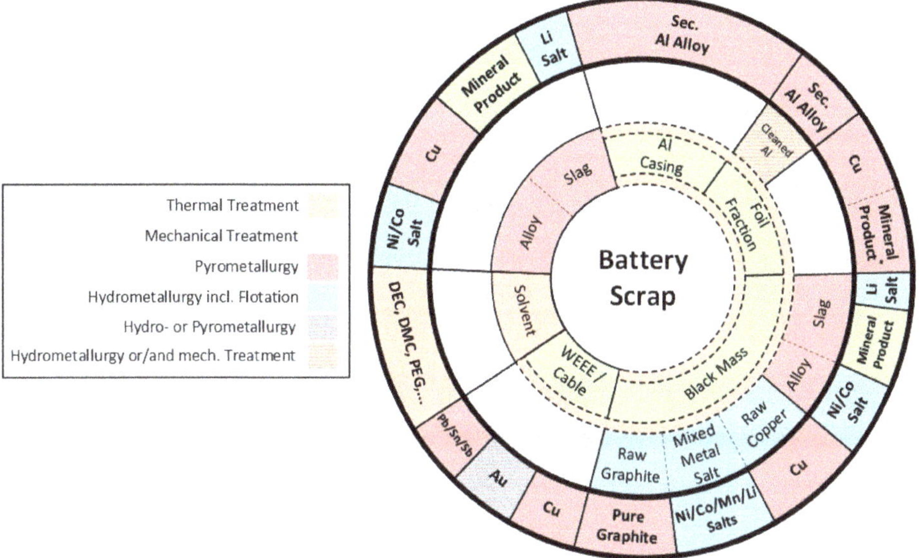

Figure 3. Entry points for recycling products in current large-scale metal production value chains, based on different process modules, based on [8]. Here, the outer ring represents established large-scale operations for comparable commodity materials, also indicated as base material production. The middle ring stands for an advanced second stage in dedicated battery recycling, whereas the inner ring stands for the first process stage in dedicated battery recycling.

This wheel can be read starting from the center with end-of-life batteries, passing two rings of first and possibly second dedicated battery recycling steps into the commodity material production systems (outer ring). Hence, the two inner rings refer to battery-specific recycling facilities and unit operations, and the outer ring represents existing, large-scale metallurgical or chemical industrial production facilities. Obtained products from dedicated battery recycling facilities are channeled into existing large-scale operations for new product generation. Battery scrap, which can be either cell or module based, is treated in process modules, which obtain a specific product fraction. For example, Al casing is recoverable by the module "mechanical treatment", possibly in combination with the module "thermal pre-treatment" before or even after the mechanical process. This Al casing product is then directly transferred to existing, non-battery-specific aluminum production facilities, where it is transformed to a secondary aluminum alloy by the module "pyrometallurgy".

It is to be pointed out that the entry points of the outer ring are defined in terms of impurities by commodity material producers. Thereby, meeting these demands on battery recycling products is a pre-requisite for generating marketable products. The channeling of battery recycling products into existing infrastructure for primary production is an important tool for near zero-waste recycling. However, with all process chain optimization, it must always be kept in mind that metal losses can also increase with each process module added and that the revenues from the newly recovered products may be offset by the costs of these required additional steps. Additionally, last not but least, all recycling operations must be robust against changing battery chemistries [9,10] and impurities deriving from sorting failures or insufficient orderliness and cleanliness in the plant. One simple example is the future Si-based battery which shows a strong tendency to disable hydrometallurgical operations by gelation risk.

Based on 19 high-quality articles, this Special Issue presents methods for further improving the currently achievable recycling rate, product quality in terms of focused elements and moreover, approaches for the enhanced mobilization of lithium, graphite and electrolyte components. In particular, the target of early stage Li removal is a central point of various research approaches in the world, which has been reported, for example, under the names early stage lithium recovery (ESLR) [11] or CO_2 leaching (COOL) [12]. These processes are a strongly focusing on environmentally friendly lithium mobilization before entering pyrometallurgy or conventional hydrometallurgy. Figure 4 simplifies the effect of this approach.

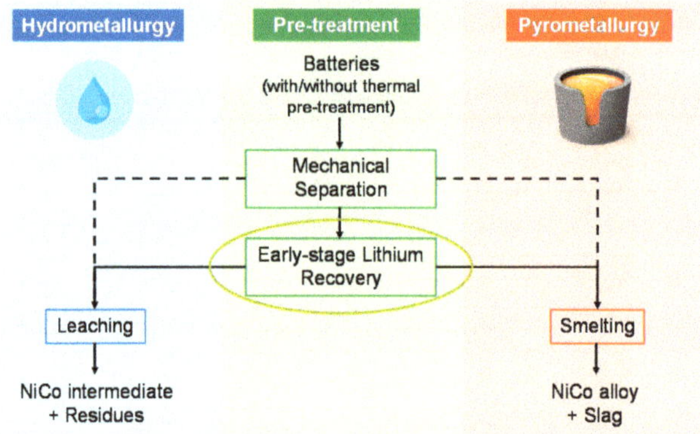

Figure 4. Early stage Li recovery (ESLR) process scheme, based on [11,13].

It has to be pointed out that early stage Li recovery is a tool which can be incorporated into all existing recycling paths, hence, it is an effective add-on for pursuing a high recycling efficiency for this critical element. Currently, the process is investigated by using (thermally treated) black mass [11], but directly shredded material may also be a practical option for future research.

Besides the topic of environmentally friendly lithium mobilization, many more approaches are present in this Special Issue, starting with robotic disassembly and dismantling of Li-ion batteries [14,15]. Moreover, the optimization of various pyro- and hydrometallurgical as well as combined battery recycling processes for the treatment of conventional Li-ion batteries, up to an evaluation of the recycling on an industrial level, and different battery recycling topics, are addressed as well. The recovery of lithium by innovative methods comes to the fore as an important component. In addition to the consideration of the Li distribution in compounds of a Li_2O-MgO-Al_2O_3-SiO_2-CaO system, the Li recovery from battery slags is also discussed. The development of suitable recycling strategies for new battery systems, such as all-solid-state batteries, but also lithium–sulfur batteries, are also taken into account in this Special Issue. Some articles also discuss the issue that battery recycling processes do not have to produce end products such as high-purity battery materials, but that they should be aimed at finding an "entry point" into existing proven large-scale industrial processes where marketable product generation is possible and cost-efficient (referring to the discussion around Figure 3).

The contributions of this issue are structured according to their research areas, as can be seen in Table 1.

Table 1. Published articles in this Special Issue, "New Science Based Concepts for Increased Efficiency in Battery Recycling" sorted by research field and given as sources in the References.

Research Field	Source in Special Issue
Dismantling	[14,15]
Shredding/Separation	[5,16]
Thermal Conditioning	[11]
Smelting	[6,7,17–19]
Hydrometallurgy/Chem. Processing	[9,10,12,20–22]
Reviews	[23–25]

Participants in this Special Issue originate from 18 research institutions from eight countries. We would hereby like to thank all of them for the high-quality research and reviews including the evaluation and derivation of recommendations for future work.

Funding: This research received no external funding.

Conflicts of Interest: The authors declare no conflict of interest.

References

1. Friedrich, B.; Schwich, L. Impact of process chain design on metal recycling efficiency—A critical view on the actual total weight-based target. In Proceedings of the 24th International Congress for Battery Recycling ICBR, Lyon, France, 18–20 September 2019. Available online: https://www.researchgate.net/publication/335889260_Impact_of_process_chain_design_on_metal_recycling_efficiency_-A_critical_view_on_the_actual_total_weight-based_target (accessed on 16 March 2021). [CrossRef]
2. European Commission. Batteries—Modernising EU Rules. Proposal for a Regulation. 2020. Available online: https://ec.europa.eu/info/law/better-regulation/have-your-say/initiatives/12399-Modernising-the-EU-s-batteries-legislation (accessed on 17 March 2021).
3. Friedrich, B.; Schwich, L. Status and Trends of industrialized Li-Ion battery recycling processes with qualitative comparison of economic and environmental impacts. In Proceedings of the 22nd International Congress for Battery Recycling ICBR, Lisbon, Portugal, 20–22 September 2017. Available online: https://www.researchgate.net/publication/319964237_Status_and_Trends_of_industrialized_Li-Ion_battery_recycling_processes_with_qualitative_comparison_of_economic_and_environmental_impacts (accessed on 16 March 2021). [CrossRef]
4. European Commission. Communication from the Commission to the European Parliament, the Council, the European Economic and Social Committee and the Committee of the Regions. On the 2017 List of Critical Raw Materials for the EU. 2017. Available online: https://ec.europa.eu/transparency/regdoc/rep/1/2017/EN/COM-2017-490-F1-EN-MAIN-PART-1.PDF (accessed on 17 March 2021).
5. Ruismäki, R.; Rinne, T.; Dańczak, A.; Taskinen, P.; Serna-Guerrero, R.; Jokilaakso, A. Integrating Flotation and Pyrometallurgy for Recovering Graphite and Valuable Metals from Battery Scrap. *Metals* **2020**, *10*, 680. [CrossRef]
6. Sommerfeld, M.; Vonderstein, C.; Dertmann, C.; Klimko, J.; Oráč, D.; Miškufová, A.; Havlík, T.; Friedrich, B. A Combined Pyro- and Hydrometallurgical Approach to Recycle Pyrolyzed Lithium-Ion Battery Black Mass Part 1: Production of Lithium Concentrates in an Electric Arc Furnace. *Metals* **2020**, *10*, 1069. [CrossRef]
7. Holzer, A.; Windisch-Kern, S.; Ponak, C.; Raupenstrauch, H. A Novel Pyrometallurgical Recycling Process for Lithium-Ion Batteries and Its Application to the Recycling of LCO and LFP. *Metals* **2021**, *11*, 149. [CrossRef]
8. Friedrich, B.; Sabarny, P.; Stallmeister, C. Process Flow Alternatives for LIB Recycling. In *Berliner Recycling- und Sekundärrohstoffkonferenz*; Berlin, Germany, 2021. Available online: https://www.researchgate.net/publication/350071276_Process_Flow_Alternatives_for_LIB_Recycling (accessed on 24 March 2021). [CrossRef]
9. Schwich, L.; Küpers, M.; Finsterbusch, M.; Schreiber, A.; Fattakhova-Rohlfing, D.; Guillon, O.; Friedrich, B. Recycling Strategies for Ceramic All-Solid-State Batteries—Part I: Study on Possible Treatments in Contrast to Li-Ion Battery Recycling. *Metals* **2020**, *10*, 1523. [CrossRef]

10. Schwich, L.; Sabarny, P.; Friedrich, B. Recycling Potential of Lithium–Sulfur Batteries—A First Concept Using Thermal and Hydrometallurgical Methods. *Metals* **2020**, *10*, 1513. [CrossRef]
11. Schwich, L.; Schubert, T.; Friedrich, B. Early-Stage Recovery of Lithium from Tailored Thermal Conditioned Black Mass Part I: Mobilizing Lithium via Supercritical CO2-Carbonation. *Metals* **2021**, *11*, 177. [CrossRef]
12. Pavón, S.; Kaiser, D.; Mende, R.; Bertau, M. The COOL-Process—A Selective Approach for Recycling Lithium Batteries. *Metals* **2021**, *11*, 259. [CrossRef]
13. Stallmeister, C.; Schwich, L.; Friedrich, B. Early-Stage Li-Removal—Vermeidung von Lithiumverlusten im Zuge der Thermischen und Chemischen Recyclingrouten von Batterien. In *Recycling und Rohstoffe*; Holm, O., Thomé-Kozmiensky, E., Goldmann, D., Friedrich, B., Eds.; Thomé-Kozmiensky Verlag GmbH: Neuruppin, Germany, 2020; pp. 545–557. ISBN 9783944310510.
14. Marshall, J.; Gastol, D.; Sommerville, R.; Middleton, B.; Goodship, V.; Kendrick, E. Disassembly of Li Ion Cells—Characterization and Safety Considerations of a Recycling Scheme. *Metals* **2020**, *10*, 773. [CrossRef]
15. Choux, M.; Marti Bigorra, E.; Tyapin, I. Task Planner for Robotic Disassembly of Electric Vehicle Battery Pack. *Metals* **2021**, *11*, 387. [CrossRef]
16. Sinn, T.; Flegler, A.; Wolf, A.; Stübinger, T.; Witt, W.; Nirschl, H.; Gleiß, M. Investigation of Centrifugal Fractionation with Time-Dependent Process Parameters as a New Approach Contributing to the Direct Recycling of Lithium-Ion Battery Components. *Metals* **2020**, *10*, 1617. [CrossRef]
17. Schirmer, T.; Qiu, H.; Li, H.; Goldmann, D.; Fischlschweiger, M. Li-Distribution in Compounds of the Li_2O-MgO-Al_2O_3-SiO_2-CaO System—A First Survey. *Metals* **2020**, *10*, 1633. [CrossRef]
18. Wittkowski, A.; Schirmer, T.; Qiu, H.; Goldmann, D.; Fittschen, U.E.A. Speciation of Manganese in a Synthetic Recycling Slag Relevant for Lithium Recycling from Lithium-Ion Batteries. *Metals* **2021**, *11*, 188. [CrossRef]
19. Li, Y.; Yang, S.; Taskinen, P.; Chen, Y.; Tang, C.; Jokilaakso, A. Cleaner Recycling of Spent Lead-Acid Battery Paste and Co-Treatment of Pyrite Cinder via a Reductive Sulfur-Fixing Method for Valuable Metal Recovery and Sulfur Conservation. *Metals* **2019**, *9*, 911. [CrossRef]
20. Vieceli, N.; Reinhardt, N.; Ekberg, C.; Petranikova, M. Optimization of Manganese Recovery from a Solution Based on Lithium-Ion Batteries by Solvent Extraction with D2EHPA. *Metals* **2021**, *11*, 54. [CrossRef]
21. Klimko, J.; Oráč, D.; Miškufová, A.; Vonderstein, C.; Dertmann, C.; Sommerfeld, M.; Friedrich, B.; Havlík, T. A Combined Pyro- and Hydrometallurgical Approach to Recycle Pyrolyzed Lithium-Ion Battery Black Mass Part 2: Lithium Recovery from Li Enriched Slag—Thermodynamic Study, Kinetic Study, and Dry Digestion. *Metals* **2020**, *10*, 1558. [CrossRef]
22. Gerold, E.; Luidold, S.; Antrekowitsch, H. Selective Precipitation of Metal Oxalates from Lithium Ion Battery Leach Solutions. *Metals* **2020**, *10*, 1435. [CrossRef]
23. Doose, S.; Mayer, J.K.; Michalowski, P.; Kwade, A. Challenges in Ecofriendly Battery Recycling and Closed Material Cycles: A Perspective on Future Lithium Battery Generations. *Metals* **2021**, *11*, 291. [CrossRef]
24. Werner, D.; Peuker, U.A.; Mütze, T. Recycling Chain for Spent Lithium-Ion Batteries. *Metals* **2020**, *10*, 316. [CrossRef]
25. Brückner, L.; Frank, J.; Elwert, T. Industrial Recycling of Lithium-Ion Batteries—A Critical Review of Metallurgical Process Routes. *Metals* **2020**, *10*, 1107. [CrossRef]

 *metals*

Article

Cleaner Recycling of Spent Lead-Acid Battery Paste and Co-Treatment of Pyrite Cinder via a Reductive Sulfur-Fixing Method for Valuable Metal Recovery and Sulfur Conservation

Yun Li [1,2], Shenghai Yang [1], Pekka Taskinen [2], Yongming Chen [1,*], Chaobo Tang [1,*] and Ari Jokilaakso [1,2,*]

[1] School of Metallurgy and Environment, Central South University, Changsha 410083, China
[2] Department of Chemical and Metallurgical Engineering, School of Chemical Engineering, Aalto University, P.O. Box 16100, 02150 Espoo, Finland
* Correspondence: csuchenyongming@163.com (Y.C.); chaobotang@163.com (C.T.); ari.jokilaakso@aalto.fi (A.J.); Tel.: +86-731-8883-0470 (C.T.); +358-50-3138-885 (A.J.); Fax: +86-731-8883-0470 (Y.C.)

Received: 28 July 2019; Accepted: 16 August 2019; Published: 20 August 2019

Abstract: This study proposes a cleaner lead-acid battery (LAB) paste and pyrite cinder (PyC) recycling method without excessive generation of SO_2. PyCs were employed as sulfur-fixing reagents to conserve sulfur as condensed sulfides, which prevented SO_2 emissions. In this work, the phase transformation mechanisms in a $PbSO_4$-Na_2CO_3-Fe_3O_4-C reaction system were studied in detail. Furthermore, the co-treatment of spent LAB and PyCs was conducted to determine the optimal recycling conditions and to detect the influences of different processing parameters on lead recovery and sulfur fixation. In addition, a bench-scale experiment was carried out to confirm the feasibility and reliability of this novel process. The results reveal that the products were separated into three distinct layers: slag, ferrous matte, and crude lead. 98.3% of lead and 99% of silver in the feed materials were directly enriched in crude lead. Crude lead with purity of more than 98 wt.% (weight percent) was obtained by a one-step extraction. Lead contents in the produced matte and slag were below 2.7 wt.% and 0.6 wt.%, respectively. At the same time, 99.2% total sulfur was fixed and recovered.

Keywords: lead-acid battery recycling; pyrite cinder treatment; lead bullion; sulfide matte; SO_2 emissions; pilot plant

---

1. Introduction

Lead-acid batteries (LABs) are widely applied in automobiles and electric bicycles. Recently, advances in energy, transportation, and telecommunication industries have increasingly expanded its demand, and its scrap volume grows worldwide [1]. It has been estimated that the number of spent LABs would be multiplied annually based on the mean lifetime of 2–3 years and will continue to grow, especially in China and other emerging economies [2]. As a result, it has become the most significant secondary lead source worldwide [3]. About 80–85% of total secondary lead is recycled from lead paste [4]. Typically, a spent LAB consists of four main components: lead paste (30–40%), lead alloy grids (24–30%), polymeric materials (22–30%), and waste electrolytes (11–30%) [5,6]. Of these, lead paste is the most difficult to deal with [7].

Spent LABs are of great environmental concern [8,9] because of the toxicity of lead [10]. It will be a serious problem to environmental protection and human health if disposed and treated improperly [11]. In many countries, spent LAB is classified as a hazardous waste. In China, plants without a certificate for hazardous waste treatment are not allowed to handle or process spent LABs [10,12]. Currently, pyrometallurgy is the predominant methodology for LAB recycling worldwide [13]; spent lead paste is recovered as metallic lead through an energy-intensive decomposition process using traditional pyrometallurgical processes [14]. Reverberatory furnaces, shaft furnaces, electric furnaces, rotary furnaces [15], and bottom blowing methods [16] are usually selected to smelt lead-containing residues and wastes [15]. Although the traditional recycling technology of spent LABs through smelting is relatively mature, occasional and serious pollution is still possible [2], for example, in form of lead particulates, discharge of toxic and unstable sludge, slag, and wastewater [14]. Since 2009, there has been more than 10 major lead poisoning accidents recorded in China, and nearly 4000 children have been affected by these accidents [10].

Some emerging technologies [9,17,18], including electrowinning, solid phase electrolysis, biological technique, vacuum methods [19], and so on, are being developed and applied. However, in most methods, some problems still exist and need to be solved (e.g., sulfur footprint removal; lead-containing dust, industrial SO_2 emissions and pollution in pyrometallurgy [20,21]; tedious procedures; large amounts of generated unmanageable waste water; and high electricity consumption [22] in hydrometallurgy).

In addition, pyrite cinder (PyC), another hazardous waste, is produced in large quantities in industrial sulfuric acid manufacture. Only in China, the production of PyCs annually exceeds 12 million tons [23]. It often contains toxic metals, including Pb, Cd, As, and so on, as well as appreciable quantities of valuable metals such as Cu, Co, In, Au, and Ag [24]. The recycling or harmless treatment of PyCs is a huge challenge, even in developed countries. The recycling status of PyCs is not encouraging: the recycling rate in USA is about 80%–85% and about 70%–80% in Japan. China recycles only around 50% of PyCs, and large quantities of PyCs are simply landfilled or dumped without any treatment [23]. A limited amount of PyCs is recycled to produce cement, bricks, iron oxide pigments, or iron sponges [23,24]. This leads to serious consequences including occupation of considerable land resources, dust problems, and contamination of soil and ground water [24].

In view of environmental regulations, treatment costs, and limited availability of landfill/disposal sites, the search for new and cost-effective practices for the recycling of LABs and, in recent years, valorization of PyCs has become increasingly important [10]. In this article a novel, integrated resourcing treatment method of LABs and PyCs, a reductive sulfur-fixing recycling process [25,26], was proposed to recover valuable metals, recycle iron and sulfur values, as well as to co-treat hazardous materials. The novelties of this process are SO_2-free generation, sulfur fixation and conservation, a much shorter flowsheet, absence of harmful by-products, and a wide adaptability for different secondary materials. This work investigated the experimental feasibility and reliability of this novel process. The fundamental phase conversion mechanisms and reaction paths in the treatment were further examined.

2. Experimental Parameters

2.1. Materials

The materials employed in the reaction mechanism investigations included $PbSO_4$, Na_2CO_3, Fe_3O_4, and carbon powder. They were of analytical grade (≥99.9 wt.%) and purchased from Aladin Industrial Corporation, China. Nitrogen with a purity of 99% was applied as a protective gas. LAB paste, pyrite cinder, and metallurgical coke were adopted in the batch processing tests. The lead paste used was separated from scrap LABs and supplied by Yuguang Gold & Lead Co., Henan, China. The coke and sulfur-fixing agent, pyrite cinder, were supplied by Jiuquan Iron & Steel Co., Ltd., Gansu, China. Their chemical compositions were analyzed by inductively coupled plasma-atomic emission spectrometry (ICP-AES, Perkin Elmer, Optima 3000, Norwalk, CT, USA). The assays are shown in Table 1. The concentration of the various metal oxides in LAB and PyC and phase compositions of lead in LAB paste were determined by EDTA (Ethylene Diamine Tetracetic Acid) titration. The contents of PbO, PbO_2, and metallic Pb components presented in Table 1 were determined by dissolving each individual component in turn into solution, followed by EDTA titration. Firstly, lead paste was added to 5% (mass fraction) acetum to dissolve the PbO component. Secondly, the residue was moved and further dissolved in HNO_3 solution (1 + 1 volume fraction) to separate metallic Pb. Then, the remaining residue was dissolved in the mixed solution of 1 + 1 HNO_3 and 1 + 40 H_2O_2 to separate PbO_2. Thus, the contents of PbO, PbO_2, and metallic Pb can be titrated in the corresponding solution by means of EDTA (0.05 mol/L) titration. The $PbSO_4$ component in lead paste was determined by sulfur concentrate. It suggests that the lead paste comprised 54.7 wt.% $PbSO_4$, 22.1 wt.% PbO_2, 8.5 wt.% Pb, and 8.5 wt.% PbO.

Table 1. Chemical composition of the corresponding materials (wt.%), * g/t.

Materials	Pb	S	Fe	SiO$_2$	CaO	Na	Mg	Al	Ba	Sb
Lead paste	72.90	5.77	0.02	5.48	0.22	0.33	0.06	0.03	0.14	0.09
	PbSO$_4$	PbO	PbO$_2$	Pb	-	-	-	-	-	-
	54.68	8.49	22.05	8.53	-	-	-	-	-	-
Pyrite cinder	Pb	S	Fe	SiO$_2$	CaO	Na	MgO	Al$_2$O$_3$	Cu	Ag*
	0.02	1.31	58.23	9.68	1.02	0.01	0.19	2.08	0.05	324
Coke	Industrial analysis (%)				Chemical composition of the ash (%)					
	Fixed Carbon		Volatile		Ash	Fe$_{total}$	SiO$_2$	Al$_2$O$_3$	CaO	MgO
	84.05		1.13		13.94	15.96	30.96	18.18	4.05	1.45

2.2. Apparatus

The equipment used in this investigation is shown in Figure 1. Equipment included a horizontal tube furnace equipped with a gas controller and temperature controller (SR3-8P-N-90-100Z, SHIMADEN Co., Ltd., Tokyo, Japan, accuracy ±1 °C) and two S-type thermocouples. Silicon carbide (Si-C) heating elements were used to heat the furnace. An alumina crucible was used to carry the reaction mixture. The solid products were quenched in liquid N_2. The tail gas was absorbed by a NaOH solution.

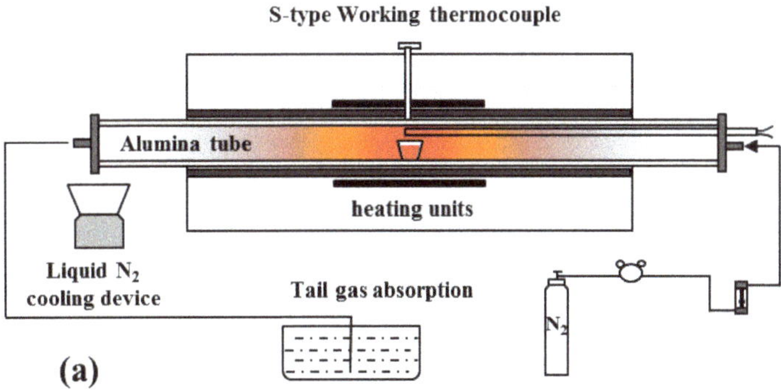

Figure 1. *Cont.*

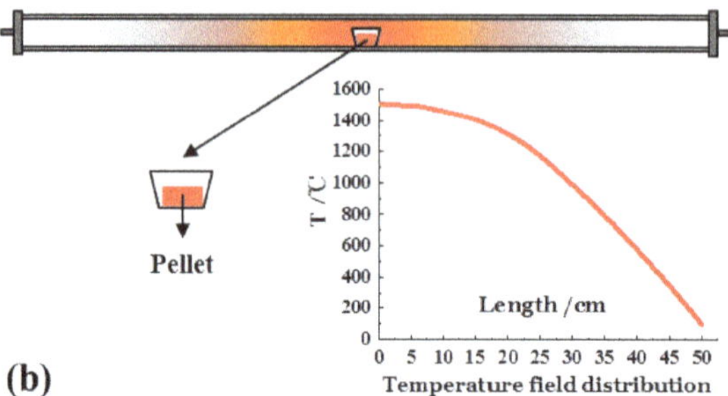

Figure 1. (**a**) Schematic of the experimental apparatus and (**b**) temperature profile of the furnace.

2.3. Procedure

When investigating the reaction mechanism, pure $PbSO_4$ was selected as the model compound because it is the major and most difficult component to deal with in lead paste. In order to reveal the reaction paths, the $PbSO_4$-Fe_3O_4-Na_2CO_3-C reaction mixture was synthesized using pure chemicals according to their potential reaction stoichiometry. Five grams of $PbSO_4$ was employed in each synthesized specimen. The specimens were pressed uniaxially under 15 MPa to obtain cylindrical samples 10 mm in diameter, loaded into the alumina crucible, and pushed into the constant temperature zone of the furnace at the desired temperature for 5, 10, 15, 20, or 30 min. The protective N_2 gas flow rate was fixed at 0.5 L/min. After the preset reaction time, the product was taken out rapidly and quenched in liquid nitrogen. The phase compositions were characterized by XRD (D/max 2550PC, Rigaku Co., Ltd., Tokyo, Japan) with Cu-Ka radiation. The XRD data were collected in the range of $2\theta = 10°$–$80°$ with a 2θ step width of $1°$. The recorded spectra were evaluated by comparison with entries from the PDF-2 database [27].

In the batch experiments of real lead paste, 200 g of lead paste was mixed evenly with a given amount of coke, pyrite cinder, Na_2CO_3, and other fluxes (CaO and SiO_2) for every test. The mixture was placed into an alumina crucible and positioned in the furnace. After the required treatment time, the crucible was rapidly quenched in liquid nitrogen and weighed. Next, the crucible was broken to carefully separate and weigh crude lead, ferrous matte, and slag. Each sample was well prepared for ICP-AES analysis. The microstructures of matte and slag samples were characterized by a scanning electron microscope and energy diffraction spectrum (SEM-EDS, Carl Zeiss LEO 1450, Oberkochen, Germany; EDS, INCA Wave 8570, Oxford Instruments, Oxford, UK). The direct Pb recovery rate (η) and sulfur-fixing rate (γ) were calculated based on following Equations (1) and (2), respectively:

$$\eta = \frac{Mass\ of\ Pb\ in\ the\ crude\ lead}{Mass\ of\ Pb\ in\ the\ initial\ feed\ materials} \times 100\%; \tag{1}$$

$$\gamma = \frac{Mass\ of\ sulfur\ in\ the\ ferrous\ matte\ and\ slag}{Mass\ of\ sulfur\ in\ the\ inital\ feed\ materials} \times 100\% . \tag{2}$$

3. Results and Discussion

3.1. Phase Transformation Mechanisms

The XRD patterns of the $PbSO_4$-Na_2CO_3-Fe_3O_4-C reaction system are presented in Figure 2. It is observed from Figure 2a that exchange reactions between $PbSO_4$ and Na_2CO_3 occurred above 500 °C. At 650 °C, $xPbO·PbSO_4$ (x can be 1, 2, or 4), PbO, and $Na_4CO_3SO_4$ were detected. This indicates that the reactions between $PbSO_4$ and Na_2CO_3 were a multistage process. PbO was generated from $PbSO_4$ in the presence of Na_2CO_3 through the following sequence: $PbSO_4 \rightarrow PbO·PbSO_4 \rightarrow 2PbO·PbSO_4 \rightarrow 4PbO·PbSO_4 \rightarrow PbO$. A schematic of the reaction path is shown in

Figure 3a. Thus, SO_3 in $PbSO_4$ transferred to Na_2SO_4. With increased temperature, PbS and metallic Pb emerged at 750 °C, and the main products after 30 min reaction were Pb, PbS, PbO, Fe_3O_4, and Na_2SO_4. It indicates that a part of $PbSO_4$ was reduced to PbS. At the same time, PbO was reduced to metallic Pb. When temperature rose to 850 °C, $NaFeS_2$ was generated after 30 min lead time. This suggests that iron oxide was involved in the sulfur-fixation reactions. Sulfur in PbS and Na_2SO_4 was transferred to $NaFeS_2$. Metallic Pb emerged and settled from the reaction mixture as a metal layer. When temperature was increased to 1000 °C, intermediate products—sodium iron oxides, sodium lead oxides, and lead iron oxides—were detected, and as temperature rose from 1100 °C to 1200 °C, their generation grew. However, the main products were Pb, $NaFeS_2$, and FeS.

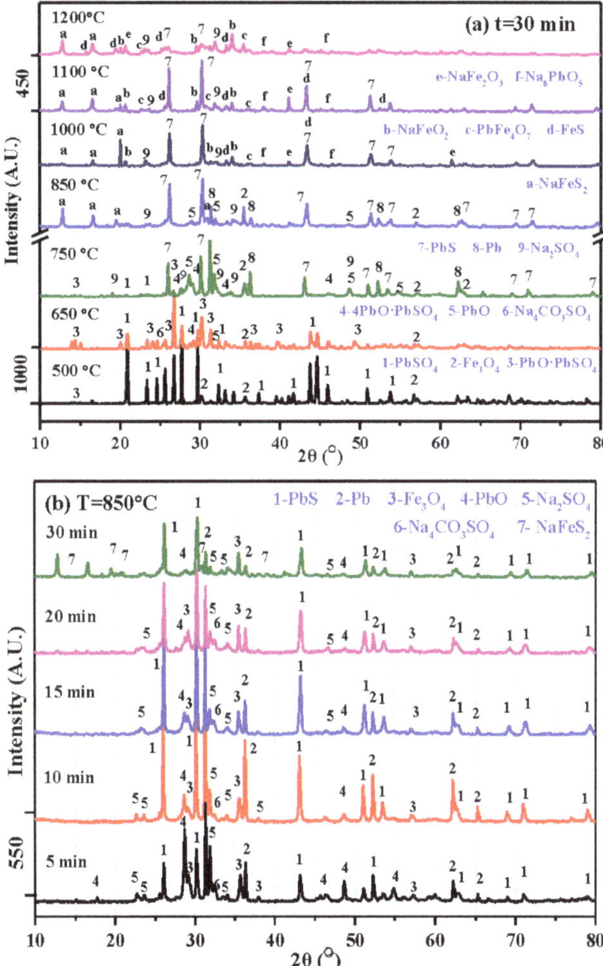

Figure 2. XRD pattern of the $PbSO_4$-Na_2CO_3-Fe_3O_4-C mixture (molar ratio 3:3:1:18) at (**a**) different temperatures and (**b**) reaction times.

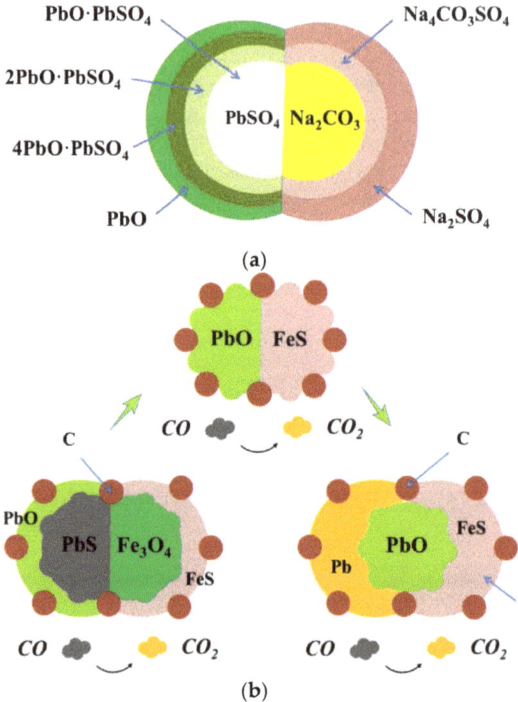

Figure 3. Schematic of (**a**) the multistage process between $PbSO_4$ and Na_2CO_3 and (**b**) lead extraction reactions between PbS and Fe_3O_4.

Figure 2b suggests that Na_2CO_3 first reacted with $PbSO_4$ to convert it to PbO. At the same time, PbO can be reduced to metallic Pb at 850 °C. The unconverted $PbSO_4$ will also be reduced to PbS. After 5 min reaction, the products comprised Pb, PbS, Na_2SO_4, PbO, and Fe_3O_4. As reaction time extended, the diffraction intensity of metallic Pb increased steadily, and that of PbO decreased. Fe_3O_4 did not react with PbS within around the first 20 min but 30 min later when $NaFeS_2$ emerged.

The above results indicate that the presence of Na_2CO_3 and reductant is a critical factor to ensure sufficient sulfur and lead recovery. At low temperatures and weakly reductive atmospheres, as Figure 3a shows, $PbSO_4$ first reacted with Na_2CO_3 to convert to PbO and Na_2SO_4, which helped the decomposition of $PbSO_4$ and ensured that sulfur was conserved in the recycling system in the solid state (as Na_2SO_4) without generating and emitting SO_2 gas. As temperature increased, unconverted $PbSO_4$ was selectively reduced to PbS. Then, the sulfur-fixing agent, Fe_3O_4, reacted with PbS, as shown in Figure 3b. The sulfur in PbS was further immobilized and finally recycled as FeS and $NaFeS_2$.

The entire reaction path can be described as follows, where the Gibbs free energies ΔG_T^θ of the reactions below are calculated by HSC 9.2.6 and its database [28] (unit of ΔG_T^θ is kJ/mol, temperature T is °C):

$$PbSO_4 + Na_2CO_3 = PbO + Na_2SO_4 + CO_{2(g)}; \ \Delta G_T^\theta = -0.149\,T + 7.586; \tag{3}$$

$$C + CO2_{(g)} = CO_{(g)}; \ \Delta G_T^\theta = -0.173\,T + 121.52; \tag{4}$$

$$PbSO_4 + 4CO_{(g)} = PbS + 4CO_{2(g)}; \ \Delta G_T^\theta = -0.681\,T + 166.39; \tag{5}$$

$$3PbS + Fe_3O_4 + 4CO_{(g)} = 3Pb + 3FeS + 4CO_{2(g)}; \ \Delta G_T^\theta = -0.066\,T - 24.284; \tag{6}$$

$$PbS + Na_2CO_3 + CO_{(g)} = Na_2S + Pb + 2CO_{2(g)}; \ \Delta G_T^\theta = -0.132\,T + 128.00; \tag{7}$$

$$PbO + CO_{(g)} = Pb + CO_{2(g)}; \Delta G_T^{\theta} = -0.013\,T - 67.25,$$
$$T \leq 900\,°C;\ \Delta G_T^{\theta} = 0.014\,T - 90.546,\ T \geq 900\,°C; \tag{8}$$

$$Na_2SO_4 + 4CO_{(g)} = Na_2S + 4CO_{2(g)};\ \Delta G_T^{\theta} = 0.034\,T - 127.88; \tag{9}$$

$$4Na_2SO_4 + Fe_2O_3 + 16CO_{(g)} = 2NaFeS_2 + 3Na_2O + 16CO_{2(g)};\ \Delta G_T^{\theta} = 0.207T - 126.14; \tag{10}$$

$$6Na_2SO4 + Fe_3O_4 + 23.5CO_{(g)} = 3NaFeS_2 + 4.5Na_2O + 23.5CO_{2(g)};\ \Delta G_T^{\theta} = 0.339\,T - 164.05; \tag{11}$$

$$2Na_2SO4 + 2FeS + 7CO_{(g)} = 2NaFeS_2 + Na_2O + 7CO_{2(g)};\ \Delta G_T^{\theta} = 0.1884\,T - 122.67. \tag{12}$$

Figure 4 illustrates the ΔG_T^{θ} versus T, Log(pS$_2$) versus T, and Log(pSO$_2$) versus T diagrams of the reaction system. The thermodynamic calculation results agreed well with the experimental reaction mechanism. $PbSO_4 \xrightarrow{Na2CO3} PbO$ and $PbSO_4 \xrightarrow{CO} PbS$ reactions tended to take place firstly at a low temperature. Then, lead was extracted from PbS and PbO with the help of Fe$_3$O$_4$ [29] and Na$_2$CO$_3$ [30]. Sulfur was conserved as FeS, Na$_2$S, and NaFeS$_2$.

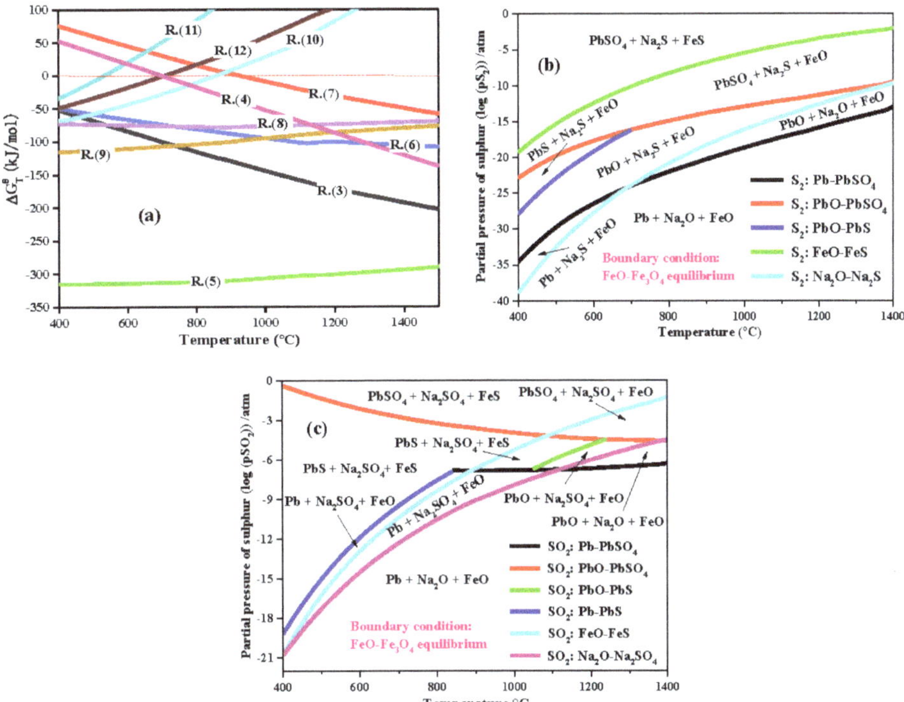

Figure 4. (a) The ΔG_T^{θ} and T diagram, (b) Log(pS$_2$) and T equilibrium diagram, and (c) Log(pSO$_2$) and T equilibrium diagram in the boundary condition of FeO-Fe$_3$O$_4$ equilibria (data from HSC 9.2.6 database).

The reaction systems in Figure 4b,c were defined by oxygen pressures based on the FeO/Fe$_3$O$_4$ boundary conditions. Thus, the activity of elemental carbon, gaseous sulfur polymers, and CO$_{(g)}$ can be calculated from the results obtained, and they will not violate the equilibria by any means. It was found from Figure 4b,c that the Pb, Na$_2$S, and Na$_2$SO$_4$ equilibrium domains formed before FeS. It indicated that Na$_2$CO$_3$ will first react with PbSO$_4$ to extract metallic Pb. The results coincide with previous experimental results. Additionally, the Na$_2$O/Na$_2$SO$_4$ equilibrium generated the lowest SO$_2$

partial pressure (i.e., allows the smallest SO_2 losses to gas). Therefore, it is essential for sulfur (dioxide) emissions to have Na_2CO_3 in the raw materials.

3.2. Batch Experiments of End of Life (EoL) Lead Paste

3.2.1. Influence of Coke Addition

Lab-scale batch experiments using real EoL (end of life) lead battery paste were carried out to investigate the effect of major recycling parameters on direct lead recovery and sulfur fixation efficiency. The results of increasing coke addition are presented in Figure 5. It reveals that the direct Pb recovery and sulfur-fixation rates increased with increasing coke dosage and were stable after 12% coke addition, where 95.9% lead was recovered and enriched in crude lead, and 97.7% sulfur was recovered in the ferrous matte and slag. This indicates that the reductive atmosphere was a significant factor for lead recovery and sulfur fixation. As the results shown in Figure 2 indicate, lead extraction from $PbSO_4$, PbS, and PbO, as well as sulfur conversion from $PbSO_4$, PbS, and Na_2SO_4, should rely on carbothermal reduction. A suitable coke addition can notably improve lead recovery and sulfur fixation according to Equations (1) and (2), respectively.

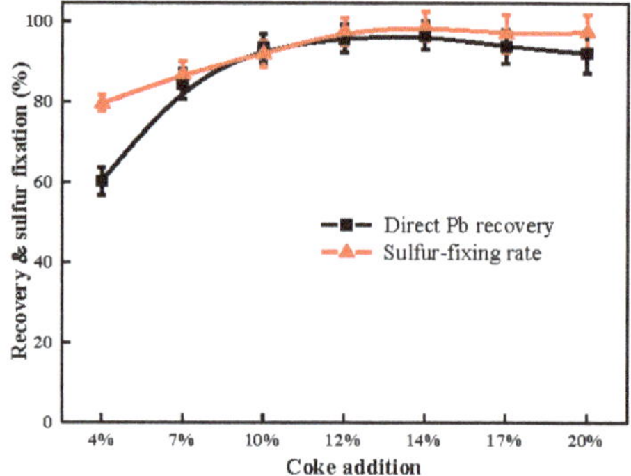

Figure 5. The effects of coke dosage on lead recovery and sulfur fixation. ($W_{\text{lead paste}}$:W_{Na2CO3}:$W_{\text{pyrite cinder}}$ = 200:16:50 g, FeO/SiO_2 = 1.3, CaO/SiO_2 = 0.4, 1200 °C, 2 h).

3.2.2. Influence of Na_2CO_3 Addition

The effects of Na_2CO_3 addition on direct Pb recovery and sulfur fixation are depicted in Figure 6. It illustrates that lead recovery and sulfur fixation rates gradually increased as Na_2CO_3 dosage increased from 0% to 4%, where 97.0% lead was recovered, and 99.2% sulfur was made immovable. Without Na_2CO_3 addition, $PbSO_4$ could be directly reduced to PbS, and it further was converted to PbO with the help of sulfur-fixing agent Fe_3O_4. As a result, metallic Pb was extracted and recovered from lead oxide. However, when the Na_2CO_3 addition exceeded 4%, direct Pb recovery and sulfur-fixing rates no longer increased and were maintained at around 96% and 98%, respectively. This means that the active Na_2CO_3 reached saturation. Excess Na_2CO_3 was unable to further increase the capacities of lead recovery and sulfur fixation.

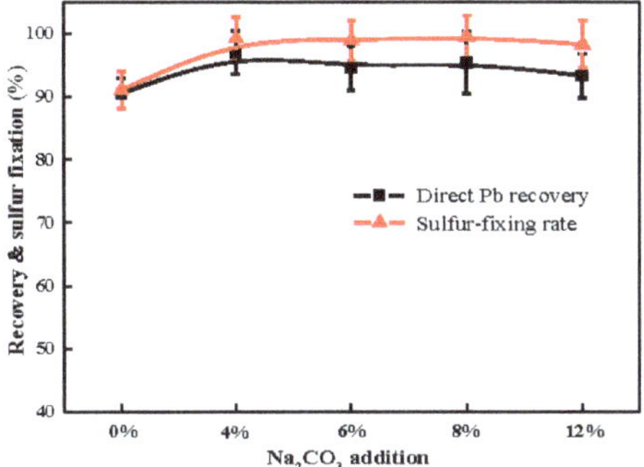

Figure 6. The effects of Na_2CO_3 dosage on lead recovery and sulfur fixation. ($W_{lead\ paste}$:$W_{pyrite\ cinder}$: W_{coke} = 200:50:24 g, FeO/SiO$_2$ = 1.3, CaO/SiO$_2$ = 0.4, 1200 °C, 2 h).

3.2.3. Influence of Treatment Temperature

The influences of temperature on lead recovery and sulfur-fixing rates are presented in Figure 7. The results show that 1200 °C was the optimal recycling temperature to obtain a high lead and sulfur recovery, where 97.0% lead was recovered in crude lead. The sulfur fixation rate was maintained at around 98% when temperature increased from 1150 °C to 1350 °C. This indicates that the sulfur-fixation reactions had completed before 1150 °C. Excessive temperature was unbeneficial for lead enrichment because volatilization of lead will intensify at high temperatures. The above presented XRD results of the $PbSO_4$-Na_2CO_3-Fe_3O_4-C mixtures also indicate that the intermediate products, lead oxides, tended to combine with sodium oxide and iron oxides at high temperatures, which limited further reduction of lead oxide to metallic lead.

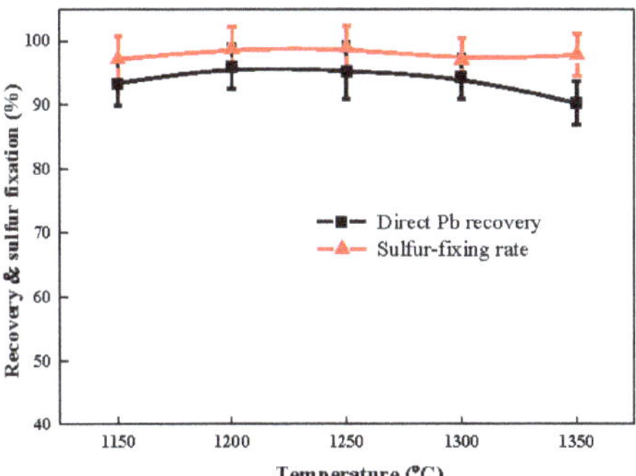

Figure 7. The effects of treatment temperature on lead recovery and sulfur fixation. ($W_{lead\ paste}$:W_{Na2CO3}: $W_{pyrite\ cinder}$:W_{coke} = 200:8:50:24 g, FeO/SiO$_2$ = 1.3, CaO/SiO$_2$ = 0.4, 2 h).

3.2.4. Influence of Treatment Time

The direct Pb recovery and sulfur-fixation rates at different treatment times are presented in Figure 8. It implies that 1.5 h was an acceptable recycling time for sulfur fixation, lead alloy settling, and enrichment, where more than 96.7% of lead and 99.5% of sulfur were recovered. A declining trend was observed when the treatment time exceeded 1.5 h because of the increasing volatilization of lead and sodium salt. The results in Figure 2b reveal that the lead extraction and sulfur fixing reactions could take place in minutes. However, an acceptable lead recovery and sulfur fixation rate relied on reactions between PbS and Fe_3O_4. Adequate completion of these reactions would take time.

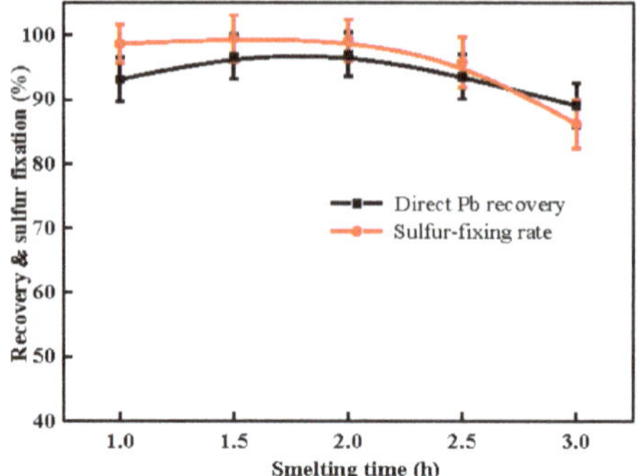

Figure 8. The effects of treatment time on lead recovery and sulfur fixation. ($W_{\text{lead paste}}$:W_{Na2CO3}:$W_{\text{pyrite cinder}}$:Wcoke = 200:8:50:24 g, FeO/SiO$_2$ = 1.3, CaO/SiO$_2$ = 0.4, 1200 °C).

3.3. Comprehensive Experiments and Characterization of Products

3.3.1. Comprehensive Experimental Results

Comprehensive bench-pilot experiments with 1500 g lead paste were conducted to detect the repeatability and reliability of this novel process. The above obtained optimal recycling conditions were used—$W_{\text{lead paste}}$:W_{Na2CO3}:$W_{\text{pyrite cinder}}$:W_{coke} = 100:4:25:12 g, treatment temperature 1200 °C, and treatment time 1.5 h. SiO$_2$ and CaO were added in the initial feed materials to form slag and adjust FeO/SiO$_2$ = 1.3 and CaO/SiO$_2$ = 0.4. Figure 9 shows a physical macrograph and the corresponding XRD patterns of recycling products. It is clear that the products were separated into three distinct layers: slag, ferrous matte, and crude lead bullion. Table 2 shows the chemical compositions of different products obtained.

Figure 9. Physical macrograph (**a**) and corresponding XRD patterns (**b**) of the recycling products. ($W_{\text{lead paste}}$:W_{Na2CO3}:$W_{\text{pyrite cider}}$:W_{coke} = 1500:60:375:180 g, FeO/SiO$_2$ = 1.3, CaO/SiO$_2$ = 0.4, 1200 °C, 1.5 h).

Table 2. Chemical compositions of recycling products in the comprehensive expansion experiments (wt.%), * g/t.

No.	Product	Chemical Compositions (wt.%)								
		Pb	Fe	S	Na	Sb	Ag *	SiO_2	CaO	Al_2O_3
1	Crude lead	98.87	0.26	0.04	-	0.075	136	-	-	-
2		98.02	0.38	0.07	-	0.089	115	-	-	-
Average		98.45	0.32	0.06	-	0.082	126	-	-	-
1	Ferrous matte	2.37	53.25	22.54	9.56	-		-	-	-
2		2.73	52.66	24.39	9.68	-		-	-	-
Average		2.55	53.09	23.47	9.62	-		-	-	-
1	Slag	0.39	22.96	3.24	9.79	-		31.22	18.86	6.83
2		0.59	24.31	3.19	9.88	-		30.72	18.50	6.14
Average		0.49	23.64	3.22	9.84	-		30.97	18.68	6.49

The results show that 98.3% lead and 99% silver were directly enriched in crude lead in the pilot test. More than 98% purity crude lead was obtained by a one-step extraction. Lead contents in the matte and slag were 2.7% and 0.6%, respectively. At the same time, 99.2% of total sulfur was fixed and recovered in the treatment system (85.9% in ferrous matte and 13.3% in slag), which helped to prevent the generation and emission of SO_2. The main phases in the solidified matte were FeS, Fe_3O_4, and $NaFeS_2$. Some unreacted PbS and entrained gangue materials, such as $Ca_3Al_2(SiO_4)_3$, $FeSiO_3$, $Ca_2(Al(AlSi)O_7)$, and $CaSiO_3$, were also detected. The matte can be sold directly or used for sulfuric acid manufacture and regenerate the sulfur-fixing agent. The slag comprised $Ca_2(Al(AlSi)O_7)$, Fe_2SiO_4, Fe_3O_4, $Na_2Si_2O_5$, $NaAlSiO_4$, $CaSiO_3$, $Ca_3Al_2(SiO_4)_3$, and some entrained FeS. It is harmless and can be used as raw material for cement production after water-quenching and granulation.

3.3.2. SEM-EDS Characterization of Products

The recycling products obtained, ferrous matte and slag, were characterized by SEM-EDS techniques. The results are presented in Figures 10 and 11. Figure 10 illustrates that ferrous oxide FeO and few metallic Fe droplets were embedded in the ferrous sulfide and its FeS minerals. At the same time, iron spinel (magnetite) Fe_3O_4 was adjacent to FeS. Some metallic lead and lead oxide particles were embedded and entrained in FeS minerals. It helps to confirm that lead extraction from PbS was carried out by exchange reactions between iron oxide (Fe_3O_4 and FeO) and PbS.

The SEM-EDS results presented in Figure 11 show that magnetite Fe_3O_4 tended to combine with sodium silicate $Na_2Si_2O_5$. Carnegieite Na_2AlSiO_4 particles were found embedded in magnetite and fayalite. Mackinawite FeS was found to be mainly entrained on the interfaces between gehlenite $Ca_2(Al(AlSi)O_7)$ and fayalite Fe_2SiO_4.

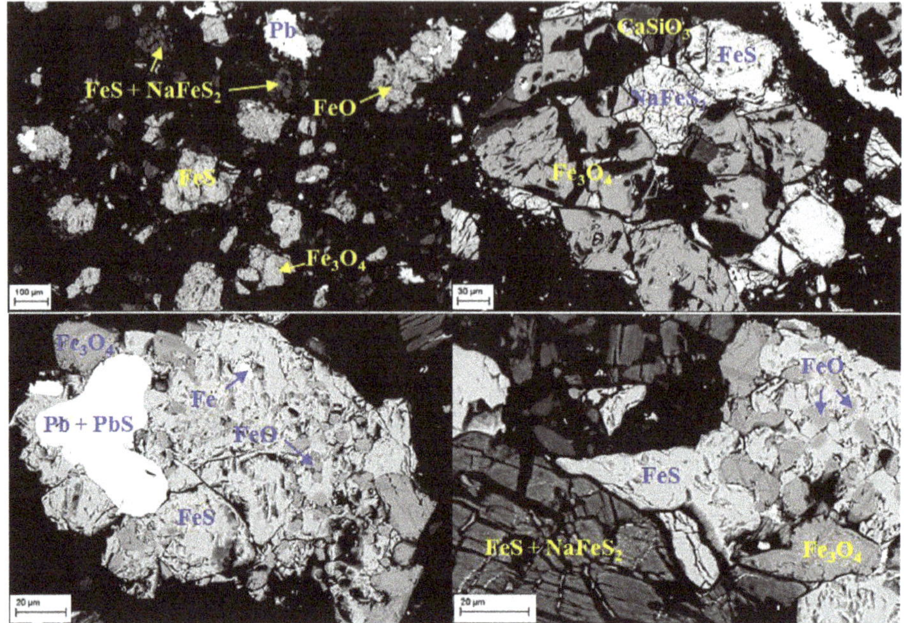

Figure 10. SEM-EDS results of the matte produced from conditions of $W_{lead\ paste}$:W_{Na2CO3}:$W_{pyrite\ cinder}$:W_{coke} = 1500:60:375:180 g, FeO/SiO_2 = 1.3, CaO/SiO_2 = 0.4, 1200 °C, and 1.5 h.

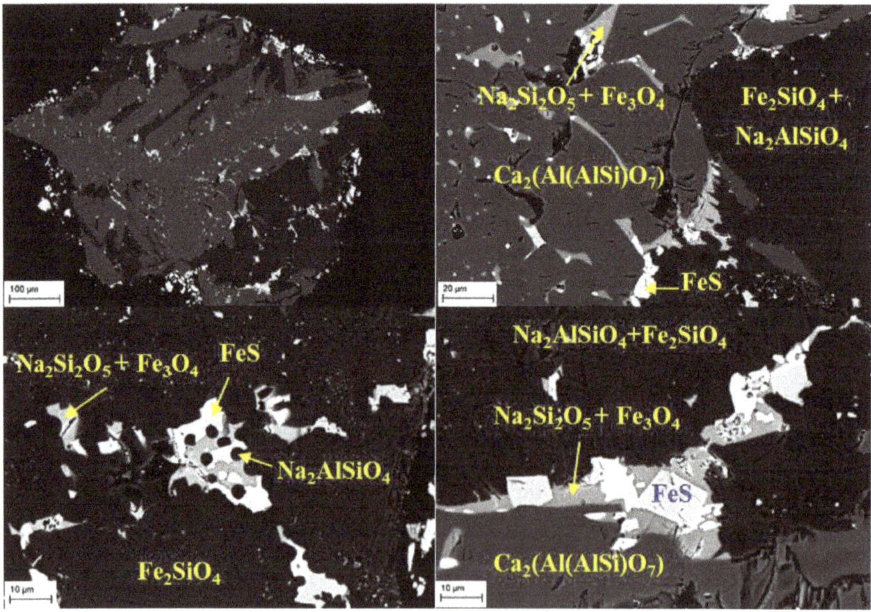

Figure 11. SEM-EDS results of the slag produced from conditions of $W_{lead\ paste}$:W_{Na2CO3}:$W_{pyrite\ cinder}$:W_{coke} = 1500:60:375:180 g, FeO/SiO_2 = 1.3, CaO/SiO_2 = 0.4, 1200 °C, and 1.5 h.

4. Conclusions

A cleaner recycling of lead from EoL LAB paste and co-treatment of PyCs by a reductive sulfur-fixing recycling technique was experimentally confirmed to be feasible. The optimal treatment conditions were determined as $W_{\text{lead paste}}:W_{\text{Na2CO3}}:W_{\text{pyrite cinder}}:W_{\text{coke}} = 100:4:25:12$ g, $FeO/SiO_2 = 1.3$, $CaO/SiO_2 = 0.4$, treatment temperature 1200 °C, and time 1.5 h. Under these conditions, 98.3% lead in the raw materials was directly enriched and recovered in crude lead, and 99.2% total sulfur was fixed and recovered to matte and slag. Crude lead of 98% purity was obtained in a one-step extraction. Lead contents in matte and slag were 2.7% and 0.6%, respectively.

The phase transformation mechanisms were clarified. $PbSO_4$ originally reacted at low temperatures with Na_2CO_3 to convert to PbO and Na_2SO_4, which avoided decomposition of $PbSO_4$ and ensured that sulfur was conserved in the recycling system in the solid state (Na_2SO_4) without generating and emitting SO_2 gas. As temperature increased, unconverted $PbSO_4$ was selectively reduced to PbS. Then, Fe_3O_4 reacted with PbS. The sulfur in PbS was further transferred and finally fixed as FeS and $NaFeS_2$. This new process provides a promising alternative recycling and treatment method for various secondary lead-containing materials and iron-bearing industrial wastes.

Author Contributions: Y.L. organized the research plan, performed the experiments, and wrote and revised the manuscript according to the comments from the co-authors. S.Y., Y.C., and C.T. were in charge of project administration and assisted with the supervising work; P.T. modified the thermodynamic section; A.J. and P.T. reviewed and checked the language of the manuscript.

Funding: This research was funded by Specialized Research Project of Guangdong Provincial Applied Science and Technology, China, grant number 2016B020242001; Hunan Provincial Science Fund for Distinguished Young Scholars, China, grant number 2018JJ1044; National Natural Science Foundation of China, China, grant number 51234009 and 51604105; and CMEco by Business Finland, Finland, grant number 2116781.

Conflicts of Interest: The authors declare no conflict of interest.

References

1. Tian, X.; Wu, Y.; Gong, Y.; Zuo, T. The lead-acid battery industry in China: outlook for production and recycling. *Waste Manage. Res.* **2015**, *33*, 986–994. [CrossRef] [PubMed]
2. Zhu, X.; He, X.; Yang, J.; Gao, L.; Liu, J.; Yang, D.; Sun, X.; Zhang, W.; Wang, Q.; Kumar, R.V. Leaching of spent lead acid battery paste components by sodium citrate and acetic acid. *J. Hazard. Mater.* **2013**, *250*, 387–396. [CrossRef] [PubMed]
3. Zhang, Q. The current status on the recycling of lead-acid batteries in China. *Int. J. Electrochem Sci.* **2013**, *8*, 6457–6466.
4. Huggins, R.A. *Lead-Acid Batteries*; Springer: Boston, MA, USA, 2016; pp. 237–250.
5. Li, Y.; Chen, Y.; Tang, C.; Yang, S.; Klemettinen, L.; Rämä, M.; Wan, X.; Jokilaakso, A. A New Pyrometallurgical Recycling Technique for Lead Battery Paste without SO_2 Generation—A Thermodynamic and Experimental Investigation. *Extraction 2018* **2018**, 1109–1120. [CrossRef]
6. Li, Y.; Tang, C.; Chen, Y.; Yang, S.; Guo, L.; He, J.; Tang, M. One-Step Extraction of Lead from Spent Lead-Acid Battery Paste via Reductive Sulfur-Fixing Smelting: Thermodynamic Analysis. In *8th International Symposium on High-Temperature Metallurgical Processing*; Springer: Cham, Switzerland, 2017; pp. 767–777. [CrossRef]
7. Lin, D.; Qiu, K. Recycling of waste lead storage battery by vacuum methods. *Waste Manage.* **2011**, *31*, 1547–1552. [CrossRef] [PubMed]
8. Van der Kuijp, T.J.; Huang, L.; Cherry, C.R. Health hazards of China's lead-acid battery industry: A review of its market drivers, production processes, and health impacts. *Environ. Health* **2013**, *12*, 61. [CrossRef] [PubMed]
9. Tian, X.; Wu, Y.; Hou, P.; Liang, S.; Qu, S.; Xu, M.; Zuo, T. Environmental impact and economic assessment of secondary lead production: Comparison of main spent lead-acid battery recycling processes in China. *J. Clean. Prod.* **2017**, *144*, 142–148. [CrossRef]
10. Sun, Z.; Cao, H.; Zhang, X.; Lin, X.; Zheng, W.; Cao, G.; Sun, Y.; Zhang, Y. Spent lead-acid battery recycling in China—A review and sustainable analyses on mass flow of lead. *Waste Manage.* **2017**, *64*, 190–201. [CrossRef]

11. Liu, T.; Qiu, K. Removing antimony from waste lead storage batteries alloy by vacuum displacement reaction technology. *J. Hazard. Mater.* **2018**, *347*, 334–340. [CrossRef]

12. Ma, Y.; Qiu, K. Recovery of lead from lead paste in spent lead acid battery by hydrometallurgical desulfurization and vacuum thermal reduction. *Waste Manage.* **2015**, *40*, 151–156. [CrossRef]

13. Kreusch, M.A.; Ponte, M.J.J.S.; Ponte, H.A.; Kaminari, N.M.S.; Marino, C.E.B.; Mymrin, V. Technological improvements in automotive battery recycling. *Resour. Conserv. Recy.* **2007**, *52*, 368–380. [CrossRef]

14. Lassin, A.; Piantone, P.; Burnol, A.; Bodénan, F.; Chateau, L.; Lerouge, C.; Crouzet, C.; Guyonnet, D.; Bailly, L. Reactivity of waste generated during lead recycling: An integrated study. *J. Hazard. Mater.* **2007**, *139*, 430. [CrossRef] [PubMed]

15. Queneau, P.B.; Leiby, R.; Robinson, R. Recycling lead and zinc in the United States. *World of Metallurgy-ERZMETALL* **2015**, *68*, 149.

16. Li, W.F.; Jiang, L.H.; Zhan, J.; Zhang, C.F.; Li, G.; Hwang, J.Y. Thermodynamics Analysis and Industrial Trials of Bottom-Blowing Smelting for Processing Lead Sulfate-Containing Materials. In *6th International Symposium on High-Temperature Metallurgical Processing*; Springer: Cham, Switzerland, 2015; pp. 507–514.

17. Zhang, X.; Li, L.; Fan, E.; Xue, Q.; Bian, Y.; Wu, F.; Chen, R. Toward sustainable and systematic recycling of spent rechargeable batteries. *Chem. Soc. Rev.* **2018**, *47*, 7239–7302. [CrossRef] [PubMed]

18. Dutta, T.; Kim, K.H.; Deep, A.; Szulejko, J.E.; Vellingiri, K.; Kumar, S.; Kwon, E.E.; Yun, S.T. Recovery of nanomaterials from battery and electronic wastes: A new paradigm of environmental waste management. *Renew. Sustain. Energ. Rev.* **2018**, *82*, 3694–3704. [CrossRef]

19. Liu, K.; Yang, J.; Liang, S.; Hou, H.; Chen, Y.; Wang, J.; Liu, B.; Xiao, K.; Hu, J.; Wang, J. An Emission-Free Vacuum Chlorinating Process for Simultaneous Sulfur Fixation and Lead Recovery from Spent Lead-Acid Batteries. *Environ. Sci. Technol.* **2018**, *52*, 2235–2241. [CrossRef] [PubMed]

20. Wei, J.; Guo, X.; Marinova, D.; Fan, J. Industrial SO_2 pollution and agricultural losses in China: evidence from heavy air polluters. *J. Clean. Prod.* **2014**, *64*, 404–413. [CrossRef]

21. Mohanty, C.; Adapala, S.; Meikap, B. Removal of hazardous gaseous pollutants from industrial flue gases by a novel multi-stage fluidized bed desulfurizer. *J. Hazard. Mater.* **2009**, *165*, 427–434. [CrossRef] [PubMed]

22. Zhang, W.; Yang, J.; Wu, X.; Hu, Y.; Yu, W.; Wang, J.; Dong, J.; Li, M.; Liang, S.; Hu, J. A critical review on secondary lead recycling technology and its prospect. *Renew. Sustain. Energ. Rev.* **2016**, *61*, 108–122. [CrossRef]

23. Zhang, J.; Yan, Y.; Hu, Z.; Fan, X.; Zheng, Y. Utilization of low-grade pyrite cinder for synthesis of microwave heating ceramics and their microwave deicing performance in dense-graded asphalt mixtures. *J. Clean. Prod.* **2018**, *170*, 486–495. [CrossRef]

24. Alp, I.; Deveci, H.; Yazıcı, E.; Türk, T.; Süngün, Y. Potential use of pyrite cinders as raw material in cement production: Results of industrial scale trial operations. *J. Hazard. Mater.* **2009**, *166*, 144–149. [CrossRef] [PubMed]

25. Hu, Y.; Tang, C.; Tang, M.; Chen, Y.; Yang, J.; Yang, S.; He, J. Reductive smelting of spent lead-acid battery colloid sludge in molten salt of sodium at low temperature. *Chin. Nonferr. Metall.* **2014**, *43*, 75–79.

26. Ye, L.; Tang, C.; Chen, Y.; Yang, S.; Yang, J.; Zhang, W. One-step extraction of antimony from low-grade stibnite in Sodium Carbonate–Sodium Chloride binary molten salt. *J. Clean. Prod.* **2015**, *93*, 134–139. [CrossRef]

27. Toby, B. CMPR-a powder diffraction toolkit. *J. Appl. Crystallogr.* **2005**, *38*, 1040–1041. [CrossRef]

28. Roine, A. Outotec HSC Chemistry. Available online: www.hsc-chemistry.com (accessed on 19 August 2019).

29. Li, Y.; Yang, S.; Taskinen, P.; He, J.; Liao, F.; Zhu, R.; Chen, Y.; Tang, C.; Wang, Y.; Jokilaakso, A. Novel recycling process for lead-acid battery paste without SO_2 generation-Reaction mechanism and industrial pilot campaign. *J. Clean. Prod.* **2017**, *217*, 162–171. [CrossRef]

30. Li, Y.; Yang, S.; Taskinen, P.; He, J.; Chen, Y.; Tang, C.; Wang, Y.; Jokilaakso, A. Spent Lead-Acid Battery Recycling via Reductive Sulfur-Fixing Smelting and Its Reaction Mechanism in the $PbSO_4$-Fe_3O_4-Na_2CO_3-C System. *JOM* **2019**, *71*, 1–12. [CrossRef]

 *metals*

Review

Recycling Chain for Spent Lithium-Ion Batteries

Denis Werner *, Urs Alexander Peuker and Thomas Mütze

Institute of Mechanical Process Engineering and Mineral Processing, TU Bergakademie Freiberg, 09599 Freiberg, Germany; Urs.Peuker@mvtat.tu-freiberg.de (U.A.P.); Thomas.Muetze@mvtat.tu-freiberg.de (T.M.)
* Correspondence: Denis.Werner@mvtat.tu-freiberg.de; Tel.: +49-3731-393023

Received: 14 February 2020; Accepted: 26 February 2020; Published: 28 February 2020

Abstract: The recycling of spent lithium-ion batteries (LIB) is becoming increasingly important with regard to environmental, economic, geostrategic, and health aspects due to the increasing amount of LIB produced, introduced into the market, and being spent in the following years. The recycling itself becomes a challenge to face on one hand the special aspects of LIB-technology and on the other hand to reply to the idea of circular economy. In this paper, we analyze the different recycling concepts for spent LIBs and categorize them according to state-of-the-art schemes of waste treatment technology. Therefore, we structure the different processes into process stages and unit processes. Several recycling technologies are treating spent lithium-ion batteries worldwide focusing on one or several process stages or unit processes.

Keywords: environmental technologies; waste treatment; recycling; spent lithium-ion batteries; recycling chain; process stages; unit processes; industrial recycling technologies

---

1. Introduction

The development of information technologies, electrically powered vehicles, stationary energy storage systems, and consumer electronics will further increase the consumption of lithium-ion batteries (LIBs) over the next few years [1,2]. As a result, there will be also an increasing amount of spent batteries in the future, which will have to be treated by suitable processes [3,4]. Review articles and research activities, which focus mainly on the valuable active materials of the cathodes, define process steps involved in recovering these active materials as pretreatment. From this point of view, most of the recycling technologies, independent from the technological scale, cannot be clearly structured, represented, compared, and differentiated. Due to the increasing importance of the individual process steps prior to metallurgical treatment, this article classifies the steps for the recycling of spent LIBs along the generic process chain for waste materials providing uniform nomenclature. Moreover, it compiles an overview of the known industrial processes.

LIBs are rechargeable electrochemical energy converters in which chemical energy is converted into electrical energy by reversible redox reactions during the discharge process and vice versa during charging [5]. LIBs are one of the most important mobile energy storage devices for electrical and electronic applications. Moreover, this battery type has a great potential of success in the international market due to its beneficial properties like high energy density, no memory effect, and low self-discharge. Another fact is that the production of LIBs requires a considerable amount of metallic resources which represent a potential risk to the environment if disposed to landfills and which have to be returned to the material cycle [6].

The recycling of LIBs is of great importance not only from an economic and environmental perspective, but also from a geostrategic point of view and some health aspects [7]. Spent LIBs contain geographically unevenly distributed rare and valuable materials and generate large quantities of metal-containing waste [8]. High voltage and current from residual energy can lead to severe injuries. Furthermore, active materials containing nickel oxide have a carcinogenic potential [9,10]. For these reasons, recycling of spent LIBs enhances environmental protection and the idea of a circular economy by separating the valuable metallic constituents into different products. These products are secondary raw materials for the production of metal or metal composite products [3].

Besides the intrinsic value of the battery materials and components, national or supranational legislation like the Battery Act or the Directive 2006/66 EC provide the framework for the recycling of LIB. These directives prescribe responsibilities and procedures for each contributor included in the life cycle of a battery. Moreover, the directives classify different battery types in respect to their implemented application and address waste management. Especially, collection and recycling targets are formalized to promote material instead of energy recovery or disposal. Therefore, recycling is defined as the reprocessing of waste materials in a production process either for their original purpose, i.e., material or raw material recovery, or for other purposes, i.e., other recovery like backfilling, but excluding energetic recovery [11]. Recycling efficiency is the result of the ratio of output material to feed material and targets 50 mass percentage. However, the different purposes of the output materials are not further distinguished. Moreover, recycling efficiency can be already achieved by a beneficial ratio of the cell mass in a battery system, when the battery system is dismantled to cell level.

Since their market launch in 1990, the energy and power densities of LIBs increased continuously and new areas of application have been opened up [12]. This led to a high and still increasing variety of battery types with different material compositions [13]. In this initial stage, robust pyrometallurgical recycling technologies are mainly used on industrial scale for waste treatment. These technologies reach the recycling efficiency by backfilling the slag. In addition, operating above the smelting temperatures of the contained metals pyrometallurgical technologies are very energy-intensive at the same time [14], since they are mainly focusing on high value materials like cobalt, nickel and copper. Therefore, and with regard to the increasing number of battery types and the global energy consumption, international research also focuses on the development of alternative or supplementary mechanical processes in combination with hydrometallurgical refining [15,16].

2. Material Recycling of Lithium-Ion Batteries

The smallest functional unit of a LIB is called a battery cell. It generally consists of two electrodes, a separator, electrolyte, and cell housing. Each electrode is composed of a metallic conductor foil and a coating, the so-called active material. By definition, the designation of the negative and the positive electrode is assigned during the discharge process. Therefore, the anode is mostly a copper foil with graphite coating and typically, the cathode is an aluminum foil coated with an intercalated lithium compound. The separator is mostly a porous polyolefin, the electrolyte a mixture of organic solvents and a lithium salt. The cell housing is a sealed container made of aluminum, steel, special plastics, or highly refined aluminum composite foils [17]. Battery cells are cylindrical, prismatic, or bag-shaped. The cells can be connected in series or parallel to form either a single block or a module as subunit of a bigger battery system [5,9,18].

If the recycling chain for waste materials is applied to LIBs, the treatment has to be subdivided into process stages on the one hand which are characteristic in terms of process technology and into unit processes on the other hand which limit the amount of waste material and the material conversion processes. Therefore, the recycling chain consists of four process stages with two unit processes each [19,20]:

1.   Preparation: waste logistics and presorting,
2.   Pretreatment: dismantling and depollution,
3.   Processing: liberation and separation,
4.   Metallurgy: extraction and recovery.

Typical recycling processes do not mention the process stages' preparation and pretreatment [21,22]. However, both of them have a significant influence on the efficiency of the downstream steps, i.e., processing and metallurgy. Furthermore, the differentiation between recycling concepts becomes more precise if their combinations with preparation and pretreatment is indicated as well. As a result, both process stages are included into the holistic characterization of the recycling processes of spent LIB here.

2.1. Process Stages of the Recycling Chain for Lithium-Ion Batteries

Figure 1 shows the generic recycling chain for lithium-ion batteries with its four process stages and the associated unit processes. Different mixtures of batteries are collected and sorted in the preparation stage either according to battery types and/or to active materials. During the subsequent pretreatment and processing, the first secondary raw materials are produced to be fed into established recycling processes. The treatment of enriched fractions e.g., concentrates of active materials, occurs within the framework of metallurgy. The recovery of organic solvents as well as electronic and auxiliary components or casing and support materials, i.e., plastics and metals, are not considered in detail, here (cf. Figure 1). For most of these materials conventional recycling chains already exist.

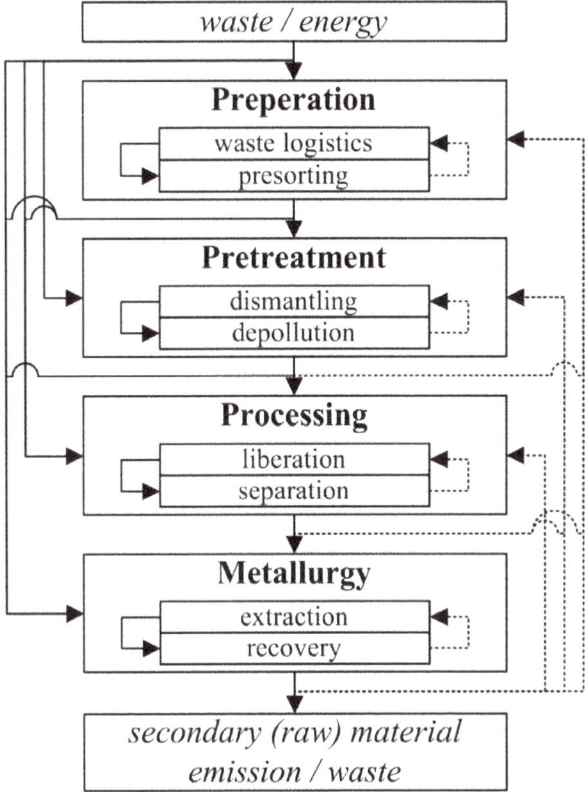

Figure 1. Recycling chain with process stages and unit processes for spent lithium-ion batteries.

Since it focusses on the central challenges of battery recycling, this generalized recycling chain can be used to evaluate all of the present recycling concepts, as well as in laboratory scale to whole industrial facilities. Some of the unit processes within the process stages are used iteratively, others are omitted altogether (cf. Figure 1) [20].

2.2. Unit Processes

The individual unit processes are composed of unit subprocesses and unit operations that are linked in different complexes. Unit subprocesses describe the character of the implicated state of aggregation, whereas unit operations treat a material within a single device [23,24]. Martens and Goldmann [19] present the general objectives of the respective unit processes in detail. These are explained below for spent LIB.

2.2.1. Preparation

The unit processes of preparation as the first process stage in LIB recycling are waste logistics and presorting as shown in Figure 2.

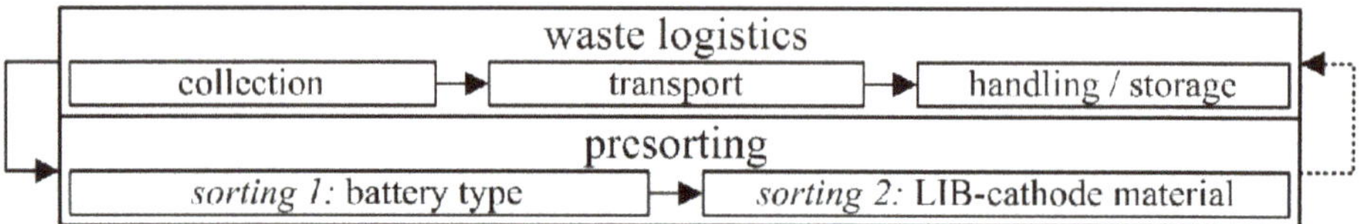

Figure 2. Process stage 1 for recycling of lithium-ion batteries: preparation.

Waste Logistics

In the past, LIBs were mainly produced for portable and mobile applications. Currently, different collection systems collect the batteries depending on the legislation of the respective country [25,26]. For example, according to the European Directive (2006/66/EC) on waste batteries and accumulators, distributors of industrial batteries from electric vehicles have to take back the batteries themselves. Therefore, collection systems for those battery types are either already available or under development. In this field, car manufacturers do often closely cooperate with local recycling companies due to the low sales figures and correspondingly few returns now [27–31]. Afterwards, spent LIBs are either transported directly to the subsequent treatment facility or handled and stored before they are handed over [32,33].

Nevertheless, there is a great potential for the collection of spent LIB. One example is the LIB collection rate of Germany for portable systems e.g., smartphones or laptops which amount to only 45%. In a global comparison, even this figure in many countries is not achieved and more often no collection systems exist [34,35]. However, a large proportion of portable batteries are subject to the hoarding effect [36] or stockpiling [9,37]. Besides, they end up in incinerators, landfills, or other wastes due to a lack of consumer or sorting personal awareness, respectively [29,37–43]. In addition, electrical and electronic scrap containing LIBs is illegally exported from the developed to African and Asian countries, which also reduces the collection rates [44,45].

Due to the considerable high hazard potential (cf. Depollution), both the collection and transport of LIBs require special measures against short circuits and leakage of the electrolyte [46]. The classification of spent lithium-ion batteries as a hazardous good [13] demands special transport containers, warning signs, and packaging [46–52]. Therefore, the waste logistics cause a significant share of the total cost in battery recycling due to the high safety requirements and the resulting low specific transport weight [53].

Presorting

Spent batteries of any origin are usually not collected separately. They are accumulated as battery mixtures of different battery types or of lithium-ion batteries of different composition [29,36,54]. Due to the large number of battery types, it is not possible yet to recover all the constituents from the mixture using a single recycling process [26,55,56]. Therefore, presorting by battery type is necessary for recycling technologies specializing in lithium-ion batteries to define material flows for further treatment.

The sorting technologies used for presorting of batteries are assigned to picking. Characteristic features of each individual battery are analyzed and evaluated using trained personnel or sensors. Then, the batteries are separated manually or automatically [29,47,55]. Recycling technologies based on picking are designed for complex battery materials, so high recycling rates can be achieved [34,56]. In addition, it is possible to sort by different cathode active materials (e.g., Figure 2) within a single LIB type. However, picking requires an initial dismantling (cf. Dismantling) of the battery cells into their functional unit and an elemental analysis of the active material previously. As this type of preparation is associated with a significant processing time and hazard potential (cf. Depollution), it is currently not used industrially.

Currently, most of the technologies for recycling of spent LIBs operate without presorting, especially technologies based on pyrometallurgical unit operations [16] (cf. Pyrometallurgy). The processes were derived from the recycling of nickel-based battery systems [57,58], the production of primary and secondary metallic raw materials [16], or the recycling of completely different wastes [35]. The changing composition of these mixed material systems causes low recycling rates and poor quality of the secondary raw materials. Therefore, only metals like nickel, cobalt, and copper are recovered and by-metals like aluminum or manganese are discharged into the furnace slag [34]. The latter ones can only be reintroduced into the material cycle at great expense.

2.2.2. Pretreatment

After the preparation stage, the batteries need to be dismantled to a defined level. Furthermore, different hazard potentials are deactivated thermally, electrically, or cryogenically (cf. Figure 3). Depending on the size or original purpose of the battery system, both linear and iterative pretreatment is performed within this process stage and its unit processes.

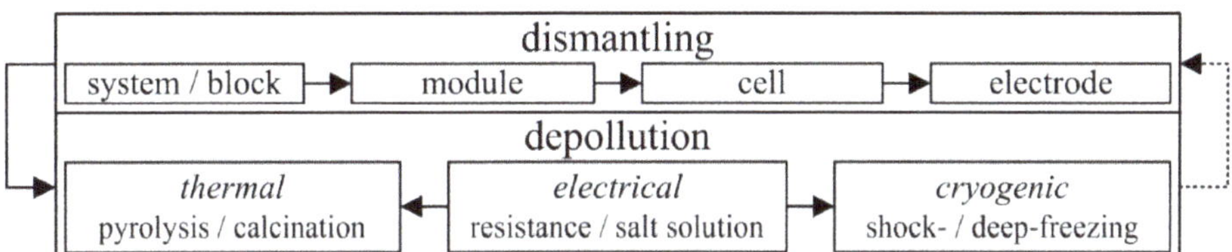

Figure 3. Process stage 2 for recycling of lithium-ion batteries: pretreatment.

Dismantling

The increasing market penetration of electric cars and the resulting increase of spent LIBs in respect to amount, mass, and size makes dismantling reasonable and in some cases urgently necessary. Dismantling is a time-consuming and thus cost-intensive process step due to the complexity of spent batteries [9,56]. Often, technological equipment such as furnaces or crushing devices are limited regarding to the maximum size or mass of the feed in order to achieve process ability, sufficient process stability, and efficiency [59]. In addition, it is indispensable for some methods of depollution to access the battery modules or cells (see Depollution). Besides the battery cell, a battery system consists of a high proportion of periphery components like the battery management system (BMS) or cooling parts. Therefore, dismantling of LIBs from system to module, cell, or even electrode level generates products made of metals, plastics, and electronic components that can be fed to established recycling routes increasing the overall recycling efficiency [14,56,60,61]. Furthermore, functional components or reusable assemblies can also be obtained for Second Life applications [9,62]. From an economic point of view, dismantling is an optimization problem between the dismantling level, i.e., the revenue from recovered individual parts or energy savings in mechanical and/or metallurgical processing, and the costs for the equipment and operating expenditure [9].

Several dismantling concepts are known, such as manual, semi-automatic (hybrid), and fully automatic dismantling [9,56,63]. From the economical and safety perspective, it is difficult to implement only manual dismantling [54]. Hybrid concepts try to overcome this by combining manual activities with robots [64]. The implementation of automated dismantling by industrial robots is subject of controversial discussions about the manifold battery designs and the rapid technical development of batteries [9,46,54,56,64–66]. Currently, battery systems are manually dismantled to different dismantling levels within the framework of research projects and some industrial applications [39,54,66,67]. Occupational safety requires that the DC voltage is reduced at least below 60 V [54,68].

Depollution

The depollution within the recycling of LIBs prevents carry-over of critical or hazardous components into subsequent process steps and avoids the release of harmful emissions into the environment [19]. The hazard potentials of spent LIBs can be summarized as electrical, chemical, and thermal ones which interact with one another [56,69]. Depending on the individual recycling process, its intermediate and final products, the depollution utilizes different methods like discharge, cryogenic treatment, and/or thermal treatment to remove hazardous substances or conditions [70]. As a result, depollution is also called as deactivation [10], passivation, or stabilization [9].

Discharging is a method that lowers the electrochemical energy content of the battery [9]. This method is primarily used for recycling processes using dismantling (cf. Dismantling) or mechanical liberation (cf. Liberation) [31,71]. Various methods for discharging were subject of the research project LithoRec II [10,18,63], whereas discharging using an external circuit with resistor is the most common and a practical method for large battery cells with high capacities [16,71]. In addition, cells are discharged in salt brines [9,72–74], in powders from metallic conductor foils or graphite [75], or in stainless-steel containers with stainless-steel chips [70]. Especially, when using brines, undesired side reactions are also mentioned leading to corrosion of electrical contacts or housing components as well as to the release of hydrogen [10,31,75] or other gases [9]. The discharge in salt brines (mainly NaCl) in particular is currently a common method for discharging for low capacity batteries [70,74]. If high voltage batteries are discharged, this method can lead to leaks in the battery housing and has its limits regarding to discharging process and efficiency as well as contamination of electrode materials [75]. According to Zhao [31], only high discharge currents and professional technical equipment can achieve the safe and complete discharge of LIBs. However, this results in an enormous expenditure of time and money, especially with regard to high-performance batteries from electric vehicles [9]. This does not apply to batteries with internal damage anyway.

Cryogenic treatment is one method that avoids exothermic reactions, especially during liberation (cf. Liberation). If LIBs are exposed to temperatures around −200 °C, the ion mobility decreases significantly [10,69]. Appropriate safety measures must be taken (e.g., heat-resistant conveyor units and exhaust gas purification systems), since the chemical reactions after liberation occur later compared to the treatment at normal temperatures.

Thermal methods such as pyrolysis or calcination easily remove flammable electrolyte components and decompose the electrolyte components by breaking down the organic compounds thermochemically. The fact that these thermal processes partially decompose components such as the separator and the binder of the electrode coatings and that they delaminate the coating from the metallic conductor foils is in turn used in a number of processes [54,76]. These processes are carried out in vacuum induction furnaces [76], rotary kilns [34], or blast furnaces [77]. Appropriately designed dedusting and flue gas cleaning systems then separate the resulting decomposition products [6,10,54,76]. At laboratory scale, pyrolysis in tube furnaces with a vacuum environment is also applied as pretreatment of mixed active materials for further lithium recovery [78].

2.2.3. Processing

The third process stage in the recycling chain of LIBs is the processing with the unit processes liberation and separation (cf. Figure 4). The aim of this process stage is to break up the bonds between the individual components, i.e., materials, in order to separate them into defined concentrates. Liberation also includes size control to influence adaptability and efficiency of physical separation technologies. Especially, if the active materials are separated from the metallic conductor foils, impurities must be minimized in the corresponding fraction in order to unburden or even enable the subsequent refining (cf. 2.2.4) [3].

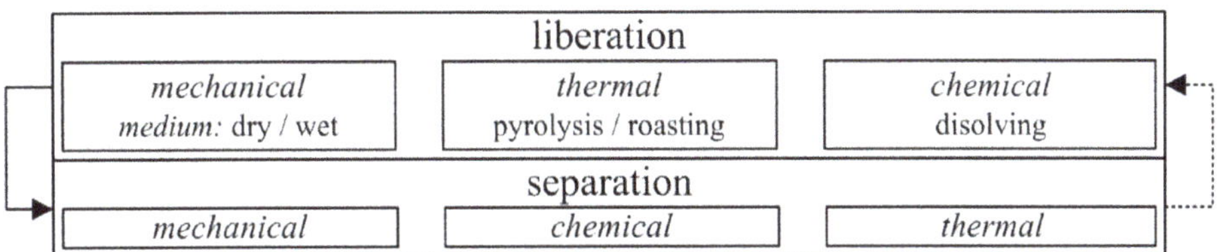

Figure 4. Process stage 3 for recycling of lithium-ion batteries: processing.

A separation efficiency depends generally on the mechanical degree of liberation of the individual components that need to be separated [79]. Secondly, it depends on the efficiency of the separation technology used. A combination of crushing, size classification, and sorting is intended to either produce secondary raw materials in the sense of material recycling, to prepare the material recycling by metallurgical processes, or to achieve waste treatment and disposal [33]. In addition, thermal, chemical, and mechanical separation processes can be used for special purposes e.g., electrolyte separation by drying, de-coating of the metallic conductor foils, or shape modification of crushed and enriched materials [18,80], respectively.

Liberation

The mechanical liberation of LIBs is mainly achieved by shear, cutting, and tearing stresses since their material behavior is ductile [81]. Therefore, mechanisms based on slow or fast compression are more likely to trap materials or create new compounds [82]. The liberation of LIB modules is usually carried out in two stages [34,82], rarely in one stage [10,16,54,83]. The first step is precrushing of the feed and the second liberation of the components itself. During this mechanical treatment it is essential to protect the tools and process chambers of the equipment against corrosion caused by electrolytes [31]. Furthermore, there are approaches for electrohydraulic defragmentation of LIBs, but so far, this could not be converted to a large technical scale successfully [84–87]. Liberation by using diamond saws, water, or laser beams is described in the literature as well, but it is very time consuming and lacks efficiency [31]. In addition, there is always a fire hazard when diamond saws and laser beams are used.

Depending on the type of pretreatment and depollution (cf. 2.2.2), it is necessary to adopt adequate safety measures to provide a safe work environment during the liberation. For example, the medium in the process chamber can cause strong exothermic decomposition reactions up to explosions [3]. Ideally, each single battery cell is completely discharged for liberation [20,31,80]. The hazard potential deriving from volatile electrolyte components can be reduced by thermal or cryogenic treatment in order to use ambient air during crushing. Charged low capacity batteries have to be crushed either under a protective gas [3,9], such as carbon dioxide [88], nitrogen [63], argon [88,89], or helium [69], or in liquid media, such as water or salt solutions [88]. Using water, further safety precautions are required due to undesirable side reactions of the electrolyte with the process medium (e.g., formation of hydrofluoric acid). Regardless of the process medium, the volatile electrolytes and dust must be separated from the medium afterwards by adequate purification systems.

Thermal and chemical liberation methods are applied to delaminate the coating material from the current collector foils, especially for cathodes. Thermal methods like pyrolysis or roasting decompose the binder, whereas chemical methods either dissolve the aluminum or detach the cathode active materials from the aluminum foil in a special solution [9].

Separation

Mechanical sorting according to electromagnetic, electrostatic, density, and granulometric properties mainly separates liberated components and materials. At times, hydrometallurgical treatment follows separation, hydrometallurgical processes require a high purity of the intermediate products for sufficient process stability and selectivity. Commonly used processes are magnetic and eddy current separation, screening, gravity sorting in flow fields [31,69] or with pneumatic shaking tables [88] as well as flotation, to either enrich valuable materials or to deplete impurities in fractions [3,88,90]. Furthermore, it is necessary to install appropriate measures for the dedusting and separation of the electrolyte from the process media, latter thermally or chemically [18,31].

The combination of separation steps is extremely material and process specific. Nevertheless, in most recycling processes adopting processing technologies the materials are enriched in products such as casing materials, plastics (mainly the separator foil), a mixture or separated electrode foils, and a mixture or the individual active materials [9,18,80,91]. If the electrode foils still contain active materials after the mechanical liberation [9], coating and metallic foil are liberated and separated by further mechanical [14,80], thermal [92] or chemical treatment [3,77,93,94] (cf. Liberation).

Active materials can be sorted by flotation [95] or multistage magnetic separation after appropriate preparation steps [96]. This mechanical processing of active materials is the preferred option from the energetic and geostrategic point of view (cf. 1). Challenges arise from the small particle sizes at sufficient liberation and the small differences in the material properties there [79]. Therefore, the active material fractions are mostly not sorted mechanically but fed to either established or newly developed pyro- and/or hydrometallurgical processes (cf. 2.2.4).

2.2.4. Metallurgy

The final stage of the LIB recycling feeds pretreated, i.e., dismantled, batteries or material fractions from processing into extractive metallurgical processes for the production of sufficiently pure intermediate materials. The unit processes are extraction and recovery. Both unit processes are always carried out in a coupled manner. Consequently, the refining process is classified into hydro- and pyrometallurgical processes (cf. Figure 5). For materials used in the housing of the batteries, in peripheral components, and the conductor foils established metallurgical processes exist, e.g., for aluminum, copper, and alloyed steel. These components are treated in pure secondary smelters using specifically designed and optimized processes [19]. The metals cobalt, nickel, copper, manganese, and iron bear an outstanding position in the recycling of LIBs due to their high intrinsic value and the comparatively low cost of recovery [56]. Here, pyrometallurgy can be combined with hydrometallurgical processes but pure hydrometallurgical processing is becoming more and more important [39,58,97,98]. When discussing different metallurgical technologies for the processing of spent LIBs, the pyro- and hydrometallurgical unit processes are compared as stand-alone.

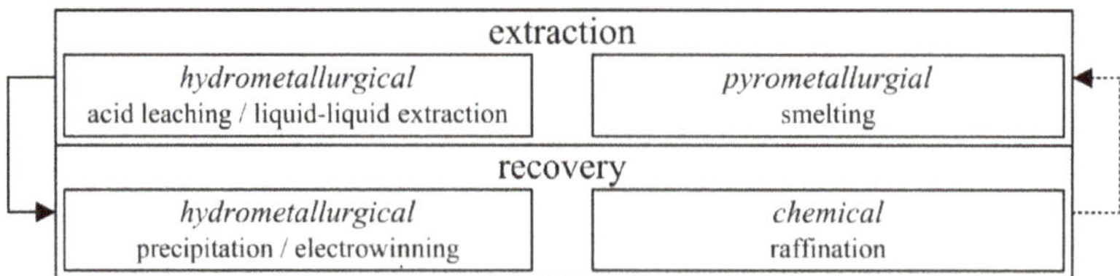

Figure 5. Process stage 4 for recycling of lithium-ion batteries: metallurgy.

Pyrometallurgy

In industrial pyrometallurgical processes, the solid feed is melted down, typically in an electric arc furnace [99] or shaft furnace [7], and transferred into an alloy, a slag, and/or a matte and the off gas with dust particles [9]. Organic components of the feed are pyrolyzed or entirely burned depending on the furnace technology. Graphite can be generally used as a reducing reagent for carbo-reductive melting processes. If the graphite content of the feed material is too high, problems occur during pyrometallurgical processing influencing reaction kinetics due to low reactivity of graphite, properties of the melt like melting point and viscosity, slag formation, and consequent metal recovery as well as the overall process efficiency [7,99]. Depending on the composition of the feed and the process specifications, cobalt, nickel, copper, and iron can be collected as metal alloy or matte. The metal alloy is sold as a secondary raw material, such as alloyed steel [100], whereas the matte is further processed hydrometallurgically. Unfortunately, due to its principle, pyrometallurgical processes transfer lithium, manganese, and aluminum into the slag which is currently used as filler material, e.g., in road construction [34,37,92] or in concrete [101], or is deposited in landfills [40]. In principle, the recovery of these three substances from slag is technically feasible via hydrometallurgical processes [9,40] but not economically [6,35,88,102,103]. Furthermore, pollutants such as carbon dioxide, dioxins, and furans occur [3,9,93,104].

Pyrometallurgical technologies exhibit comparatively low flexibility due to the required high economic investments and the complex processes involved in the extraction of metals [93]. Other disadvantages of such technologies are low capacities, high energy consumption, as well as limited recycling efficiency [9,37,93]. An advantage of this technology is its robustness which requires only minor pretreatment and conditioning of the feed since many hazard potentials are eliminated automatically during smelting [9].

Hydrometallurgy

One option in pyrometallurgical processing is to produce intermediate instead of final products which are refined by hydrometallurgical processes [105]. Currently, the favorable option is to directly process active material concentrates with a combination of several hydrometallurgical unit operations [16,70]. Hydrometallurgical processing is mainly applied for the metals coated on the cathode. The processing can be subdivided into dissolving and concentrating of the feed as well as cleaning and recovering the metal salts [106]. Various organic and inorganic acids were examined as solvents often mixed with deoxidizing agents to increase the recovery rate [9,93,106]. Furthermore, bio-organisms are able to dissolve and convert the metals by caustic methods as well [4,9,77].

Once dissolved, the metals are extracted from the solvent by liquid–liquid extraction, ion exchange, or chemical precipitation [69,77,97]. If the resulting metal salts meet the quality requirements of the corresponding raw materials, subsequent recovery can be forgone. Otherwise, further precipitation, crystallization, or electrochemical processes such as extraction electrolysis or electro winning are used. Thereby, corresponding metal compounds or impurities are separated selectively or deposited on electrodes, respectively [4,77,106].

Hydrometallurgical processes require material preparation and size control provided by manual dismantling and/or liberation and separation. The direct manual dismantling of batteries to electrode level provides highly pure feed materials on one hand [61] and decreases the required piece sizes for further hydrometallurgical processing on the other. For industrial implementation, however, manual removal of the valuable cathode materials does not appear to be expedient from both an economic and occupational safety point of view [70] (cf. Dismantling and Depollution). If the electrode coatings are separated mechanically (cf. 2.2.3), the coating fraction contains some copper and aluminum particles of the metallic conductor foils as contaminants [9,80]. Their particle size distribution and material composition are thus significantly influenced by the type of processing, which in turn determines the metallurgical effort and yield of recyclable materials [9,106]. Further impurities are residues of organic solvents, which influence the pH value of the solution and the performance of other solvent based processes. Therefore, these impurities have to be considered in the design stage of hydrometallurgical processes as well.

As a more general rule, hydrometallurgical processes require strong organic acids and expensive additives producing considerable amounts of waste liquids [107] and harmful or toxic emissions [22]. In contrast, metals from the cathode coating can be recovered in an energy-efficient and selective way [4,69].

3. Industrial Recycling Technologies

Numerous recycling companies worldwide treat spent LIBs of different types and forms (cf. Table 1). The recycling technologies differ according to the used process stages and unit processes as well as the generated final products. This is due to the historical development of the individual companies, the environmental conditions and regulations, as well as the relevant market situation. In principle, it is possible to characterize each technology on the basis of the process stages and unit processes introduced in Section 2. Literature often shows only the unit processes for the last process stage (metallurgy) in tabular form (cf. Table 1—data from tables). In some cases, there are still indications of pretreatment and mechanical processing from which the overall recycling technology can be guessed, but which cannot be reliably traced. Detailed descriptions and explanations are available for certain process technologies (cf. Table 1—process described). However, the low information density prevents a clear and complete characterization according to the process stages and unit processes [6].

Table 1. Overview of industrial recycling technologies for spent lithium-ion batteries (mech = mechanical processing; hydro = hydrometallurgy; pyro = pyrometallurgy; n. d. = no data).

Company	Data from Tables	Process Described
ACCUREC GmbH	mech [108], pyro [3,26,69], pyro and hydro [10,59], pyrolysis and hydro [109,110], disassembly, pyrolysis, mech [7], n. d. [111,112]	[3,7,10,13,98,113]
AEA Technology Batteries	hydro [59,69]	[98,114]
AERC Recycling Solutions	pyro [59] n. d. [69]	-
AFE Group (Valdi)/ ERAMET	pyro [59,69]	-
AkkuSer	mech [59], mech and hydro [110], n. d. [111]	[98,113]
American Manganese	n. d.	-
Anhua Taisen Recycling Technology Co. Ltd.	mech and hydro	-
Battery Resourcers LLC	n. d.	[13]
Battery Safety Solutions	collection, discharge and disassembly	-
Batrec Industrie AG	mech [108], mech and pyro [59], mech and hydro [26], pyro [3,69], hydro [10], pyrolysis and pyro [109,110], mech, pyrolysis, mech, hydro [7], n. d. [111,112,115]	[3,7,10,13,98,113,115]
BDT	n. d. [111]	-
Brunp Recycling Technology Co.	hydro [3,59,116]	[13]
Cawleys	n. d.	-
Chemetall	n. d. [1,2]	-
DOWA Eco-Systems Co. Ltd.	pyro [59,69], n. d. [111,115]	[117]
DK Recycling und Roheisen GmbH	pyro [69]	-
Düsenfeld GmbH	mech and hydro [118]	(for LithoRec [13])
Earthtech	disassembly	-
Erlos/Nickelhütte Aue	disassembly [112], pyro and hydro	-
Euro Dieuze Industrie/ SARP	hydro [59,69,109,110], n. d. [111]	-
Farasis Energy	-	[13]
Fuoshan Bangpu Ni/Co High-Tech Co.	n. d. [111]	-
GHTECH	-	[13]
G&P Batteries (Ecobat Technologies Ltd.)	pyro and hydro [109,110], n. d. [59,111,115]	-
GRS Batterien	pyro [109,110], n. d. [111]	-
Guangdong Guanghua Sci-Tech Co., Ltd.	disassembly	-
Highpower International Inc.	disassembly, pyro and hydro	[13]
Huayou Cobalt New Material Co Ltd.	pyro	-
Inmetco	pyro [3,7,69], n. d. [115]	[7,10,13,113,119]
Japan Recycling Center	pyro [69]	-
JX Nippon Mining and Metals Co.	pyro [59], pyro and hydro [69], n. d. [111]	-
KYOEI Steel	pyro	-
Li-Cycle US	mech and hydro	-
Lithion Recycling	hydro	-
Metal-Tech Ltd.	n. d. [111,115]	-
Neometals	hydro	-

Table 1. *Cont.*

Company	Data from Tables	Process Described
Nippon Recycle Center Corp.	pyro [59]	-
OnTo Technology Oregon US	hydro [69], mech [116], presorting, disassembly and hydro [7]	[7,13]
Pilagest	mech and hydro [109,110], n. d. [69]	-
PROMESA GmbH & Co. KG	mech	-
Recupyl S.A.S	mech [108], mech and hydro [7,110], hydro [3,10,26,59,69,109], n. d. [111,115]	[7,10,13,98,113,119]
REDUX GmbH	pyro [69], pyrolysis and mech	[54]
REVATECH	n. d. [115], n. d. [111]	-
SAFT. AB	pyro [69]	-
Salesco Systems	pyro [69]	-
Shenzhan BAK Battery Co.	disassembly [31]	-
Shenzhen Green Eco Manufacturer Hi-Tech. Co., Ltd.	mech and hydro [116], hydro [3,59], n. d. [111,115]	[3,13]
Shenzhen Tele Battery Recycling Co.	hydro [31]	-
SK Innovation Co	n. d.	-
S.N.A.M.	mech, pyrolysis and pyro [110], pyro [3,69,109], pyro and hydro [59], n. d. [111,115]	[98,113]
Sony Corp. & Sumitomo Metals and Mining Co.	pyro [3,59,69], n. d. [111,115]	[3,13,98,113]
Soundon New Energy Tech. Co. Ltd.	-	-
SungEel Hitech Ltd.	mech and hydro [116]	[13]
Technologies Inc	n. d. [59]	-
TES-AMM China	n. d. [111]	[13]
Toxco/ Retriev Tech.	hydro [26,59], mech [69,108], disassembly, cryogenic pretreatment, mech and hydro [3,7,10], n. d. [111,112,115]	[3,7,10,98,113,115,119]
Umicore	pyro [10,59], pyro and hydro [3,7,26,108–110], n. d. [69,111,112,115]	[3,7,10,13,98,113,119]
Xstrata/ Glencore	pyro [7,26], pyro and hydro [3,59,69,108], n. d. [111,115]	[7,10,13,98,113]
4R Energy Corp.	n. d. [116]	-

The production of LIBs has so far taken place almost exclusively in China, South Korea, and Japan [39,58,77,120]. Hence, battery waste is mainly recycled in Asian and only a few European and North American plants [26,29,82,91,111,121]. American and European recycling companies show a wide variety of technologies but lack the volumes of spent batteries for profitable operation [70,121]. Avoiding high investments for dedicated process equipment, spent LIBs are also fed as secondary feed in existing metallurgical plants [7] (cf. Presorting).

Lv et al. [6] and De-Leon [116] published information on the capacities of specialized technologies for certain feed materials or material mixtures. Currently, the industrial recycling approaches focus primarily on the recovery of the valuable metals cobalt and nickel from portable and industrial batteries [29,40]. Therefore, the entire LIB is broken down either thermally or mechanically in order to be recovered by pyro- and/or hydrometallurgical processes. Aiming to increase the total recycling efficiency of current industrial recycling processes, also aluminum and organic battery components, such as the electrolyte and plastics, should be considered for recovery [20,80,99]. Hence, unit operations like thermal pretreatment or separation have to be adjusted in order to avoid decomposition of the organic battery components.

In general, three industrial process routes can be identified for material recovery of spent LIBs depending on the temperature depolluting the batteries, effort for preparation and processing and overall recycling efficiency (cf. Figure 6).

- high temperature route with optional presorting and calcination as deactivation, no processing but direct pyrometallurgical treatment (cf. Figure 6A)-p), and optional hydrometallurgical refining (cf. Figure 6A)-h)
- moderate temperature route with pyrolysis as thermal pretreatment, multistage mechanical processing (cf. Figure 6B)-n; n ≥ 1), and pyro- and (cf. Figure 6B)-p-h)/or hydrometallurgical (cf. Figure 6B)-h) refining
- low temperature route (often called direct recycling process [13]) with electrical and no/or cryogenic depollution, multistage processing, and hydrometallurgical refining (cf. Figure 6C)-n; n ≥ 1)

In general, the material, characterized by the arrows with continuous lines, flows top down through the different process stages and unit processes transforming spent batteries into secondary (raw) materials, emissions, and waste. Arrows with dotted lines show on the one hand an optional hydrometallurgical treatment after pyrometallurgical unit operations (cf. Figure 6A,B) or an optional iteration of the respective unit operations. Due to generalization and low information density of the industrial technologies, the material transformation within the process stages and unit processes cannot be further distinguished in concentrates for further treatment and secondary materials.

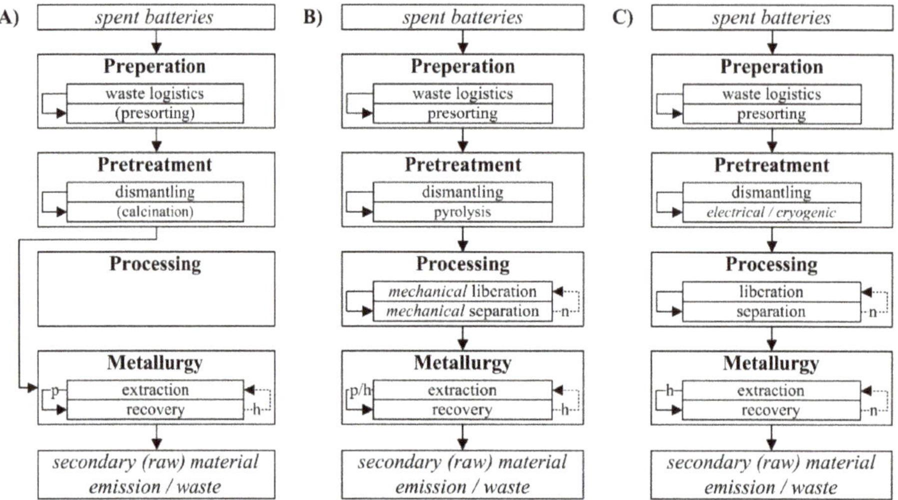

Figure 6. Schematic flow chart of the three most common industrial process routes: high temperature route (**A**), moderate temperature route, (**B**) and low temperature route (**C**); p—pyrometallurgical, h—hydrometallurgical, n—number of iterations.

It can be stated that the possible overall recycling efficiency decreases with higher process temperatures due to the decomposition of organic components and later due to the downgrading of metals like aluminum and manganese into slag. A detailed discussion of the different process routes is presented by Harper et al. [9], Lv et al. [6], Chen et al. [13], and Pinegar and Smith [7]. Nevertheless, an overall quantitate process analysis for the three main and other industrial recycling technologies in respect to recycling efficiency, energy, and auxiliary material consumption are scarce in the literature [101]. For example, the energy demand to recover the different metals depends on the specific recycling technology employed, as well as the forms of the final secondary (raw) materials [101]. Only one life cycle assessment can be found in literature for technologies containing two process routes. On the one hand, a combination of the unit processes dismantling, depollution, pyrometallurgical extraction, and recovery represents the high temperature route [60], and on the other hand a combination of the unit processes dismantling, depollution, liberation, separation, multistep hydrometallurgical extraction, and recovery [122] characterize the low temperature route.

4. Conclusions

The global search of recycling technologies for spent LIBs has led to various nomenclature and confusing classifications of relevant processes. The common goal is a high recycling efficiency and the suitability in terms of recovery of material and energy. In contrast, the generic process chain for waste from Martens and Goldmann [19] offers the possibility of dividing the individual processes into four process stages and associated unit processes. The different recycling technologies can thus be clearly classified, differentiated, and process specifications addressed.

It is conspicuous that, currently, no company is carrying out the entire process chain for spent LIBs. They rather specialize in certain process stages, combinations of process stages, or only unit processes. Advantages and disadvantages of the unit operations discussed in this review address the existing technologies knowing that an optimized entire process chain does not exist yet. Most probably, the optimum from an energetic, material, ecological, and economic point of view will be found from a combination of different unit processes and operations. It is noticeable that some of the unit processes are partly iterated. For example, the process stage of processing can have multiple steps of liberation and subsequent separation. Then, other stages or groups are completely skipped, especially in technologies with pyrometallurgical processes, which partly dispense with depollution or processing. Although further and deeper information about the industrial technologies are missing, a rough assessment and understanding of these technologies can be gained in respect to the currently used process stages and unit operations. Nevertheless, a reliable comparison and evaluation of the several recycling technologies cannot be done.

So far, research has focused on metallurgical processing with pyro- and increasingly hydrometallurgical processes. On a laboratory scale, manual dismantling of the battery systems and subsequent dismantling of the battery cells to electrode level delivers usually the relevant feed materials (cathode coating). Extrapolating this to industrial plants, mechanical liberation with subsequent sorting will become necessary. Different approaches are available for this mechanical treatment. However, they are not distinguishable based on their unit processes alone. In such a case, the unit processes have to be further subdivided into unit subprocesses and unit operations like liberation dependent on the stresses applied (kind, intensity, speed of stressing tools) process medium, specification of the machines, and apparatus used, etc. [88]. The same applies to the other process stages, but primarily hydrometallurgy, where parameters such as solid–liquid ratio, solvents used, and process conditions (temperature, residence time) influence the material conversion processes. Necessary information is also lacking due to the economic competition situation and the current spirit of optimism that will not change in the near future.

The article shows that production of secondary raw materials or materials from spent batteries is in principle technically possible. Moreover, waste treatment of spent LIBs is currently carried out already worldwide [13]. Though the collection of spent batteries and the revenue-generating sale of the secondary raw materials produced remain decisive for the recycling [123]. Additionally, the political framework for an economically and ecologically reasonable collection system need to be created for long-term success [9]. Finally, recycling of spent LIB can contribute to the global supply of metallic resources for LIB production in the long term, but for the ongoing increasing demand and battery wastes, primary resources will stay inevitable [124] and present global recycling capacity has to be expanded [70], respectively.

Author Contributions: T.M. created the initial idea for the article. D.W. produced the original concept of the Review, and wrote the article, integrating contributions from T.M. and U.A.P., editing and shaping the Review. T.M. and U.A.P. contributed to the nomenclatures and critically revised the article, the former multiple times. All authors have read and agreed to the published version of the manuscript.

Funding: This research received no external funding.

Acknowledgments: The author would like to acknowledge the technical support and contributions in kind of materials used for prior experiments by BMW Group Munich, especially Recycling und Demontagezentrum of BMW Group in Unterschleißheim, Munich.

References

1. Gu, F.; Guo, J.; Yao, X.; Summers, P.A.; Widijatmoko, S.D.; Hall, P. An investigation of the current status of recycling spent lithium-ion batteries from consumer electronics in China. *J. Clean. Prod.* **2017**, *161*, 765–780. [CrossRef]

2. Melin, H.E. *State-of-the-Art in Reuse and Recycling of Lithium-Ion Batteries—A Research Review*; Circular Energy Storage: Stockholm, Sweden, 2019.

3. Li, L.; Zhang, X.; Li, M.; Chen, R.; Wu, F.; Amine, K.; Lu, J. The Recycling of Spent Lithium-Ion Batteries: A Review of Current Processes and Technologies. *Electrochem. Energy Rev.* **2018**, *1*, 461–482. [CrossRef]

4. Zhao, S.; He, W.; Li, G. Recycling Technology and Principle of Spent Lithium-Ion Battery. In *Recycling of Spent Lithium-Ion Batteries—Processing Methods and Environmental Impacts*; An, L., Ed.; Springer Nature Switzerland AG: Cham, Switzerland, 2019.

5. Korthauer, R. *Handbuch Lithium-Ionen-Batterien*; Korthauer, R., Ed.; Springer: Berlin/Heidelberg, Germany, 2013.

6. Lv, W.; Wang, Z.; Cao, H.; Sun, Y.; Zhang, Y.; Sun, Z. A Critical Review and Analysis on the Recycling of Spent Lithium-Ion Batteries. *ACS Sustain. Chem. Eng.* **2018**, *6*, 1504–1521. [CrossRef]

7. Pinegar, H.; Smith, Y.R. Recycling of End-of-Life Lithium Ion Batteries, Part I: Commercial Processes. *J. Sustain. Met.* **2019**. [CrossRef]

8. Mathieux, F.; Ardente, F.; Bobba, S.; Nuss, P.; Blengini, G.; Alves Dias, P.; Blagoeva, D.; Torres De Matos, C.; Wittmer, D.; Pavel, C.; et al. *Critical Raw Materials and the Circular Economy—Background Report*; Publications Office of the European Union: Luxembourg, 2017; ISBN 978-92-79-74282-8.

9. Harper, G.; Sommerville, R.; Kendrick, E.; Driscoll, L.; Slater, P.; Stolkin, R.; Walton, A.; Christensen, P.; Heidrich, O.; Lambert, S.; et al. Recycling lithium-ion batteries from electric vehicles. *Nature* **2019**, *575*, 75–86. [CrossRef] [PubMed]

10. Hanisch, C.; Diekmann, J.; Stieger, A.; Haselrieder, W.; Kwade, A. Recycling of Lithium-Ion Batteries. In *Handbook of Clean Energy Systems*; Yan, J., Cabeza, L.F., Sioshansi, R., Eds.; John Wiley & Sons, Ltd.: Hoboken, NJ, USA, 2015; Volume 5, pp. 2865–2888.

11. Europäische Union. *RICHTLINIE 2006/66/EG DES EUROPÄISCHEN PARLAMENTS UND DES RATES*; Europäische Union: Brussels, Belgium, 2006.

12. Hettesheimer, T.; Thielmann, A.; Neef, N.; Möller, K.-C.; Wolter, M.; Lorentz, V.; Gepp, M.; Wenger, M.; Prill, T.; Zausch, J.; et al. *ENTWICKLUNGSPERSPEKTIVEN FÜR ZELLFORMATE VON LITHIUM-IONENBATTERIEN IN DER ELEKTROMOBILITÄT*; Fraunhofer-Institut für System- und Innovationsforschung ISI: Karlsruhe, Germany, 2017.

13. Chen, M.; Ma, X.; Chen, B.; Arsenault, R.; Karlson, P.; Simon, N.; Wang, Y. Recycling End-of-Life Electric Vehicle Lithium-Ion Batteries. *Joule* **2019**. [CrossRef]

14. Wuschke, L.; Jäckel, H.-G. Zur mechanischen Aufbereitung von Li-Ionen-Batterien. *Berg Hüttenmännische Mon.* **2016**, *161*, 267–276. [CrossRef]

15. Pinegar, H.; Smith, Y.R. Mechanical Beneficiation of End-of-Life Lithium-Ion Battery Components. In *Energy Technology 2020: Recycling, Carbon Dioxide Management, and Other Technologies*; Chen, H., Zhong, Y., Zhang, L., Howarter, J.A., Baba, A.A., Wang, C., Sun, Z., Zhang, M., Olivetti, E., Luo, A., et al., Eds.; Springer Nature: Cham, Switzerland, 2020; pp. 259–267.

16. Pinegar, H.; Smith, Y.R. End-of-Life Lithium-Ion Battery Component Mechanical Liberation and Separation. *J. Miner. Met. Mater. Soc.* **2019**, *71*, 4447–4456. [CrossRef]

17. Lowe, M.; Tokuoka, S.; Trigg, T.; Gereffi, G.; Abayechi, A. *Lithium-Ion Batteries for Electric Vehicles: THE U.S. VALUE CHAIN*; Center on Globalization, Governance & Competitiveness, Duke University: Durham, NC, USA, 2010.

18. Kwade, A.; Diekmann, J. *Recycling of Lithium-Ion Batteries—The LithoRec Way*; Kwade, A., Diekmann, J., Eds.; Springer: New York, NY, USA, 2018; p. 312.

19. Martens, H.; Goldmann, G. *Recyclingtechnik*; Martens, H., Goldmann, G., Eds.; Springer Vieweg: Wiesbaden, Germany, 2016; Volume 2.

20. Werner, D.; Mütze, T.; Jäckel, H.-G.; Peuker, U.A. Mechanical processing of spent lithium-ion-batteries from electric vehicles. In Proceedings of the International V4 Waste Recycling 21 Conference, Miscolc, Hungary, 22–23 November 2018; pp. 20–28.

21. BMJV. *Gesetz zur Förderung der Kreislaufwirtschaft und Sicherung der umweltverträglichen Bewirtschaftung von Abfällen (Kreislaufwirtschaftsgesetz—KrWG)*; Bundestag: Bonn, Germany, 2017.

22. Yun, L.; Linh, D.; Shui, L.; Peng, X.; Garg, A.; Le, M.L.P.; Asghari, S.; Sandoval, J. Metallurgical and mechanical methods for recycling of lithium-ion battery pack for electric vehicles. *Resour. Conserv. Recycl.* **2018**, *136*, 198–208. [CrossRef]

23. Schubert, H.; Heidenrein, E.; Lippe, F.; Neeße, T. *Mechanische Verfahrenstechnik*; Deutscher Verlag für Grundstoffindustrie: Leipzig, Germany, 1990; Volume 3.

24. Sattler, K. *Thermische Trennverfahren—Grundlagen, Auslegung, Apparate*; WILEY-VCH: Weinheim, Germany, 2001; Volume 3.

25. Mancha, A. A Look at Some International Lithium Ion Battery Recycling Initiatives. *J. Undergrad. Res.* **2016**, *9*, 1–5. [CrossRef]
26. Heelan, J.; Gratz, E.; Zheng, Z.; Wang, Q.; Chen, M.; Apelian, D.; Wang, Y. Current and Prospective Li-Ion Battery Recycling and Recovery Processes. *JOM* **2016**, *68*, 2632–2638. [CrossRef]
27. Zeng, X.; Li, J.; Liu, L. Solving spent lithium-ion battery problems in China: Opportunities and challenges. *Renew. Sustain. Energy Rev.* **2015**, *52*, 1759–1767. [CrossRef]
28. Lewis, H. *Lithium-Ion Battery Consultation Report*, Helen Lewis Research (Consultation Report). 3 August 2016.
29. Scrosati, B.; Garche, J.; Sun, Y.-K. Recycling lithium batteries. In *Advances in Battery Technologies for Electric Vehicles*, 1st ed.; Scrosati, B., Garche, J., Tillmetz, W., Eds.; Woodhead Publishing: Sawston, UK, 2015; pp. 503–516.
30. Larouche, F.; Demopoulos, G.P.; Amouzegar, K.; Bouchard, P.; Zaghib, K. Recycling of Li-Ion and Li-Solid State Batteries: The Role of Hydrometallurgy. In *Extraction 2018, Proceedings of the First Global Conference on Extractive Metallurgy (The Minerals, Metals & Materials Series); Ottawa, ON, Canada*; Davis, B.R., Moats, M.S., Wang, S., Gregurek, D., Kapusta, J., Battle, T.P., Schlesinger, M.E., Alvear Flores, G.R., Jak, E., Goodall, G., et al., Eds.; Springer: New York, NY, USA, 2018; pp. 2541–2553.
31. Zhao, G. Resource Utilization and Harmless Treatment of Power Batteries. In *Reuse and Recycling of Lithium-Ion Power Batteries*; Zhao, G., Ed.; John Wiley & Sons: Singapore, 2017; pp. 335–378.
32. Petrikowski, F.; Kohlmeyer, R.; Jung, M.; Steingrübner, E.; Leuthold, S. *Batterien und Akkus. Umweltbundesamt—Fachgebiet III 1.2*; Rechtsangelegenheiten, V.E.u.B., Ed.; Umweltbundesamt: Dessau, Germany, 2012.
33. Rudolph, A. *Altproduktentsorgung aus Betriebswirtschaftlicher Sicht*; Rudolph, A., Ed.; Physica-Verlag: Heidelberg, Germany, 1999; Volume 1.
34. Weyhe, R. Recycling von Lithium-Ion-Batterien. In *Recycling und Rohstoffe*; Thomé-Kozmiensky, K.J., Goldmann, D., Eds.; TK Verlag Karl Thomé-Kozmiensky: Neuruppin, Germany, 2013; Volume 6, pp. 505–525.
35. Wang, X.; Gaustad, G.; Babbitt, C.W. Targeting high value metals in lithium-ion battery recycling via shredding and size-based separation. *Waste Manag.* **2016**, *51*, 204–213. [CrossRef] [PubMed]
36. Pistoia, G. *Spent Battery Collection and Recycling*; Pistoia, G., Ed.; Elsevier Science: Amsterdam, The Netherlands, 2005; Volume 1.
37. Ellis, T.W.; Howes, J.A. ROADMAP FOR THE LIFECYCLE OF ADVANCED BATTERY CHEMISTRIES. In Proceedings of the REWAS 2016: Towards Materials Resource Sustainability, Nashville, TN, USA, 14–18 February 2016; pp. 51–56.
38. Cai, G.; Fung, K.Y.; Ng, K.M. Process Development for the Recycle of Spent Lithium Ion Batteries by Chemical Precipitation. *Ind. Eng. Chem. Res.* **2014**, *53*, 18245–18259. [CrossRef]
39. Zeng, X.; Li, J.; Singh, N. Recycling of Spent Lithium-Ion Battery: A Critical Review. *Crit. Rev. Environ. Sci. Technol.* **2014**, *10*, 1129–1165. [CrossRef]
40. Weyhe, R.; Friedrich, B. *Demonstrationsanlage für Ein Kostenneutrales, Ressourceneffizientes Processing Ausgedienter Li-Ionen Batterien aus der Elektromobilität—EcoBatRec—Abschlussbericht zum Verbundvorhaben*; IME Metallurgische Prozesstechnik und Metallrecycling: Aachen, Germany, 2016.
41. Nigl, T.; Pomberger, R. Entwicklungen und Problematik von Lithium-Batterien. In Proceedings of the Recy&DepoTech 2016, Leoben, Austria, 8–11 November 2016.
42. Wang, M.-M.; Zhang, C.-C.; Zhang, F.-S. Recycling of spent lithium-ion battery with polyvinyl chloride by mechanochemical process. *Waste Manag.* **2017**, *67*, 232–239. [CrossRef]
43. Kim, H.; Jang, Y.-C.; Hwang, Y.; Ko, Y.; Yun, H. End-of-life batteries management and material flow analysis in South Korea. *Front. Environ. Sci. Eng.* **2018**, *12*. [CrossRef]
44. Hagelüken, C. Die neue Circular Economy Strategie der EU—Ein Durchbruch zur Sicherung der Rohstoffbasis? In Proceedings of the Dechema Kolloquium: Anorganische Rohstoffe—Sicherung der Rohstoffbasis von morgen, Frankfurt, Germany, 4 February 2016.
45. Kaya, M. Recovery of metals and nonmetals from electronic waste by physical and chemical recycling processes. *Waste Manag.* **2016**, *57*, 64–90. [CrossRef]
46. Ay, P.; Markowski, J.; Pempel, H.; Müller, M. Entwicklung eines innovativen Verfahrens zur automatisierten Demontage und Aufbereitung von Lithium-Ionen-Batterien aus Fahrzeugen. *Recycl. Rohst.* **2012**, *5*, 443–456.
47. Beckmann, J. *Batterien und Akkumulatoren*; Bayerisches Landesamt für Umwelt: Augsburg, Germany, 2017.

48. ZVEI; EPTA. *Versand von Lithium-Ionen-Batterien für Elektrowerkzeuge und Elektrische Gartengeräte: Umsetzung der Gefahrgut-Vorschriften*; EPTA–European Power Tool Association: Frankfurt, Germany, 2017.

49. LAGA. *Umsetzung des Elektro- und Elektronikgerätegesetzes–Anforderungen an die Entsorgung von Elektro- und Elektronikaltgeräten*; Bund/Länder-Arbeitsgemeinschaft Abfall: Berlin, Germany, 2017.

50. BDE. *Praxisleitfaden: Lithiumbatterien und -Zellen (Auch in Elektroaltgeräten): Sammlung, Verpackung und Transport Gemäß ADR*; Bundesverband der Deutschen Entsorungs: Berlin, Germany, 2017.

51. Richa, K. Sustainable Management of Lithium-Ion Batteries after Use in Electric Vehicles. Ph.D. Thesis, Rochester Institute of Technology, Rochester, NY, USA, 2016.

52. DSLV. *ADR 2017: Die Wichtigsten Änderungen der Vorschriften für die Beförderung Gefährlicher Güter auf der Straße im Überblick*; DSLV Deutscher Speditions- und Logistikverband e. V.: Bonn, Germany, 2016; p. 32.

53. Nickel, W. *Recyclinghandbuch: Strategien—Technologien—Produkte*; Nickel, W., Ed.; VDI-Verlag: Düsseldorf, Germany, 1996.

54. Arnberger, A.; Coskun, E.; Rutrecht, B. Recycling von Lithium-Ionen-Batterien. In *Recycling und Rohstoffe*; Thomé-Kozmiensky, K.J., Goldmann, D., Eds.; TK Verlag Karl Thomé-Kozmiensky: Neuruppin, Germany, 2018; Volume 11, pp. 583–599.

55. Sziegoleit, H. Sortierung von Gerätebatterien. In *Recycling und Rohstoffe*; Thomé-Kozmiensky, K.J., Goldmann, D., Eds.; TK Verlag Karl Thomé-Kozmiensky: Neuruppin, Germany, 2013; Volume 6, pp. 495–504.

56. Elwert, T.; Römer, F.; Schneider, K.; Hua, Q.; Buchert, M. Recycling of Batteries from Electric Vehicles. In *Behaviour of Lithium-Ion Batteries in Electric Vehicles—Battery Health, Performance, Safety, and Cost*; Pistoia, G., Liaw, B., Eds.; Springer International Publishing AG: Cham, Switzerland, 2018.

57. Bernardes, A.M.; Espinosa, D.C.R.; Tenório, J.A.S. Recycling of batteries: A review of current processes and technologies. *J. Power Sources* **2004**, *130*, 291–298. [CrossRef]

58. Zhang, X.; Xie, Y.; Lin, X.; Li, H.; Cao, H. An overview on the processes and technologies for recycling cathodic active materials from spent lithium-ion batteries. *J. Mater. Cycles Waste Manag.* **2013**, *15*, 420–430. [CrossRef]

59. Lebedeva, N.; Di Persio, F.; Boon-Brett, L. *Lithium Ion Battery Value Chain and Related Opportunities for Europe*; European Commission: Petten, The Netherlands, 2016.

60. Buchert, M.; Jenseit, W.; Merz, C.; Schüler, D. *Verbundprojekt: Entwicklung Eines Realisierbaren Recyclingkonzepts für die Hochleistungsbatterien Zukünftiger Elektrofahrzeuge—LiBRi*; Öko-Institut e.V.: Darmstadt, Germany, 2011.

61. Li, L.; Zheng, P.; Yang, T.; Sturges, R.; ELLIS, M.W.; Li, Z. Disassembly Automation for Recycling End-of-Life Lithium-Ion Pouch Cells. *J. Miner. Met. Mater. Soc.* **2019**, *71*, 4457–4464. [CrossRef]

62. Idjis, H.; Attias, D.; Bocquet, J.-C.; Sophie, R. Designing a Sustainable Recycling Network for Batteries from Electric Vehicles. Development and Optimization of Scenarios. In Proceedings of the 14th Working Conference on Virtual Enterprises, Dresden, Germany, 30 September–2 October 2013; pp. 609–618.

63. Steinbild, M. *Recycling von Lithium-Ionen-Batterien—LithoRec II: Abschlussbericht der Beteiligten Verbundpartner*; Bundesministeriums für Umwelt, Naturschutz, Bau und Reaktorsicherheit: Bonn, Germany, 2017.

64. Steinbild, M. *Recycling von Lithium-Ionen-Batterien—LithoRec II*; Rockwood Lithium GmbH: Frankfurt, Germany, 2012.

65. Treffer, F. *Entwicklung Eines Realisierbaren Recyclingkonzeptes für die Hochleistungsbatterien Zukünftiger Elektrofahrzeuge*; Umicore AG & Co. KG: Hanau, Germany, 2011.

66. Zhao, G. Assessment Technology Platform and Its Application for Reuse of Power Batteries. In *Reuse and Recycling of Lithium-Ion Power Batteries*; Zhao, G., Ed.; John Wiley & Sons: Singapore, 2017; pp. 37–260.

67. Wegener, K.; Andrew, S.; Raatz, A.; Dröder, K.; Herrmann, C. Disassembly of Electric Vehicle Batteries Using the Example of the Audi Q5 Hybrid System. In Proceedings of the Conference on Assembly Technologies and Systems, Dresden, Germany, 13–14 November 2014.

68. Plaßmann, W.; Schulz, D. *Handbuch Elektrotechnik: Grundlagen und Anwendungen für Elektrotechniker*; Plaßmann, W., Schulz, D., Eds.; Vieweg+Teubner: Wiesbaden, Germany, 2009; Volume 5.

69. Gama, M. *European Li-Ion Battery Advanced Manufacturing For Electric Vehicles-ELIBAMA*, online. 2014.

70. Pinegar, H.; Smith, Y.R. Recycling of End-of-Life Lithium-Ion Batteries, Part II: Laboratory-Scale Research Developments in Mechanical, Thermal, and Leaching Treatments. *J. Sustain. Metall.* **2020**. [CrossRef]

71. Wuschke, L.; Jäckel, H.-G.; Leißner, L.; Peuker, U.A. Crushing of large Li-ion battery cells. *Waste Manag.* **2019**, *85*, 317–326. [CrossRef]

72. Nan, J.; Han, D.; Zuo, X. Recovery of metal values from spent lithium-ion batteries with chemical deposition and solvent extraction. *J. Power Sources* **2005**, *152*, 278–284. [CrossRef]

73. Choubey, P.K.; Chung, K.-S.; Kim, M.-S.; Lee, J.-C.; Srivastava, R.R. Advance review on the exploitation of the prominent energy-storage element Lithium. Part II: From sea water and spent lithium ion batteries (LIBs). *Miner. Eng.* **2017**, *110*, 104–121. [CrossRef]

74. Shaw-Stewart, J.; Alvarez-Reguera, A.; Greszta, A.; Marco, J.; Masood, M.; Sommerville, R.; Kendrick, E. Aqueous solution discharge of cylindrical lithium-ion cells. *Sustain. Mater. Technol.* **2019**, *22*, e00110. [CrossRef]

75. Li, J.; Wang, G.; Xu, Z. Generation and detection of metal ions and volatile organic compounds (VOCs) emissions from the pretreatment processes for recycling spent lithium-ion batteries. *Waste Manag.* **2016**, *52*, 221–227. [CrossRef]

76. Träger, T.; Friedrich, B.; Weyhe, R. Recovery Concept of Value Metals from Automotive Lithium-Ion Batteries. *Chem. Ing. Tech.* **2015**, *87*, 1550–1557. [CrossRef]

77. Vezzini, A. Manufacturers, Materials and Recycling Technologies. In *Lithium-Ion Batteries Advances and Applications*; Pistoia, G., Ed.; Elsevier: Amsterdam, The Netherlands, 2014.

78. Xiao, J.; Li, J.; Xu, Z. Recycling metals from lithium ion battery by mechanical separation and vacuum metallurgy. *J. Hazard. Mater.* **2017**, *338*, 124–131. [CrossRef] [PubMed]

79. Rao, S.R. *Resource Recovery and Recycling from Metallurgical Wastes*; Rao, S.R., Ed.; Elsevier: Amsterdam, The Netherlands, 2006; Volume 7.

80. Wuschke, L. Mechanische Aufbereitung von Lithium-Ionen-Batteriezellen. Ph.D. Thesis, TU Bergakademie Freiberg, Freiberg, Germany, 2018.

81. Schubert, G. Comminution Equipment for Non-Brittle Waste and Scrap. *AUFBEREITUNGSTECHNIK* **2002**, *43*, 6–23.

82. Wuschke, L.; Jäckel, H.-G.; Peuker, U.A.; Gellner, M. Recycling of Li-ion batteries—A challenge. *Recovery* **2015**, *4*, 48–59.

83. Arnberger, A.; Gresslehner, K.-H.; Pomberger, R.; Curtis, A. Recycling von Lithium Ionen Batterien aus EVs & HEVs. In Proceedings of the DepoTech 2012, Leoben, Austria, 6 November 2012.

84. Kilo, M.; Bittner, A. *Chemikalienfreie Wertstoffabtrennung für Hocheffiziente Recyclingprozesse*; ISCF: London, UK, 2016.

85. Berg, N.; Eisert, S. Bessere Verwertungsmöglichkeiten von Verbundmaterialien mittels Schockwellentechnologie. In *Ressourceneffizienz vor Ort*; Freiberg, Germany, 2016.

86. Bokelmann, K.; Horn, D.; Zimmermann, J.; Gellermann, C.; Stauber, R. Recycling von Li-Ionen-Batterien. *Chem. Unserer Zeit* **2018**, *52*, 2–3. [CrossRef]

87. Leißner, T.; Hamann, D.; Wuschke, L.; Jäckel, H.-G.; Peuker, U.A. High voltage fragmentation of composites from secondary raw materials—Potential and limitations. *Waste Manag.* **2018**, *74*, 123–134. [CrossRef] [PubMed]

88. Valio, J. Critical Review on Lithium Ion Battery Recycling Technologies. Master's Thesis, Aalto University, Helsinki, Finland, 2017.

89. Tedjar, F.; Foudraz, J.-C. Method for the mixed recycling of lithium-based anode batteries and cells. U.S. Patent 7,820,317, 26 October 2010.

90. Zhan, R.; Oldenburg, Z.; Pan, L. Recovery of active cathode materials from lithium-ion batteries using froth flotation. *Sustain. Mater. Technol.* **2018**, *1*, e00062. [CrossRef]

91. Hanisch, C.; Haselrieder, W.; Kwade, A. Recycling von Lithium-Ionen-Batterien—Das Projekt LithoRec. In *Recycling und Rohstoffe*; Thomé-Kozmienski, K.J., Goldmann, D., Eds.; TK Verlag: Neuruppin, Germany, 2012; Volume 5, pp. 691–698.

92. Hanisch, C.; Loellhoeffel, T.; Diekmann, J.; Markley, K.J.; Haselrieder, W.; Kwade, A. Recycling of lithium-ion batteries: A novel method to separate coating and foil of electrodes. *J. Clean. Prod.* **2015**, *108*, 301–311. [CrossRef]

93. Gao, W.; Zhang, X.; Zheng, X.; Lin, X.; Cao, H.; Zhang, Y.; Sun, Z. Lithium Carbonate Recovery from Cathode Scrap of Spent Lithium-Ion Battery: A Closed-Loop Process. *Environ. Sci. Technol.* **2017**, *51*, 1662–1669. [CrossRef]

94. Shuva, A.H.; Kurny, A.S.W. Hydrometallurgical Recovery of Value Metals from Spent Lithium Ion Batteries. *Am. J. Mater. Eng. Technol.* **2013**, *1*, 8–12.

95. Vanderbruggen, A. Characterization and Beneficiation of Pyrolyzed Black Mass: Increasing the Recycling Rate of Spent Lithium Ion Battery. Master'Thesis, TU Bergakademie Freiberg, Freiberg, Germany, 2018.

96. Moradi, B.; Botte, G.G. Recycling of graphite anodes for the next generation of lithium ion batteries. *J. Appl. Electrochem.* **2016**, *46*, 123–148. [CrossRef]

97. Chen, W.-S.; Ho, H.-J. Recovery of Valuable Metals from Lithium-Ion Batteries NMC Cathode Waste Materials by Hydrometallurgical Methods. *Metals* **2018**, *8*. [CrossRef]

98. Chagnes, A. *Lithium Process Chemistry—Resources, Extraction, Batteries and Recycling*; Chagnes, A., Swiatowska, J., Eds.; Elsevier: Amsterdam, The Netherlands, 2015.

99. Georgi-Maschler, T.; Friedrich, B.; Weyhe, R.; Heegn, H.; Rutzc, M. Development of a recycling process for Li-ion batteries. *J. Power Sources* **2012**, *207*, 173–182. [CrossRef]

100. Hagelüken, C. Cobalt for Li-Ion Batteries–Future demand expectations and actions for sustainable sourcing. In Proceedings of the World Resource Forum, Geneva, Switzerland, 24–25 October 2017.

101. Huang, B.; Pan, Z.; Su, X.; An, L. Recycling of lithium-ion batteries: Recent advances and perspectives. *J. Power Sources* **2018**, *399*, 274–286. [CrossRef]

102. Wang, X.; Gaustad, G.; Babbitt, C.W.; Bailey, C.; Ganter, M.J.; Landi, B.J. Economic and environmental characterization of an evolving Li-ion battery waste stream. *J. Environ. Manag.* **2014**, *135*, 126–134. [CrossRef] [PubMed]

103. Wanger, T.C. The Lithium future—Resources, recycling, and the environment. *Conserv. Lett.* **2011**, *4*, 202–206. [CrossRef]

104. King, S.; Boxall, N.J.; Bhatt, A.I. *Australian Status and Opportunities for Lithium Battery Recycling*; CSIRO: Canberra, Australia, 2018.

105. Friedrich, B.; Träger, T.; Peters, L. Lithium Ion Battery Recycling and Recent IME Activities. In Proceedings of the AABC Advanced Automotive Battery Conference, Mainz, Germany, 30 January–2 February 2017.

106. Pagnanelli, F.; Moscardini, E.; Altimari, T.; Abo Atia, T.; Toro, L. Leaching of electrodic powders from lithium ion batteries: Optimization of operating conditions and effect of physical pretreatment for waste fraction retrieval. *Waste Manag.* **2017**, *60*, 706–715. [CrossRef]

107. Xu, J.; Thomas, H.R.; Francis, R.W.; Lumb, K.R.; Wang, J.; Liang, B. A review of processes and technologies for the recycling of lithium-ion secondary batteries. *J. Power Sources* **2008**, *177*, 512–527. [CrossRef]

108. Friedrich, B.; Träger, T.; Weyhe, R. Recyclingtechnologien am Beispiel Batterien. In Proceedings of the 25. Aachener Kolloquium Abfallwirtschaft, Forum M, Aachen, Germany, 29 November 2012.

109. Kushnir, D. *Lithium Ion Battery Recycling Technology 2015—Current State and Future Prospects*; Chalmers University of Technology: Chalmers, Sweden, 2015; p. 56.

110. Romare, M.; Dahllöf, L. *The Life Cycle Energy Consumption and Greenhouse Gas Emissions from Lithium-Ion Batteries*; IVL Swedish Environmental Research Institute: Stockholm, Sweden, 2017.

111. Knights, B.D.H.; Saloojee, F. *Lithium Battery Recycling*; CM Solutions—Metallurgical Consultancy and Laboratories: Modderfontein, RSA, 2015.

112. Rahimzei, E.; Sann, K.; Vogel, M. *Kompendium: Li-Ionen-Batterien—Grundlagen, Bewertungskriterien, Gesetze und Normen*; Elektronik Informationstechnik e. V.: Frankfurt am Main, Germany, 2015.

113. Marinos, D.; Mishra, B. Processing of Lithium-Ion Batteries for Zero-Waste Materials Recovery. In *Sustainability in the Mineral and Energy Sectors*; Devasahayam, S., Dowling, K., Mahapatra, M.K., Eds.; CRC Press: Boca Raton, FL, USA, 2016.

114. Lain, M.J. Recycling of lithium ion cells and batteries. *J. Power Sources* **2001**, *97–98*, 736–738. [CrossRef]

115. Nigl, T. Zwischenlagerung, Aufbereitung und Verwertung von Lithium-Ionen-Batterien und -akkumulatoren. In *Der Umgang mit High-Tech-Produkten in der Abfallwirtschaft*; Montanuniverstität Leoben Lehrstuhl für Abfallverwertungstechnik und Abfallwirtschaft: Wien, Austria, 2016.

116. De-Leon, S. Lithium Ion Battery Recycling Market 2018. Available online: https://www.sdle.co.il/download/ (accessed on 21 November 2018).

117. DOWA Holdings Co. Ltd. *Dowa Eco-System Commercializes Lithium-Ion Battery Recycling Business*; DOWA Holdings Co. Ltd: Tokyo, Japan, 2010.

118. Burkert, A. Effective Recycling of Electric-vehicle Batteries. *ATZ Worldw.* **2018**, *120*, 10–15.

119. Saloojee, F.; Llyod, J. *LITHIUM BATTERY RECYCLING PROCESS*; CRUNDWELL MANAGEMENT SOLUTIONS: Modderfontein, South Africa, 2015.

120. Jiang, J.; Zeng, X.; Li, J. Feasibility Analysis of Recycling and Disposal of Spent Lithium-ion Batteries in China. *Appl. Mech. Mater.* **2015**, *768*, 622–626. [CrossRef]

121. Melin, H.E. *The Lithium-Ion Battery End-of-Life Market—A Baseline Study*; Circular Energy Storage: London, UK, 2018.

122. Buchert, M.; Jenseit, W.; Merz, C.; Schüler, D. *Ökobilanz zum "Recycling von Lithium-Ionen-Batterien" (LithoRec)*; Öko-Institut e.V.: Darmstadt, Germany, 2011.

123. Jäckel, H.-G.; Peuker, U.A.; Wuschke, L. Probleme der Aufbereitung metallhaltiger Werkstoffverbunde aus lithiumhaltigen Geräten und Batterien. *Chem. Ing. Tech.* **2014**, *86*, 806–813. [CrossRef]

124. Konietzko, S.; Gernuks, M. *Ressourcenverfügbarkeit von Sekundären Rohstoffen -Potenzialanalyse für Lithium und Kobalt*; Volkswagen AG: Wolfsburg, Germany, 2011.

Article

Integrating Flotation and Pyrometallurgy for Recovering Graphite and Valuable Metals from Battery Scrap

Ronja Ruismäki, Tommi Rinne, Anna Dańczak, Pekka Taskinen, Rodrigo Serna-Guerrero and Ari Jokilaakso *

Department of Chemical and Metallurgical Engineering, School of Chemical Engineering, Aalto University, 02150 Espoo, Finland; ronja.ruismaki@aalto.fi (R.R.); tommi.rinne@aalto.fi (T.R.); anna.danczak@aalto.fi (A.D.); pekka.taskinen@aalto.fi (P.T.); rodrigo.serna@aalto.fi (R.S.-G.)
* Correspondence: ari.jokilaakso@aalto.fi; Tel.: +358-50-313-8885

Received: 29 April 2020; Accepted: 19 May 2020; Published: 21 May 2020

Abstract: Since the current volumes of collected end-of-life lithium ion batteries (LIBs) are low, one option to increase the feasibility of their recycling is to feed them to existing metals production processes. This work presents a novel approach to integrate froth flotation as a mechanical treatment to optimize the recovery of valuable metals from LIB scrap and minimize their loss in the nickel slag cleaning process. Additionally, the conventional reducing agent in slag cleaning, namely coke, is replaced with graphite contained in the LIB waste flotation products. Using proper conditioning procedures, froth flotation was able to recover up to 81.3% Co in active materials from a Cu-Al rich feed stream. A selected froth product was used as feed for nickel slag cleaning process, and the recovery of metals from a slag (80%)–froth fraction (20%) mixture was investigated in an inert atmosphere at 1350 °C and 1400 °C at varying reduction times. The experimental conditions in combination with the graphite allowed for a very rapid reduction. After 5 min reduction time, the valuable metals Co, Ni, and Cu were found to be distributed to the iron rich metal alloy, while the remaining fraction of Mn and Al present in the froth fraction was deported in the slag.

Keywords: mechanical treatment; slag cleaning; cobalt; nickel; manganese; lithium-ion battery; recycling; circular economy

---

1. Introduction

Since their commercial launch in 1991, lithium-ion batteries (LIBs) have become the dominant power source technology for a variety of electronic devices, from electric vehicles (EVs) to laptops, due to their superior electrochemical properties such as low self-discharge rate and high energy density [1]. The prospected demand of LIBs is expected to grow annually by 25% from 180 GWh in 2018 to 2600 GWh in 2030 [2]. The major driving factor for the growing demand is attributed to the transportation sector shifting to a low-emission fleet [2]. Consequently, an increasing demand of LIBs sets pressure on both the upstream processes (e.g., mining and refining) to extract raw materials and manufacture components, and downstream processes (e.g., second life and recycling) to maximize the recovery of secondary raw materials. Depending on the vehicle model, the LIB in an EV can make up 40% of the total costs, making it the most valuable component [3]. Therefore, the development of cost-efficient EVs is strongly focused on the value chain of batteries [3]. This in turn advocates for integrating recyclability as a design feature in the development phase [4].

The estimated lifetime of EV batteries is eight years during first life and five years for second life [5]. In consumer electronics, the lifespan is generally less than three years [6]. In 2035, 104 GWh of battery capacity is expected to reach end-of-life (EoL) [5], thus making the relatively large amount of metal reserves in EoL batteries attractive for recycling [7]. The benefits of recycling are multifold in terms of economics, regulatory perspective, securing raw material supply, new business creation opportunities [7], and environmental protection. Developing cost-efficient technologies and processes for LIB recycling is thus a necessity.

Despite this, only a small portion of the EoL LIBs are currently properly collected [8], while the vast majority are either hoarded in households, or end up in landfills [9]. This not only raises environmental concerns but is also a waste of valuable resources [10,11]. Additionally, the current EU Battery Directive 2006/66/EC in place does not sufficiently reflect the integration of the life cycle concept or the growing importance of LIBs as the recycling efficiencies are not defined for specific components or elements (e.g., Li and Co) [4,12].

The major components of a LIB include its casing, separator membrane, electrolyte, current collectors, a polymeric binder, and the active materials found in the electrodes. The active material usually comprises of graphite (anode) and some of various lithium metal oxides (cathode), and the mixture of the two is commonly referred to as black mass in industrial jargon [13]. The different components, relative amounts, and currently the most commonly applied materials were recently summarized by Velazquez-Martinez et al. [14]. Due to cost savings, supply risks [15], and increasing energy density requirements, cathode chemistries such as NMC 622 and 811, with less cobalt compared to NMC 111 or LCO chemistries, have entered the market. [7] In waste LIBs, the black mass hosts the elements with the highest economic value, and recovering these elements has, consequently, been the main focus of the industrial LIB recycling operations. [13,14] In the current state-of-the-art industrial LIB recycling processes, mechanical unit operations are first performed to recover the macroscopic components, and to separate the black mass fraction. The black mass is then further treated with either hydro- or pyrometallurgical processes (or a combination of both), and the valuables are recovered as either alloys or salts [16].

The main goal of the initial mechanical processing is to produce a sufficiently pure stream of black mass for the subsequent chemical purification, while providing a high enough throughput to ensure the economic profitability of the process. Therefore, a trade-off exists between the throughput and the grade/recovery of the valuables [17]. For example, it has been found that the different LIB components have distinctive particle size distributions after crushing, with the black mass consisting of considerably finer particles than the other components [17]. Consequently, after crushing the waste LIBs, screening is usually applied to separate the black mass in the underflow, while retaining the majority of the coarser components (Cu, Al, plastics) in the overflow [14,17]. However, due to incomplete liberation of the active particles [18], in order to increase the black mass recovery, the industrially applied sieve opening sizes are relatively large (~500 μm) [14]. This, however, results in quantifiable amounts of Al, Cu, and plastics reporting in the underflow. In addition, the Cu-Al rich sieve overflow has been shown to retain a considerable amount of the black mass [18].

Generally, it is seen that one of the advantages of applying pyrometallurgy in battery recycling is to minimize mechanical pre-treatment [19,20]. However, a relatively high fraction of metals is lost to the slag, which in some cases is further refined to recover metals or utilized by the construction industry [11,14,19]. These include metals, such as Li and Mn, with high oxide stability at high temperatures [21]. The industrial Umicore ValÉas™ and Sumitomo-Sony recycling concepts include pre-processing steps, such as dismantling and sorting, but no mechanical treatment [14]. In laboratory scale, Ren et al. [22] utilized two waste streams, namely (1) LIBs with Al cans and (2) copper slag from an industrial electric arc furnace cleaning process to produce an Fe-Co-Ni-Cu alloy and slag with fayalite (Fe_2SiO_4) and hernycite ($FeAl_2O_4$). Guoxing et al. [23] proposed a smelting reduction process at 1475 °C based on a MnO-SiO_2-Al_2O_3 slag system resulting in a Co-Ni-Cu-Fe alloy and manganese rich slag to recover valuable metals from LIBs.

The EoL LIBs will start gradually returning to metals production as their amount increases and the recycling efficiency improves. Currently, the volumes are rather low and the most feasible way to recycle them is to feed them to existing metals production processes instead of developing and building new production plants. In this approach, the waste LIB would be partially functioning as a chemical reagent, applied as a secondary feed for the industrial process, and ideally eliminating the need for new chemical reagents. Consequently, decreasing the loss of valuable metals to the slag must be researched by introducing suitable mechanical pre-treatment. However, at the same time, new flexible processes should also be developed as the chemistries and compositions of the LIBs (or other batteries) are continuously developing.

In the present investigation, the aim is to integrate existing unit operations (sieving, grinding, froth flotation) and a unit process (EF slag cleaning) for recovering Co and Ni from the black mass fraction of LIB scrap and replacing the conventional reducing agent coke with graphite. Figure 1 shows a flowchart of the proposed process investigated in this article. In previous studies by the authors, the integration of battery recycling to nickel slag cleaning was reported successful for the first time [24]. Unlike these previous studies and many industrial pyrometallurgical processes for battery recycling, this approach emphasizes the importance of mechanical pre-treatment to minimize the loss of valuable metals to the slag. Coupling froth flotation to a pyrometallurgical unit process is introduced as a novel method for improving the LIB recycling efficiency, by allowing the selective recovery of leftover active materials from the Cu-Al rich sieve overflow, which is produced as a side stream during the initial mechanical processing of the waste LIBs. This approach is fundamentally different compared to the conventional black mass flotation studies [25–30] that have targeted to separate the anodic graphite in the froth phase, and the cathode active components in the tailings. Furthermore, the recovered froth flotation products are utilized as the only source of reducing agent in the nickel slag cleaning process replacing the fossil coke.

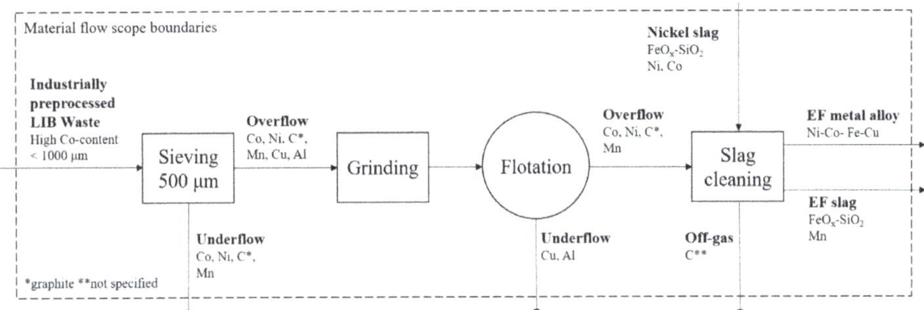

Figure 1. A simplified flowchart of the investigated LIBs recycling process. * graphite, ** not specified.

2. Materials and Methods

2.1. Froth Flotation

2.1.1. Sample Preparation and Characterization of the Waste Lithium Ion Battery Feed

LIB waste was obtained from Akkuser Oy (Nivala, Finland). In the industrial process, the waste LIBs are sorted, crushed, and sieved. Firstly, sorting was performed based on the cathode chemistry of the EoL LIBs, dividing the waste into fractions of high Co content and low Co content. The sorted fractions were subsequently crushed and sieved with a ca. 1000 µm sieve, and magnetic separation was applied to recover the steel casing materials. The present work utilized the underflow (<~1000 µm) of the industrially sieved high Co content fraction.

A sample of waste LIB of ca. 1.3 kg was taken and sieved in two batches for one hour. Sieving was performed using a vibratory sieve shaker (Fritsch Analysette 3, Idar-Oberstein, Germany) using a stack of four sieves with opening sizes of 1250 µm, 400 µm, 200 µm, and 100 µm. Since the exact sieve opening size of the industrial sieving process was unknown, the coarsest sieve selected was 1250 µm. To enable characterization with an X-ray fluorescence spectrometer (XRF, Niton XL3t, Thermo Fisher

Scientific, Waltham, MA, US), a sample of the fractions with sizes > 100 μm was pulverized for 10 min in a disc mill (Retsch RS 200, Haan, Germany), and characterized with XRF for elements with atomic number ≥ 12.

As will be further discussed in Section 3, based on the composition results of the size fractions, the samples for froth flotation experiments were prepared by dry sieving ca. 190 g of LIB waste with a 500 μm sieve, at an amplitude of 5.5 mm over 10 min. The overflow (>500 μm) of this laboratory sieving process was used as feed for the flotation experiments.

2.1.2. Experimental Procedure of Froth Flotation

As a first conditioning step, the feed material was treated in a vibratory micro mill (Fritsch, Pulverisette 0, Idar-Oberstein, Germany), with an amplitude of 6 mm, and a batch size of ca. 13.33 g. The energy input provided by the micro mill under these conditions is low, as the objective of this step was solely to improve the particle liberation, by purifying the Cu/Al/plastic film surfaces, and not to further comminute the active materials. This procedure was repeated thrice to obtain a total feed mass of 40 g. In the experimental series, seven different milling times (0 min, 5 min, 10 min, 15 min, 20 min, 30 min, and 40 min) were studied.

After milling, the material was placed in a one-liter laboratory-scale flotation cell (Lab Cell—60mm FloatForce mechanism, Outotec, Espoo, Finland), and one liter of tap water was added. The solid-water suspension was then subjected to 1000 rpm of agitation for 3 min. After this period, 150 g/t of kerosene (collector) were added, allowing 3 min of conditioning under stirring. Finally, 8 ppm of MIBC (frother) were added, followed by two minutes of conditioning. Subsequently, air was fed to the cell at a flowrate of 2 L/min, marking the beginning of the flotation experiment. Throughout conditioning and flotation stages, the agitation was kept constant at 1000 rpm.

After the air flow was turned on, the system was given approximately 10 s to form a stable froth fraction, after which the froth was manually scooped at defined time intervals for a total of 25 min. To monitor the kinetics of the separation, froth products were collected into three different fractions, namely 0–1 min, 1–10 min, and 10–25 min from the beginning of the experiment. Since froth production was at its most hectic in the earlier stages of the experiment, for the 0–1 min fraction, the froth was scooped in a continuous fashion, for the 1–10 min fraction, froth was collected every 30 s, and for the 10–25 min fraction, sampling was carried once per minute. To maintain a constant froth height, tap water was added into the flotation cell as needed.

Each froth fraction was subsequently filtered and dried for 48 hours in a convection oven (Memmert UE400, Büchenbach, Germany). The drying temperature was kept at 40 °C, to avoid the formation of hydrofluoric acid (HF) via decomposition of any leftover electrolyte or other fluorine compounds. During all working stages of the flotation experiments, the laboratory atmosphere was monitored with a portable HF (g) detector (GfG Micro IV, Dortmund, Germany), but no peaks of HF release were recorded. This suggests that the majority of the electrolyte was already recovered during the industrial processing.

2.1.3. Characterization of the Flotation Products

The dry froth products and tailings were weighed, and the elemental compositions for atomic numbers ≥ 20 were determined by a portable XRF gun (Oxford Instruments, X-MET 5000, Abingdon, UK). To account for the heterogeneous nature of the LIB waste, five 30 s measurements were performed for each froth/tailings sample, and an average value was calculated. To improve the accuracy in the analysis of composition, inductively coupled plasma mass spectrometry (ICP-MS) was performed on selected samples, and a linear regression model was built based on the observed correlations, for each individual element of interest. The regression models were then used to adjust the results obtained with the XRF.

Based on the measured mass and the elemental compositions of the fractions, the elemental mass per fraction, head grade, cumulative grade, and cumulative recovery were calculated for each component of interest. Furthermore, separation efficiency (SE) values were calculated for Cu as shown in Equation (1) [31]:

$$SE = \frac{100Cm(c - f)}{(m - f)f} \tag{1}$$

where,

C = Fraction of feed weight reporting to the froth at time t (1, 10, 25 min) [wt%]
m = Metal content of Cu in the mineral [wt%]
c = Fraction of Cu reporting to the froth at time t (1, 20, 25 min) [wt%]
f = Fraction of Cu reporting to the feed at time t (1, 10, 25min) [wt%].

Since all Cu content in the waste LIB feed was assumed to be in a metallic phase, the value for m was assumed to be 100 wt%. The active materials, however, were assumed to be present in various chemistries (as discussed in Section 3), and determining a single m value for them would have thus been difficult. For this reason, SE values were not calculated for Co, Ni, and Mn.

Based on the results of the characterization, an optimal milling time was determined, and an additional flotation experiment was performed under the optimal parameters (as discussed in Section 3) to collect a sufficient amount of material for the subsequent nickel slag cleaning experiments.

2.2. Slag Cleaning

2.2.1. Materials

The slag was acquired from Boliden Harjavalta (Harjavalta, Finland). Table 1 shows the elemental composition of the slag, which was analyzed with X-Ray Fluorescence (XRF) spectrometry. Most of the metals in the slag were in oxidic form, and therefore they were converted into oxides. For iron, the oxidation state can be 2+ and/or 3+, so it was left in the metallic state. The sum in Table 1 is less than 100 wt% mostly due to the unknown amount of oxygen in iron oxide. Additionally, the slag contained entrained metallic or sulfidic droplets [24].

Table 1. The elemental composition of the industrial nickel slag.

Compound	Fe	SiO$_2$	MgO	Al$_2$O$_3$	Ni	CaO	Co	Cu	S	Zn
wt%	35.86	33.86	7.14	3.12	3.46	1.65	0.46	0.52	0.15	0.06

In order to investigate whether graphite from recycled LIBs could be used as a reducing agent in nickel slag cleaning, the battery material from the selected flotation fraction was added to the slag. Supposedly, this fraction should contain the highest amount of graphite and cobalt. The slag and flotation fraction were mixed in a mortar with a 4:1 ratio to ensure that there is excessive carbon available. The elemental composition of the starting mixture with slag and battery fraction from flotation for the experiments is presented later.

2.2.2. Experimental Procedure

The furnace settings and the experimental procedure for high-temperature heating and quenching were described in detail earlier [32,33]. The experiments were conducted in an LTF 16/450 single-phase vertical tube furnace (Lenton, Parsons Lane, Hope, UK). A schematic of the furnace can be seen in Figure 2. The furnace was equipped with four silicon carbide heating elements, an alumina working tube (Ø 35 mm ID, Frialit AL23, Friatec AG, Mannheim, Germany) and an inner tube (Ø 22 mm ID) installed inside the working tube to ensure the correct position of the sample in the hot zone. A Kanthal A1 wire (Ø 0.65 mm) was used in the experiments for lifting the sample to the hot zone.

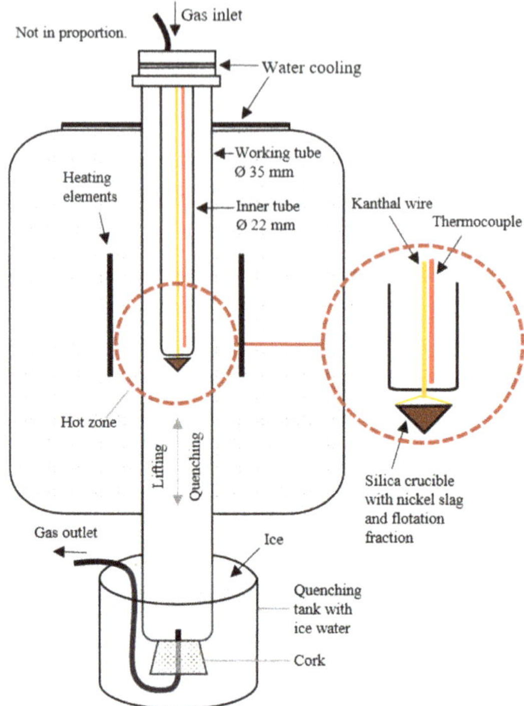

Figure 2. Furnace settings.

The experiments were conducted in inert atmosphere with a gas flow rate of 400–500 mL/min argon (99.999% purity, AGA Linde, Espoo, Finland). The end of the working tube was sealed with a cork allowing gas to flow out of the furnace through a tube installed inside it. The silica crucible with the nickel slag and flotation fraction mixture was placed into a basket made of Kanthal wire. To ensure an inert atmosphere in the experiments, the furnace was flushed with argon gas for 15 minutes before lifting the sample to the hot zone from the cold zone. Time zero (t = 0) was set to be the moment when the crucible was lifted and positioned in the hot zone. The samples were held in the hot zone at 1350 °C and 1400 °C for 5, 10, 20, 30, and 60 min. The end of the working tube was immersed in ice water before the end of the contact time and after reaching full contact time, the cork was taken out and the Kanthal wire was pulled sharply upwards to release the basket with the sample to be quenched in ice water.

After quenching, the samples were dried, mounted in epoxy, ground, polished, and coated with carbon. The SEM-EDS (Scanning Electron Microscopy-energy Dispersive X-ray Spectroscopy) analyses were carried out at Aalto University Department of Chemical and Metallurgical Engineering with a MIRA 3 SEM (Tescan, Brno, Czech Republic) equipped with an UltraDry Silicon Drift Energy Dispersive X-Ray Spectrometer and NSS Microanalysis Software (Thermo Fisher Scientific, Waltham, MA, USA). An accelerating voltage of 15 kV and a beam current of 9.8 nA were used in the analyses. The counting time was set to 20 s per spectrum. The standards (Astimex Standards Ltd. Toronto, Ontario, Canada) in EDS analyses were as follows: pure metals (Co Kα, Cu Kα, Ni Kα, Zn Kα, Mg Kα, Mn Kα, Al Kα), quartz (Si Kα and O Kα), hematite (Fe Kα), marcasite (S Kα), calcite (Ca Kα), and sanidine (K Kα).

3. Results & Discussion

3.1. Characterization of LIB Waste

The size distribution of the industrially produced waste LIB sample used in this study is reported in Figure 3. As presented in the figure, the highest mass fractions are those at sizes <100 µm and 400–1250 µm with ca. 30 wt% each, whereas the middle size fractions represent around 20 wt% each. The fact that the fraction > 1250 µm only hosts 0.3 wt% of the total mass indicates that the industrial sieving was efficient.

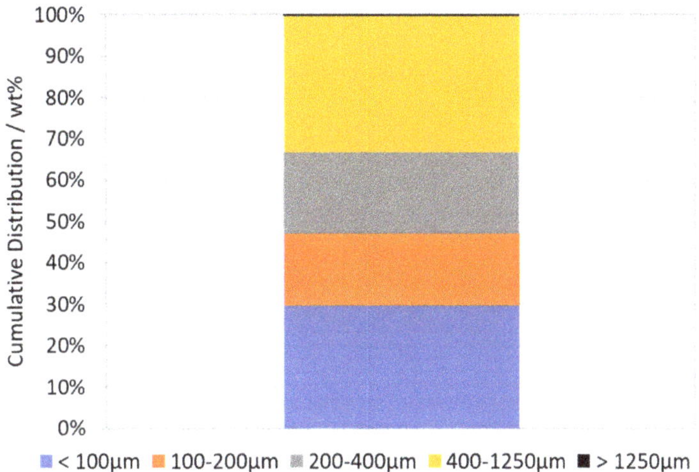

Figure 3. The mass balance of the waste LIB sieving process.

For the most abundant elements (Fe, Al, Cu, Ni, Mn, Co), the elemental compositions of the sieved fractions are presented in Figure 4, as measured by XRF. Other components, including e.g., graphitic carbon and Li, are collectively represented by the "other" column.

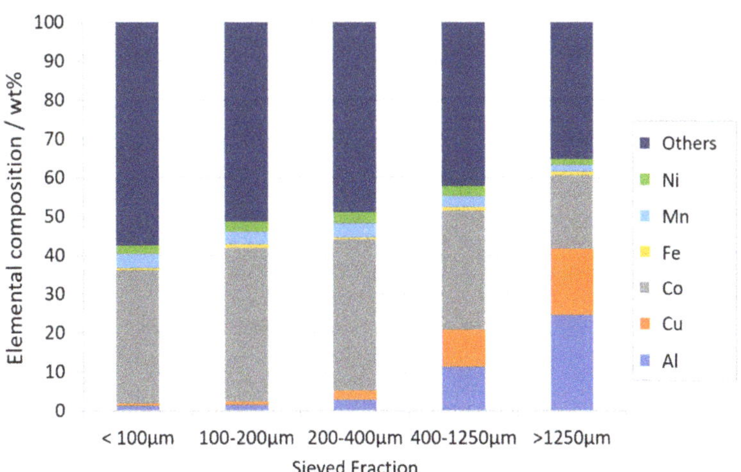

Figure 4. The elemental compositions of the sieved waste LIB feed.

The cumulative distributions of the main components are presented in Figure 5. From this figure, it is seen that Cu and Al report to the > 400 µm fractions, whereas Fe, Co, Ni, Mn, and other components are dominant in the finer fractions. For Cu and Al, this behavior is expected, as these metals likely originate from the large components in the LIBs, such as current collector plates and casing materials. However, it is noteworthy that both Al and Cu are also present in the finer fractions, as approximately 20 wt% of these elements are recovered in the fractions < 400 µm. Interestingly, Al is more abundant than Cu in all of the < 400 µm fractions, which could be explained by its presence in some cathode chemistries ($LiNiCoAlO_2$). In any case, these results indicate that an industrial crushing process will result in fine-sized Al and Cu components, thus emphasizing the need for the LIB recycling studies to be conducted with real battery waste, instead of artificially produced anode-cathode mixtures, as has been the case in various publications [25–30]. Indeed, the presence of these metals in the black mass and its repercussion in subsequent metallurgical processes has been overlooked so far by other authors.

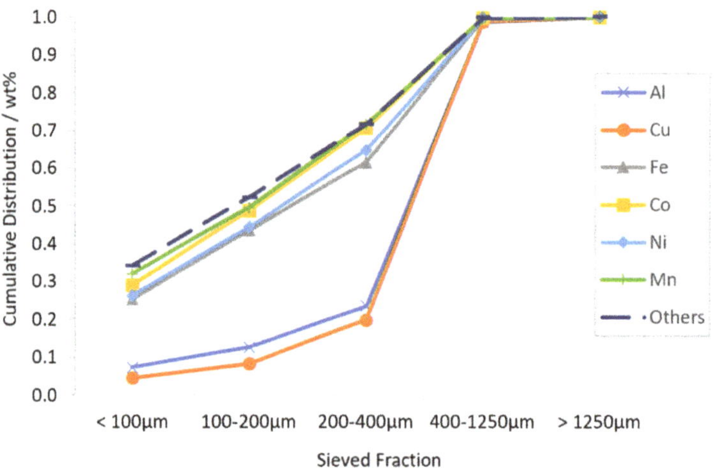

Figure 5. The cumulative distribution of the most abundant metals in the sieved waste LIB fractions. Non-metals and less abundant metals are collectively represented in the "others" chart.

The cathode active components (Co, Ni, and Mn) are expected to report in the finer fractions, since these materials are applied as fine powders (<100 µm) in the electrodes of the battery. However, the fact that all of these components are also present in the size fractions > 100 µm suggests that the currently used combination of crushing and dry sieving is inefficient in separating the active particles. The majority (~70 wt%) of the Co, Ni, and Mn particles are carried on to the coarser fractions, possibly adhered on the surfaces of the coarser components (Al, Cu, plastics). This phenomenon is also reported in other studies conducted with industrial LIB waste [18].

Fe seems to follow a similar trend with the cathode active components, indicating that the waste battery feed might involve LIBs with Fe-based cathode chemistry ($LiFePO_4$). Further evidence for the Fe content in the feed being at least partially of-cathodic-origin rather than as residue of casing materials is the fact that magnetic separation had been performed by the industrial operator for the separation of any metallic Fe. As the industrial operator claims to have sorted the waste LIB feed for high Co cathode chemistries only, the assumed $LiFePO_4$-based Fe content is most likely either due to inefficiency in the sorting process, or the presence of mixed cathode chemistries.

The major constituents of the unidentified "other" materials are expected to be graphite (anode active material), carbon black (conductive additive), Li (cathode active material), F (PVDF), O (cathode active material), and polymers (PVDF and the separator films). The cumulative distribution of these materials seems to follow a very similar trend with Co, Ni, and Mn. Since the majority of the elements in the "others" fraction are associated with the active materials in the battery structure, they can be expected to be distributed in a similar manner with the cathode active elements.

Based on the cumulative distributions of the components, it can be argued that the dominant cathode chemistry in the waste LIB feed is most likely $LiCoO_2$, including some $LiNi_xMn_yCo_zO_2$ variants (various NMC-ratios) and $LiNiCoAlO_2$. Furthermore, it is likely that small amounts of batteries with $LiFePO_4$ cathodes have ended up in the waste stream as well.

3.2. Flotation Experiments

3.2.1. Cumulative Grades and Recoveries

The froth flotation results for the metallic components of waste LIBs after various activation times are compiled in Figures 6–8.

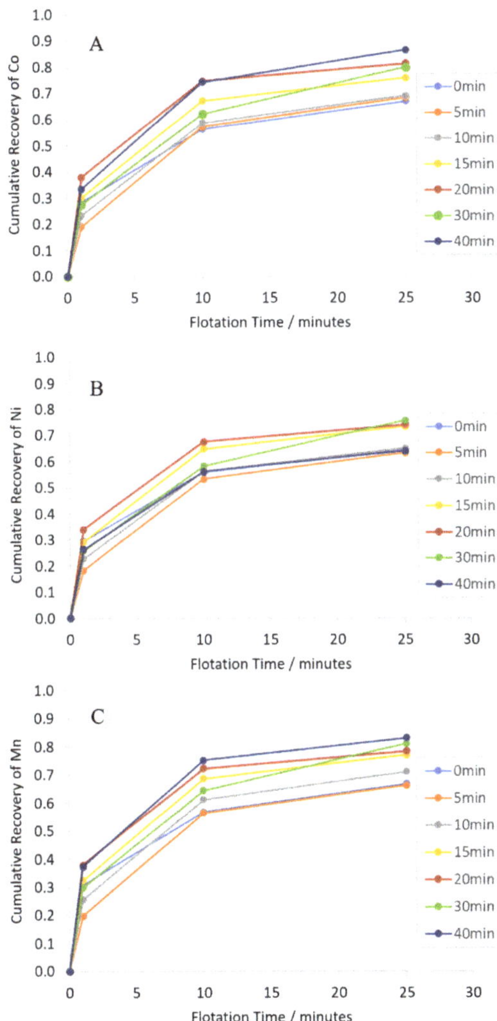

Figure 6. The cumulative recoveries of selected metals (**A**) Co, (**B**) Ni, and (**C**) Mn from cathode active materials in waste LIB after various milling times.

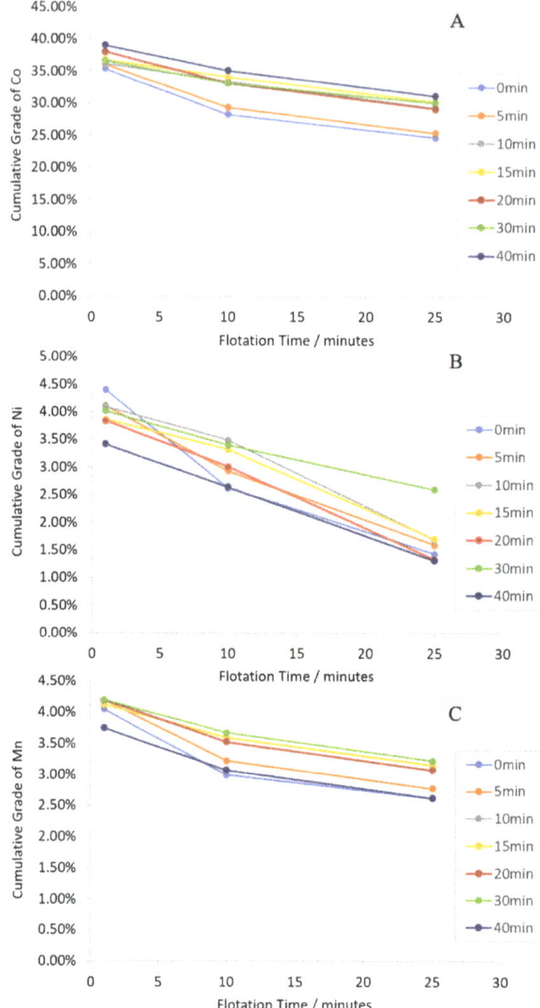

Figure 7. The cumulative grades of selected metals (**A**) Co, (**B**) Ni, and (**C**) Mn from cathode active materials in waste LIB after various milling times.

As Figures 6–8 show, when the binder surfaces are not removed, cathode active material recoveries of ~70–80% can be achieved, while maintaining low (~10%) recovery of Cu. Low energy milling seems to improve the separation up to the 20-min mark, after which the separation efficiency stabilizes. In this experimental series, the optimal separation was achieved using 20-min milling time, with cumulative recoveries (after 25 min) of 81.3%, 67.6%, and 78.4%, for Co, Ni, and Mn, respectively. The cumulative recovery for Cu in the overflow was only 10% (i.e., 90% recovery in the tailings). Furthermore, the separation efficiency of Cu, with the optimal 20-min milling time, was reported to be −54.5%, indicating an efficient enrichment to the underflow. It is to be noted that even though only 7.4 wt% of Cu was floated with 5-min milling time (92.6% recovery in tailings, Figure 8A), the enhanced separation of the active materials using longer milling times ultimately results in an increased Cu separation efficiency. Milling enhances the Cu separation efficiency noticeably, as the SE value for Cu in the absence of milling was −40.3%.

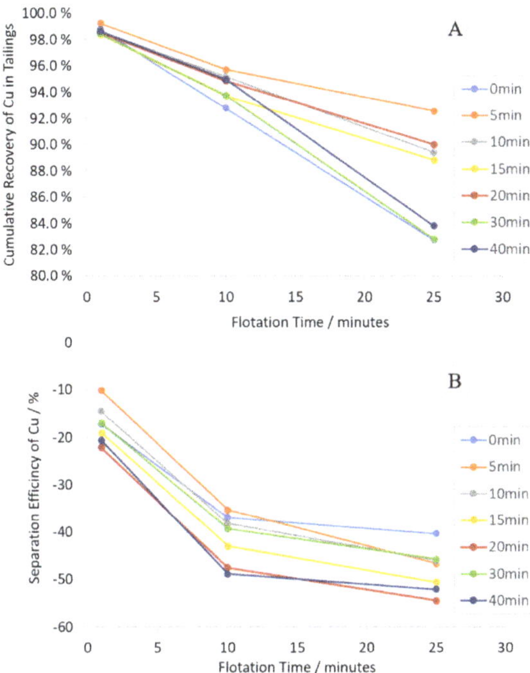

Figure 8. The (**A**) cumulative recovery in the tailings and the (**B**) separation efficiency for Cu, after various milling times.

The results hereby presented indicate that flotation can be applied for the selective recovery of the leftover black mass in the Cu-Al rich sieve overflow, by extracting the active particles in the froth phase and simultaneously enriching Al and Cu in the tailings. The Al-Cu enriched tailings are a valuable material stream on its own, and it could be further separated relatively easily via density-based separation, for example.

The exact mechanism of the extraction of the cathode active components in a seemingly true flotation manner is unknown. A reasonable explanation is that the floatability of the cathode materials is a result of the PVDF binder coating, which arguably hydrophobizes the Li-salt particles. The observed trend of improved floatability with respect to grinding time could also be a result of the different grinding response of graphite and the anode materials, as graphite is softer than the crystal structures of the cathode. Yu et al. [25] described the floatability of ground black mass, and reported that short grinding times (<5 min) resulted in an increment in the floatability of the cathode active components. This was hypothesized to be a result of anode-cathode aggregate formation. In the aforementioned article, the purpose of the milling was to wear down the binder surfaces, and it is likely that a higher energy mill was used compared to the one applied in this article. Consequently, it is plausible that the increased floatability of the cathode components reported in this article after longer grinding times (20 min and above) could also be a result of the anode–cathode aggregate formation. Furthermore, when milled with a low energy input, the adhered black mass particles are expected to be liberated from the Cu, Al, and plastic surfaces, allowing for a larger amount of the active particles to be extracted via true flotation. Nevertheless, as Figure 6 shows, grinding is not necessary for the floatability of the cathode, as significant Co, Ni, and Mn recoveries (~60%) were reported in this study, even in the absence of milling (0 min milling time).

The recovery of Cu and the other macroscopic components (Al, plastic separator) in the froth fractions, however, is likely a result of the hydrodynamic conditions in the flotation cell, and the density of these materials. The plastic films, for example, are so lightweight that they float readily on top of the suspension surface even before the airflow was started. Cu and Al, on the other hand, tend to sink at first, but the strong agitation in the flotation cell continuously carries a certain amount of these particles towards the froth fraction. This behavior is more akin to entrainment than true flotation and could be tested in future works by changing the parameters that effect the hydrodynamic conditions of the flotation process, such as the agitation speed, and the volume/geometry of the flotation cell.

3.2.2. Characterization of the Froth Fraction Used for the Nickel Slag Cleaning

The 0–1 min froth fraction was selected for further pyrometallurgical treatment, due to its observed highest graphite content. The average cumulative grade of the combined 0–1 min froth fractions (two experiments with 20 min milling) are presented in Table 2, along with the average head grade of the two experiments. The amounts of oxygen and lithium in Table 2 were calculated based on the assumption that O and Li are present with a relative molar ratio of 2:1 and 1:1 with respect to Co, respectively, according to commercial Co-containing cathode chemistries. In the 0–1 min fraction, the other materials are expected to be predominantly graphite, Al, F, P, and Fe. In the calculated head grade, the other materials are expected to be primarily Al, graphite, and the separator film plastics.

Table 2. The composition of the combined 0–1 min froth fractions (obtained from two experiments with 20 min milling time), and the average head grade of the experiments.

Grades	Co	Ni	Mn	Cu	O	Li	Other
Head Grade [wt%]	19.00	2.26	2.39	24.29	10.32	2.24	39.50
0–1 min Grade [wt%]	37.30	3.97	4.16	1.95	20.25	4.39	27.98

Characterizing the froth fraction quantitatively for graphite is difficult due to the heterogeneous composition of the industrial LIB waste. One main complication is that, in addition to the graphite anode, other sources of carbon are present in the material, namely carbon black, PVDF, and separator films. To achieve an accurate characterization with Fourier-transform infrared spectroscopy, for example, these carbon sources need to be thoroughly removed, possibly via roasting/acid leaching pretreatment. Other traditional carbon characterization techniques such as Raman spectroscopy are vulnerable to the fluorescent response of the metallic Al and Cu particles that are present in the froth fraction. Likewise, the multiple cathode chemistries, and different elemental ratios within similar cathode chemistries, hinder the applicability of X-ray diffraction for quantitative characterization. Therefore, and considering that the aim of the present manuscript is to evaluate the use of flotation products as smelter feed rather than their separation into high purity components, the graphite concentration hereby presented is an estimate based on the concentrations of the other components that can be accurately measured.

To estimate the amount of graphite in the 0–1 min fraction, the following series of assumptions are made. Firstly, Al is expected to be present in a similar or slightly higher concentration than Cu, as its lower density makes it more likely for the Al particles to end up in the froth fraction. Furthermore, as indicated by Figures 4 and 5, Al-containing cathode chemistries are expected to be present in the feed, likely increasing the Al content of the froth fraction. Secondly, F is expected to be present in a similar concentration range as Li. Thirdly, taking into account the possible LFP cathode content (indicated by Figures 4 and 5), the concentration of Fe, P, and other elements is expected to be in the range of 1.5–2 wt%. This results in a conservative estimate of 19–20 wt% for the graphite grade in the 0–1 min fraction.

As Table 2 shows, in the 0–1 min fraction, the grades of the elements associated with the cathode active components (Co, Ni, Mn, O, Li) were nearly doubled when compared to the head grade, and a drastic reduction in the Cu grade was reported (from 24.29 wt% to 1.95 wt%). This suggests that, during the first minute of the flotation process, a very selective separation of the active material takes place.

These high grades were associated with recovery values ranging from 34% (Ni) to 38% (Co and Mn), as indicated by Figure 6. Even though the separation efficiency was improved towards the later stages of the experiment (Figures 6 and 8B), the chemical composition and especially the assumed high graphite grade made the 0–1 min fraction particularly attractive for the pyrometallurgical slag cleaning procedure. It is noteworthy, that even though the head grade of "others" was higher compared to the assumed graphite grade in the 0–1 min fraction, graphite could still be expected to be enriched to the froth. This is because the Al grade was expected to drop in the froth fraction, following a similar trend to Cu.

Elemental mapping at the microstructure of 0–1 min froth fraction by scanning electron microscopy (SEM) with energy dispersive X-ray spectrometry (EDS) is presented in Figure 9. The results confirmed the assumptions described above. It can be seen that the cobalt, nickel, and manganese were concentrated together in the grains of cathode material, whereas aluminum was present separately in the form of wires. Carbon presence between the grains of cathode material was detected.

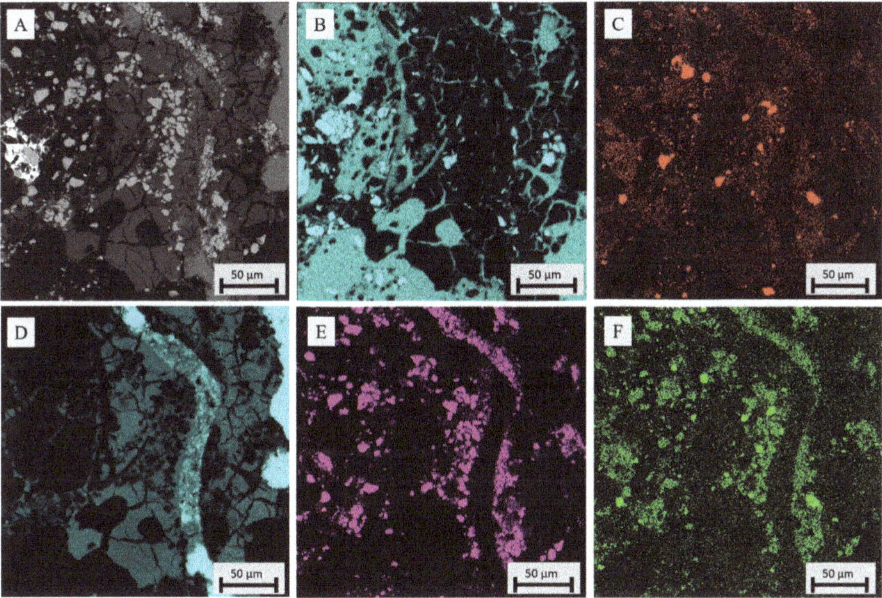

Figure 9. EDS-mapping of 0–1 min froth fraction, (**A**) overview, (**B**) carbon, (**C**) manganese, (**D**) aluminum, (**E**) cobalt, and (**F**) nickel.

3.3. Reduction of Metals

3.3.1. Sample Microstructure

Figure 10 shows SEM-backscattered electron (BSE) micrographs of different samples. Figure 10A shows the microstructure of molten slag without any additions after 20 min contact time in argon atmosphere, whereas Figure 10B–D show the microstructures of samples with slag-battery scrap system for 5, 30, and 60 min contact times, respectively. The visual observations did not indicate much difference between the sample structures after different contact times. Even after 5 min contact time, the metal alloy (the lightest color in the micrograph) had already formed in the sample, as is seen in Figure 10B. This was expected based on earlier findings by Ruismäki et al. [24].

The typical microstructure of a sample with slag-battery scrap system consisted of a glassy slag and metal alloy, presented in Figure 11. The metal alloy was mainly homogenous, but contained some sulfides (matte), which are also visible in the micrographs and seemed to concentrate on grain boundaries. In most samples, the metal alloy was concentrated into metal droplet and was located on top of the slag. The metal alloy had not yet settled to the bottom of the crucible as opposed to industrial nickel slag cleaning process. The chemical composition of the slag and metal are presented in Sections 3.3.2 and 3.3.3, respectively.

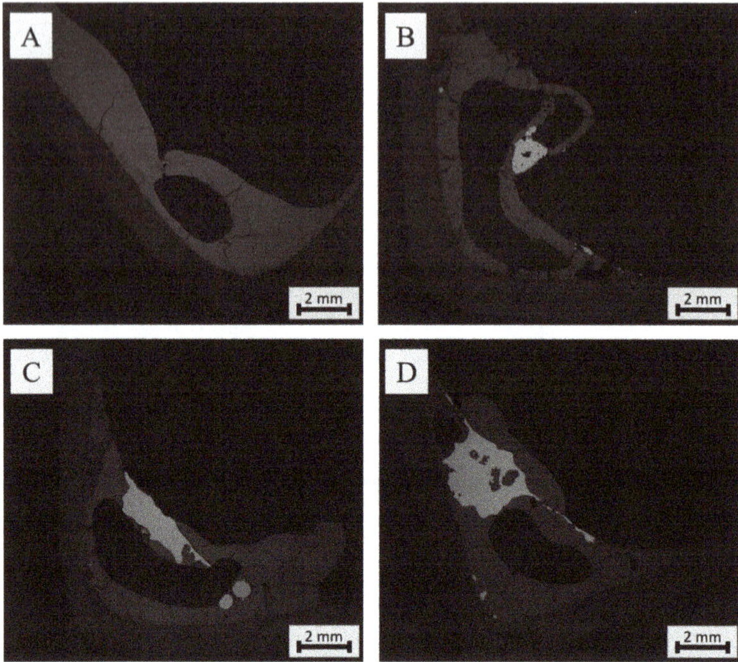

Figure 10. SEM-BSE micrographs of different samples: (**A**) molten slag without additions after 20 min contact time in argon atmosphere at 1400 °C, (**B–D**) samples with slag-battery slag mixtures after different contact time 5, 30 and 60 min, respectively, in argon atmosphere at 1350 °C.

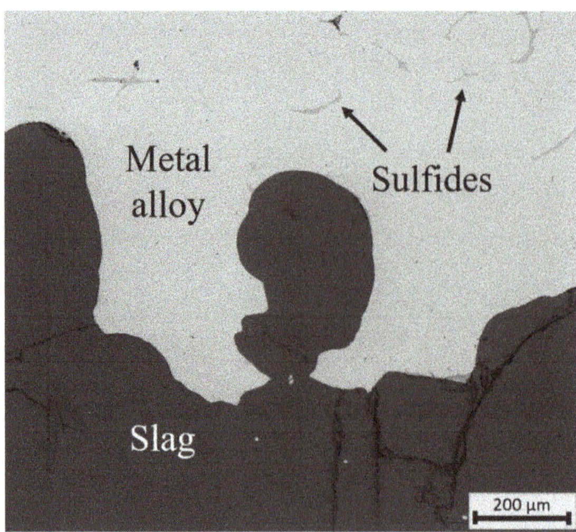

Figure 11. A typical microstructure of a quenched sample. This specific sample was held for 60 min at 1350 °C.

3.3.2. Chemical Composition of Slag

The starting slag mixture, used in the simulated conditions of an electric arc furnace, consisted of 80 wt% of industrial nickel-slag and 20 wt% battery fraction from 0–1 min flotation time. The chemical composition of the prepared mixture, seen in the Table 3, was calculated based on chemical composition of the nickel-slag (Table 1) and the 0–1 min froth fraction (Table 2). The composition of starting mixture was selected based on a previous study by Ruismäki et al. [24]. In this study, it was indicated that the addition of battery scrap rich in cobalt has an increasing effect of cobalt recovery in nickel slag cleaning.

Table 3. Chemical composition of the starting mixture.

Substance	Co	Cu	Li	Fe	Mn	Ni	Zn	S	SiO$_2$	MgO	Al$_2$O$_3$	Graphite
wt%	7.83	0.81	0.88	28.69	0.83	3.56	0.05	0.12	27.09	5.71	3.23	3.8–4.0 *

* graphite concentration is based on assumptions presented in Section 3.2.2.

Figure 12 presents the concentrations of major elements: Fe, Si, Mg, and Al in the slag as a function of contact time in argon atmosphere. It was observed that the most significant change in concentrations of major elements occurred in the first 5 min. The concentrations of elements do not seem to vary between 1350 °C and 1400 °C.

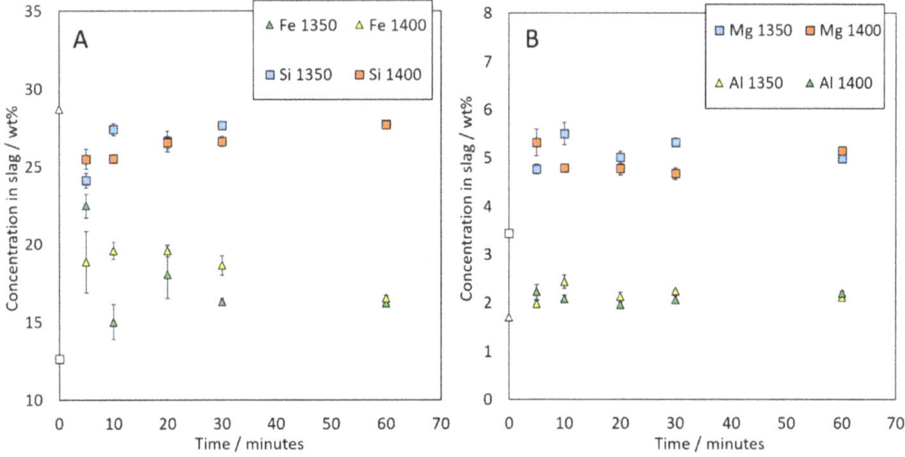

Figure 12. Concentrations of major elements (**A**) iron and silicon, (**B**) magnesium and aluminum in the slag phase at 1350 °C and 1400 °C as a function of contact time in argon atmosphere. The values at 0 min correspond to the elemental concentrations in the starting mixture.

After 5-min contact time, the concentration of Fe in the molten slag decreased significantly in comparison to its concentration in the starting mixture (Table 3). Iron concentration in the slag increased slightly as the contact time increased.

The concentrations of Si, Mg, and Al in the slag were significantly higher already after 5 min in comparison to their concentrations in the starting mixture. Si concentration in slag increased with the increasing contact time and reached the concentration of 27.7 wt% after 60 min contact time, which corresponds to 59.3 wt% of SiO$_2$. The increase in silica concentration in the slag was connected with dissolution of silica crucible used in the experiments and reduction of metal oxides to the metal alloy.

The concentrations of Mg and Al in the slag remained stable between 5 and 60 min, about 5 wt% and 2 wt%, respectively, corresponding to 8.3 wt% of MgO and 3.8 wt% of Al$_2$O$_3$.

Figure 13 shows concentrations of cobalt and manganese in the metal alloy. Nickel and copper concentrations in the slag were under detection limit of the EDS. Extremely low concentrations (under detection limit) of nickel and copper in the slag phase suggest that their oxides were successfully reduced to metallic form.

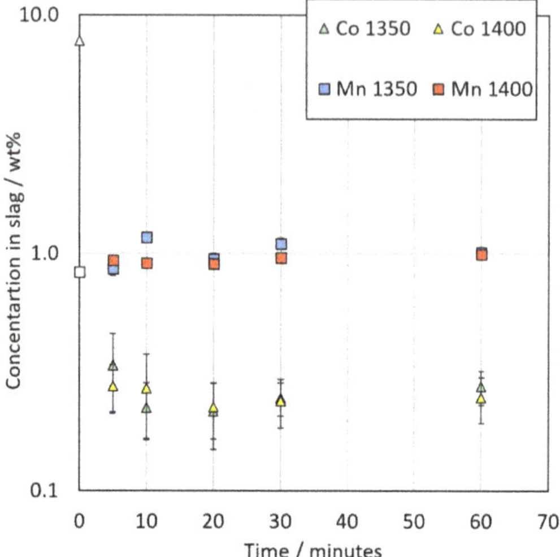

Figure 13. Cobalt and manganese concentrations (logarithmic scale) in the slag as a function of contact time in argon atmosphere. The values at 0 min correspond to the elemental concentrations in the starting mixture.

Manganese concentration in the slag varies slightly between 0.9 and 1.2 wt%. As was presented in a previous study [21], manganese oxide is characterized by a high thermodynamic stability at high temperatures and is therefore sometimes used as a deoxidizing agent. The feasibility of recovering manganese from slags should be researched in case an increasing amount of manganese ends up in slags. The precipitation of nanomanganese ferrite from industrial metallurgical slags by the oxidation route has been proposed [34].

Cobalt concentration in the slag decreased significantly in comparison to its concentration in the starting mixture (7.8 wt%). After 5-min contact time, its concentration in slag was about 0.2 wt% and seemed to be stabilized between 10 and 60 min contact time. This information suggests that a significant portion (about 93%) of cobalt was reduced to metal alloy. Similarly, as for the major elements, the concentrations of the minor elements do not seem to vary between experiments conducted at 1350 °C and 1400 °C.

3.3.3. Chemical Composition of Metal Alloy

A Co-Ni-Fe-Cu metal alloy was formed in all samples, even after the shortest contact time of 5 min. The elemental composition of the metal alloy is shown in Figure 14. Concentrations of cobalt, nickel, copper, and iron did not seem to change between 5 and 60 min contact times or at different temperatures. The formed alloy consisted of about 2.5 wt% of copper, 9 wt% of nickel, 18 wt% of cobalt, and 70 wt% of iron. Based on the results, the reduction of metal oxides by using graphite as a reductant is possible and the kinetics are extremely fast.

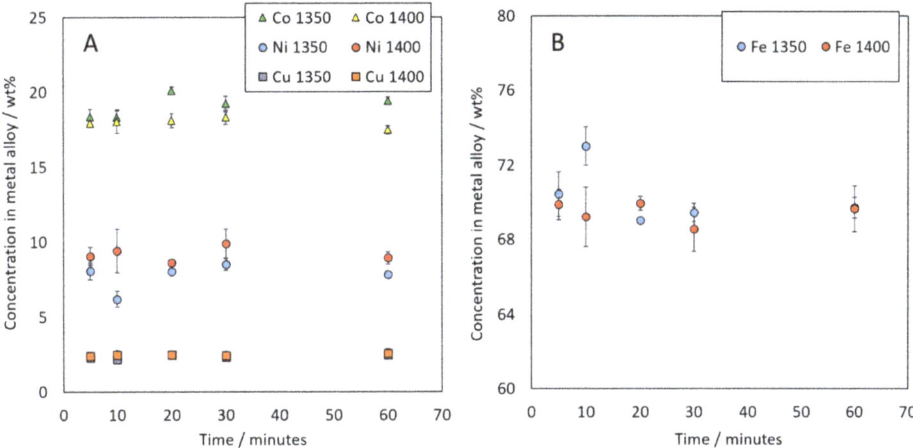

Figure 14. Concentrations of (**A**) cobalt, nickel, and copper and (**B**) iron in the metal alloy as a function of time in argon atmosphere.

The reason for the fast reduction of metal oxides might be caused by the relatively small particle size of the graphite and the metal fraction increasing its density. Additionally, the flotation fraction containing graphite was evenly mixed with the nickel-slag making the reductant readily available for the metal oxides within short diffusion distances. Compared to conventional slag cleaning, the coke floats on top of the slag resulting in the top layer of the slag to be reduced first. Instead, in this approach, the reduction proceeds supposedly quite evenly in the molten slag-battery scrap mixture. Furthermore, an even reduction behavior enables the coalescence of forming metal droplets into larger droplets relatively quickly. Thus, further analysis on the optimum charging procedure e.g., whether the reductant should be charged as mixed, on the bottom or on the top of the slag or even injected should be conducted.

3.3.4. Distribution of Cobalt between Metal Alloy and the Slag

In a system with two phases, the distribution equilibrium of elements between e.g., a slag (s) and a metal or metal sulfide phase (m) can be described with distribution coefficients. The distribution coefficient of an element Me between two phases is expressed as the ratio of weight concentrations of Me dissolved in these phases in equilibrium [35].

Figure 15 presents the distribution coefficient of cobalt between metal alloy and the slag. The temperatures of 1350 and 1400 °C do not have an impact on the cobalt distribution coefficient. This is in agreement with the results of Piskunen et al. [36], which showed that temperature range between 1350 °C and 1450 °C had relatively small effect on the matte-slag distribution coefficients. The results indicate that the distribution coefficient of cobalt reached its maximum during the first 5–10 min. The mass distribution should be studied in order to see whether maximum recovery was reached or not. Previous studies in similar conditions (5 wt.% of Co in the starting mixture, added as synthetic oxide) reported distributions coefficients of cobalt between metal and slag of $10^{1.0}$ after 10-min reduction time and approximately $10^{1.4}$ after 20 min [24]. In the present study, the reaction rate was higher by the reduction with graphite, as the coefficient was about $10^{1.9}$ after 10 min and stabilizes as opposed to the previous studies. The high concentration of cobalt in the starting mixture seems to result in a higher distribution coefficient, and the end concentration of cobalt in the slag was lower compared to the previous studies [24]. Therefore, it is suggested that the reduction kinetics within the first five minutes should be researched in more detail.

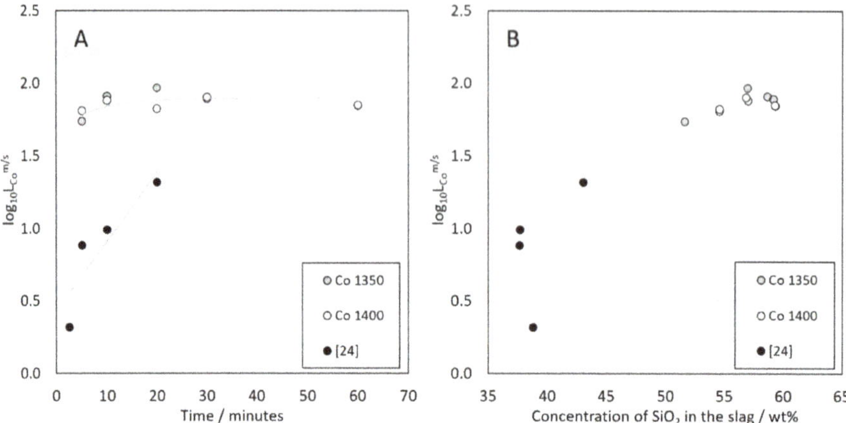

Figure 15. Apparent distribution coefficient of cobalt between the forming metal alloy and the slag (**A**) as a function of reduction time, (**B**) as a function of SiO$_2$ concentration in the slag.

Figure 15B shows the distribution coefficient of cobalt between metal alloy and the slag as a function of silica concentration in the slag. It was observed that the increasing silica concentration in the slag has an increasing effect on cobalt distribution coefficient. This information is in agreement with the study of Hellstén et al. [37] on slag cleaning equilibria in iron silicate slag-copper system. According to their observations, Fe/SiO$_2$ ratio has a significant effect on trace elements distribution between metal alloy and the slag. Hellstén et al. concluded that reduction of metal oxides can be optimized by adjusting SiO$_2$ concentration in slag and the goal should be to maintain the slag compositions higher than about 28 wt% [37]. The previous study of Ruismäki et al. together with the present study confirmed these findings, even though the silica concentration in the slag in this study was much higher, 35–43 wt% [24] versus 51–59 wt%, respectively.

However, at high silica concentrations, the viscosity of the slag increases, resulting in the decreasing reaction rates and the settling rate of particles through the slag layer [37,38]. As seen in Figure 10, the metal alloy is collected on top of the slag even after longer reduction times, as opposed to industrial practices, implying that the silica concentration may be too high for settling. The addition of basic fluxing agents (e.g., CaO, MgO) releases strong silicate formers as trace elements from the slag and increases the viscosity [37,38]. The solubility of weak silicate formers e.g., NiO, FeO and CoO in the slag, on the other hand, decreases with increasing SiO$_2$ concentration [37]. A high viscosity could also slow down the reduction rate by entraining the remaining oxides of the valuable metals in the slag as the concentration of the metals in the slag or metal alloy did not change significantly after 10 min. This is supported by the increase of the silica concentration after 10 min. Experiments with longer reduction times would give additional information on the equilibrium concentrations of the elements.

In order to separate the valuable metals, such as Co, Ni, and Cu, further refinement by pyrometallurgical or hydrometallurgical processes is essential. The concentration of the metals in the product, i.e., the metal alloy, should the adjusted partially according to the limitations of the subsequent processes. The reductant to metal oxide ratio and reduction time play major roles in adjusting the final concentrations of the metal alloy product. In many cases, Fe is considered as a non-valuable metal or impurity, whereas Co, Ni, and Cu are recovered selectively [39]. If the subsequent processing requires lower iron concentration in the metal alloy, iron can be selectively oxidized from the metal alloy by using air or oxygen and adding silica flux enables the formation of a fayalitic slag [40]. Optionally, the metal alloy from slag cleaning can be granulated and leached in order to separate Fe, Co, Cu, and Ni in a controlled way similarly as in the direct Outotec nickel flash smelting process [40]. Benefits of lower iron concentration in the metal alloy include lower processing costs of hydrometallurgical separation [41]. For maximizing resource efficiency, the recovery of metals from pyrometallurgical slags and finding greener options for replacing landfilling of slags has gained an increasing attention lately [11]. For instance, sodium roasting and subsequent water leaching has been proposed by Li et al. [42] for recovering lithium from pyrometallurgical slag, while the recovery of manganese from slags has been studied by Ayala and Fernández [43] and Baumgartner and Groot [44].

As the Co, Ni, Cu, and Fe oxides are seemingly easily and rapidly reduced with the presented parameters, reducing the amount and accessibility of the reductant, i.e., changing the charging procedure of the reductant could slow down the reaction rate and allow to adjust the final concentration. Therefore, optimizing the product concentration while keeping in mind further refinement possibilities is of industrial importance.

4. Conclusions

According to the results, the integration of industrial nickel slag cleaning and LIB recycling was successful. This study presents a novel multifold approach in which:

- Froth flotation is introduced as a mechanical pre-treatment for pyrometallurgical battery recycling
- Graphite in the flotation fraction is utilized to replace coke in the nickel slag cleaning.

The laboratory-scale flotation experiments show that when the black mass is not pretreated for the PVDF binder removal, both the anode and the cathode display a hydrophobic response and can be recovered in the froth phase. This allows the recovery of the left-over black mass from the Cu-Al rich sieve overflow in a purity comparable to the underflow of the sieve, thus increasing the overall black mass recovery of the initial mechanical processing. Low-energy milling was demonstrated to improve the black mass separation, and the optimal separation efficiency was achieved with 20-min milling time.

A proof-of-concept for the processing of industrial nickel slag using solely graphite obtained from the flotation of LIB waste as a reductant was reported for the first time. When the 0–1 min flotation fraction was mixed with industrial nickel slag in a 1:4 ratio, the metals deportment into the matte/alloy phase started immediately when the mixture reached the process temperature. The results indicated that the distribution coefficient of valuable metals reached maximum within 5 min reduction time.

The reason for a fast reduction of metals might be caused by the relatively small particle size of graphite and its even distribution in the mixture prior to melting. In conventional nickel slag cleaning, the coke is charged on top of the slag as the coke is lighter than the slag, resulting in the top parts of the slag to be reduced first. In practice, various measures are taken for improving phase contact and slag mixing in the electric furnace in order to enhance reduction of ferric oxide to ferrous after which the valuable metals can start to reduce. Instead, in this approach, the reduction appears to proceed quite evenly in the molten slag. Thus, further analysis on the optimum charging procedure e.g., whether the reductant should be charged mixed, on the bottom or on the top of the slag should be conducted. It should be noted, that in industrial slag cleaning, the slag is tapped molten from the nickel smelting furnace and in this approach, it was not the case.

For optimizing the process parameters further, the ratio of Ni-slag and battery scrap is of importance as the reductant concentration has a great impact on the forming metal alloy and its metal concentration. Therefore, mixing varying flotation fractions with the nickel slag should be studied further. Additionally, the elemental concentrations in the metal alloy should be optimized keeping in mind the further refining steps. This study suggests preliminarily that any excess carbon available after the reduction of Co, Ni, and Cu will increase the concentration of iron in the metal alloy. Due to thermodynamic constraints, it is not possible to avoid the co-reduction of iron when recovering valuable metals. However, for further refining requirements, the possibility of decreasing Fe concentration in metal alloy should be investigated. A thorough comparison with graphite and other reductants, such as coke, biochar, and methane, used in nickel slag cleaning should be conducted as well.

Author Contributions: Conceptualization, A.J. and R.S.-G.; methodology, R.R., T.R., A.D., P.T., A.J., and R.S.-G.; software, R.R., T.R., and A.D.; validation, R.R., T.R., A.D., A.J., and R.S.-G.; formal analysis, R.R., T.R., and A.D.; investigation, R.R., T.R., and A.D.; resources, A.J. and R.S.-G.; data curation, R.R., T.R., and A.D.; writing—original draft preparation, R.R., T.R., and A.D.; writing—review and editing, P.T., A.J., and R.S.-G.; visualization, R.R., T.R., and A.D.; supervision, A.J. and R.S.-G.; project administration, A.J. and R.S.-G.; funding acquisition, A.J. and R.S.-G. All authors have read and agreed to the published version of the manuscript.

Funding: This research was funded by BATCircle project, grant number 4853/31/2018, SYMMET project, grant number 3891/31/2018, and ReVolt project, grant number 08_2018_IP167_ReVolt.

Acknowledgments: The authors are grateful to Boliden Harjavalta (Finland) for providing industrial nickel slag and Akkuser Oy (Finland) for providing battery scrap. The authors are grateful for Lassi Klemettinen for his SEM-EDS expertise and proofreading the article. This study utilized the RawMatTERS Finland infrastructure (RAMI, Academy of Finland) in Aalto University, VTT Espoo and GTK Espoo. The authors would like to express their gratitude to the Helmholtz Institute Freiberg for Resource Technology, for allowing the use of their research infrastructure, and Anna Vanderbruggen for her expertise in the waste battery characterization. Tommi Rinne is grateful for the doctoral study grant provided by the Finnish Steel and Metal Producers' Fund.

Conflicts of Interest: The authors declare no conflict of interest.

References

1. Huang, B.; Pan, Z.; Su, X.; An, L. Recycling of lithium-ion batteries: Recent advances and perspectives. *J. Power Sources* **2018**, *399*, 274–286. [CrossRef]
2. A Vision for a Sustainable Battery Value Chain in 2030, Insight Report, September 2019. Unlocking the Full Potential to Power Sustainable Development and Climate Change Mitigation. Available online: http://www3.weforum.org/docs/WEF_A_Vision_for_a_Sustainable_Battery_Value_Chain_in_2030_Report.pdf (accessed on 4 March 2020).
3. Wentker, M.; Greenwood, M.; Leker, J. A bottom-up approach to lithium-ion battery cost modeling with a focus on cathode active materials. *Energies* **2019**, *12*, 504. [CrossRef]
4. Commission Staff Working Document on the Evaluation of the Directive 2006/66/EC on Batteries and Accumulators and Waste Batteries and Accumulatros and Repealing Directive 91/157/EEC. Available online: https://ec.europa.eu/environment/waste/batteries/pdf/evaluation_report_batteries_directive.pdf (accessed on 23 April 2020).
5. Shi, J.; Chen, M.; Li, Y.; Eric, H.; Klemettinen, L.; Lundström, M.; Taskinen, P.; Jokilaakso, A. Sulfation roasting mechanism for spent lithium-ion battery metal oxides under SO2-O2-Ar atmosphere. *JOM* **2019**, *71*, 4473–4482. [CrossRef]
6. Wang, X.; Gaustad, G.; Babbitt, C.W.; Richa, K. Economies of scale for future lithium-ion battery recycling infrastructure. *Resour. Conserv. Recycl.* **2014**, *83*, 53–62. [CrossRef]
7. Bernhart, W. Recycling of lithium-ion batteries in the context of technology and price developments. *Atzelectron. Worldw.* **2019**, *14*, 38–43. [CrossRef]
8. Gu, F.; Guo, J.; Yao, X.; Summers, P.A.; Widijatmoko, S.D.; Hall, P. An investigation of the current status of recycling spent lithium-ion batteries from consumer electronics in China. *J. Clean. Prod.* **2017**, *161*, 765–780. [CrossRef]
9. Werner, D.; Peuker, U.A.; Mütze, T. Recycling chain for spent lithium-ion batteries. *Metals* **2020**, *10*, 316. [CrossRef]
10. Zheng, X.; Zhu, Z.; Lin, X.; Zhang, Y.; He, y.; Cao, H.; Sun, Z. A mini-review on metal recycling from spent lithium ion batteries. *Engineering* **2018**, *4*, 361–370. [CrossRef]
11. Rämä, M.; Nurmi, S.; Jokilaakso, A.; Klemettinen, L.; Taskinen, P.; Salminen, J. Thermal processing of jarosite leach residue for a safe disposable slag and valuable metals recovery. *Metals* **2018**, *8*, 744. [CrossRef]
12. Directive 2006/66/EC of the European Parliament and of the Council of 6 September 2006 on Batteries and Accumulators and Waste Batteries and Accumulators and Repealing Directive 91/157/EEC. Available online: https://eur-lex.europa.eu/legal-content/EN/TXT/PDF/?uri=CELEX:02006L0066-20131230&rid=1 (accessed on 30 March 2020).
13. Harper, G.; Sommerville, R.; Kendrick, E.; Driscoll, L.; Slater, P.; Stolkin, R.; Walton, A.; Christensen, P.; Heidrich, O.; Lambert, S.; et al. Recycling lithium-ion batteries from electric vehicles. *Nature* **2019**, *575*, 75–86. [CrossRef]

14. Velázquez-Martínez, O.; Valio, J.; Santasalo-Aarnio, A.; Reuter, M.; Serna-Guerrero, R. A critical review of lithium-ion battery recycling processes from a circular economy perspective. *Batteries* **2019**, *5*, 68. [CrossRef]

15. Nitta, N.; Wu, F.; Lee, J.T.; Yushin, G. Li-ion battery materials: Present and future. *Mater. Today* **2015**, *18*, 252–264. [CrossRef]

16. Lv, W.; Wang, Z.; Cao, H.; Sun, Y.; Zhang, Y.; Sun, Z. A critical review and analysis on the recycling of spent lithium-ion batteries. *ACS Sustain. Chem. Eng.* **2018**, *6*, 1504–1521.

17. Or, T.; Gourley, S.W.D.; Kaliyappan, K.; Yu, A.; Chen, Z. Recycling of mixed cathode lithium-ion batteries for electric vehicles: Current status and future outlook. *Carbon Energy* **2020**, *2*, 6–43. [CrossRef]

18. Porvali, A.; Aaltonen, M.; Ojanen, S.; Velazquez-Martinez, O.; Eronen, E.; Liu, F.; Wilson, B.P.; Serna-Guerrero, R.; Lundström, M. Mechanical and hydrometallurgical processes in HCl media for the recycling of valuable metals from Li-ion battery waste. *Resour. Conserv. Recycl.* **2019**, *142*, 257–266. [CrossRef]

19. Mayyas, A.; Steward, D.; Mann, M. The case for recycling: Overview and challenges in the material supply chain for automotive li-ion batteries. *SMT* **2018**, *19*. [CrossRef]

20. Chen, M.; Ma, X.; Chen, B.; Arsenault, R.; Karlson, P.; Simon, N.; Wang, Y. Recycling end-of-life electric vehicle lithium-ion batteries. *Joule* **2019**, *3*, 2622–2646. [CrossRef]

21. Dańczak, A.; Klemettinen, L.; Kurhila, M.; Taskinen, P.; Lindberg, D.; Jokilaakso, A. Behavior of battery metals lithium, cobalt, manganese and lanthanum in black copper smelting. *Batteries* **2020**, *6*, 16. [CrossRef]

22. Ren, G.; Xiao, S.; Xie, M.; Pan, B.; Chen, J.; Wang, F.; Xia, X. Recovery of valuable metals from spent lithium ion batteries by smelting reduction process based on FeO-SiO2-Al2O3 slag system. *Trans. Nonferr. Metals Soc.* **2017**, *27*, 450–456. [CrossRef]

23. Ren, G.; Xiao, S.; Xie, M.; Pan, B.; Fan, Y.; Wang, F.; Xia, X. Recovery of valuable metals from spent lithium-ion batteries by smelting reduction process based on $MnO-SiO_2-Al_2O_3$ slag system. In *Advances in Molten Slags, Fluxes, and Salts, Proceedings of the 10th International Conference on Molten Slags, Fluxes and Salts (MOLTEN16), 22–25 May 2016*; Reddy, R.G., Chaubal, P., Pistorius, P.C., Pal, U., Eds.; TMS (The Minerals, Metals & Materials Society): Pittsburgh, PA, USA, 2016.

24. Ruismäki, R.; Dańczak, A.; Klemettinen, L.; Taskinen, P.; Lindberg, D.; Jokilaakso, A. Integrated battery scrap recycling and nickel slag cleaning with methane reduction. *Minerals* **2020**, *10*, 435. [CrossRef]

25. Yu, J.; He, Y.; Ge, Z.; Li, H.; Xie, W.; Wang, S. A promising physical method for recovery of $LiCoO_2$ and graphite from spent lithium-ion batteries: Grinding flotation. *Sep. Purif. Technol.* **2018**, *190*, 45–52. [CrossRef]

26. Wang, F.; Zhang, T.; He, Y.; Zhao, Y.; Wang, S.; Zhang, G.W.; Zhang, Y.; Feng, Y. Recovery of valuable materials from spent lithium-ion batteries by mechanical separation and thermal treatment. *J. Clean. Prod.* **2018**, *185*, 646–652. [CrossRef]

27. Liu, J.; Wang, H.; Hu, T.; Bai, X.; Wang, S.; Xie, W.; Juan, H.J.; He, Y. Recovery of $LiCoO_2$ and graphite from spent lithium-ion batteries by cryogenic grinding and froth flotation. *Miner. Eng.* **2020**, *148*, 106223. [CrossRef]

28. Zhang, G.; He, Y.; Feng, Y.; Wang, H.; Zhu, X. Pyrolysis-ultrasonic-assisted flotation technology for recovering graphite and $LiCoO_2$ from spent lithium-ion batteries. *ACS Sustain. Chem. Eng.* **2018**, *6*, 10896–10904.

29. Zhang, G.; He, J.; Wang, H.; Feng, Y.; Xie, W.; Zhu, X. Application of mechanical crushing combined with pyrolysis-enhanced flotation technology to recover graphite and $LiCoO_2$ from spent lithium-ion batteries. *J. Clean. Prod.* **2019**, *231*, 1418–1427. [CrossRef]

30. He, Y.; Zhang, T.; Wang, F.; Zhang, G.; Zhang, W.; Wang, J. Recovery of $LiCoO_2$ and graphite from spent lithium-ion batteries by Fenton reagent-assisted flotation. *J. Clean. Prod.* **2017**, *143*, 319–325. [CrossRef]

31. Wills, B.A.; Finch, J. *Wills' Mineral Processing Technology: An Introduction to the Practical Aspects of Ore Treatment and Mineral Recovery*; Butterworth-Heinemann: Oxford, UK, 2015; ISBN 978-0-08-097053-0.

32. Wan, X.; Jokilaakso, A.; Fellman, J.; Klemettinen, L.; Marjakoski, M. Behavior of Waste Printed Circuit Board (WPCB) materials in the copper matte smelting process. *Metals* **2018**, *8*, 887. [CrossRef]

33. Wan, X.; Jokilaakso, A.; Iduozee, I.; Eric, H.; Latostenmaa, P. Experimental research on the behavior of WEEE scrap in flash smelting settler with copper concentrate and synthetic slag. In Proceedings of the EMC 2019, Düsseldorf, Germany, 23–26 June 2019; Volume 3, pp. 1137–1150.

34. Semykina, A.; Seetharaman, S. Recovery of manganese ferrite in nanoform from the metallurgical slags. *Metals Mater. Trans. B* **2011**, *42*, 2–4. [CrossRef]

35. Sukhomlinov, D.; Klemettinen, L.; Avarmaa, K.; O'Brien, H.; Taskinen, P.; Jokilaakso, A. Distribution of Ni, Co, precious and platinum group metals in copper making process. *Metall. Mater. Trans. B* **2019**, *50*, 1752–1765. [CrossRef]

36. Piskunen, P.; Avarmaa, K.; O'Bren, H.; Klemettinen, L.; Johto, H.; Taskinen, P. Precious metal distributions in direct nickel matte smelting with low-Cu mattes. *Metall. Mater. Trans. B* **2017**, *49*, 98–112. [CrossRef]

37. Hellstén, N.; Klemettinen, L.; Sukhomlinov, D.; O'Brien, H.; Taskinen, P.; Jokilaakso, A.; Salminen, J. Slag cleaning equilibria in iron silicate slag-copper systems. *J. Sustain. Metall.* **2019**, *5*, 463–473. [CrossRef]

38. Shlesinger, M.E.; King, M.J.; Sole, K.C.T.G.; Davenport, W.G. *Extractive Metallurgy of Copper—Chapter 5: Matte Smelting Fundamentals*; Elsevier: Oxford, UK, 2011; pp. 73–88. [CrossRef]

39. Stefanova, V.P.; Iliev, P.K.; Stefanov, B.S. Copper, nickel and cobalt extraction from FeCuNiCoMn alloy obtained after pyrometallurgical processing of deep sea nodules. In Proceedings of the Tenth ISOPE Ocean Mining and Gas Hydrates Symposium, Szczecin, Poland, 22–26 September 2013.

40. Riekkola-Vanhanen, M. *Finnish Expert Report on Best Available Techniques in Nickel Production*; Finnish Environment Institute: Helsinki, Finland, 1999; ISBN 952-11-0507-0. Available online: https://helda.helsinki.fi/bitstream/handle/10138/40684/FE_317.pdf?sequence=1&isAllowed=y (accessed on 13 May 2020).

41. Jones, R.T.; Denton, G.M.; Reynolds, Q.G.; Parker, J.A.L.; van Tonder, G.J.J. Recovery of cobalt, nickel, and copper from slags, using DC-arc furnace technology. In Proceedings of the International Symposium on Challenges of Process Intensification 35th Annual Conference of Metallurgists, Montreal, QC, Canada, 25–29 August 1996. Available online: http://www.mintek.co.za/Pyromet/Cobalt/Cobalt.htm (accessed on 30 March 2020).

42. Li, N.; Guo, J.; Chang, Z.; Dang, H.; Zhao, X.; Ali, S.; Li, W.; Zhou, H.; Sun, C. Aqueous leaching of lithium from simulated pyrometallurgical slag by sodium sulfate roasting. *RSC Adv.* **2019**, *9*, 23908–23915. [CrossRef]

43. Ayala, J.; Fernández, B. Recovery of manganese from silicomanganese slag by means of a hydrometallurgical process. *Hydrometallurgy* **2015**, *158*, 68–73. [CrossRef]

44. Baumgartner, S.J.; Groot, D.R. The recovery of manganese products from ferromanganese slag using a hydrometallurgical route. *J. S. Afr. Inst. Min. Metall.* **2014**, *114*, 2411–9717. Available online: http://www.scielo.org.za/scielo.php?script=sci_arttext&pid=S2225-62532014000400011&lng=en&nrm=iso (accessed on 14 May 2020).

Article

Disassembly of Li Ion Cells—Characterization and Safety Considerations of a Recycling Scheme

Jean Marshall [1], Dominika Gastol [2], Roberto Sommerville [2], Beth Middleton [1], Vannessa Goodship [1] and Emma Kendrick [2,*]

[1] WMG, University of Warwick, International Manufacturing Centre, Coventry CV4 7AL, UK; Jean.Marshall@warwick.ac.uk (J.M.); B.Middleton@warwick.ac.uk (B.M.); V.Goodship@warwick.ac.uk (V.G.)

[2] School of Metallurgy and Materials, University of Birmingham, Edgbaston, Birmingham B15 2TT, UK; d.a.gastol@bham.ac.uk (D.G.); R.Sommerville@bham.ac.uk (R.S.)

* Correspondence: e.kendrick@bham.ac.uk; Tel.: +44-121-414-6730

Received: 15 May 2020; Accepted: 5 June 2020; Published: 9 June 2020

Abstract: It is predicted there will be a rapid increase in the number of lithium ion batteries reaching end of life. However, recently only 5% of lithium ion batteries (LIBs) were recycled in the European Union. This paper explores why and how this can be improved by controlled dismantling, characterization and recycling. Currently, the favored disposal route for batteries is shredding of complete systems and then separation of individual fractions. This can be effective for the partial recovery of some materials, producing impure, mixed or contaminated waste streams. For an effective circular economy it would be beneficial to produce greater purity waste streams and be able to re-use (as well as recycle) some components; thus, a dismantling system could have advantages over shredding. This paper presents an alternative complete system disassembly process route for lithium ion batteries and examines the various processes required to enable material or component recovery. A schematic is presented of the entire process for all material components along with a materials recovery assay. Health and safety considerations and options for each stage of the process are also reported. This is with an aim of encouraging future battery dismantling operations.

Keywords: batteries; reuse; recycling; disassembly; safety

---

1. Introduction

Predicted sales of electric vehicles will create large volumes of end-of-life (EoL) lithium ion batteries (LIBs). Within the European Union, in 2016 it was reported that just 5% of LIBs were being recycled. [1] It is a significant challenge to create an economically viable process for the reclamation of all materials from used batteries and to re-use all of the recovered materials and components. [2,3] The automotive manufacturer remains responsible for the disposal of EoL battery packs, which is complicated by the fact that these are both expensive to ship due to safety issues and may end up in the waste stream in varying states of health.

It is, therefore, imperative to develop a well-understood and safe process for the efficient dismantling of LIBs in such a manner that the valuable materials they contain may be re-used or reclaimed. This is part of the larger circular economy picture for battery recycling and addresses some of the hierarchy of recycling process decisions illustrated previously by Harper et al [4].

In order to develop a safe recycling process, the battery must first be stabilized by being discharged to a known state of charge (SOC) [5] The SOC is defined according to the capacity that can be delivered by the battery/cell with respect to its value in the charged state. The SOC is not defined by the cell voltage, however the cell voltage can be used to infer the SOC. The exact end-of-discharge voltage required to attain SOC = 0% varies with the internal resistance and the chemistry of the cell but is usually quoted between 2.5–3V for a layered oxide–graphite chemistry [6]. It is possible to over-discharge a cell below 0% SOC to an open circuit voltage of 0 V. This can be achieved through discharge via a resistor or external short circuit, care must be taken in performing this, as any remaining energy will be delivered as heat. If over-discharge is desired, it must be performed carefully and slowly, to prevent this heat build-up. It should be noted that over-discharging a cell to an end-of-discharge voltage of 0 V will also change the chemistry of the cell [7,8]. From our experience, copper from the current collectors in the battery tends to partially dissolve in the electrolyte when the battery is discharged to 0 V, this is due to the high potential observed by the anode at 0 V and oxidation of the copper [9]. After end of discharge voltage is reached, the cell is allowed to rest to open circuit voltage (OCV). Battery discharge can be accomplished by simply connecting a load across the battery terminals, this allows for potential energy collection and reuse. An alternative that can be used for cells (not modules and packs), is a salt-water electrochemical discharge method. This does not allow energy reclamation but can render the cells safe. A recent study analyzed this technique in some detail and concluded that several different aqueous salt solutions were capable of efficiently discharging the battery without damage [10]. In the case of damaged cells where discharge cannot be performed, more safety precautions will be necessary, often these are first "made safe" by short circuiting the cell through nail penetration or discharging in a brine solution before disassembly. This does however introduce greater potential for contamination of the materials streams during processing.

Once discharged, the battery is transferred to a controlled environment in order that it may be opened safely, because some chemicals inside the battery can react with water and with oxygen. This is often done in a glove box filled with argon [11,12]. Parts of the battery can then be separated. Systems for disassembling the battery have been described previously [13]. To date, in most cases these systems involve the separation of some battery parts manually, followed by shredding [4,14,15] or crushing [16,17] to attempt to recover useful materials. Due to the hazardous nature of the battery components it is essential that engineering controls such as glove boxes or fume hoods are used for handling battery materials. If the electrolyte is not removed from the reclaimed components, then hazardous materials can be released from them at a later stage in the process [18]. The electrolyte may be leached out of the components into water and procedures have been described for the optimization of this process, for example, using flotation tanks or including additives in the water for efficient leaching. [19–21] As was the case for the discharge process, the optimal procedures for battery dismantling may depend on the reason for which the battery is being dismantled; the requirements for an industrial recycling process may be different from the requirements of academic researchers who want to open the battery to characterize its components. Such studies can give important insight into how the material inside the battery changes during its useful lifetime [22,23].

An average of the chemical cell composition is given by Mossali et al., [24] however this was based upon an average of several cell types. We compare the several chemistries used in electric vehicle cells using BATPAC© Argonne National Labs, IL, USA. [25]–see Figure 1.

One of the principal challenges in Li-ion battery recycling is the sheer complexity of the battery itself. A typical battery is enclosed in a large pack housing, within which there is a number of modules (each containing several pouch cells), circuitry and the battery management system [30,31]. The exact layout of each of these components is different between manufacturers. Even at the pouch cell level, each cell contains many chemical components (outer pouch material, aluminum and copper sheets, anode and cathode material) and separating these is not a trivial problem. A particular challenge is presented by the anode and cathode materials, which consist of a mixture of various chemical components and require advanced chemical and physical methods, such as such as ultrahigh shear de-agglomeration, calcination and soxhlet extraction, froth flotation, selective leaching, direct regeneration and mechanochemical recovery, to be separated into 'pure' materials; their constituent materials parts with no or limited contamination. [32–38] The benefits of recycling and materials recovery are, however, substantial; in a truly circular economy, we should aim to recover and protect the critical materials contained in LIBs.

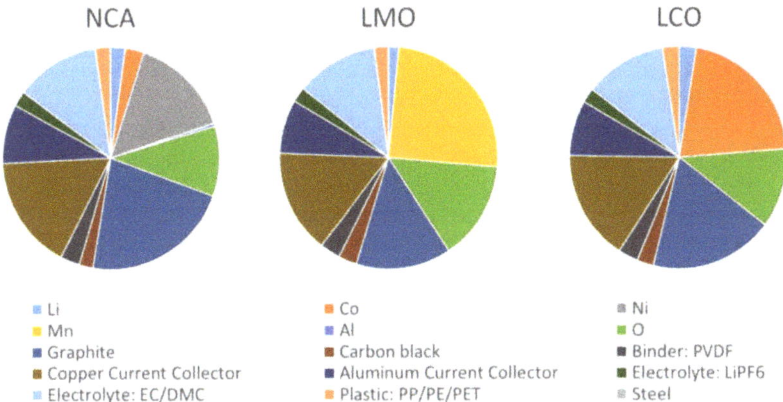

Figure 1. Percentage breakdown of components of a typical lithium ion battery (LIB) cell with different cathode materials LiNi$_{0.8}$Co$_{0.15}$Al$_{0.05}$O$_2$ (NCA), LiMn$_2$O$_4$ (LMO) and LiCoO$_2$ (LCO); identifying the key materials and critical and strategic materials [26,27] (values of each component given as a percentage by weight) [28]. As calculated for a prismatic cell using a basic cell model [29].

Figure 1 gives a visual representation of the materials present in three different prismatic LIB chemistries; LiCoO$_2$ (LCO), LiMn$_2$O$_4$ (LMO) and LiNi$_{0.8}$Co$_{0.15}$Al$_{0.05}$O$_2$ (NCA). In particular, it is imperative to find better ways to recover and re-use metals such as lithium and cobalt, due to the precarious nature of the global supply for these materials; over 50% of global lithium production is located in a few regions of South America [39], while over 60% of global cobalt production is located in one country, the Democratic Republic of the Congo [40,41].

There are concerted ongoing efforts in both academia and industry to improve the recovery levels of these critical materials but this is hindered by the fast moving nature of the industry. Recycling rates have been slow to increase, complicated by changing battery chemistries, changes in composition of materials and the lack of focus on prior material recovery stages.

As an alternative to current systems, this paper reports on work looking at dismantling and characterization of the components produced by hand dismantling systems. In order for disassembly processes to become part of the commercial recycling procedure for LIBs, there must be a potential for them to be automated, as has been discussed in the case of 'test' pouch cells [42]; we discuss how our findings can be used to identify where problems and opportunities may lie for automated disassembly of LIBs in the future, particularly in terms of the health and safety aspects of the cell opening procedures. The aim of this work was initially to produce high quality transition metal containing black mass for hydrometallurgical extraction for our project partners. However during the development of this disassembly process we also optimized the process to reclaim clean and pure materials waste streams from other components; separators, pouch material, current collectors. The composition, morphology and the change in properties are investigated for two types of different scrap cells—Quality Control (QC) Rejected and End of Life (EOL). The potential for re-use is discussed for the different components.

2. Materials and Method Development

The batteries used for this study were automotive pouch cells from a 1st-generation Nissan Leaf. The dimensions of each cell are 215 mm × 256 mm, each cell has a nominal voltage of 3.75 V. A single Nissan Leaf car contains 192 pouch cells (with 4 cells in each of 48 modules). Each pack stores an electrical energy of 24 kW.h. The cells have cathodes which are approximately 75% Lithium Manganese oxide spinel (LMO) with 25% Lithium Nickel Cobalt aluminum Oxide (NCA) on aluminum current collectors, this is similar to reported previously [43]. The mass ratio of LMO:NCA is calculated from the Inductively coupled plasma-optical emission spectrometry (ICP-OES) data as shown in Supplementary S1. The anode is comprised of graphite anodes on copper current collectors. The electrolyte is $LiPF_6$ in an organic carbonate solvent. Our process is tailored for this battery with this chemistry. Although some aspects of the procedure are general, some will have to be tailored in order to be applicable to other battery chemistries in other car models.

The teardown procedure is as described in Figure 2. We begin with cells that have been discharged to a suitable OCV; for safety we check the voltage with a multimeter before beginning their disassembly. Then, in a fume hood we manually cut the pouch open with a ceramic scalpel; during this whole process, metallic tools are not used due to the risk of electrical discharge. When opening the cell with a scalpel, care is taken to avoid damaging personal protective equipment such as gloves. Incisions into the pouch cell are made around the edges of the cell, avoiding cutting the stack of electrodes and current collectors; this stack is then separated into individual components using tweezers. This minimizes contact between the gloves of the person carrying out the disassembly and the components that are impregnated with electrolyte. All tools are thoroughly cleaned after this procedure. We envisage that for a future automated process, the pouch incision could be made using a laser. Current investigations using laser-opening methods are ongoing; the operation of which is being optimized to reduce the heat transfer to the cell components whilst still cutting open the packaging, to ensure that thermal runaway does not occur. It is noted that the settings for a laminated aluminum pouch are different from a steel can. In addition, the laminated aluminum pouch is easily cut with a blade if necessary, whereas the steel can require greater power or energy to pierce. The pouch cell is the subject of this study. The cell components are physically separated into three groups; anode, cathode and 'other' materials. This physical separation minimizes the possibility of cross-contamination during further processing, this is currently by hand, however we foresee the ability to automate this process and some steps towards automatic separation have been reported [42,44,45]. It is noted however that upon opening of the cells, the $LiPF_6$ salt in the electrolyte can hydrolyze to produce HF and end of life cells may contain lithium dendrites. Therefore, processes need to be developed such that lithium dendrites are pacified (such as with CO_2 or water mist) and gaseous HF is not released to atmosphere. To remove the electrolyte and salt from the electrode, the three components are washed in separate solvent baths of isopropan-2-ol (IPA) in a fume cupboard at ambient temperature, before drying under vacuum, (the electrolyte remains in the bath). The procedure was attempted using a number of solvents, including acetone, diethyl carbonate, dimethyl carbonate and IPA. In some of these solvents, the electrode black mass delaminated from the metal films from the cell. IPA proved to be the most effective solvent for this process because it caused the least delamination and therefore was chosen as the washing solvent. In this case, we have not separated the electrolyte any further; the carbonates remain dissolved in the IPA or water and a white lithium containing precipitate starts to form over time. The aim of washing was to eliminate the possibility of HF forming from hydrolysis of the $LiPF_6$ salt during the further component processing and handling. After washing, the outer part of the pouch and the separator can be recycled. Then, the electrode 'black mass' can be separated from the copper and aluminum foil sheets. There are several ways of liberating the materials from the current collector; thermal liberation to break down the binder, [46,47] solvent delamination [21,48], physical methods such as ultrasonic and agitation. [49] Binder calcination forms HF due to decomposition of the PVDF and solvent delamination (in N-methyl Pyrrollidone (NMP)) produces a secondary waste which falls under current REACH regulations. [50] In this work, our focus was upon an environmentally conscious solution for scale-up to industry and therefore we chose to investigate a combination of ultrasonic delamination in a green solvent. Initially water and DMC were used with limited delamination effect; upon the addition of an acid (2 M HNO_3) we noticed significant delamination. From previous work on copper and aluminum corrosion, it is known that the mineral acids and alkali's will cause pitting of aluminum, particularly if the pH is below 4 or above 7 [51]. Copper will dissolve slowly into strong acids and in aqueous solutions above pH 3 will slowly oxidize. [52] Interestingly, previous work with oxalic acid shows a passivating effect on both aluminum [53] and copper [54]. We therefore chose oxalic acid, (pH 3), in water, as a greener delamination solvent alternative to NMP. Initially the oxalic acid concentration was optimized for delamination of the QC rejected cells, where we observed the least damage and maximum delamination using weak organic solutions in an ultrasonic bath. Therefore, due to delamination efficiencies and secondary waste concerns we chose to concentrate and optimize the processes using oxalic acid in this study, greater efficiency or optimization may be possible utilizing alternatives. The black mass was extracted by sonication in a bath of oxalic acid at a respective optimized concentration for the copper and aluminum delamination process. It is possible to further reclaim the black mass components; active materials, conductive additive and binder (PVDF). These components are present in very small quantities (typically <10% by weight of the black mass). Heat treatments can break down and burn off the PVDF or binder [55], however we can only recover the active material components; graphite or the metal oxides. This process also delithiates the cathode materials [56] and also has the disadvantage of being quite energy-intensive. In this work we show as an example for the re-use case, materials which have been heat treated in this way. If we wish to recover all materials including the PVDF binder and to do this without such aggressive heating, it is necessary to consider a route that involves solvents to strip the binder out of the anode and cathode materials. We show as an example the possibility to remove the PVDF and carbon black from the black mass using NMP as a solvent. This however is not sustainable and further work for investigations of green solvents or other removal methods are required.

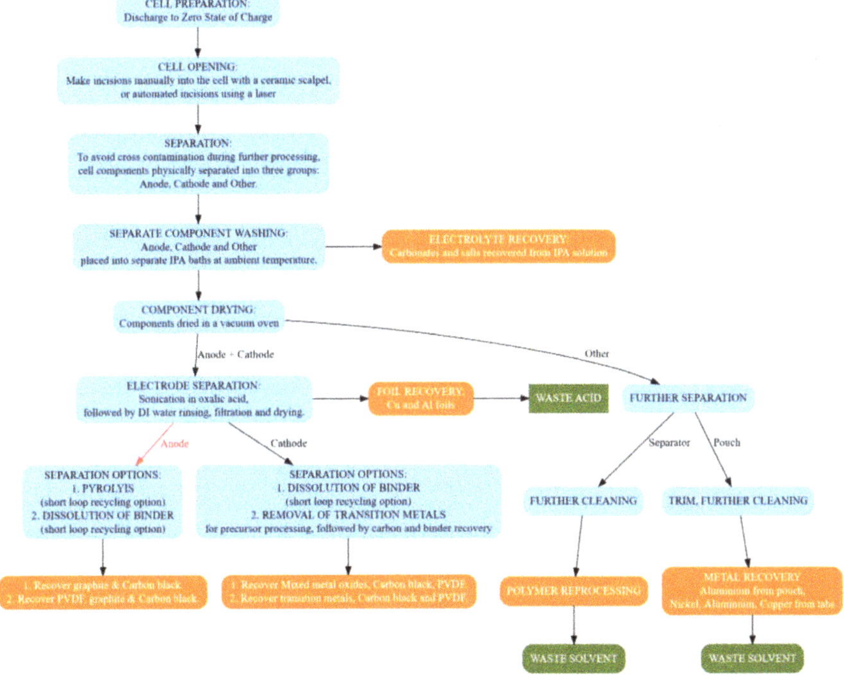

Figure 2. Schematic diagram describing our procedure for the disassembly of a Li-ion battery. Steps marked in blue are our procedure steps for each stage of the cell teardown. Boxes marked in orange represent the recovered materials. Boxes marked in green represent waste materials.

3. Experimental Details

Delamination—The anodic and cathodic "black mass" was separated from the copper and aluminum sheets present in the cell. The individual components were first washed in IPA or water (as above) and then dried for 24 h at 50–75 °C and 100 mBar. This was followed by a sonication process (40 Hz, 50 W) in a solution of oxalic acid, to remove the "black mass" from the metal foils. The anodic "black mass" was separated from the copper foils in an oxalic acid bath concentration 0.02 M, for 30 min at ambient temperature; the cathodic "black mass" was separated from the aluminum foils in an oxalic acid bath at a concentration of 0.5 M, for 5 min at 50 °C. The subsequent composite black mass powders were dried under vacuum at 60 °C before further processing or re-use. The black mass materials can be deconstructed still further and the binder and conductive additives removed for example by either calcining or by chemical extraction methods, this is currently the subject of further study.

T-Peel tests were carried out on an Instron 30 kN test frame, in tension with a 500 N load cell and 1 kN wedge grips. Specimens were pre-loaded to 1.5 N to remove slack from the specimen and tested at 2 mm/min. Each specimen comprised of 2 adherends (25 mm × 120 mm) thermally bonded at one end. Specimens were bonded using the following parameters—temperature 160 °C, time 3 s, bond width 8 mm, pressure 40 kg·cm^{-2}. SEM analysis was carried out using a Hitachi Tabletop Microscope TM3030Plus (Hitachi High-Tech Corporation, Japan). For analysis of the pouch materials, each sample was coated in AuPd, sputtered at 80 mA for 80 s under vacuum (0.01 mBar).

ICP-OES analysis—The cathodes were characterized using ICP-OES (Optima 8000, Perkin Elmer Inc., Waltham, MA, USA). Prior to analysis, the cathodes were removed from a quality control reject cell, dried at 50–75 °C and 100 mBar overnight, washed in distilled water and dried again at 50–75 °C and 100 mBar overnight. Samples were cut from the electrode sheet and dissolved in in aqua regia.

The morphological analysis of the reclaimed material was conducted with the use of Scanning Electron Microscope (SEM) (JCM-7000, JEOL, 1930 Zaventem, Belgium) equipped with a secondary electron detector and under the acceleration voltage of 15 kV. The working distances applied for the SEM study were 6.0 mm and 10.0 mm with the latter applied for the EDS analysis as well.

Electrochemical testing of the black mass—The obtained black mass was ground, sieved and subsequently utilized for the slurry preparation. The anode ink contained—reclaimed graphite, carbon black (Super C65, Timcal, Imerys, 6804 Bironica, Switzerland), carboxymethyl cellulose (Bondwell BVH8, Ashland Industries Europe GMbH, Schaffhausen, Switzerland) and styrene butadiene 40 wt. % suspension in water (Zeon Europe GmbH, 40549 Düsseldorf, Germany) in a weight ratio—92:3:2:3. The cathode coating was prepared in a dry room (dew point of −50 °C) with the following constituents—reclaimed cathode material, carbon black (Super C65), PVDF (Solef® 5130, Solvay SA, 1120 Brussels, Belgium)–prepared as 8 wt.% solution in NMP with the weight ratios of the materials—94:3:3. Both slurries were prepared with the use of centrifugal mixer (Thinky ARE 250, Intertronics, Oxfordshire, UK,) and coated onto the aluminum and copper current collectors for positive and negative electrodes, respectively. The prepared coatings were dried on a hot plate at 50 °C–anode and 80 °C–cathode. Then, they were transferred to the vacuum oven (Binder) and dried overnight at 120 °C prior to the cell assembly.

Coin cells were assembled. The electrodes and separator were cut with a TOB electrode cutter, with the disc diameter adjusted to 14.8 mm, 15.0 mm and 16.0 mm for positive electrode, negative electrode and separator (Celgard 2325, Charlotte, NC, USA) respectively. Cells were constructed with a lithium metal counter electrode and filled with 70 µL electrolyte (1 M LiPF$_6$ in 3EC:7EMC PuriEl R&D261, Soulbrian, MI, USA). The assembled cells were tested using Bio-Logic, Seyssinet-Pariset, 38170 France, BCS cycler applying the following protocol for the anode half-cells—discharge at 0.05 C to 5 mV then charged back to 1.5 V at the same rate using constant current (CC) repeated twice. Followed by 0.2 C discharge and charge steps with the same voltage limits. The cathode half-cells were tested according to the following protocol—CC charge at 0.05 C to 4.2 V, CV step at 4.2 V and CC discharge to 2.8 V.

4. Results

We herein describe how each step of the procedure is to be followed safely and how each component may be reclaimed and possibly re-used.

4.1. Discharge and Cell Opening

Prior to opening the cells it is necessary to ensure that the Open Circuit Voltage (OCV) is low enough to commence a safe teardown. All relevant risk assessment and COSHH documents should be completed to ensure a safe scheme of work is followed due to the potentially toxic nature of these materials and the risk of sparking and/or fire. There is still some debate as to what the discharge level actually is on cell opening and how to accurately capture discharge data due to charge recovery effects. Furthermore, once discharged, there can be some minor charge recovery which may vary from battery to battery.

For the purposes of this study, batteries were provided to us already discharged to a suiTable SOC, in this case we discharged to 2.5 V cell voltage. At this level of discharge and as we are concerned with materials recovery rather than degradation analysis, we were able to open the pouch cells and separate the components in a fume cupboard (without an inert atmosphere) with no extreme adverse reactions. The fume hood must be well-ventilated to prevent release of any toxic materials into the working environment.

We examined cells in two different end states;

(1)   Modules from EoL batteries that had previously been discharged to a low SOC (2.5 V) as recommended on a 2nd life data sheet [57]. Figure 3a is a photograph of one of these modules just as the outer module casing is opened. Inside are two plastic sheets (the blue layers in the photograph) and between these are 4 pouch cells, as shown in Figure 3b. Since the outer casing is mostly aluminum, it can be sent directly for metal recycling. The pouches are separated from the casing for further processing.

(2)   Pouch cells that had been rejected as part of routine Quality Control (QC) procedures. Having been rejected during QC, these cells had never been placed in a module and were supplied as individual cells, discharged to 0 V. Such cells may be considered as part of the manufacturing waste associated with making LIBs. Figure 3c is a photograph of one of these cells. Figure 3d is a photograph of this cell after the pouch has been cut open to expose the layers of electrode materials in the pouch.

In both cases, before opening the actual pouches the external tabs on each cell were removed. The photograph in Figure 3c clearly shows these tabs, on the right-hand side of the photograph as viewed here. Note that the tabs are larger in the QC rejects, as they have never been welded and cut shorter during the process of being fitted into a module. These external tabs are made of Copper, coated in Nickel.

Once the tabs are removed, an incision is made through the outer wall of the pouch; this is done using a ceramic blade in order to prevent any potential risk of shock or short circuit between electrode layers (we envisage that in a future 'automated' version of this procedure, the incision could be made using a laser or automated knife). The pouch is slit open on three sides as shown in Figure 3d. The outer casing of the pouch is made of aluminum with a polymer coating, while inside each pouch there are multiple layers of copper, aluminum, anode and cathode 'black mass' materials and polypropylene (PP) separator layers. In addition, the electrode layers are suffused with the electrolyte solution.

Once the pouch is open, separate stacks are formed of copper foils (coated in anode black mass), aluminum foils (coated in cathode black mass) and the other components of the cell (aluminum pouch and PP separator. For the characterization of the cell components, these three stacks are kept separate in order to prevent cross-contamination. The following sections will deal with the separation and re-use of these components.

4.2. Materials Separation and Recovery

Figure 4 describes the average mass of material collection from the different components as the cell is split down into its various components, a complete analysis is shown in Table 1. A total of 80% and 77% of material is recovered from the QC rejects and EoL cells, respectively. The electrolyte is not currently being reclaimed from the water and solvent waste and therefore this is the major loss in these numbers. These figures were obtained as an average from teardowns of three cells. From these data, we can conclude that the amounts of material recovered are similar at most stages of the process. Exceptions to this are the tabs (which are more readily recovered from QC reject cells, having not been welded into a module, as discussed above) and the electrode black mass, which is more heavily weighted towards the anodic black mass in the case of EoL cells. A number of factors could contribute to this result, including Li uptake into the anode, dendrite formation and Cu dissolution, in the case of EoL cells.

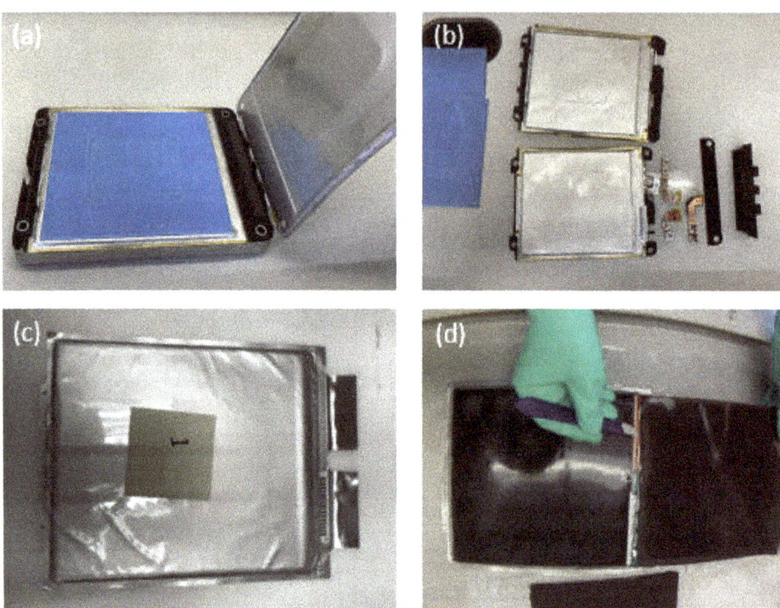

Figure 3. (**a**) Photograph of a module of an end of life (EoL) battery, straight after the casing is opened. (**b**) Photograph of the open module, showing the pouch cells inside. (**c**) Photograph of an individual Quality Control (QC) reject pouch cell. (**d**) Photograph of an open cell pouch, showing the layers of electrode materials within.

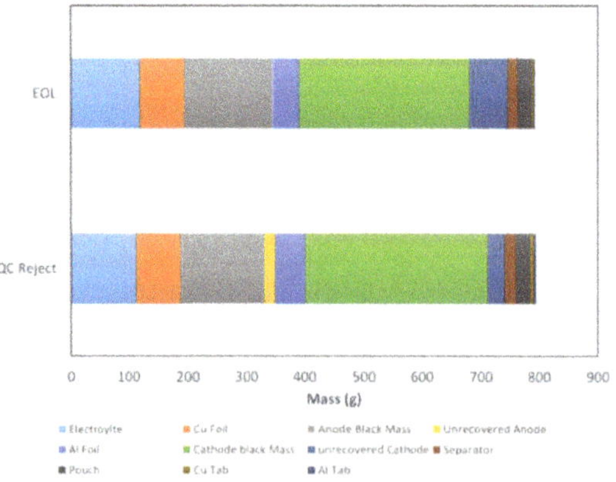

Figure 4. Shows the quantity of material reclaimed from the teardown methodology. Values are given for both QC reject and EoL cells. The values given are averaged over 3 cells.

Table 1. Summary of tear down and reclaimed masses of cell components from a QC rejected cell and EOL.

Cell, Components and Materials				QC Reject		EOL	
				Component Mass (g)		**Component Mass (g)**	
Full Cell				791.0(6)		795(1)	
			Electrolyte	113(20)		120(1)	
Full Cell	Electrodes	Anodes	Cu Foil	237(2)	74(2)	226.67(3)	74.6(2)
			Anode Black Mass		143(1)		152(2)
			Unrecovered Anode		19(1)		0.07(1)
		Cathodes	Al Foil	390(9)	51(2)	399.4(2)	44.41(5)
			Cathode black Mass		310(4)		290(1)
			Unrecovered Cathode		29.12		65.26
	Other	Plastics	Separator	44.0(4)	19.0(4)	42 (2)	17(1)
			Pouch		25.00(2)		25.1(2)
	Metals	Tabs	Cu Tab	1.01(1)	6.21(6)	5.23(6)	3.15(5)
			Al Tab		3.9(1)		2.08(2)

4.2.1. Electrodes: Current Collectors

The current collectors are easily reclaimed after the ultrasonic delamination process, the delamination is nearly 100% effective and little of the black mass is retained on the current collector. The current collectors which are obtained through purely removal of the electrode coatings using NMP solvent are shown in Figure 5. We can observe that for both the cathode and the anode significant indentations into the current collectors are observed. This is from the calendaring of the active material components onto the current collector in the manufacturing process. The image of the aluminum current collector (Figure 5b) has some remaining metal oxides which are clearly embedded into the current collector. The SEM images show insignificant changes between the reclaimed current collectors from solvent removal and the delamination process indicating minimal damage from the ultrasonic oxalic acid solution bath. The copper current collectors from the QC rejected cells have additional pitting due to the oxidation of the copper when discharged to 0 V, (Figure 5a). It should be noted the practice of taking QC rejected cells to 0 V is a safety measure to prevent these cells being used in any other application. The dissolution of the copper is confirmed by the presence of copper oxide in the cathode material as illustrated in the next section.

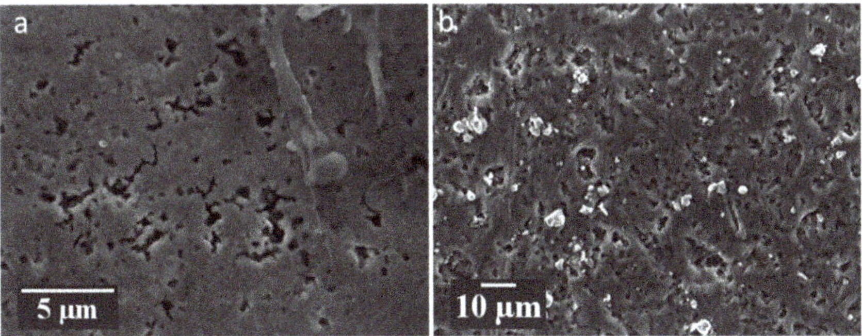

Figure 5. Scanning electron microscopy (SEM) images of copper (**a**) and aluminum (**b**) current collectors obtained from the QC reject electrodes.

4.2.2. Electrodes: Black Mass

The black mass separated from the metallic films by this process can then be filtered out of the oxalic acid bath and rinsed with water or with IPA. This black mass is comprised of the active component of the anode and cathode, with a conductive carbon and PVDF, the SEM images of the cathode from the QC rejected (a,b) and the EOL (c,d) cells are shown in Figure 6. The EDX analysis shows that the QC rejected cells have copper contamination from the discharge process to 0 V (Table S1.2). From the ICP-OES analysis (Table S1.1) the transition metal ratio Mn:Co:Ni is 74.7:2.8:22.5, which is similar to that observed by EDX. The undissolved content was calculated to be 6.7% by mass, this is due to the conductive carbon and PVDF content of the electrode coating. Indicating that there is approximately 93.3% current collector and active mass content for the cathode, which is reasonable. ICP-OES also showed 3.3% by weight of the total metal content. From EDX, the transition metal composition varied between the QC reject and the EOL, a greater percentage of cobalt was present in the EOL cell, possibly indicating low levels of manganese dissolution during the life-time of the cell. Ratios of Mn:Co:Ni are 74:3:23 and 73:6:22 for QC reject and EOL respectively. It should also be noted the Mn is preferentially leached by oxalic acid [58].

The anode black mass was analyzed using SEM-EDX, before and after washing in either IPA or in water. The elemental composition of the black mass varied across the electrode, therefore several elemental maps were taken over the electrode and averaged. Figure 7 shows the SEM images of the black mass from the QC reject and EOL cells. The elemental composition of and the morphology of the surface of the pieces of reclaimed electrode are slightly different depending where within the electrode they have come from. The sample shown in Figure 7a is the surface of the electrode, nearest the separator, whereas Figure 7b is that closest to the current collector. EDS mapping of the elements showed that fluorine concentrated in the spider-like covering shown in Figure 7b and initial XPS results indicate that this is from PVDF binder (Supplementary S2).

The EDS analysis of the black mass from the QC rejects and EOL cells was compared for samples which were not washed, washed in IPA and subsequently delaminated using oxalic acid solution (Table S2.1). We analyzed and compared the particles which were closest to the separator (S) and the current collector (C). The main elemental components in the EOL cell are carbon, fluorine, sulphur, phosphorous and aluminum, the samples obtained from close to the current collector contained a greater percentage of fluorine from the PVDF binder. The ratios of the washed electrodes are very similar to that of the unwashed electrodes, indicating very little change of the black mass composition upon washing. For the QC rejected cells, the black mass has a similar elemental composition to the EOL cells, except for copper being present. The quantity of copper was also variable depending upon whether the electrode was closest to the current collector or to the separator, as is expected the greater copper and fluorine percentage was observed closest to the current collector.

At this stage, we now have a separate stack of anode black mass and cathode black mass. The anode black mass contains graphite, carbon black, PVDF binder and other conductive additive. The cathode black mass contains Lithium Manganese Oxide (LMO) and Lithium Nickel Cobalt Aluminum Oxide (NCA powder), PVDF binder and conductive additive. Also, subsequent generations of automotive cells involve different chemistries, as well as binder polymer, so the separation process will need to be adapted for each model.

As an example; the SEM images shown in Figure 8a,b show large graphite particles (10–30 microns) connected by binder material. These particles are packed together fairly closely in the anode layer. After suspending some of this anode material in NMP (50 °C with ultra-sonication for 30 min) a suspension was formed; passing this liquid through a standard cellulose filter (Whatman grade 1) results in most (>90% of the material remaining on the filter. The SEM images in Figure 8c,d show discrete separated graphite particles resulting from graphite reclamation via this route. The PVDF binder then remains in the solvent used to strip it from the graphite and can be recovered either by evaporation of the solvent or by addition of a non-solvent to remove it from the solution. This process however uses the solvent NMP and therefore alternative routes are required to ensure a more sustainable process.

The exact processes used to separate the electrode components will depend on the specific requirements for the reclaimed material. From this process, the copper and aluminum metal sheets can be cleanly recovered for potential reuse. The waste acid produced in the oxalic acid bath may be contaminated with heavy metals or particulates and will require further treatment for recovery. The cathodic black mass produces a mixed metal oxide material which can be sent for hydrothermal or pyro metallurgical extraction but still contains the PVDF and carbon black at this stage and the anodic black mass contains mainly graphite with PVDF and carbon black. We have shown the potential for future post processing, such as solvent extraction of the PVDF and the carbon black or potentially heat treatments. Significantly more work is required to understand the effects of the post processing upon the active materials and hence the potential for re-use. For the purposes of demonstration of black-mass re-use (Section 4), we have utilized heat treatment for post-processing the black mass after reclamation, this is because the as reclaimed materials did not perform. Heat treatments have safety implications as HF is produced as the PVDF decomposes.

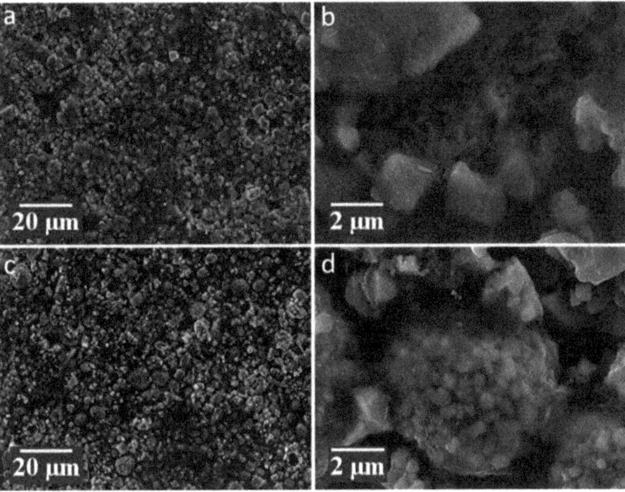

Figure 6. SEM images of the reclaimed cathode black mass for (**a**,**b**) QC rejected cells and (**c**,**d**) end of life cells.

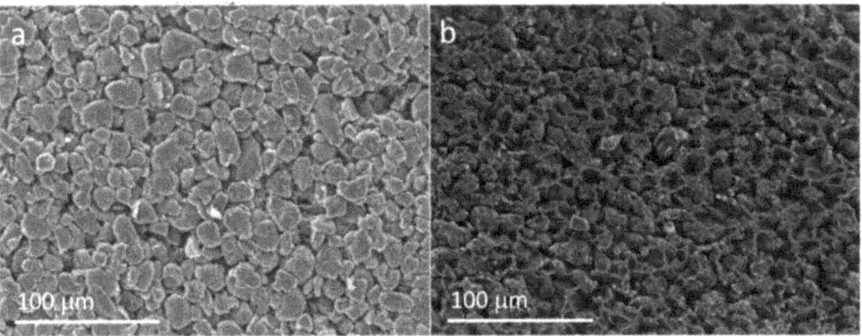

Figure 7. SEM images of anode black mass for (**a**) QC rejected cell and (**b**) EOL cell. Both electrodes have been washed in IPA (**a**) is the surface of the electrode nearest the separator and (**b**) is towards the current collector.

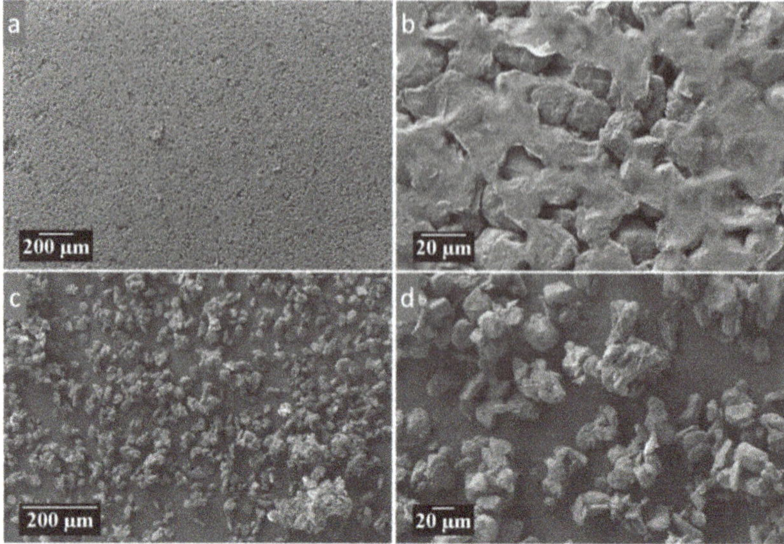

Figure 8. SEM images showing (**a**), (**b**) the anode material after being separated from the copper film (**c**), (**d**) graphite particles from the anode, after the binder is removed by dissolving in NMP (N-methyl Pyrrollidone) followed by filtration.

4.2.3. Electrolyte

The procedure to be followed in order to collect and characterize the electrolyte depends on the aim of the cell teardown. If the aim is simply to collect the electrodes for recycling and in this case the cathodic black mass for hydrometallurgical processing, it may not be necessary for separation of the solvent and electrolyte salt. Here we have not reclaimed the solvent or the salt from the washings. The quantity of solvent and salt has been calculated from the difference in the weights of the total cell and the disassembled and washed components. However, the solvent and the salt can be reclaimed from the washings through precipitation and solvent evaporation. The salt however will not be collected as $LiPF_6$ as it is unstable in water and LiF will precipitate over time as the $LiPF_6$ reacts [59,60].

4.2.4. Outer Pouch

The outer pouch, separator and trim materials make up roughly 7% by weight of a dismantled battery cell. This particular waste stream is, however, of considerably lower economical value than the critical materials (particularly metals) present in the electrodes and electrolyte. Once collected, the pouch is placed in a water bath and subsequently tested for re-use/recycling.

The unused pouches in this system are made of aluminum covered by a polymer coating; the inner coating is polypropylene (PP) while the outer coating contains polyethylene terephthalate (PET) and nylon 6 (PA6). We will use the as-supplied pouches as a 'baseline,' to which we can compare the reclaimed pouches. Figure 9 presents SEM images from the inner surface of the 'baseline' pouch material compared to that of a reclaimed pouch from a QC reject cell.

T-Peel tests were carried out on 14 specimens, using a method based on ISO 11339. Each specimen comprised 2 adherends (25 mm × 120 mm) thermally bonded at one end. There were 5 pristine pouch material specimens and 9 specimens made from reclaimed pouch (from a QC reject cell). Figure 10a gives example load vs extension curves for one example of a "baseline" pouch and a "reclaimed" pouch. In the 0–5 mm extension regime, both curves show an initial maximum load; between 10–15 mm the load variation is flatter. Figure 10b gives average results across all samples tested, for the initial peak load, the maximum load obtained and the mean load seen across the "flatter" 5–15 mm region. It is clear that the reclaimed pouches retain much of their mechanical strength; while there is some decrease in the initial peak load and overall maximum load, the load variation across the 5–15 mm region has not changed significantly. Therefore, it appears likely that pouch materials could be reused (in smaller format sizes) if the weldability of the reused PP can be determined to be safe. Alternatively, it may be beneficial to re-use the pouch material in a less hazardous environment as part of an alternative supply chain.

Figure 9. SEM images of the inner surface of (**a**) a "pristine" pouch, compared to (**b**) and (**c**) a surface from a reclaimed pouch (from a QC reject cell) that has been washed and dried. It is clear that some modification of the surface has occurred.

4.2.5. Separator Films

The separator is a porous polypropylene (PP) film held between the two layers of black mass forming the cathode and anode. It is peeled away from the layers of black mass during the cell disassembly stage.

In order to investigate the contamination of the separator, we cleaned pieces of separator by ultra-sonication in solvents for 30 min (ultrasound at 50 W and 40 KHz was used; the ultrasound bath was held at 50 °C during this time). The separator films were then dried and re-weighed. The difference in mass between the initially reclaimed separator and the separator after washing and drying is shown in Figure 11. In all of these solvents, a significant decrease in mass was seen after the washing process; we conclude that the solvents abstracted a significant amount of impurity from each film.

In addition, we note that the separator films were very weak and easily tore during the battery dismantling process. It therefore seems unlikely that they can be re-used as separator films. PP is a commonly recycled polymer, however and (depending on the cleaning procedures) it could be recycled in products requiring lower grade PP.

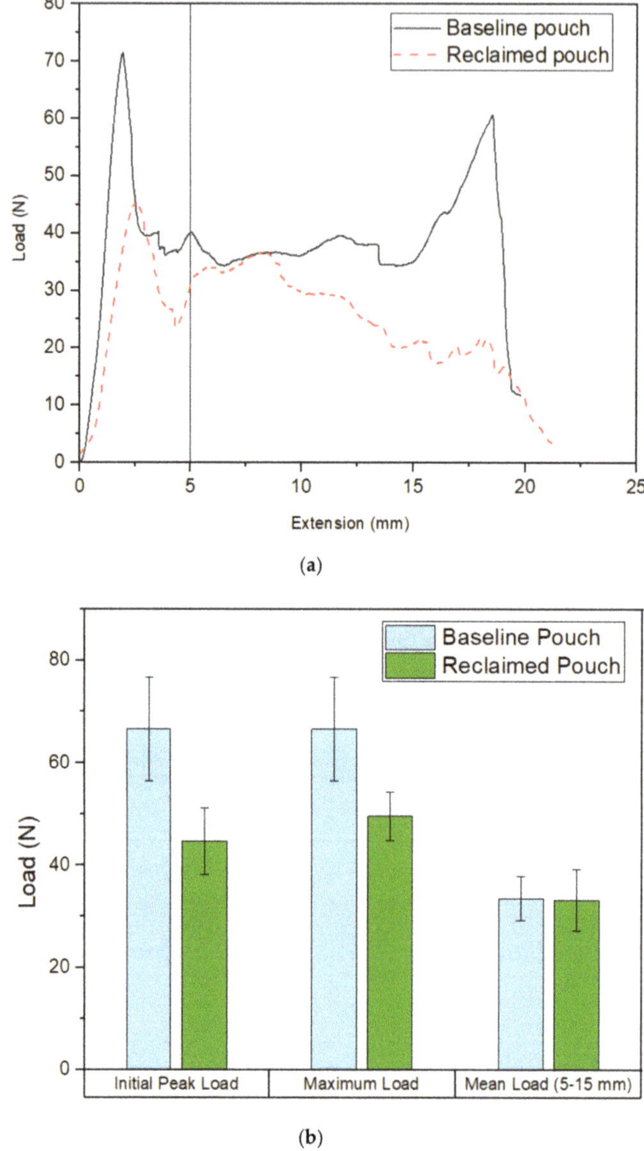

Figure 10. Graphs to show load vs extension curves for a number of pouch specimens. (**a**) Gives an example load vs extension graph for a baseline pouch compared to a reclaimed pouch. We see an initial peak load below 5 mm and then the graph appears flatter between 5–15 mm. (**b**) compares the average peak load, maximum load and mean load over a number of baseline and reclaimed pouches.

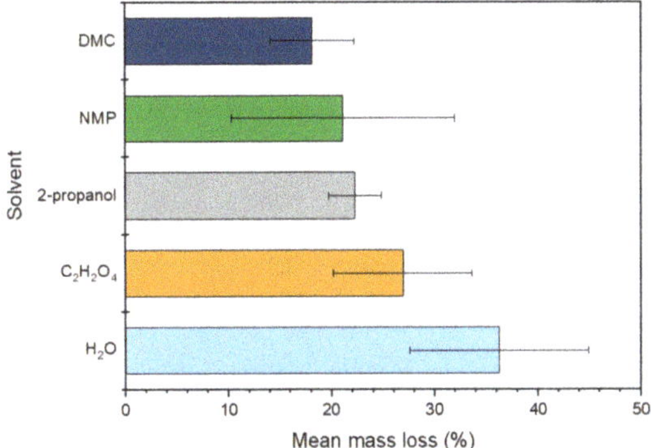

Figure 11. Mass loss of the separator on washing with various solvents (DMC, NMP, IPA, oxalic acid and water).

5. Reuse and Recycling of the Components

After reclamation we must consider the re-use cases for the cell components that we have extracted. We consider the current collectors, separator, pouch and black mass.

5.1. Pouch Materials

Used pouch material still has comparable mechanical strength to pristine pouch material. T-Peel tests show a slight decrease in the initial peak load and overall maximum load but the results are broadly comparable. If it can be demonstrated that the weldability of the pouch material is suitable, there is potential for used pouch material to be reused in new pouch cells, albeit of a smaller format. Used pouch material could also find applications in less hazardous environments if mechanical strength is paramount.

5.2. Current Collectors

The Aluminum current collector is affected by the processing and is more delicate than the copper. Figure 5 shows SEM images of current collectors from an uncycled (QC reject) cell, with the electrode coating removed using NMP. The current collectors are pitted despite not being processed using ultrasound. The copper current collector is affected by the state of charge it is stored at and ICP-OES and EDX data in Supplementary Data Tables S1.1 and S1.2 show Cu contamination in the cathode and anode black mass which is attributed to the dissolution of copper in the overdischarge process applied to these cells. The pitting and possible dissolution further weaken the foils, making reuse in a new cell difficult. Large scale electrode manufacture prints onto long rolls of foil which are cut into discrete foils after printing. Laboratory scale operations can use draw down coaters for individual foils. Recovering Al in the metallic form allows it to be recycled. Recycling Al saves 95% of the energy invested in primary production of Al [61].

5.3. Black Mass

The black mass for the negative and positive electrodes were collected separately, Cathode Black Mass was analyzed using SEM-EDS, compared to the untreated electrode from supplementary data Table S1.2 and is shown in Table 2. These results show a similar Ni:Mn and Ni:Co ratio to the starting material, although the Co content of the end of life material has decreased over the lifetime of the cell.

Graphite anode black mass content was analyzed to be variable as shown by XPS and EDX (Table S2.1) was dependent upon the position in the electrode, however if we assume the total carbon is an average of the surfaces, we observe 91.9% carbon by weight for the end of life and 91.95% by weight for the QC rejected cell. The major additional content was from fluorine and in the case of the QC rejected copper. It should be noted that the lithium content was not possible to analyze.

Table 2. Metal Ratios of an untreated electrode (Data from Table S2.1) and two electrodes treated with the ultrasonic separation technique.

Metal Ratios	Untreated Electrode	Ultrasonicated QC Reject	Ultrasonicated EOL
Ni:Mn	0.30	0.31	0.30
Ni:Co	8.1	7.8	3.9
Co:Mn	0.04	0.04	0.08

To investigate the re-use case for the anode and cathode, the respective black mass was made directly into an electrode after drying (Figure 12). Initial results were poor where both anode and cathode exhibited high polarization and low capacities. In the case of the cathodic black mass, the processing in oxalic acid solution has affected the reclaimed cathode. Although not observable from X-ray diffraction measurements, it is likely that the surface of the materials have reacted to form carbonates, hydroxides or oxalates, which will form a resistive coating. We therefore investigated post treatment of the black mass. The cathode black mass was heat treated at 300 °C for 4 h. whereas, the anode black mass was heat treated at 600 °C for 2 h.

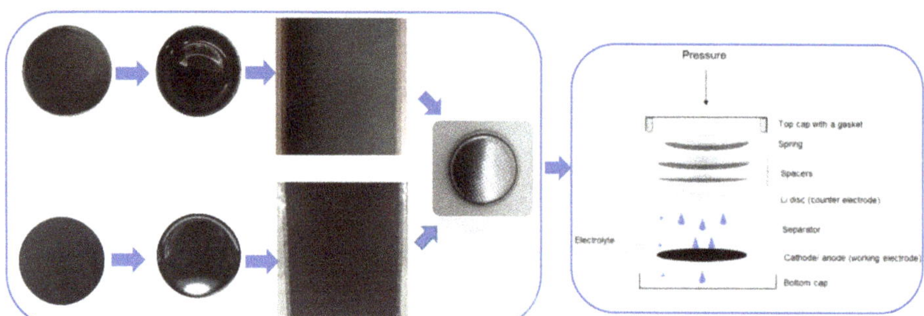

Figure 12. Anode and cathode coating preparation from the reclaimed material, top diagram presents anode, bottom cathode process steps: reclaimed active material, mixed slurry, electrode coating and coin cell.

The obtained results from the formation cycles of the half-cells testing are presented in Figure 13. The measured specific capacities of the graphite are 360 mAh·g^{-1} and 370 mAh·g^{-1} for the material reclaimed from QC reject and EoL cells, respectively. The specific capacity of the reclaimed cathode material is of 66.2 mAh·g^{-1} for the cathode black mass. The first cycle efficiencies for all the half-cells were measured to be above 90%. From the results shown from the EOL cathode material, we can see that the water and oxalic acid negatively affects the performance of the cathode. In order to "reclaim" any capacity from the materials they must first be heat-treated. Even with this heat treatment, the specific capacities are still significantly lower than we would expect. This indicates a likely loss of lithium during the reclamation process and relithiation steps require investigating for short loop recycling. If we consider however the black mass for hydrometallurgical and pyro metallurgical processes, there is very little loss of transition metal during the reclamation processes. The cathodic black mass waste stream has an extremely high purity of transition metal Mn:Ni:Co which can be leached and separated for synthesizing into a new cathode material or utilized in other industries. Cobalt is used in animal feeds and pigments, manganese has applications in steel and metallurgical alloys, primary alkaline batteries, chemical feed for water purification and treat waste water, Nickel is used in metallurgical alloys for example in coins and turbine blades.

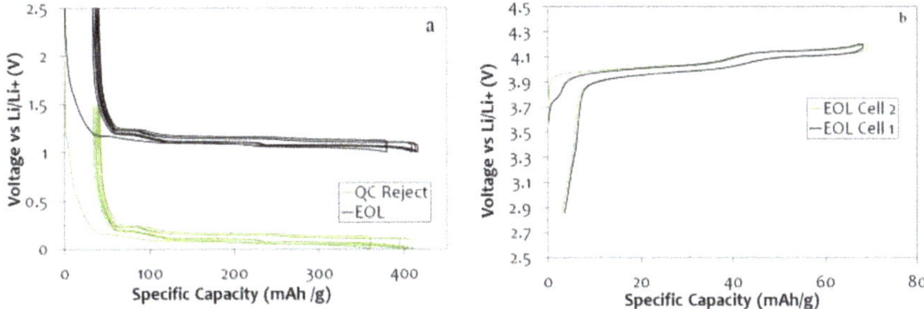

Figure 13. Formation cycle of the (**a**) anodes half-cells from QC reject and end of life anodes (**b**) and two cathode half-cells for end of life cathode after heat treatment of the reclaimed black mass.

From the above analysis of the re-use of the components, there is a possibility to re-use each component apart from the electrolyte. Each component has been compromised in some way from the use, disassembly and reclamation of the components. The current collectors are dented and pitted. We expect that if re-used the level of contact between the current collector and active components reduced, which will increase the resistance. The separator and pouch material have both been damaged in the cell and the strength reduced. The black mass requires post processing and heat treatment before it can be used. We demonstrate that we can short loop recycle the graphite and we obtain good specific capacities during low rate cycling, the cathodic black mass requires further re-lithiation, we observe useable capacities but with a large resistance. However, the cathodic black mass is suitable for further hydrometallurgical separation and is present in a very pure materials stream for reclamation.

6. Conclusions

As discussed above, there are significant challenges inherent in the recycling of Li-ion batteries, not least of which is the variety of different chemistries in use for the cathode and anode materials. Any one approach cannot be completely general and must be adapted for the particular system being used. However, valuable insight is gained by dismantling a battery of one type and some parts of the procedure can be generalized. This is part of the hierarchy of recycling, and we have investigated the potential for re-use or remanufacturing with materials, as we further refine the waste streams. The work was originally focused upon reclamation of pure cathodic black mass waste stream for further hydrometallurgical extraction of the transition metal components, and subsequently the disassembly route developed to reclaim most of the component parts in decontaminated waste streams. We have developed a methodology with a sustainable disassembly route in mind, using sustainable solvents, low cost routes and no toxic chemicals. We also discuss the safety aspects of each process, and the methodology we have adopted to ensure chemical and electrical safety.

The principal limitation of this approach is that it is rather labor-intensive, as it requires a person to manually make incisions in the cell and separate the internal components. However, some steps of the procedure could be automated, such as the opening of the pouch in order to separate the pouch components into separate stacks. Automation of the procedure would allow this process to be scaled up. The bottle necks in the automation are in cell opening and component separation. In both of these steps consideration around the types of chemicals potentially formed during the process or with exposure to air is required. In addition, the components material composition for the equipment needs consideration to ensure no corrosion of the parts. In addition, there are many types of cells; cylindrical, pouch and prismatic which are constructed in different manners, with different materials and joining mechanisms. Intelligent designs of equipment and identification tools will be required in order to identify the different cell types and therefore knowledge upon how to dismantle them.

In this article, we have demonstrated a workable method for the safe dismantling of pouch cells and have compared differences in method for the dismantling of EoL cells and QC reject cells. This "disassembly" approach allows us to recover most of the valuable materials present in each cell and to separate them into material types with greater purity than is usually possible using conventional methods that involve shredding the battery. This approach, then, has a considerable environmental benefit, as it can allow for potential re-use of some components and recovery of others for recycling.

In terms of the materials that are usefully recovered using this process, we straightforwardly obtain a significant quantity of aluminum and copper in the form of the current collectors, polymer from the separator films and the 'black mass' from the electrodes. Further processing is required in order to extract useful materials from the black mass. For some of the low-concentration chemical components of the black mass, it may be uneconomic to separate them completely and a short-loop recycling process may be more appropriate; this will be the subject of a future article. The polymer separators are unsuitable for simple re-use but have the potential to be recycled to form lower-grade polymer materials.

It is apparent that not all the components can be re-used, and if they are re-used they may have life-time and contamination issues. We have demonstrated the electrochemical performance of the anodic and cathodic black mass from the reclaimed and reprocessed materials, and show that the graphite can be short loop recycled, but the reclaimed cathode requires further relithiation. This is part of the circular economy for battery materials; some components can be re-used at different points of reclamation and some require remanufacturing. The ability to produce a completely recycled cell from the reclaimed components is unrealistic at present, but it may be possible to remanufacture the component parts from the reclaimed materials to produce a recycled cell. This is a great example of the circular economy picture for lithium ion battery recycling.

Supplementary Materials: The following are available online at http://www.mdpi.com/2075-4701/10/6/773/s1, Supplementary S1: Analysis of Cathode Black Mass; Supplementary S2: Analysis of Anode Black Mass. Figure S1. 1 EDX mapping analysis showing distribution of the elements (a), the transition metals Mn (b) and Ni (c) in the SEM images, there appear to be two separate phases the first is Mn rich, the second is Ni rich. Table S1. 1 ICP-OES analysis of the extracted cathode electrode, washed with distilled water, not delaminated. Table S1. 2. EDX analysis of the elemental composition of the QC reject and EOL after washing and delamination. Figure S2. 1 SEM (a) and EDS elemental mapping (b) of QC rejected anode, showing copper contamination throughout the electrode. Figure S2. 2 SEM (a) and EDS elemental mapping (b) of EOL cell, showing carbon particles and patches of high fluorine content due to the PVDF binder. Figure S2. 3 XPS analysis of the surface of the QC rejected anode black mass, and the fluorine analysis. Table S2. 1 Elemental composition analysed by EDX of the anode black mass with different treatments and positions within the cell, C–closest to current collector, S–closest to separator.

Author Contributions: Conceptualization, E.K. and V.G.; methodology, D.G., R.S., B.M., J.M.; investigation, D.G., R.S., J.M., B.M.; writing-original draft preparation, J.M., E.K, R.S. writing-review and editing, D.G., R.S., B.M., V.G., E.K.; supervision, E.K. and V.G.; funding acquisition, E.K. and V.G.; project administration, E.K. and V.G. All authors have read and agreed to the published version of the manuscript.

Funding: We acknowledge funding from UKRI project "Reclamation and Re-manufacture of lithium ion batteries," reference number 104425, and Faraday Institution ReLIB FIRG005 and FIRG006.

Acknowledgments: We would like to acknowledge the help of Marc Walker and the use of the XPS facilities at the Warwick Photoemission Facility (University of Warwick). We acknowledge Marny Bharndal (University of Warwick) for assistance with the T-Peel tests. We also acknowledge the ReLIB team at Birmingham, particularly Anton Zorin, Newcastle, and Leicester for helpful discussions and advice.

Conflicts of Interest: The authors declare no conflict of interest. The funders had no role in the design of the study; in the collection, analyses or interpretation of data; in the writing of the manuscript or in the decision to publish the results.

References

1. Heelan, J.; Gratz, E.; Zheng, Z.; Wang, Q.; Chen, M.; Apelian, D.; Wang, Y. Current and Prospective Li-Ion Battery Recycling and Recovery Processes. *JOM* **2016**, *68*, 2632–2638.
2. Velázquez-Martínez, O.; Valio, J.; Santasalo-Aarnio, A.; Reuter, M.; Serna-Guerrero, R. A Critical Review of Lithium-Ion Battery Recycling Processes from a Circular Economy Perspective. *Batteries* **2019**, *5*, 68. [CrossRef]
3. Tarascon, J.M.; Armand, M. Issues and challenges facing rechargeable lithium batteries. *Nature* **2001**, *414*, 359–367.
4. Harper, G.; Sommerville, R.; Kendrick, E.; Driscoll, L.; Slater, P.; Stolkin, R.; Walton, A.; Christensen, P.; Heidrich, O.; Lambert, S.; et al. Recycling lithium-ion batteries from electric vehicles. *Nature* **2019**, *575*, 75–86. [CrossRef]
5. Waldmann, T.; Iturrondobeitia, A.; Kasper, M.; Ghanbari, N.; Aguesse, F.; Bekaert, E.; Daniel, L.; Genies, S.; Gordon, I.J.; Löble, M.W.; et al. Review—Post-mortem analysis of aged lithium-ion batteries: Disassembly methodology and physico-chemical analysis techniques. *J. Electrochem. Soc.* **2016**, *163*, A2149–A2164. [CrossRef]
6. Kobayashi, Y.; Kobayashi, T.; Shono, K.; Ohno, Y.; Mita, Y.; Miyashiro, H. Decrease in Capacity in Mn-Based/Graphite Commercial Lithium-Ion Batteries. *J. Electrochem. Soc.* **2013**, *160*, A1181–A1186. [CrossRef]
7. Golubkov, A.W.; Scheikl, S.; Planteu, R.; Voitic, G.; Wiltsche, H.; Stangl, C.; Fauler, G.; Thaler, A.; Hacker, V. Thermal runaway of commercial 18650 Li-ion batteries with LFP and NCA cathodes - Impact of state of charge and overcharge. *RSC Adv.* **2015**, *5*, 57171–57186. [CrossRef]
8. Sonoc, A.; Jeswiet, J.; Soo, V.K. Opportunities to improve recycling of automotive lithium ion batteries. *Procedia CIRP* **2015**, *29*, 752–757.
9. Hendricks, C.E.; Mansour, A.N.; Fuentevilla, D.A.; Waller, G.H.; Ko, J.K.; Pecht, M.G. Copper Dissolution in Overdischarged Lithium-ion Cells: X-ray Photoelectron Spectroscopy and X-ray Absorption Fine Structure Analysis. *J. Electrochem. Soc.* **2020**, *167*, 090501. [CrossRef]
10. Shaw-Stewart, J.; Alvarez-Reguera, A.; Greszta, A.; Marco, J.; Masood, M.; Sommerville, R.; Kendrick, E. Aqueous solution discharge of cylindrical lithium-ion cells. *Sustain. Mater. Technol.* **2019**, *22*, e00110. [CrossRef]
11. Aurbach, D.; Markovsky, B.; Rodkin, A.; Cojocaru, M.; Levi, E.; Kim, H.J. An analysis of rechargeable lithium-ion batteries after prolonged cycling. *Electrochim. Acta* **2002**, *47*, 1899–1911. [CrossRef]
12. Williard, N.; Sood, B.; Osterman, M.; Pecht, M. Disassembly methodology for conducting failure analysis on lithium-ion batteries. *J. Mater. Sci. Mater. Electron.* **2011**, *22*, 1616–1630.
13. Wegener, K.; Andrew, S.; Raatz, A.; Dröder, K.; Herrmann, C. Disassembly of electric vehicle batteries using the example of the Audi Q5 hybrid system. *Procedia CIRP* **2014**, *23*, 155–160.
14. Wang, X.; Gaustad, G.; Babbitt, C.W. Targeting high value metals in lithium-ion battery recycling via shredding and size-based separation. *Waste Manag.* **2016**, *51*, 204–213. [CrossRef]
15. Sloop, S.; Crandon, L.; Allen, M.; Koetje, K.; Reed, L.; Gaines, L.; Sirisaksoontorn, W.; Lerner, M. A direct recycling case study from a lithium-ion battery recall. *Sustain. Mater. Technol.* **2020**, *25*. [CrossRef]
16. Diekmann, J. Ecological recycling of lithium-ion batteries from electric vehicles with focus on mechanical processes. *J. Electrochem. Soc.* **2017**, *164*, A6184–A6191.
17. Smith, W.N.; Swoffer, S. Recovery of lithium ion batteries. U.S. Patent 8616475B1, 31 December 2013.
18. Grützke, M.; Krüger, S.; Kraft, V.; Vortmann, B.; Rothermel, S.; Winter, M.; Nowak, S. Investigation of the Storage Behavior of Shredded Lithium-Ion Batteries from Electric Vehicles for Recycling Purposes. *ChemSusChem* **2015**, *8*, 3433–3438. [CrossRef]
19. Marinos, D.; Mishra, B. An Approach to Processing of Lithium-Ion Batteries for the Zero-Waste Recovery of Materials. *J. Sustain. Metall.* **2015**, *1*, 263–274. [CrossRef]
20. Wang, M.M.; Zhang, C.C.; Zhang, F.S. Recycling of spent lithium-ion battery with polyvinyl chloride by mechanochemical process. *Waste Manag.* **2017**, *67*, 232–239. [CrossRef]
21. Meshram, P.; Pandey, B.D.; Mankhand, T.R. Extraction of lithium from primary and secondary sources by pre-treatment, leaching and separation: a comprehensive review. *Hydrometallurgy* **2014**, *150*, 192–208.

22. Kovachev, G.; Schröttner, H.; Gstrein, G.; Aiello, L.; Hanzu, I.; Wilkening, H.M.R.; Foitzik, A.; Wellm, M.; Sinz, W.; Ellersdorfer, C. Analytical Dissection of an Automotive Li-Ion Pouch Cell. *Batteries* **2019**, *5*, 67. [CrossRef]

23. Qian, D.; Ma, C.; More, K.L.; Meng, Y.S.; Chi, M. Advanced analytical electron microscopy for lithium-ion batteries. *NPG Asia Mater.* **2015**, *7*.

24. Mossali, E.; Picone, N.; Gentilini, L.; Rodrìguez, O.; Pérez, J.M.; Colledani, M. Lithium-ion batteries towards circular economy: A literature review of opportunities and issues of recycling treatments. *J. Environ. Manage.* **2020**, *264*, 110500. [CrossRef]

25. BatPaC Model Software|Argonne National Laboratory. Available online: www.anl.gov/cse/batpac-model-software (accessed on 24 November 2019).

26. *Study on the Review of the List of Critical Raw Materials—Critical Raw Materials Factsheets*; EU Publications, European Commission: Brussels, Belgium, 2017; pp. 1–93.

27. Fortier, S.M.; Nassar, N.T.; Lederer, G.W.; Brainard, J.; Gambogi, J.; McCullough, E.A. *Draft Critical Mineral List—Summary of Methodology and Background Information—U.S. Geological Survey Technical Input Document in Response to Secretarial Order No. 3359*; U.S. Geological Survey: Reston, VA, USA, 2018. [CrossRef]

28. Gaines, L.; Cuenca, R. *Costs of Lithium-Ion Batteries for Vehicles*; U.S. Department of Energy-Office of Scientific and Technical Information: Argonne, IL, USA, 2000.

29. Dai, Q.; Spangenberger, J.; Ahmed, S.; Gaines, L.; Kelly, J.C.; Wang, M. *EverBatt: A Closed-loop Battery Recycling Cost and Environmental Impacts Model*; Department of Energy-Office of Scientific and Technical Information: Argonne, IL, USA, 2019.

30. Gaines, L. Lithium-ion battery recycling processes: research towards a sustainable course. *Sustain. Mater. Technol.* **2018**, *17*, e00068.

31. Gaines, L. The future of automotive lithium-ion battery recycling: charting a sustainable course. *Sustain. Mater. Technol.* **2014**, *1–2*, 2–7.

32. Zhan, R.; Payne, T.; Leftwich, T.; Perrine, K.; Pan, L. De-agglomeration of cathode composites for direct recycling of Li-ion batteries. *Waste Manag.* **2020**, *105*, 39–48. [CrossRef]

33. Paulino, J.F.; Busnardo, N.G.; Afonso, J.C. Recovery of valuable elements from spent Li-batteries. *J. Hazard. Mater.* **2008**, *150*, 843–849.

34. Zhan, R.; Oldenburg, Z.; Pan, L. Recovery of active cathode materials from lithium-ion batteries using froth flotation. *Sustain. Mater. Technol.* **2018**, *17*, e00062.

35. Gao, W. Lithium carbonate recovery from cathode scrap of spent lithium-ion battery: a closed-loop process. *Environ. Sci. Technol.* **2017**, *51*, 1662–1669.

36. Li, X.; Zhang, J.; Song, D.; Song, J.; Zhang, L. Direct regeneration of recycled cathode material mixture from scrapped LiFePO4 batteries. *J. Power Sources* **2017**, *345*, 78–84.

37. Dolotko, O.; Hlova, I.Z.; Mudryk, Y.; Gupta, S.; Balema, V.P. Mechanochemical recovery of Co and Li from LCO cathode of lithium-ion battery. *J. Alloys Compd.* **2020**, *824*. [CrossRef]

38. Lestriez, B. Functions of polymers in composite electrodes of lithium ion batteries. *Comptes Rendus Chim.* **2010**, *13*, 1341–1350.

39. Egbue, O.; Long, S. Critical issues in the supply chain of lithium for electric vehicle batteries. *EMJ-Eng. Manag. J.* **2012**, *24*, 52–62. [CrossRef]

40. Banza Lubaba Nkulu, C.; Casas, L.; Haufroid, V.; De Putter, T.; Saenen, N.D.; Kayembe-Kitenge, T.; Musa Obadia, P.; Kyanika Wa Mukoma, D.; Lunda Ilunga, J.M.; Nawrot, T.S.; et al. Sustainability of artisanal mining of cobalt in DR Congo. *Nat. Sustain.* **2018**, *1*, 495–504. [CrossRef]

41. Turcheniuk, K.; Bondarev, D.; Singhal, V.; Yushin, G. Ten years left to redesign lithium-ion batteries. *Nature* **2018**, *559*, 467–470.

42. Li, L.; Zheng, P.; Yang, T.; Sturges, R.; Ellis, M.W.; Li, Z. Disassembly Automation for Recycling End-of-Life Lithium-Ion Pouch Cells. *JOM* **2019**, *71*, 4457–4464. [CrossRef]

43. Cell, Module, and Pack for EV Applications | Automotive Energy Supply Corporation. Available online: http://archive.is/h0BZL (accessed on 2 February 2020).

44. Schmitt, J.; Haupt, H.; Kurrat, M.; Raatz, A. Disassembly automation for lithium-ion battery systems using a flexible gripper. In Proceedings of the IEEE 15th International Conference on Advanced Robotics: New Boundaries for Robotics, ICAR 2011, Tallin, Estonia, 20–23 June 2011; pp. 291–297.

45. Herrmann, C.; Raatz, A.; Mennenga, M.; Schmitt, J.; Andrew, S. Assessment of automation potentials for the disassembly of automotive lithium ion battery systems. In Proceedings of the Leveraging Technology for a Sustainable World—Proceedings of the 19th CIRP Conference on Life Cycle Engineering, Berkeley, CA, USA, 23–25 May 2012; Springer: Berlin/Heidelberg, Germany; pp. 149–154.

46. Hanisch, C.; Loellhoeffel, T.; Diekmann, J.; Markley, K.J.; Haselrieder, W.; Kwade, A. Recycling of lithium-ion batteries: A novel method to separate coating and foil of electrodes. *J. Clean. Prod.* **2015**, *108*, 301–311. [CrossRef]

47. Zhang, G.; He, Y.; Feng, Y.; Wang, H.; Zhang, T.; Xie, W.; Zhu, X. Enhancement in liberation of electrode materials derived from spent lithium-ion battery by pyrolysis. *J. Clean. Prod.* **2018**, *199*, 62–68. [CrossRef]

48. Xin, Y.; Guo, X.; Chen, S.; Wang, J.; Wu, F.; Xin, B. Bioleaching of valuable metals Li, Co, Ni and Mn from spent electric vehicle Li-ion batteries for the purpose of recovery. *J. Clean. Prod.* **2016**, *116*, 249–258. [CrossRef]

49. Li, J.; Shi, P.; Wang, Z.; Chen, Y.; Chang, C.C. A combined recovery process of metals in spent lithium-ion batteries. *Chemosphere* **2009**, *77*, 1132–1136. [CrossRef]

50. Regulation (EC) No 1907/2006 of the European Parliament and of the Council of 18 December 2006 concerning the Registration, Evaluation, Authorisation and Restriction of Chemicals (REACH), establishing a European Chemicals Agency, amending Directive 1999/45/EC and repealing Council Regulation (EEC) No 793/93 and Commission Regulation (EC) No 1488/94 as well as Council Directive 76/769/EEC and Commission Directives 91/155/EEC, 93/67/EEC, 93/105/EC and 2000/21/EC. Available online: https://www.desitek.dk/sites/default/files/media/files/DEHN-REACH-Certificate.pdf (accessed on 8 June 2020).

51. Tait, W.S. Controlling corrosion of chemical processing equipment. In *Handbook of Environmental Degradation Of Materials: Third Edition*; Elsevier Inc., William Andrew Publishing: Norwich, NY, USA, 2018; pp. 583–600. ISBN 9780323524735.

52. Feng, Y.; Siow, K.S.; Teo, W.K.; Tan, K.L.; Hsieh, A.K. Corrosion Mechanisms and Products of Copper in Aqueous Solutions at Various pH Values. *Corrosion* **1997**, *53*, 389–398. [CrossRef]

53. Satyabama, P.; Rajendran, S.; Nguyen, T.A. Corrosion inhibition of aluminum by oxalate self-assembling monolayer. *Anti-Corrosion Methods Mater.* **2019**, *66*, 768–773. [CrossRef]

54. Pernel, C.; Farkas, J.; Louis, D. Copper in organic acid based cleaning solutions. *J. Vac. Sci. Technol. B Microelectron. Nanom. Struct.* **2006**, *24*, 2467–2471. [CrossRef]

55. Nguyen, T. Degradation of Polyfvinyl Fluoride) and Poiyfvinylidene Fluoride). *J. Macromol. Sci. Part C* **1985**, *25*, 227–275. [CrossRef]

56. Diaz, F.; Wang, Y.; Moorthy, T.; Friedrich, B. Degradation Mechanism of Nickel-Cobalt-Aluminum (NCA) Cathode Material from Spent Lithium-Ion Batteries in Microwave-Assisted Pyrolysis. *Metals (Basel)* **2018**, *8*, 565. [CrossRef]

57. Datasheet Gen2 Cell Module, Second life EV Batteries. Available online: https://www.secondlife-evbatteries.com/nissan-leaf-1670wh-module.html (accessed on 7 February 2020).

58. Sahoo, R.N.; Naik, P.K.; Das, S.C. Leaching of manganese from low-grade manganese ore using oxalic acid as reductant in sulphuric acid solution. *Hydrometallurgy* **2001**, *62*, 157–163. [CrossRef]

59. Gorman, S.F.; Pathan, T.S.; Kendrick, E. The 'use-by date' for lithium-ion battery components. *Philos. Trans. R. Soc. A Math. Phys. Eng. Sci.* **2019**, *377*, 20180299. [CrossRef]

60. Lux, S.F.; Lucas, I.T.; Pollak, E.; Passerini, S.; Winter, M.; Kostecki, R. The mechanism of HF formation in $LiPF_6$ based organic carbonate electrolytes. *Electrochem. commun.* **2012**, *14*, 47–50. [CrossRef]

61. Rombach, G. Raw material supply by aluminium recycling-Efficiency evaluation and long-term availability. *Acta Mater.* **2013**, *61*, 1012–1020. [CrossRef]

 *metals* 

Article

A Combined Pyro- and Hydrometallurgical Approach to Recycle Pyrolyzed Lithium-Ion Battery Black Mass Part 1: Production of Lithium Concentrates in an Electric Arc Furnace

Marcus Sommerfeld [1,*], Claudia Vonderstein [1], Christian Dertmann [1], Jakub Klimko [2], Dušan Oráč [2], Andrea Miškufová [2], Tomáš Havlík [2] and Bernd Friedrich [1]

[1] IME Process Metallurgy and Metal Recycling, Institute of RWTH University; Intzestraße 3, 52056 Aachen, Germany; cvonderstein@ime-aachen.de (C.V.); cdertmann@ime-aachen.de (C.D.); bfriedrich@ime-aachen.de (B.F.)

[2] Institute of Recycling Technologies, Faculty of Materials, Metallurgy and Recycling, Technical University of Košice, Letna 9, 042 00 Košice, Slovakia; jakub.klimko@tuke.sk (J.K.); dusan.orac@tuke.sk (D.O.); andrea.miskufova@tuke.sk (A.M.); tomas.havlik@tuke.sk (T.H.)

* Correspondence: msommerfeld@ime-aachen.de; Tel.: +49-241-809-5200

Received: 15 July 2020; Accepted: 5 August 2020; Published: 7 August 2020

Abstract: Due to the increasing demand for battery raw materials such as cobalt, nickel, manganese, and lithium, the extraction of these metals not only from primary, but also from secondary sources like spent lithium-ion batteries (LIBs) is becoming increasingly important. One possible approach for an optimized recovery of valuable metals from spent LIBs is a combined pyro- and hydrometallurgical process. According to the pyrometallurgical process route, in this paper, a suitable slag design for the generation of slag enriched by lithium and mixed cobalt, nickel, and copper alloy as intermediate products in a laboratory electric arc furnace was investigated. Smelting experiments were carried out using pyrolyzed pelletized black mass, copper(II) oxide, and different quartz additions as a flux to investigate the influence on lithium-slagging. With the proposed smelting operation, lithium could be enriched with a maximum yield of 82.4% in the slag, whereas the yield for cobalt, nickel, and copper in the metal alloy was 81.6%, 93.3%, and 90.7% respectively. The slag obtained from the melting process is investigated by chemical and mineralogical characterization techniques. Hydrometallurgical treatment to recover lithium is carried out with the slag and presented in part 2.

Keywords: lithium-ion battery; recycling; cobalt; nickel; circular economy; lithium minerals; lithium slag characterization; thermochemical modeling; critical raw materials; smelting

---

1. Introduction

Lithium-ion batteries (LIBs) are currently considered as one of the most important energy storage systems, which is reflected in a wide range of applications, especially for portable devices [1–7]. Due to the extensive electrification expected in the field of electromobility, batteries will have another key role in the future, ensuring the transition towards a climate neutral economy [8]. In addition to the implementation of electromobility and their widespread use for portable applications, lithium-ion batteries are also indispensable as intermediate storage for the stabilization of decentralized power systems [2–5,9,10]. Compared to other battery types, LIBs have advantageous technical properties that substantiate their dominance as energy storage systems, including, e.g., high energy density and low self-discharge [10,11]. As a result of increasing applications of lithium-ion batteries, a significantly higher demand for batteries containing critical or strategic raw materials, such as cobalt, lithium, and nickel, is to be expected. Those crucial metals are only available in limited quantities and currently obtained mainly from primary sources [2]. Recycling is an essential aspect of closing the entire substance cycle of LIBs and securing the supply of raw materials for new battery production. To meet the increasing demand for strategic metals, the development of a raw-material recycling economy, in addition to the expansion of mining capacities, is therefore unavoidable [5].

In the European Union, Directive 2006/66/EC applies to the recycling of LIBs, which has been implemented into national law in Germany by the Battery Law (BattG). This directive requires the recycling of fifty percent of the average weight of used batteries, including spent LIBs [12]. The extensive recycling of battery components that exceed the recycling quota of fifty percent is of central importance to ensure the supply of materials for the new battery production and consequently the transition to a climate-neutral economy. This also requires a consideration of battery components such as lithium.

Although lithium does not represent a critical metal, its recovery, especially regarding future battery systems, where lithium will be manifested as an indispensable cathode component, becomes essential. In the field of battery recycling, research projects have been carried out for several years dealing with both single and combined mechanical, pyrometallurgical and hydrometallurgical processes as well as pyrolysis to recover battery components [5,10,11,13–23]. However, the focus is set mainly on critical and valuable metals, which is the reason why lithium as a component has not been sufficiently considered [24]. Overall, the recovery of lithium from active material has not been solved satisfactorily since the recovery is made more difficult by the ignoble character of the metal. Currently, only one percent of the total end-of-life lithium is recycled [25,26]. In pyrometallurgical processes, lithium is converted into slag, which is either used as a construction material, undergoes further hydrometallurgical treatment, or can be sold, e.g., for the cement industry [27]. Hydrometallurgical processes allow lithium to be recovered from black mass, for example as lithium carbonate [28].

Within the scope of this work, a combined pyro- and hydrometallurgical process was designed, which enables a complete recovery of the valuable metals present in the black mass of spent LIBs. In a first process step, the production of artificial lithium concentrates will enable the recovery of lithium, while more precious components, such as cobalt and nickel, can be recovered via the generated metal alloy.

To obtain a maximized lithium yield through a combined pyro- and hydrometallurgical process, in this study the preconditions of lithium-containing slag for the subsequent hydrometallurgical recovery are designed. This slag design aims to ensure a stable process in the presence of varying metal contents in the black mass resulting from fluctuating scrap input materials. A comparable approach was followed by Georgi-Maschler et al. [29], whereby the deviation in the present study is a slag design adapted to the subsequently performed hydrometallurgy. Additionally, a different slag system was used to generate a SiO_2-Al_2O_3-Li_2O slag. A delimitation is also given by the addition of copper oxide, which enables the use of excess graphite in the black mass as a reducing agent.

2. Materials and Methods

2.1. Used Materials

The black mass used as initial material for pyrometallurgical treatment was provided by Accurec Recycling GmbH (Germany). To obtain the pelletized black mass for subsequent pyrometallurgical treatment it has to pass several pretreatment process steps which are shown in Figure 1. In the dismantling step, a manual removal of Cu cables, the steel casing, plastics, and electrical components is implemented. The subsequent pyrolysis step enables a deactivation of batteries and vaporization of the electrolyte. At last, the pyrolyzed batteries pass further mechanical treatment steps, such as comminution and sieving to separate the coarse fraction from the black mass-containing fine fraction <100 μm, which is further pelletized and applied as raw material for the electric arc furnace (EAF) smelting process.

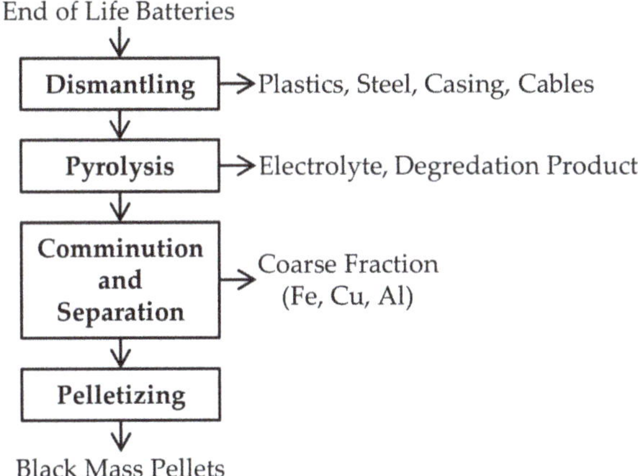

Figure 1. Schematic process flow sheet for black mass pellet generation.

The pelletized black mass was analyzed with a "Spectro CIROS Vision" inductively coupled plasma-optical emission spectrometer (ICP-OES) made by "SPECTRO Analytical Instruments GmbH, Kleve, Germany". All ICP-OES measurements were carried out twice. Carbon analysis was carried out with an "ELTRA CS 2000" system made by "ELTRA GmbH, Haan, Germany" based on a combustion method. Carbon measurements were carried out three times per sample. Table 1 shows mean value of the analysis of the received black mass and the standard deviation of the sample set (Std. Dev.). Regarding the composition of pelletized black mass, the high cobalt content compared to manganese and nickel content must be emphasized. Further elements like halogens or phosphorus in the initial resource were not analyzed.

Table 1. Composition of the black mass in wt.% analyzed by ICP-OES and combustion.

Compound	Co	Fe	Mn	Al	Cu	Si	Zn	Ni	Ag	Li	C
Mean	22.0	6.51	0.75	3.88	4.69	0.37	0.11	2.71	0.32	2.24	20.5
Std. Dev.	0.14	0.12	0.00	0.05	0.02	0.07	0.00	0.04	0.01	0.02	0.27

Commercial grade quartz from "Quarzwerke GmbH, Frechen, Germany" was used as a flux in this research with a SiO_2-content above 98 wt.%. Copper(II) oxide from "Lomberg GmbH, Oberhausen, Germany" was used with a CuO-content above 98.9 wt.%. For calculations and simulations, the SiO_2 and CuO contents of those raw materials were assumed to be 100 wt.%.

2.2. Thermochemical Modeling

Thermochemical modeling was carried out using the "Equilib-module" of FactSage^{TM} 7.3 [30] to simulate the smelting process step. The influence of fluxing the black mass with SiO_2, Al_2O_3 and CaO and mixtures of those oxides on the lithium-slagging were investigated. Furthermore, the influence of the process temperature was studied. The distribution of lithium in the process was investigated in detail for varying amounts of SiO_2-additions.

The databases FactPS, FToxid and SGTE 2014 were used [30]. The FToxid database was used for oxidic solid solutions, the FToxid-SLAGA phase and pure oxides, the FactPS database was used for pure substances, while duplicates already included in the FToxid phase were excluded from the FactPS database, the liquid alloy phase from the SGTE 2014 database was used. Due to the fact that the FToxid-SLAGA solution does not contain lithium components in the original model, liquid lithium oxide, silicates, aluminate, and carbonate were merged into the solution model and treated as an ideal solution, therefore the activity coefficient is assumed to be one, which is not realistic and is one of the limitations of this model. In the "compound species" selection, the "ideal" option was used for the gas phase. In all cases, the normal equilibrium was calculated, the pressure was set to be one atmosphere and the molar volume of solids and liquids was assumed to be zero.

To simulate the change in Gibbs energy for reactions of pure substances, the "Reaction-module" of FactSage^{TM} 7.3 with the database FactPS was used [30]. As a step size, 10 K was used and involved in the reaction was always the most stable form of a compound at any given temperature.

A solidification simulation of the slag based on the analyzed composition of the slag was carried out using FactSage^{TM} 7.3 [30]. As only the slag was of interest for this step, only the FactPS and FToxid database [30] were used, the other settings were the same compared to the smelting simulation. The Scheil-Gulliver cooling model was used, therefore, after each temperature step solidified species are excluded from the total mass balance and equilibrium. The starting temperature was set to be the temperature, where no solids were present and only the slag phase occurred, the cooling step rate was defined as five Kelvin in the program. Transition metal oxides and oxides of minor elements were neglected in the solidification simulation, and therefore only the following oxides were included in the simulation: Li_2O, SiO_2, Al_2O_3, MgO, CaO, and BaO.

2.3. Smelting Trials in an Electric Arc Furnace

Smelting experiments were carried out in a direct-current (DC) electric arc furnace with a voltage between zero and eighty volt and a current between zero and thousand amperes. Therefore, the maximum power is eighty kilowatts. The power is infinitely variable, while the voltage is dependent on the electrical resistance of the charge and the furnace itself. The electrical current is therefore controlled by the operating system according to the set power. The position of the top electrode can be adjusted with a hydraulic system. A schematic sketch of the furnace and a picture of the furnace in operation is shown in Figure 2.

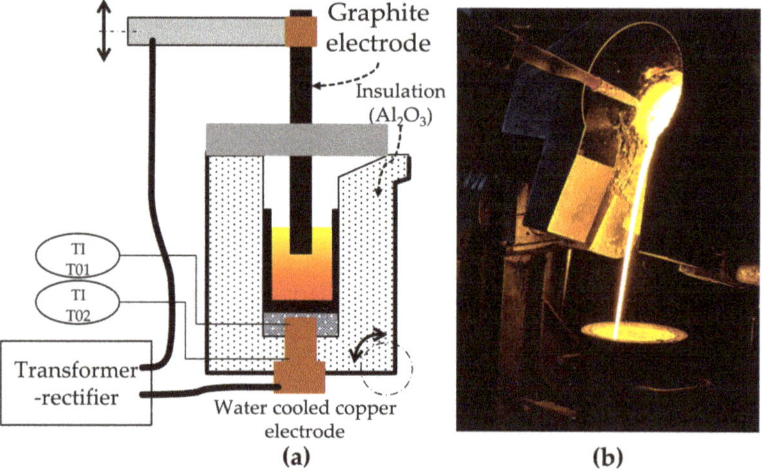

Figure 2. Laboratory electric arc furnace: (**a**) Schematic concept of the furnace; (**b**) tapping of the furnace after an experiment.

The smelting operation was carried out in a high-purity graphite crucible with an inner diameter of 120 mm. The volume of the crucible was 2 L. The graphite top electrode had a diameter of 50 mm and was immersed in the slag during smelting. Before the trial, the crucible was heated to roughly 1000 °C. Fluxes, copper(II) oxide and pelletized black mass were feed simultaneously by hand. Thereby, 3.5 kg of black mass was smelted per trial, whereas the flux-addition was varied. The copper(II) oxide addition was also varied within the trials, as the influence of the CuO-addition can vary from trial to trials. It was planned, that CuO reacts with excess carbon from the black mass, but as CuO can also react with carbon monoxide in ascending gases [31] as shown in Equations (1) and (2) or with graphite from the crucible or electrode, a controlled CuO-addition is difficult and therefore the CuO-addition was adjusted during the trial, based on the visual appearance of the slag. If excess graphite was floating on the slag, additional CuO was added.

$$CuO_{(s)} + CO_{(g)} \rightleftharpoons Cu_2O_{(s)} + CO_{2(g)} \tag{1}$$

$$Cu_2O_{(s)} + CO_{(g)} \rightleftharpoons 2Cu_{(s)} + CO_{2(g)} \tag{2}$$

The smelting temperature was 1600 °C in every trial with an estimated accuracy of ±25 °C. The temperature in the slag was measured discontinuously with type B thermocouple immersion probes made by "Heraeus Electro-Nite GmbH & Co. KG, Hagen, Germany".

The feeding time was between 90 and 110 min. After the material was completely charged into the furnace, the melt was kept at a constant temperature for ten minutes to allow further reactions between the slag and metal. Slag samples were taken during the holding time with a cold cast-iron rod, the chemical composition of the solidified slag was then analyzed. Samples for mineralogical and chemical investigation were taken from the bulk slag phase after each trial. After the holding time, the melt was either poured into a cast-iron-mould, as can be seen in Figure 2, or the melt was kept in the crucible and furnace. This was done to investigate the effect of the cooling rate on the slag mineralogy. The cooling rate for tapped trials is significantly higher, compared to the trials which solidified in the furnace since the refractory material of the furnace is also heated up during the trial and is working as further heat insulation and heat storage. Metal alloys from two trials were analyzed after solidification and a remelting operation.

3. Results

Results obtained from the thermochemical simulation are described in this section and the results of the smelting trials, including chemical and mineralogical investigations of the obtained slag and an exemplary chemical analysis of the metal phases obtained from two trials. Off-gas and dust is not analyzed from the trials.

3.1. Results of Thermochemical Modeling

A thermochemical simulation is carried out with the program FactSage^{TM} [30] to determine possible process conditions, which increase the amount of lithium being transferred into the slag. The influence of various oxide additions and the melt temperature are considered as the main variable parameters. Furthermore, the stability of lithium components is investigated and the distribution of lithium into different phases occurring in the process is shown in detail for one slag system.

3.1.1. Influence of Different Oxidic Fluxes on the Lithium Slagging

To investigate the effect of different oxides on the lithium slagging, a thermochemical simulation was carried out at a constant temperature of 1600 °C and a constant addition of copper(II) oxide of 65 g per 100 g of black mass with the composition listed in Table 1. The CuO-addition was set to 65 g, because at 1600 °C, in the absence of fluxes, there was still solid carbon present in the system, which is always the case in the laboratory trials, due to the contact of the graphite crucible and the graphite electrode with the molten phases. An additional amount of 12.09 g of Oxygen per 100 g of black mass was added to the solution, based on the assumption that the lithium is present as $LiCoO_2$ in the material and that leftover cobalt, nickel, and manganese are present as a divalent metal oxide.

The lithium slagging in this work is defined, as the amount of lithium present in the slag and any occurring solid lithium aluminate or aluminosilicate in relation to the lithium input. This assumption was made, as the regular slag solution model does not contain lithium compounds and therefore, the slag could be completely molten, even if solid lithium compounds occur according to the simulation. Equation (3) is used to calculate the slagging.

$$\text{Li} - \text{slagging} = 100\% \cdot \frac{m_{\text{Li}_{\text{Slag}}} + m_{\text{Li}_{\text{Solids}}}}{m_{\text{Li}_{\text{Black mass}}}} \tag{3}$$

Figure 3 shows the results of the simulation. Only Al_2O_3, CaO, SiO_2 were investigated in this simulation and mixtures of those oxides. Binary mixtures contained 50 wt.% of each oxide and ternary mixtures contained 33.333 wt.% of each oxide. The step size of the flux addition was 1.5 g per 100 g black mass.

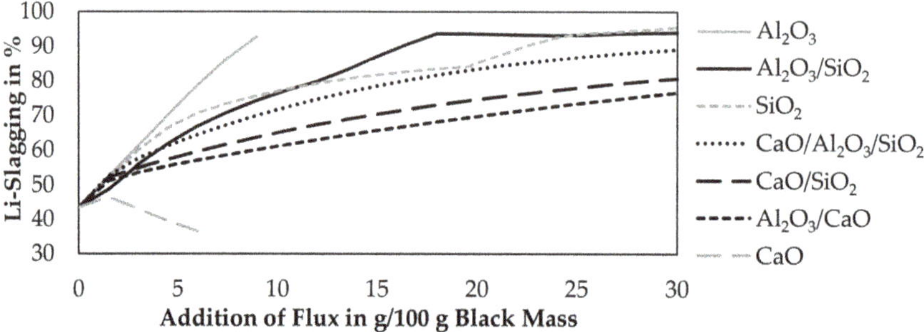

Figure 3. Influence of fluxing on the Li-slagging at 1600 °C with an addition of 65 g CuO per 100 g black mass.

According to the simulation, a CaO-addition above 6 g will lead to a saturation of the slag and the presence of solid CaO. An Al_2O_3-addition above 9 g will also lead to a saturation of the slag and the presence of solid Al_2O_3. Therefore, the addition of those fluxes is not investigated further the point of saturation.

Al_2O_3 and SiO_2 improve the lithium slagging in every investigated combination, while pure CaO is the only investigated flux, which decreases the lithium slagging. This could be explained by the lower lithium solubility in CaO-slag systems [32]. An addition of more fluxes is beneficial to achieve a higher slagging of lithium, except for lime, where the slagging increases after an addition of 1.5 g CaO and starts to decrease with higher additions again. More than 90% of slagging is obtained, if more than 22.5 g of SiO_2, 9 g of Al_2O_3, or more than 16.5 g of an Al_2O_3-SiO_2 mixture is added.

3.1.2. Influence of the Process Temperature on the Lithium Slagging

The previous investigated fluxes and mixtures were simulated at different temperatures to investigate the influence of the temperature on the lithium slagging. Pure CaO is already excluded as a possible flux, due to a decrease in lithium slagging accompanied by an addition of CaO according to Figure 3. A constant addition of 21 g fluxes per 100 g black mass was investigated while maintaining the other parameters constant compared to the previous simulation. The step size was 50 °C. Figure 4 shows the results of that simulation.

Two slag system showed solid species at lower temperatures in this simulation. An addition of Al_2O_3 leads to solid Al_2O_3 at temperatures below 1650 °C and the mixture of Al_2O_3-SiO_2 leads to solid Al_2O_3 at temperatures below 1550 °C, therefore, those results are excluded. In all cases, a lower smelting temperature leads to lower lithium losses into the gas and metal. The highest lithium slagging is obtained by an addition of SiO_2 at 1500 °C with a lithium slagging of 98.3%, followed by an addition of an Al_2O_3-SiO_2 mixture at 1550 °C with a lithium slagging of 97.3%. The mixtures containing CaO are inferior compared to SiO_2, Al_2O_3, and the mixture of those oxides, especially at higher temperatures, where the disadvantage of CaO becomes obvious.

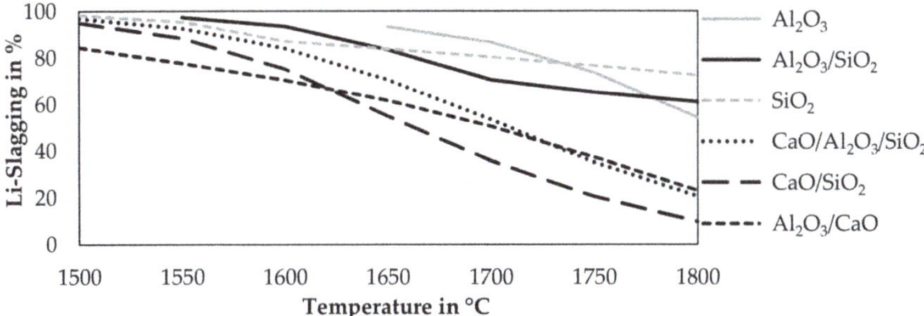

Figure 4. Influence of temperature on the Li-slagging using 21 g of fluxes and 65 g of CuO per 100 g black mass.

3.1.3. Theoretical Stability of Lithium Minerals

The positive influence of Al_2O_3 and SiO_2 can be explained, by the change of Gibbs energy for reactions including one mole Li_2O with SiO_2, Al_2O_3 or both. Figure 5 shows the change of Gibbs energy of reactions for lithium minerals available in the FactSage^{TM} databases [30] dependent on the temperature.

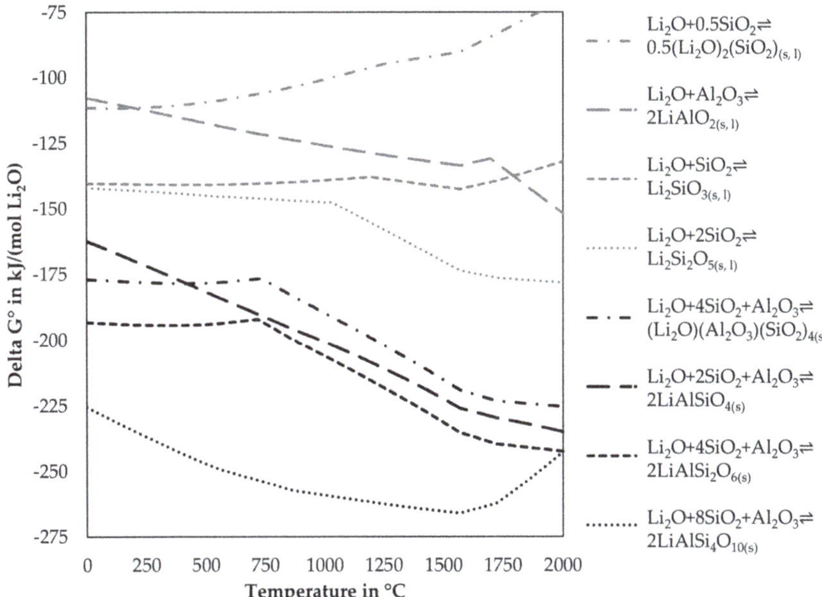

Figure 5. Simulated change of Gibbs energy for the reaction of lithium oxide with alumina, silica and alumina and silica.

As Delta G° of the reactions presented in Figure 5 is always negative in the investigated temperature range, those reactions would occur spontaneously, if all reactants are available for reactions. The lithium aluminosilicates have lower Delta G° values compared to the silicates and the aluminate and should therefore be more dominant in the slag, however, this depends also on the composition of the slag and other elements, which could react with alumina or silica, as only the elements listed in each reaction are considered. Changes in the slope of a graph are normally accompanied by phase transition. However, the dataset for lithium aluminosilicates only contains those in the solid-state, whereas for silicates and aluminates also liquid phases are included in the database. Table 2 lists the investigated lithium-containing silicates, aluminates, and aluminosilicates from Figure 5. Included in the table is the transition temperature from the low-temperature modification to the high-temperature modification, if available and also the melting point, if available in the dataset. Furthermore, the lithium content for stoichiometric compounds is listed.

Table 2. Theoretical transition temperatures of selected lithium compound and lithium content according to FactSage™ [30].

Compound Formula	Name According to Database	Solid Transition	Melting Point	Li-Content in wt.%
$(Li_2O)_2(SiO_2)$	Lithium Orthosilicate	-	1254.85 °C	23.17
$LiAlO_2$	Lithium Aluminium Oxide	-	1699.85 °C	10.53
Li_2SiO_3	Lithium Silicate	-	1200.85 °C	15.43
$Li_2Si_2O_5$	Lithium Silicate	935.85 °C	1033.85 °C	9.25
$(Li_2O)(Al_2O_3)(SiO_2)_4$	Spodumene	736.46 °C	-	3.73
$LiAlSiO_4$	Eucryptite	1026.85 °C	-	5.51
$LiAlSi_2O_6$	Spodumene	720.04 °C	-	3.73
$LiAlSi_4O_{10}$	Petalite	-	-	2.27

To avoid lithium losses due to fuming, the stability of pure lithium compounds was evaluated against volatilization. Therefore, the activity of the gas phase above pure lithium compounds listed in Table 2 was investigated for varying temperatures at a constant pressure of one atmosphere. The step size was 5 °C. For the simulation, only the listed lithium compound was allowed to form as condensed phases for each curve. There were no restrictions on the gaseous components which could form. Figure 6 shows the results of this model for seven lithium compounds.

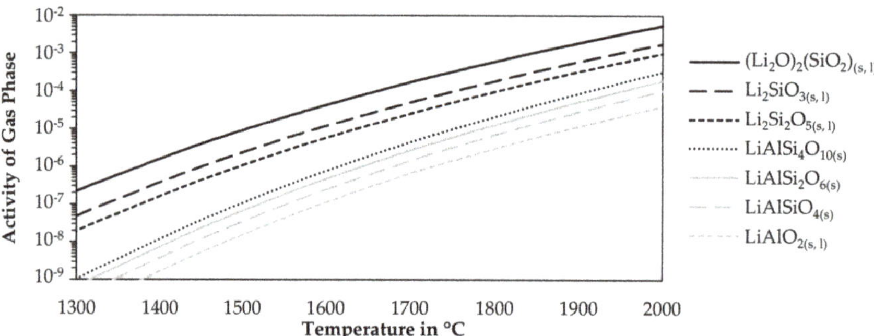

Figure 6. Activity of the gas phase in equilibrium with pure lithium compounds.

The higher the activity, the more likely it is, that a component is volatilized. As the activity is always below one, no gas was formed in equilibrium in absence of other components, however, in an open smelting process, there will always be an atmosphere above the charge or the molten phases, which would allow the uptake of lithium from the slag or solid lithium components. The figure shows that the lithium silicates are more likely to be volatilized while increasing silicon contents in the silicates decrease the activity of the gas phase. The aluminosilicates are more stable. However, a lower silicon content is beneficial in this case. The most stable compound is lithium aluminate according to the simulation.

3.1.4. Detailed Investigation of the Addition of Quartz as a Flux

Due to the abilities of quartz to promote the slagging of lithium and the prospect of lower possible smelting temperatures according to the findings in Figures 3 and 4, the behavior of lithium while varying the quartz addition was investigated further in a simulation presented in Figure 7. The CuO-addition was again fixed with 65 g per 100 g black mass. As a temperature, 1600 °C was chosen.

The step-size was decreased to an addition of one gram of quartz per 100 g black mass. Even though a lower temperature should increase the slagging of lithium and should be possible according to the simulation shown in Figure 4, a higher temperature was chosen as slags show lower viscosities at higher temperatures and therefore valuable metal losses due to entrained metal droplets in the slag should be lower at higher temperatures. Liquid lithium components in the figure are all merged as ideal solutions into the slag model FactSage^{TM} [30]. The combined amount of lithium in this idealized slag phase is expressed by the black dashed line. The solid black line is expressing the combined lithium slagging as calculated by Equation (3). Lithium contained in the gas phase is considered as a loss. Also, lithium contained in the two metal phases is considered as lost. According to the simulation, two immiscible metal phases are formed. Those are named "Copper"-phase and "Cobalt"-phase, which is in both cases the element with the highest concentration in the metal, accompanied by other mostly metallic elements as well.

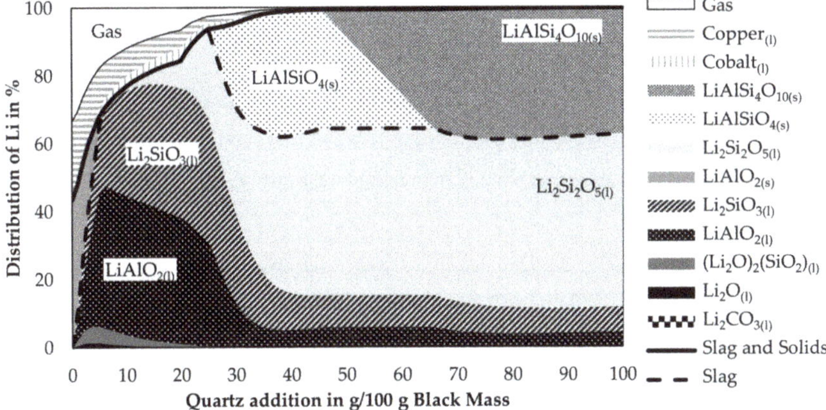

Figure 7. Distribution of lithium into different phases occurring in the process at 1600 °C with a CuO-addition of 65 g per 100 g black mass.

Lithium losses decrease while adding more quartz as a flux. Most lithium is lost due to volatilization into the gas phase, also the copper phase contains significant amounts of lithium, while losses into the cobalt phase are less significant. This can be explained by the higher amount of copper produced in the process compared to cobalt. Even though the lithium content in the cobalt phase is higher than in the copper phase. For example, an addition of 20 g quartz leads to a lithium content of 0.24 wt.% in the cobalt phase and 0.19 wt.% in the copper phase. The simulation predicts a completely molten slag for additions between 6 g and 25 g. Additions below 6 g led to solid $LiAlO_2$ and additions above 25 g led to solid lithium aluminosilicate. Most lithium in the slag is present as liquid lithium silicate or liquid lithium aluminate. The content of lithium oxide and carbonate is rather small and decreases with increasing quartz additions even further. Even though a dataset for solid $LiAlSi_2O_6$ is available in the database, it does not appear in the simulation as a stable phase, instead, an equilibrium involving $LiAlSiO_4$ and $LiAlSi_4O_{10}$ can be observed for additions of 46 g up to 65 g of quartz.

As indicated by Figures 3, 4 and 7, slagging of lithium can be increased by adding quartz, however, the lithium content in the occurring lithium phases decreases with increasing quartz additions. To ease the lithium recovery from the slag, a high lithium content in the lithium minerals is beneficial. Figure 8 shows the lithium slagging calculated by Equation (3) and the lithium content in the slag for various temperatures and quartz additions. Furthermore, the SiO_2-content in the slag is displayed. Solid lithium minerals are assumed to be part of the slag, they are therefore added to the amount of slag generated and also the lithium and silicon in solid lithium minerals are included in the calculation. The step size of the simulation is 1.5 g of quartz addition.

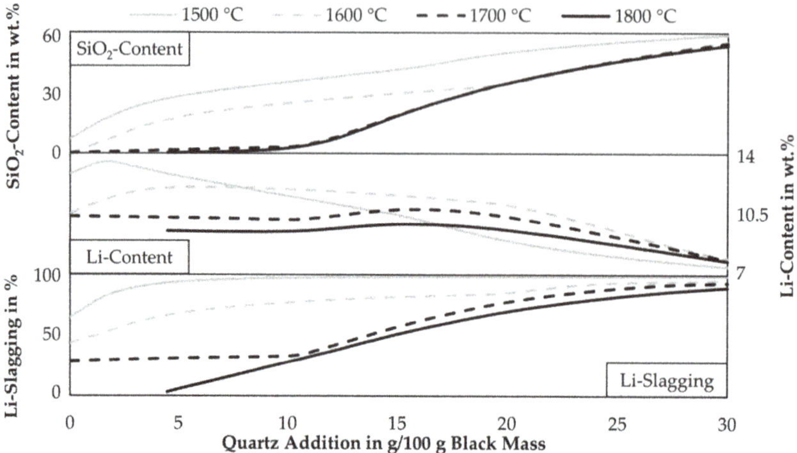

Figure 8. Influence of temperature and quartz addition on the SiO$_2$- and Li-content in slag and Li-slagging with a CuO-addition of 65 g per 100 g black mass.

Especially at temperatures above 1700 °C, the SiO$_2$-content only increases slowly while adding quartz. Due to the reductive conditions in the process, silicon is reduced into the metal phase and is not incorporated into the slag at higher temperatures until the leftover graphite is completely consumed. At 1800 °C quartz additions over 4.5 g are necessary to obtain a liquid slag, therefore, the data points below 4.5 g are excluded from the diagram. As already seen before, the lithium slagging is increased while adding quartz. The lithium content in the slag also starts to increase while adding quartz but decreases with further additions of quartz in every case. This dilution occurs already after additions of 3 g quartz at 1500 °C, higher temperatures shift the position of the highest lithium content in the slag to higher quartz additions. For a temperature of 1600 °C, which was already investigated in detail in Figure 7, a quartz addition of 10 g or more should be investigated further, as the lithium slagging is relatively high with 77%. The lithium content in the slag is 11.9 wt.% in this case. Further additions would still increase the lithium slagging but also decrease the lithium content. For example, an addition of 20 g quartz increases the lithium slagging to 84.5% while decreasing the lithium content slightly to 11.1 wt.%. Above an addition of 20 g, the lithium content decreases more rapidly. Also, it seems that the lowest shown temperature would lead to the lowest lithium content in the slag for quartz additions above 16.5 g, this can be explained by the already mentioned reduction of SiO$_2$ at higher temperatures according to the simulation and therefore, the slag at 1500 °C would be more diluted by SiO$_2$. If the SiO$_2$-reduction is as strong as predicted by the model is uncertain, FactSage^{TM} [30] calculates the equilibrium at a constant temperature for all phases, but in direct current electric arc furnaces, the slag temperature is higher than the metal temperature, due to the higher electrical resistance of the slag compared to the metal [33]. The result can be a significantly lower silicon content in the metal compared to the predicted model, which was already observed in previous trials carried out in the laboratory scale electric arc furnace used in this study [34].

Influences of the investigated parameters on the metal recovery were not shown here, as the yield for nickel, cobalt and copper are sufficient in every simulated parameter combination. The lowest yields for nickel, cobalt and copper are 99.95%, 99.91%, and 98.75% respectively. Whereas the mean value of the yields based on 1029 simulated parameter combinations is 99.99%, 99.96%, 99.60%.

3.2. Results of Smelting Trials in an Electric Arc Furnace

The analysis of the smelting trials includes a mass balance of the solid obtained products (metal and slag), the chemical analysis of both immiscible metal phases for one sample, the detailed chemical composition of all slag samples and a detailed mineralogical investigation of the slag, carried out by X-ray diffraction (XRD) and Raman analysis.

3.2.1. Mass Balance

Every trial was carried out with an input mass of 3500.0 g pelletized black mass pellets. The additions of quartz and CuO are presented in Table 3. Those components are the only additional inputs into the process. The weight of metal and slag after a manual separation is also listed in Table 3. It was not possible to obtain the mass of flue dust or gas during the trials. Instead, the total weight loss is included in the table and an adjusted weight loss, under the assumption, that oxygen bound to cobalt, copper and nickel and carbon from the black mass is subtracted from the weight loss, as those components leave the system as off-gas. Furthermore, the trial number is included in the table, which is consistently used in all tables and figures in this paper and the solidification method, as the melt is either poured into a cast iron mould or solidified in the graphite crucible after the trial.

Table 3. Input and output mass of the trials.

Trial No.	Black Mass in g	Addition in g Per 100 g Black Mass		Solidification Condition	Metal in g	Slag in g	Weight Loss in %	Adjusted Weight Loss in %
		SiO$_2$	CuO					
1	3500.0	20.0	95.0	Mould	3462.6	1015.6	40.5	18.9
2	3500.0	20.0	90.0	Crucible	3511.9	1246.8	35.3	13.7
3	3500.0	20.0	80.0	Crucible	3080.0	992.7	47.3	25.0
4	3500.0	20.0	65.0	Mould	2958.3	851.1	41.2	19.4
5	3500.0	10.0	92.3	Mould	3393.8	742.2	41.6	19.0
6	3500.0	10.0	96.3	Crucible	3401.4	714.0	43.0	20.4

Even the adjusted weight loss is considerably high, which could be due to losses of material as dust, volatilization of further components from the input materials or reduction of oxidic material in the charge not included in the corrected weight loss. Furthermore, slag samples taken during the holding time are not included in the mass balance and in the analysis of the trials. Two samples were taken per trial during the holding time with a mass of roughly 10 g per sample. This results in an error of the slag mass between 1.5% and 3.0%, the influence on the weight loss is considerably lower, compared to the total input mass between 6475 g and 7525 g. Besides the weight loss due to the slag samples, the higher share of the adjusted weight loss is due to dust consumed by the off-gas system, before the material could react with the molten metal or slag phase. Those losses are not quantifiable and cannot be avoided, because the furnace has to be used with off-gas suction at all times. Even in a technical-scale electric arc furnace, for a trial with 350 kg roasted black mass, it was not possible to obtain the whole flue dust, as part of the dust is always attached to off-gas pipes or gas cleaning equipment [29].

3.2.2. Slag Composition

The slag is the product of main interest in this investigation and was therefore investigated thoroughly. Table 4 lists the chemical composition of the bulk slag phase after the trials, samples taken during the holding time and minor elements analyzed in the bulk slag are only listed in the Supplementary Material. A "Spectro ARCOS" ICP-OES made by "SPECTRO Analytical Instruments GmbH, Kleve, Germany" was used to analyze the cobalt, nickel, lithium and copper content of slag samples. All ICP-OES measurements were carried out twice. The chemical composition of slag samples was analyzed using a wavelength dispersive X-ray fluorescence spectrometer (XRF) "Axios^{mAX}" made by "Malvern Panalytical B.V., Almelo, Netherlands". The samples were ground, sieved to a grain size below 63 µm and analyzed as fused-cast beads with the wide range oxide (WROXI) calibration from "Malvern Panalytical". The measured results were all in the calibrated composition area. All measurements were carried out twice. Carbon and sulfur analysis were carried out with an "ELTRA CS 2000" system made by "ELTRA GmbH, Haan, Germany" based on a combustion method. Carbon and sulfur measurements were carried out three times per sample.

Table 4. Chemical analysis of bulk slag generated in the trials.

Trial No. Method	Mass in g	Li	Cu	Co	Ni	C	SiO_2	Al_2O_3	Fe_2O_3	Mn_3O_4	BaO
			ICP-OES			Combustion		XRF			
1	1015.6	5.53	0.23	0.09	0.01	0.029	56.2	22.0	0.52	2.25	0.57
2	1246.8	5.18	1.4	1.44	0.05	0.035	53.4	20.2	2.25	2.23	0.50
3	992.7	5.84	0.39	0.41	0.05	0.120	49.7	27.1	0.67	1.69	0.69
4	851.1	6.24	0.40	0.25	0.03	0.285	46.5	30.2	0.22	0.84	0.76
5	742.2	6.77	0.35	0.07	0.01	0.238	43.1	32.2	0.31	2.12	0.87
6	714.0	7.40	0.10	0.06	0.01	0.184	41.8	33.4	0.14	0.70	0.83

The trial numbers are sorted, starting with the highest silicon content in the slag and decreasing in silicon content. Furthermore, the standard deviation of the sample set was determined and shown in Supplementary Table S3.

Obvious is the deviation in the copper- and cobalt content for the slag from trial number 2. The other samples show a significantly lower content of those metals. Probably, the CuO-addition was slightly too high in trial 2, even though the CuO-additions in other trials were higher, however, the effect of the CuO-addition can vary within trials, as described in the Materials and Methods. The copper, cobalt, nickel, iron, and manganese contents are not further discussed here, as they are discussed in detail later in this chapter. The lithium content in the slag is quite high with values over 5% and surpasses the lithium content of pure petalite, spodumene and in most cases also eucryptite. The lithium contents in selected minerals were already presented in Table 2. The low atomic mass of lithium makes it difficult to obtain minerals with a high content of lithium, as the lithium concentration is heavily diluted by other elements, which are heavier compared to lithium, like aluminum, oxygen and silicon, even in a stoichiometric mineral. The main component of the slag is in all cases SiO_2, followed by Al_2O_3. Except for trial 2, a decreasing SiO_2-content leads to an increased Al_2O_3-content in the slag. The aluminum-input comes from the black mass and is transferred into the slag. This can be considered to be a constant input, while the SiO_2-addition was varied and the amount of silicon transferred into the slag is also influenced by the reductive conditions during the trials. A higher amount of silicon being transferred into the slag will, therefore, dilute the constant aluminum mass in the slag and explains the correlation between the SiO_2- and Al_2O_3-content.

The lithium slagging was already simulated in detail and is evaluated based on experimental trials in Figure 9. Instead of presenting each trial individually, trials 1 and 2, 3 and 4, and 5 and 6 are combined. The quartz addition was the same for each pair of trials, even though the CuO-addition slightly varied for those trials. Therefore, the mean CuO-addition of two trials is listed in the figure and the CuO-addition per trial is presented in Table 3. Furthermore, the slagging of Mn, Fe, Co, Ni, and Cu are presented in the figure.

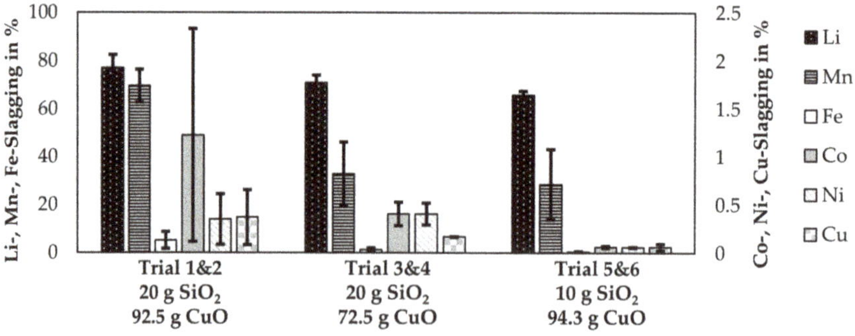

Figure 9. Slagging of metals observed in the laboratory-scale trials. Oxide addition in g per 100 g black mass.

The bars in the figure represent the mean value of two trials and the error bars represent the slagging in the individual trials. The SiO_2-content was the same for each pair of trials, whereas the mean value of CuO-addition is listed in Figure 9. The individual CuO-additions are listed in Table 3. High deviations are noticeable for the cobalt-, nickel-, and copper losses for trial 1&2 as already noticed in Table 4. The amount of cobalt, nickel and copper in the slag in relation to the input amount is 2.3%, 0.6%, and 0.7% for trial number 2. For the other trials, those values are below 0.5%, 0.5% and 0.2% for cobalt, nickel and copper respectively, while also values under 0.1% were possible. The amount of lithium transferred into the slag varied between 64.1% and 82.4% and is correlated with SiO_2-content of the slag as higher SiO_2-contents yielded a higher amount of lithium being transferred into the slag. The only other metal investigated, which was transferred into the slag with a considerable amount is manganese with 13–76%.

As the recovery of the cobalt, copper and nickel into the metal is the clear aim of pyrometallurgical processes, low contents of those metals in the slag are beneficial. For iron and manganese, the desired distribution depends on the further processing of the products. If iron and manganese are considered an impurity in the metal, high contents of those elements in the slag would be beneficial. If the slag is treated by hydrometallurgy, those elements could be also considered an impurity in the slag. Figure 10 shows, how selectively the valuable metals could be separated from iron and manganese by a smelting reduction process. For this evaluation, the contents of cobalt and nickel are related to the manganese- and iron content of the slag. The analysis of bulk slag samples presented in Table 4 are used in addition to slag samples taken during the holding time, which are only listed in the Supplementary Material.

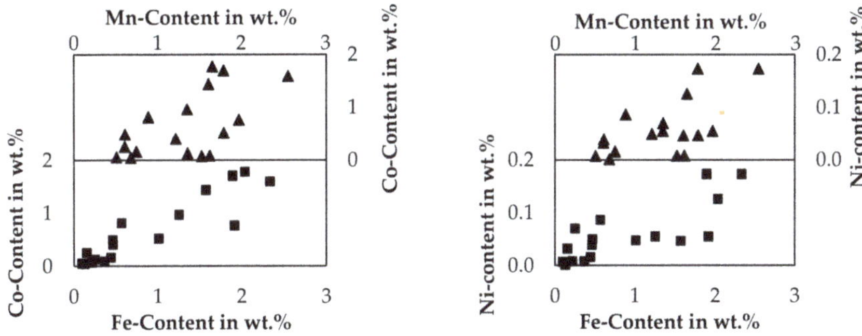

Figure 10. Relations of valuable metal content in slag samples compared to iron- and manganese content.

Due to high deviations, the limited dataset consisting only of 17 samples and the varied SiO_2-addition and CuO-addition, no interpolated graphs are presented here in the figures, linear equations and exponential equations are however listed in the Supplementary Material including the coefficient of determination, which is relatively low in all cases. Even if those simple models fail to describe the relation of valuable metals to iron or manganese, it is obvious, that the content of iron or manganese has to be considerably low to achieve high yields according to the general trend observable. The nickel content is the lowest of the metals investigated and is always under 0.2 wt.% and even significantly lower than 0.1 wt.% for samples with a low iron- and manganese content. The highest cobalt content observed is 1.8 wt.%, but with decreasing iron and manganese contents, the cobalt content decreases further and is even lower than 0.1 wt.% for samples with a low iron- and manganese content. Furthermore, also the iron- and manganese content are related. Already in this small investigated process area, a clear separation of the valuable metals nickel and cobalt from iron and manganese is difficult. If manganese and iron should not be reduced, considerably high losses of cobalt and nickel are expected. Instead, it seems plausible to reduce iron and manganese completely, to ensure high cobalt, copper and nickel yields and to generate a slag with fewer impurities, which have to be taken care of in a hydrometallurgical purification step.

3.2.3. Metal Composition

The metal samples from trial 3 and trial 4 were melted in a resistance heated furnace, slowly solidified, and separated by sawing to obtain homogenous samples and the weight of the individual cobalt- and copper phase. This was necessary, because the crucible diameter in the electric arc furnace and in the cast-iron-mould were too high, to allow a clear phase separation. This was mainly done for analytical reasons. A detailed description of the second melting operation is supplied in the Supplementary Material, together with macrographs and micrographs of the metal samples. A "Spectro ARCOS" ICP-OES made by "SPECTRO Analytical Instruments GmbH, Kleve, Germany" was used to analyze the metal samples. All ICP-OES measurements were carried out twice. Additional analytical results by arc spark optical emission spectrometry and XRF are also supplied in the Supplementary Material. After the second melting step, 62.9 wt.% of the metal can be described as a copper-rich phase and 37.1 wt.% a cobalt-rich metal phase. Figure 11 shows the chemical composition of both metal phases from the combined melting of the samples from trial number 3 and 4. As a comparison, the results from smelting the input mass of both trials according to Table 3 at 1600 °C is modeled with FactSage™ [30] and the chemical composition of both immiscible liquid phases is presented. A second model is derived by excluding the slag and gas phases from the previous model and cooling the liquid metal to 1060 °C, which is the temperature at which both metal immiscible metal phases are still liquid according to FactSage™ [30].

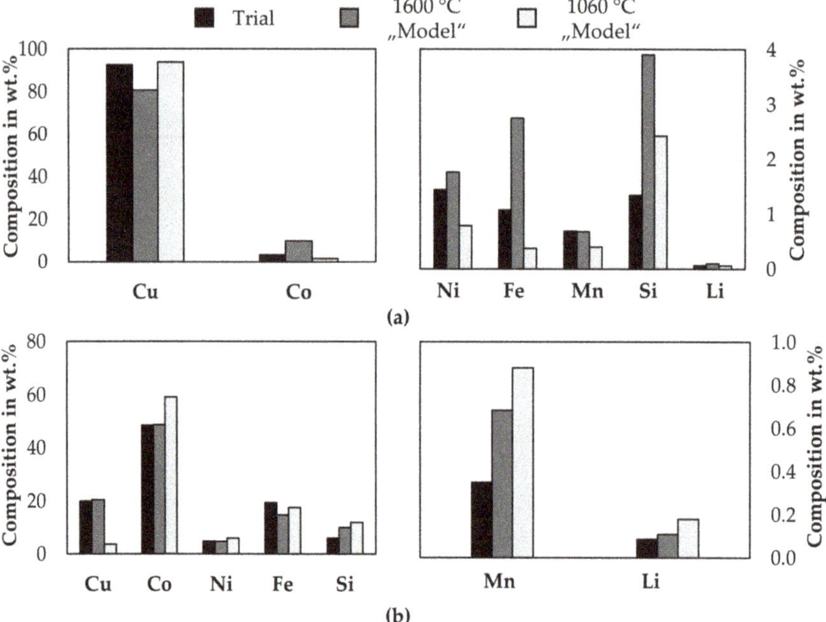

Figure 11. Comparison between analyzed metal composition and thermochemical model. (**a**): copper-rich phase; (**b**): cobalt-rich phase.

The results from the trials do not show a general alignment with the simulation, which makes a prediction solely based on a simulation difficult to evaluate the metal quality and underlines the necessity of experimental trials. Some elements differ significantly from both models, some elements show better alignment with the model at 1600 °C and for others, the model 1060 °C shows better results. In all cases, the silicon-, and lithium content is higher according to the model. The copper- and cobalt content in the copper phase is better described by the model at 1060 °C, whereas the manganese- and nickel content is better described by the model at 1600 °C, iron is significantly differing from both models. The copper-, cobalt- and nickel content in the cobalt-rich alloy is better described by the model at 1600 °C, whereas the iron content is better described with the model at 1060 °C. The measured manganese content is significantly lower compared to both models.

The separation step was mostly carried out for analytical reasons. It could also be used as a preconcentration step. In an industrial scale, metal and slag could be tapped separately and if the metal is solidified slowly enough in an iron mould, separate metal phases could be obtained. However, it is also feasible to treat water-granulated alloys containing cobalt, copper, nickel, iron and manganese [35,36]. Therefore, directly granulating the alloy obtained from a furnace could be an option as well. Based on the analysis presented in Figure 11 and the weighed amount of each metal phase. A theoretical composition of water granulated metal can be calculated. Table 5 shows the calculated composition of a single-phase metal based on the analyzed metal phases, based on the individual analysis of the samples, the calculated maximum and minimum content is listed in the table.

Table 5. Composition of quenched metal in wt.%.

Element	Cu	Co	Ni	Fe	Mn	Si	Li
wt.%	66.1–65.0	20.2–19.9	2.83–2.68	8.05–7.69	0.58–0.55	3.18–3.10	0.074–0.070

3.2.4. Distribution of Relevant Elements in the Process

Based on the results from the previous subchapters, a comparison of the distribution of elements to each phase with the FactSage™ [30] model at 1600 °C presented already in Figure 11 is shown in Figure 12. It was not possible to obtain the flue dust in the trials. Therefore, the results from the trial do not show the fraction going to the gas phase, theoretically, this should be the balance to 100% compared to the amount analyzed in the slag, copper, and cobalt phase.

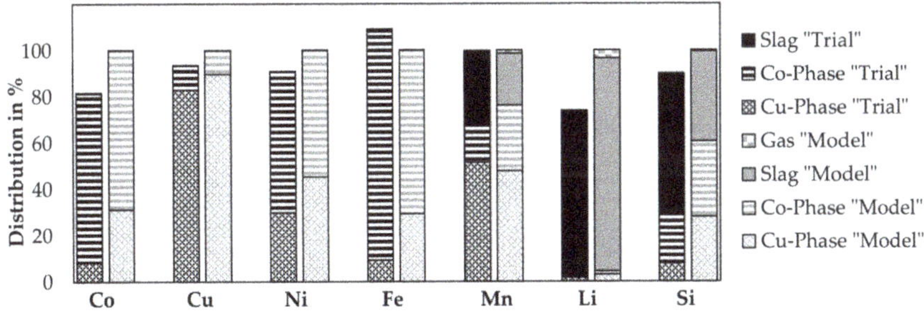

Figure 12. Distribution of elements in laboratory trials between, metal, slag and gas phase of Trial 3 and 4 (**left bars**) combined, compared with the thermochemical simulation (**right bars**).

As the simulated metal composition for both metal phases differs from experimental trials as shown by Figure 11 and was already discusses, it is not further discussed here, instead, the slag phase is considered in detail.

The results for cobalt, copper, nickel, and iron are similar, as losses to the slag phase are not observed from the experiments and neither predicted by the model. Further, 81.6% of cobalt could be identified in the products, as 73.0% is present in the cobalt-rich phase and 8.2% in the copper-phase, while the remaining amount of cobalt in the slag is considerably low with 0.4%. The model predicts that cobalt is only distributed between the metal phases and no losses occur to the gas- or slag phase.

82.8% of the copper is present in the copper phase and 10.5% in the cobalt phase, while the loss in the slag is 0.2%. In total, 93.5% of copper is in solid products. The model does not show any copper losses to the slag and 0.2% of copper losses into the gas phase. 91.1% of nickel is present in the solid products, whereas 29.9% of Nickel is in the copper phase, 60.8% is in the cobalt phase and 0.4% in the slag. The model shows no nickel losses. The iron analysis shows an unphysical result with 109.2% of iron in the solid products this could be due to inhomogeneous analyzed samples or inaccuracies with the input analyses. 9.3% of iron is present in the copper phase, 98.6% in the cobalt phase and 1.3% in the slag phase. According to the model, 0.04% is lost into the slag and 0.01% is lost into the gas phase. Manganese is the first element, with a significant distribution between all three molten phases, 99.8% of manganese is obtained in the solid samples, 51.5% in the copper phase, 15.5% in the cobalt phase and 32.8% in the slag. The model shows, that 28.3% of manganese is transferred to the slag and 1.32% to the gas phase. 73.7% of lithium is transferred to the solid products. Most of the lithium is transferred to the slag, 70.8% is enriched in that phase. 1.6% and 1.2% is present in the copper phase and cobalt phase, respectively. The model predicts 2.96% of the lithium in the copper phase, 1.52% in the cobalt phase and 92.01% in the slag phase, while the rest is lost to the gas. 89.9% of silicon could be identified in the products. 61.1% is in the slag phase, 7.8% in the copper phase and 21.0% in the cobalt phase. This differs significantly from the model, which predicts only 38.97% of the silicon in the slag phase, 27.98% in the copper phase, 32.46% in the cobalt phase and 0.58% in the gas. This proves the previous assumption, that the model predicts a higher degree of silicon reduction for DC electric arc furnace processes.

Except for iron and manganese, a considerable amount of the other elements is lost in the trials and cannot be identified in the solid products. Volatilization of cobalt, nickel and copper seems to be rather unlikely, to explain those losses. An explanation for this could be that, during charging of the material, fine material was directly taken by the off-gas suction of the furnace without reacting. Either by charging the material more carefully, which is not possible in the laboratory scale as material falls directly into the turbulent zone of the furnace or by recirculating flue dust, higher yields for valuable metals, and higher lithium slagging could be possible at a larger scale.

3.2.5. Qualitative X-ray Diffraction Phase Analysis of Slag

Qualitative X-ray diffraction (XRD) analysis was carried out and is presented in Figure 13. X-ray diffraction (XRD) of slag samples was carried out using a powder diffractometer "STADI P" made by "Stoe&Cie GmbH, Darmstadt, Germany", equipped with a copper anode (40 kV, 30 mA). A Germanium monochromator was applied to use the $K\alpha1$-radiation (wavelength: 1.540598 Å) for analysis. The scan range per sample was from 1.324° to 116.089° and the measuring time was two hours. "Match! 3.9.0.158" was used for evaluation of the pattern. Reference patterns were obtained from the "Crystallography Open Database". The version "COD-Inorg REV218120 2019.09.10" was used [37–42].

No background subtraction or smoothing of raw data was applied. For visual reasons, minor peaks of the identified minerals were indexed with vacant symbols, while the indices of the strongest peaks of each mineral are filled with color. The reference card information is supplied in the Supplementary Material. The trial numbers are the same used already in Table 3. Therefore, starting with the highest silicon content in the slag for trial 1 and decreasing in silicon content. Trial numbers 1, 4, and 5 are poured into a cast iron mould after the trial and trials 2, 3, and 6 are solidified in the crucible after the trial with a significantly lower cooling rate.

Even though the cooling condition was varied, no significant influence on the pattern is visible by comparing the trials 1 and 2, 3 and 4, and 5 and 6, which have a similar composition and variable cooling conditions.

Four minerals were identified in the samples, γ-lithium aluminate, lithium metasilicate and two lithium aluminum silicates. Furthermore, at least one mineral is present in the slags 4, 5 and 6, which could not be identified, since peaks at for example 21.4°, 31.2°, 31.7°, 45.3°, and 45.9° are not described by the previous minerals.

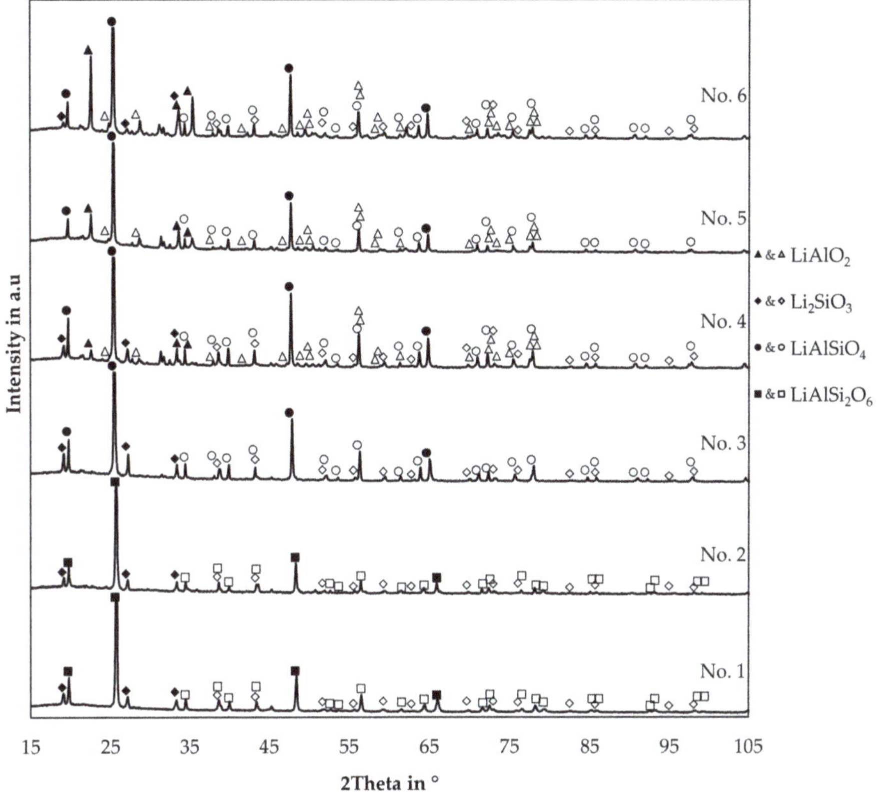

Figure 13. XRD-patterns of slag samples originated from the trials.

Lithium metasilicate is present in all samples, except for trial 5. Trial 6 contains even less silicon, but lithium metasilicate is present in that sample, even though the peak intensity is quite low compared to the slag 1, 2, 3, and 4.

γ-LiAlO$_2$ is present in sample 4, 5, and 6. A trend can be observed, as the peak intensity of γ-LiAlO$_2$ decreases with increasing content of silicon in the slag until it is not present in the slags 1, 2, and 3.

In all cases, the most intense peak belongs to the phases indexed as LiAlSiO$_4$ or LiAlSi$_2$O$_6$. Three possible candidates were examined, beta-spodumene with a tetragonal crystal system and a P4$_3$2$_1$2 space group with a simplified formula of LiAlSi$_2$O$_6$ [43]. Moreover, two different stoichiometries from the LiAlSiO$_4$-SiO$_2$ join, both with a hexagonal crystal system and a P6$_2$22 space group [44]. Distinguishing between the two hexagonal and the tetragonal mineral is difficult since all three minerals show the strongest peak slightly above 25°. In this case, the hexagonal minerals were selected as the matching phases, since the second strongest β-spodumene peak slightly above 22.5° is only observed in slag number 4, 5 and 6 and is already explained by the presence of γ-LiAlO$_2$. Also, several minor peaks better fit the hexagonal LiAlSiO$_4$-SiO$_2$ system as well.

The hexagonal LiAlSiO$_4$ mineral is called β-eucryptite and is a stuffed derivate of quartz and forms a solid-solution with SiO$_2$ [44]. For trials 3, 4, 5, and 6 a reference card with the chemical formula LiAlSiO$_4$ was used, and for trial numbers 1 and 2, a reference card with the chemical formula LiAlSi$_2$O$_6$ with the same space group was used. This was done because the lithium aluminum silicate peaks are slightly shifted to higher angles if the silicon content is increased in the slag samples.

The same observation was already made by Xu et al. and Nakagawa et al. with synthetic samples in the $LiAlSiO_4$-SiO_2 solid solution system [44,45]. To examine this further, Figure 14 shows the position of the four strongest lithium aluminum silicate peaks in the slag samples.

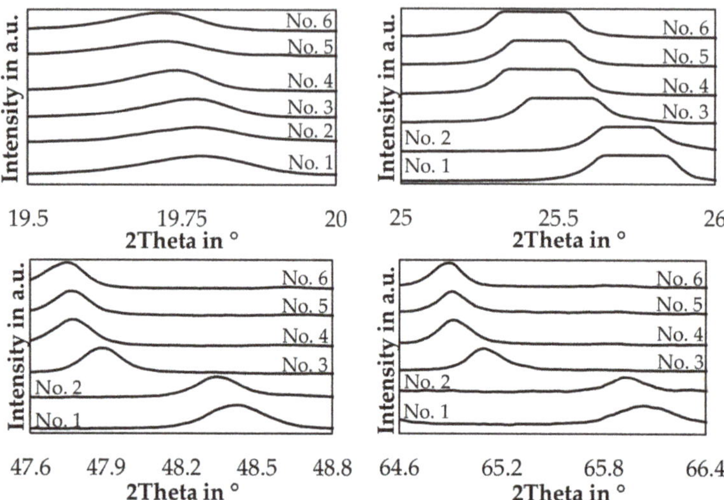

Figure 14. Detailed XRD-pattern of the four strongest β-eucryptite peaks.

The slags show a clear trend, that an increased SiO_2-content in the bulk slag, shifts the investigated peaks continuously to higher angles, especially for trial 1 and 2 a significant shift can be observed.

Based on the chemical composition of the slag samples shown in Table 4, a solidification simulation of the slag was carried out with FactSage^{TM} [30]. Figure 15 shows the amount of lithium containing minerals present in the slag after solidification. Minerals not containing lithium are expressed as "others", as those are only a minor portion of the slag.

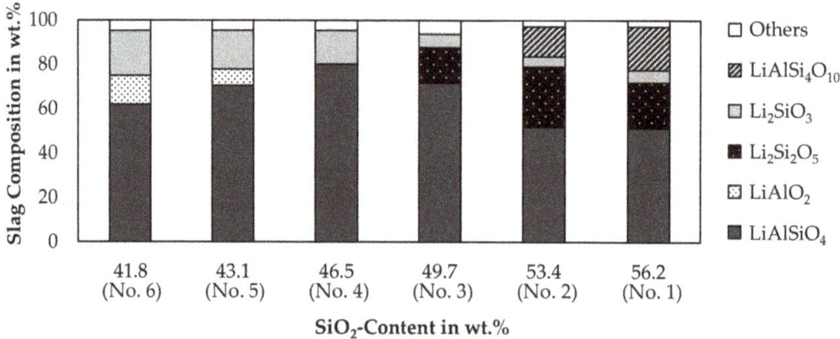

Figure 15. Simulated mineralogical composition of the slag in wt.%.

According to the simulation, the major mineral in the slag is $LiAlSiO_4$, which is also observed in the XRD-pattern. Li_2SiO_3 is predicted for every sample and in addition, $Li_2Si_2O_5$ is present for slag 1, 2 and 3. The presence of $Li_2Si_2O_5$ could not be confirmed by the XRD-results, whereas Li_2SiO_3 is present in all analyzed slag samples, except for slag number 5. The simulation proposes, that $LiAlO_2$ is formed in slag 5 and 6, which have the highest aluminum content. The presence of $LiAlO_2$ is confirmed by XRD for those two samples, but it is also confirmed by XRD for slag 4. The XRD-results do not show $LiAlSi_4O_{10}$ in any sample, even though the simulation predicts the presence of $LiAlSi_4O_{10}$ for slag number 1 and 2. As already predicted by the simulation of the smelting process shown in Figure 7, no $LiAlSi_2O_6$ is formed and instead, an equilibrium between $LiAlSiO_4$ and $LiAlSi_4O_{10}$ is predicted.

The disadvantage of the simulation is, that lithium minerals are all assumed as stoichiometric phases and no solid solutions are available in the databases, even though the $LiAlSiO_4$-SiO_2 solid solution [44] is relevant for this simulation. The absence of this solid solution could be an explanation for the predicted occurrence of $Li_2Si_2O_5$ and $LiAlSi_4O_{10}$. Instead of those minerals with a higher silicon content compared to $LiAlSiO_4$, a β-eucryptite phase with a higher silicon content than the stoichiometric $LiAlSiO_4$ included in the model probably has formed as indicated by Figure 14. Since the model has to consider the leftover silicon somehow, it predicts the formation of the non-observable $Li_2Si_2O_5$ and $LiAlSi_4O_{10}$.

3.2.6. Raman-Analysis of Slag

To verify the presence of β-eucryptite, Raman analysis was carried out using a "MA-RBE-V02" Raman microscope with a magnification of 50 made by "Stonemaster UG, Linkenheim-Hochstetten, Germany" equipped with an Nd-YAG laser. The used wavelength was 532 nm. The numerical aperture was 0.55. The accuracy of the spectral data is ±2 cm^{-1}.

The lithium aluminosilicate system is already well studied by Raman spectroscopy due to the importance in glass-ceramics. Raman spectroscopy can be used to distinguish the minerals in the Li_2O-Al_2O_3-SiO_2 ternary system and to measure indirectly the SiO_2-content in β-eucryptite [46–51]. A discussion about which rotational or vibrational state is responsible for a frequency band is omitted in this research, as it is already discussed in the literature [46–51].

To distinguish the minerals β-eucryptite, β-spodumene and γ-spodumene, literature data for peak positions and spectral characteristics are compiled in Table 6 [49,51]. Peaks below a Raman shift of 160 cm^{-1} are neglected in the table.

Table 6. Reported Raman shifts in cm^{-1} with an accuracy of 2 cm^{-1} and spectral characteristics of lithium aluminosilicates according to literature.

β-Eucryptite [51]		β-Spodumene [49]		γ-Spodumene [49]	
Raman Shift	Characteristic	Raman Shift	Characteristic	Raman Shift	Characteristic
187	m [1,2]	184	m [1]		
233	vw				
282	w, bd	288	w		
352	m				
		412	w		
466	(sh)			440	(sh) [1]
483	s	492	s	480	s
636	vw				
711	w	~720	vw, bd		
762	w	770	w, bd	742	vw, bd
		864	vw, bd		
987	w	990	(sh)		
1032	s				
1049	(sh)			1044	(sh)
1067	vw				
1086	m			1088	w, bd
1099	w	1094	w, bd		

[1] Abbreviations: v, very; w, weak; m, medium; s, strong; bd, broad; sh, shoulder; [2] Zhang et al. [51] did not list the characteristics, therefore they are derived from the published figure.

The strongest peaks were reported for Raman shifts between 480 cm^{-1} and 492 cm^{-1} for all three minerals presented in Table 6. Deviations useful to distinguish those minerals can be the peak at 187 cm^{-1} and 184 cm^{-1} reported for β-eucryptite and β-spodumene, respectively, and the peak at 352 cm^{-1} reported for β-eucryptite.

Figure 16 shows Raman spectra of slag samples originating from the bulk slag phase after a trial. The trial numbers used for labeling are the same used in Table 3 and Figure 13. Thereby, starting with the highest silicon content in the slag for trial 1 and decreasing in silicon content.

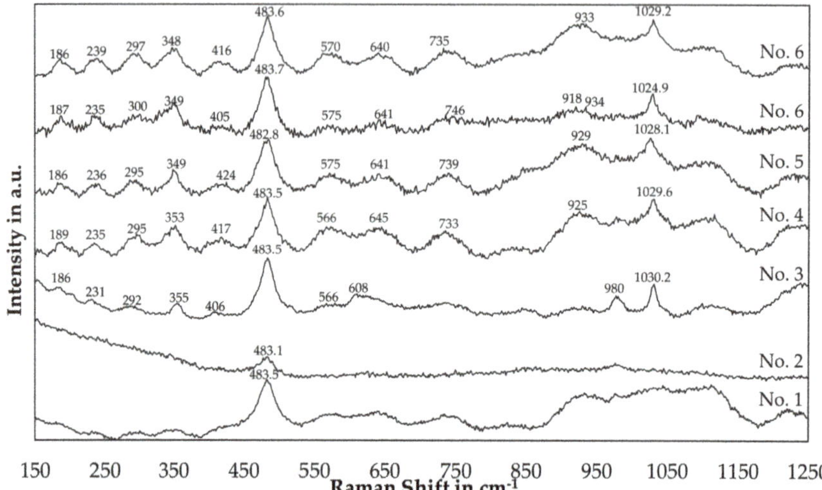

Figure 16. Raman Analysis of slag samples originated from the trials.

By comparison of Figure 16 with Table 6, a strong similarity of peak positions for sample 3 to 6 with the referenced β-eucryptite can be found. Especially the strong peaks at 482–483 cm^{-1} and 1024–1030 cm^{-1} are reported in the reference as well, however the reference peak at 1032 cm^{-1} deviated a little bit from the measured results. Furthermore, the medium-strong peaks at 187 cm^{-1} and 352 cm^{-1} can be found with small deviations in the patterns of those trials. More difficult is the evaluation of the patterns of trial 1 and 2. The peak at 483 cm^{-1} is also observed, however the other peaks are either not detectable or only weak. Furthermore, background noises below 400 cm^{-1} for slag number 2 and between 925 cm^{-1} and 1075 cm^{-1} for slag number 1 are present in the samples.

One explanation for the disappearance of peaks could be, that with an increasing silicon-content in the slag, β-eucryptite (LiAlSiO$_4$) is either replaced or partially replaced by β-spodumene (LiAlSi$_2$O$_6$) or γ-spodumene (LiAlSi$_2$O$_6$), where less peaks were observed in the references. As the main peak for β-spodumene is observed at 492 cm^{-1} and the XRD indicates, that β-spodumene is not present in sample 1 and 2, the presence of γ-spodumene or β-eucryptite seems more likely. However, a definite assignment of the spectra of samples 1 and 2 is not possible, whereas the assumption of the presence of β-eucryptite is confirmed by XRD and Raman for sample 3, 4, 5 and 6.

Similar to the peak displacement due to variations in the silicon-content already observed for lithium aluminosilicates in the XRD-analysis, Alekseeva et al. [46] proposed a linear relationship for the position of the Raman bands as a function of the silicon content. Those bands are the peaks observed at roughly 483 cm^{-1} and between 1025 cm^{-1} and 1030 cm^{-1} in our study. In our case, the peak at 483 cm^{-1} is observed at the same position for every sample besides a small deviation, which is smaller than the accuracy of the measurement device. A SiO$_2$-content of 59 (±3–4) mol.% would results in a Raman band at this position according to the linear approximation by Alekseeva et al. [46]. Since both bands have to change simultaneously and only the second band at higher Raman shifts deviates, no definite conclusion about the silicon-content in the minerals investigated can be presented in this paper.

4. Discussion

In the discussion, the results are compared with similar published investigations.

One chapter is dedicated to limitations of the research carried out in this paper and future research directions, which could be investigated even further.

4.1. Discussion of the Obtained Results with Previous Work

The comparison of previous work with the research presented in this paper is carried with a focus on the distribution of valuable metals in the process and the mineralogical investigation of the slag.

4.1.1. Comparison of Valuable Metal Distribution during Smelting

Georgi-Maschler et al. [29] carried out smelting trials in an electric arc furnace with pyrolyzed black mass as well, even though there are a few deviations. They used a considerably higher melting temperature between 1700 °C and 1750 °C and a CaO-SiO$_2$ slag as a flux. Also, the graphite content of the black mass was reduced by prior thermal treatment and not utilized as a reducing agent for another resource, as it has been done in this investigation. A cobalt yield between 60% and 100% was found in laboratory trials and a cobalt yield of 88% in a technical-scale electric arc furnace is reported. The cobalt yield was similar compared to the yield of 80% presented in this paper, even though Georgi-Maschler et al. [29] reported cobalt losses of 3.1% into the slag, whereas in our findings cobalt losses into the slag were below 0.5% except for one trial. Besides higher melting temperatures and a different slag system, the higher amount of fluxes used by Georgi-Maschler et al. [29] could be an explanation for the higher amount of cobalt being lost into the slag phase. The higher amount of fluxes also resulted in a relatively low lithium content of 1.4 wt.%. in the slag phase, which is equal to 31% of the total input lithium mass. A higher distribution of lithium into the flue dust was determined, which could be due to the fluxing by CaO or the higher melting temperature. Both parameters decrease the amount of lithium transferred into the slag, according to the simulation presented in chapter 3.1. Compared to the work from Georgi-Maschler et al. [29], a higher amount of lithium transferred into the slag and a higher lithium content in the slag could be achieved in this paper by adding quartz as the only flux. However, in our work, it was aimed to transfer lithium into the slag, whereas Georgi-Maschler et al. [29] aimed to enrich lithium in the flue dust. In both cases, it was not possible to only enrich lithium in either the slag or dust and losses occurred. Therefore, either treating the dust and slag to recover lithium is necessary or one of both by-products needs to be recirculated into the process to minimize the losses.

A more recent study by Ruismäki et al. [52] investigated an approach similar to ours to use graphite from spent batteries as a reducing agent. They smelted a concentrate generated by froth flotation of industrially pre-processed lithium-ion battery waste with nickel slag to reduce oxides in the nickel slag. A cobalt yield about 93% was reported, based on the analyzed slag. This surpasses the yield presented in this paper but can be explained by the less turbulent conditions in the laboratory tube furnace used by Ruismäki et al. [52] compared to the electric arc furnace used in this study. Furthermore, two different methods to calculate yields were used, which leads to different results. The calculated yield in this paper based on the metal output can be seen as a pessimistic baseline scenario. Yields calculated based on the losses in the slag would have been higher in this study as well than the reported values as well. A comparison of the lithium content in the slag with the paper of Ruismäki et al. [52] cannot be carried out, as only the lithium content in the starting mixture with 0.88 wt.% is reported and after reducing valuable metals, higher lithium contents in their slag seem probable.

4.1.2. Comparison of Lithium Minerals Present in Slags

Elwert et al. [53] investigated three lithium slags that originated from Umicore facilities with a high lithium content by XRD and electron probe microanalysis (EPMA). The main components of the slag were Al_2O_3, CaO, Li_2O and SiO_2 in variable amounts. Table 7 shows the chemical composition of the slags investigated by Elwert et al. [53].

Table 7. Composition of lithium-containing slags from Umicore in wt.% [53].

Slag System	Low Aluminium Content	High Manganese Content	High Aluminium Content
Al_2O_3	33.57	44.52	47.37
CaO	23.46	16.08	23.42
Li_2O	11.04	8.29	8.96
MnO_2	0.31	9.52	0.36
MgO	5.11	1.44	2.65
SiO_2	21.25	17.52	12.81

The major difference is, that the slags in our investigation contain higher amounts of SiO_2 compared to Al_2O_3 and the slags from the reference have higher Al_2O_3-contents compared to the SiO_2-content. Slags from the reference also contain calcium, which is only a minor element in our study.

All three slag systems have in common, that $LiAlO_2$ is present and that lithium aluminosilicates could not be observed [53]. In our case, lithium aluminosilicates were present in all slags, whereas lithium aluminates were only present in three slags with an aluminum content above 30 wt.%. Lithium silicate with a general formula of Li_2MeSiO_4 was found in the low aluminum slag [53]. In our case, Li_2SiO_3 was identified with XRD for five slag samples. Further components observed by the reference but not identified with XRD in our investigation were:

- Gehlenite ($Ca_2Al(AlSi)O_7$ which was identified in all slags
- Merwinite ($Ca_3Mg(SiO_4)_2$), which was present in the low aluminum slag and high aluminum slag
- Cr-Spinel, which was present in the low aluminum and high aluminum slag
- Spinels, which were present in the high manganese slag
- Silico-phosphates with a high REE content, which was present in the high manganese slag

Li et al. investigated a synthetic slag with a composition of 50 wt.% SiO_2, 35 wt.% CaO, 12 wt.% Al_2O_3 and 3 wt.% Li_2O with XRD. They identified only three phases, β-spodumene, $CaSiO_3$ and CaO [54]. Other phases were not identified, even though a considerable amount of peaks were not indexed. Even though the presence of β-spodumene for slags from this investigation seems rather unlikely, the lower amount of lithium in the slag produced by Li et al. [54] could be an explanation for the occurrence of β-spodumene in their synthetic slag.

The deviations in determined slag phases in the literature and even in this study show, how the mineralogy of the slag can be easily changed by different chemical compositions. This can have a major influence on the leaching process, as not all lithium minerals are easily leachable. α-spodumene for example is difficult to leach and is converted into β-spodumene prior to leaching [55–59].

4.2. Limitations of This Investigation and Future Research Directions

A limitation of the current work is the use of pure copper(II) oxide as a synthetic raw material. This is not feasible for an industrial process and should be replaced by an oxidic raw material like ore or oxidic industrial residues. Preferably, such a raw material contains cobalt, nickel or copper, as those elements have to be recovered from the metal alloy anyway. Further restraints are the accompanying elements of possible raw materials. Ideally, the raw material contains SiO_2, as the positive effect on the lithium slagging was proven in this work or Al_2O_3 since the simulation indicates a positive effect on the lithium slagging as well. Problems could arise if lime is included in the raw material, as the simulation shows increased lithium losses into the gas phase for lime addition.

As more than one lithium-containing mineral is present in the slag according to the XRD-analysis and the thermochemical simulation, the leachability of the slag has to be carefully investigated. If one of those minerals is not leachable, future slag design can not only focus on lithium slagging and the lithium content of the slag, as has been done in this study, it also has to focus on the formation of leachable lithium minerals. Also, the leaching behavior of impurities needs further investigation.

Since no detailed focus was put on the metal phase in this work, future work has to consider the recovery of metals from the produced metal alloy, either in the form of refined metal or pure chemicals.

It is expected that the slag and the metal are both treated by hydrometallurgical methods and the pyrometallurgical operation is used as a pre-concentration unit. Since manganese and iron could be considered an impurity in the hydrometallurgical treatment of the slag and the metal, an evaluation should determine, if those elements are easier to separate in the alloy processing or in the slag processing. Ideally, manganese and iron should be recovered as well from the intermediate products. Based on the preferred distribution of those elements for downstream processing, improvements in the pyrometallurgical process can be investigated to enrich those elements either in the slag or in the alloy. Options could be the adjustment of fluxes or the oxygen potential. Even though Figure 10 suggests, that a complete recovery of cobalt and nickel, while maintaining iron and manganese in the slag is not possible, at least for the investigated slag system. A more detailed investigation of the behavior of manganese in the process will be especially more important for newer battery generations. The manganese and nickel content in the black mass is low compared to cobalt according to Table 1 and as nickel-cobalt-manganese oxide (NCM) cathodes take a dominant role in the battery industry nowadays [60], an increased nickel and manganese content in end of life black mass can be expected in the near future.

In this project, it was only possible to analyze the metal and slag, while the flue dust could not be collected. As considerable weight losses were observed in the process and a considerable amount of lithium could not be identified in the obtained products, a flue dust analysis would enhance the accuracy of the mass balance of the process. Furthermore, an analysis of the flue dust would be necessary to evaluate whether the flue dust can be recirculated back to the electric arc furnace, or if recirculating would lead to an enrichment of volatile elements in the process. To avoid the enrichment and circulation of volatile elements, an additional treatment process of flue dust could be necessary.

Also not investigated was the influence and distribution of minor elements like phosphorous, sulfur and halogens. Halogens could be of special interest as they are commonly enriched in the flue dust in smelting processes [61,62] and halogens should not be circulated back to the smelter [63,64].

5. Conclusions

A pyrometallurgical approach was investigated to separate critical elements from pyrolyzed lithium-ion battery black mass into intermediate products by smelting in an electric arc furnace.

A thermochemical simulation was carried out to determine a fluxing strategy. Quartz was chosen as a flux and two different quartz additions were tested in six trials. To utilize excess graphite in the feed material, copper(II) oxide was fed into the furnace. The graphite was therefore used as a reducing agent in the process. Due to the experimentally proven reduction of added copper(II) oxide, carbon from black mass was utilized as a reducing agent and could therefore be included in a recycling efficiency calculation.

Cobalt, nickel, and copper were enriched in a mixed alloy, while lithium was concentrated in the slag. The yield of cobalt, nickel and copper was 81.6%, 93.3%, and 90.7% respectively for the thoroughly investigated trial with a quartz-addition of 20 g per 100 g black mass at 1600 °C based on the metal output. The reported losses for those metals into the slag were small with 0.4%, 0.2% and 0.4% respectively. Similar findings were reported by other researchers [29,52]. Besides one trial, the losses of those valuable metals in the slag were below 1% for every trial.

An enrichment of lithium into the slag was achieved in all trials with a yield between 64.1% and 82.4%. Lithium contents between 5.18% and 7.40% in the slag were achieved. Higher quartz additions increased the lithium yield, but lead to a decreased lithium content in the slag. The amount of lithium transferred into the metal alloy was below 3% compared to the lithium input.

A considerable amount of lithium, cobalt, nickel and copper from the input feed were not found in the slag or metal after the trials. Therefore, the assumption is made that they were lost as flue dust. A recirculation of flue dust into the furnace could, therefore, increase the yields significantly, as the reported losses into the slag or metal phase are considerably low. Furthermore, as the material could only be charged in the turbulent zone of the laboratory electric arc furnace, losses due to dusting of the input material could be mitigated at an industrial scale and increase the overall yield.

The slag was characterized by Raman and X-ray diffraction. Every investigated slag contains lithium aluminosilicates. Lithium aluminate and lithium metasilicate are present in three respectively five slags out of six slags in total depending on the chemical composition of the slag.

Supplementary Materials: The following are available online at http://www.mdpi.com/2075-4701/10/8/1069/s1, Figure S1: Macrographs of slag: (a) generated in trial number 5, (b) generated in trial number 6, Table S1: Chemical Formula, mineral name and information card number, Table S2: Chemical composition of slag samples taken during the holding time and after solidification, Table S3: Standard deviation of the chemical analyses presented in Table 4 in the main paper in wt.%. Table S4: Linear Equations of Metal Relations in Slag Samples and Coefficient of Determination, Table S5: Exponential Equations of Metal Relations in Slag Samples and Coefficient of Determination, Figure S2: Macrograph of metal obtained from trial number 3, Figure S3: Macrograph of metal obtained from trial number 4, Figure S4: Micrograph of the interface between the cobalt and copper phase of trial number 3, Figure S5: Micrograph of the cobalt matrix including copper inclusions of trial number 4, Figure S6: Micrograph of the copper matrix including cobalt inclusion of trial number 4, Figure S7: Micrograph of the bottom of the solidified ingot form trial number 3, Figure S8: Macrograph of slowly solidified metal from trial 3 and 4, Table S6: Comparison of selected elements in metal samples analyzed by different methods.

Author Contributions: Conceptualization, M.S., C.V., and. B.F.; methodology, M.S. and C.V.; software, M.S.; validation, M.S., C.V., C.D., J.K., D.O., B.F., T.H., and A.M.; formal analysis, M.S.; investigation, M.S. and C.V.; resources M.S. and C.V.; data curation, M.S. and C.V.; writing—original draft preparation, M.S. and C.V.; writing—review and editing M.S., C.V., C.D., J.K., D.O., B.F., T.H., and A.M.; visualization, M.S.; supervision, B.F.; project administration, C.V. and J.K.; funding acquisition, B.F., A.M., T.H., J.K., and C.V. All authors have read and agreed to the published version of the manuscript.

Funding: This research was funded by Deutscher Akademischer Austauschdienst (DAAD), grant number 57453240.

Acknowledgments: The authors are grateful to Accurec Recycling GmbH (Germany) for providing pelletized black mass. The authors would also like to express their gratitude to the DAAD for enabling the joint research in battery recycling.

Conflicts of Interest: The authors declare no conflict of interest. The funders had no role in the design of the study; in the collection, analyses, or interpretation of data; in the writing of the manuscript, or in the decision to publish the results.

References

1. Dańczak, A.; Klemettinen, L.; Kurhila, M.; Taskinen, P.; Lindberg, D.; Jokilaakso, A. Behavior of Battery Metals Lithium, Cobalt, Manganese and Lanthanum in Black Copper Smelting. *Batteries* **2020**, *6*, 16. [CrossRef]
2. Helbig, C.; Bradshaw, A.M.; Wietschel, L.; Thorenz, A.; Tuma, A. Supply Risks Associated with Lithium-Ion Battery Materials. *J. Clean. Prod.* **2018**, *172*, 274–286. [CrossRef]
3. Nitta, N.; Wu, F.; Lee, J.T.; Yushin, G. Li-Ion Battery Materials: Present and Future. *Mater. Today* **2015**, *18*, 252–264. [CrossRef]
4. Gu, F.; Guo, J.; Yao, X.; Summers, P.A.; Widijatmoko, S.D.; Hall, P. An Investigation of the Current Status of Recycling Spent Lithium-Ion Batteries from Consumer Electronics in China. *J. Clean. Prod.* **2017**, *161*, 765–780. [CrossRef]
5. Contestabile, M.; Panero, S.; Scrosati, B. A Laboratory-Scale Lithium-Ion Battery Recycling Process. *J. Power Sources* **2001**, *92*, 65–69. [CrossRef]
6. Pinegar, H.; Smith, Y.R. Recycling of End-of-Life Lithium-Ion Batteries, Part II: Laboratory-Scale Research Developments in Mechanical, Thermal, and Leaching Treatments. *J. Sustain. Metall.* **2020**, *6*, 142–160. [CrossRef]
7. Porvali, A.; Aaltonen, M.; Ojanen, S.; Velazquez-Martinez, O.; Eronen, E.; Liu, F.; Wilson, B.P.; Serna-Guerrero, R.; Lundström, M. Mechanical and Hydrometallurgical Processes in HCl Media for the Recycling of Valuable Metals from Li-Ion Battery Waste. *Resour. Conserv. Recycl.* **2019**, *142*, 257–266. [CrossRef]

8. European Commission. Report from the Commission to the European Parliament, the Council, the European Economic and Social Committee, the Committee of the Regions and the European Investment Bank: On the Implementation of the Strategic Action Plan on Batteries: Building a Strategic Battery Value Chain in Europe. Available online: https://eur-lex.europa.eu/legal-content/EN/TXT/HTML/?uri=CELEX:52019DC0176&from=EN (accessed on 16 June 2020).

9. Pinegar, H.; Smith, Y.R. Recycling of End-of-Life Lithium Ion Batteries, Part I: Commercial Processes. *J. Sustain. Metall.* **2019**, *5*, 402–416. [CrossRef]

10. Tarascon, J.; Armand, M. Issues and Challenges Facing Rechargeable Lithium Batteries. *Nature* **2001**, *414*, 359–367. [CrossRef]

11. Huang, B.; Pan, Z.; Su, X.; An, L. Recycling of Lithium-Ion Batteries: Recent Advances and Perspectives. *J. Power Sources* **2018**, *399*, 274–286. [CrossRef]

12. The European Parliament and the Council of the European Union. Directive 2006/66/EC of the European Parliament and of the Council on Batteries and Accumulators and Waste Batteries and Accumulators and Repealing Directive 91/157/EEC. Available online: https://eur-lex.europa.eu/legal-content/EN/TXT/PDF/?uri=CELEX:02006L0066-20131230&rid=1 (accessed on 6 May 2020).

13. Werner, D.; Peuker, U.A.; Mütze, T. Recycling Chain for Spent Lithium-Ion Batteries. *Metals* **2020**, *10*, 316. [CrossRef]

14. Zhong, X.; Liu, W.; Han, J.; Jiao, F.; Qin, W.; Liu, T. Pretreatment for the Recovery of Spent Lithium Ion Batteries: Theoretical and Practical Aspects. *J. Clean. Prod.* **2020**, *263*, 121439. [CrossRef]

15. Liu, J.; Wang, H.; Hu, T.; Bai, X.; Wang, S.; Xie, W.; Hao, J.; He, Y. Recovery of $LiCoO_2$ and Graphite from Spent Lithium-Ion Batteries by Cryogenic Grinding and Froth Flotation. *Miner. Eng.* **2020**, *148*, 106223. [CrossRef]

16. Zhang, G.; Du, Z.; He, Y.; Wang, H.; Xie, W.; Zhang, T. A Sustainable Process for the Recovery of Anode and Cathode Materials Derived from Spent Lithium-Ion Batteries. *Sustainability* **2019**, *11*, 2363. [CrossRef]

17. Zhang, G.; He, Y.; Feng, Y.; Wang, H.; Zhang, T.; Xie, W.; Zhu, X. Enhancement in Liberation of Electrode Materials Derived from Spent Lithium-Ion Battery by Pyrolysis. *J. Clean. Prod.* **2018**, *199*, 62–68. [CrossRef]

18. Li, J.; Lai, Y.; Zhu, X.; Liao, Q.; Xia, A.; Huang, Y.; Zhu, X. Pyrolysis Kinetics and Reaction Mechanism of the Electrode Materials During the Spent $LiCoO_2$ Batteries Recovery Process. *J. Hazard. Mater.* **2020**, *398*, 122955. [CrossRef]

19. Zhong, X.; Liu, W.; Han, J.; Jiao, F.; Qin, W.; Liu, T.; Zhao, C. Pyrolysis and Physical Separation for the Recovery of Spent $LiFePO_4$ Batteries. *Waste Manag.* **2019**, *89*, 83–93. [CrossRef]

20. Ruismäki, R.; Dańczak, A.; Klemettinen, L.; Taskinen, P.; Lindberg, D.; Jokilaakso, A. Integrated Battery Scrap Recycling and Nickel Slag Cleaning with Methane Reduction. *Minerals* **2020**, *10*, 435. [CrossRef]

21. Shi, J.; Peng, C.; Chen, M.; Li, Y.; Eric, H.; Klemettinen, L.; Lundström, M.; Taskinen, P.; Jokilaakso, A. Sulfation Roasting Mechanism for Spent Lithium-Ion Battery Metal Oxides Under SO_2-O_2-Ar Atmosphere. *JOM* **2019**, *71*, 4473–4482. [CrossRef]

22. Peng, C.; Hamuyuni, J.; Wilson, B.P.; Lundström, M. Selective Reductive Leaching of Cobalt and Lithium from Industrially Crushed Waste Li-Ion Batteries in Sulfuric Acid System. *Waste Manag.* **2018**, *76*, 582–590. [CrossRef]

23. Porvali, A.; Chernyaev, A.; Shukla, S.; Lundström, M. Lithium Ion Battery Active Material Dissolution Kinetics in Fe (II)/Fe (III) Catalyzed Cu-H_2SO_4 Leaching System. *Sep. Purif. Technol.* **2020**, *236*, 116305. [CrossRef]

24. Dewulf, J.; van der Vorst, G.; Denturck, K.; van Langenhove, H.; Ghyoot, W.; Tytgat, J.; Vandeputte, K. Recycling Rechargeable Lithium Ion Batteries: Critical Analysis of Natural Resource Savings. *Resour. Conserv. Recycl.* **2010**, *54*, 229–234. [CrossRef]

25. Reck, B.K.; Graedel, T.E. Challenges in Metal Recycling. *Science* **2012**, *337*, 690–695. [CrossRef] [PubMed]

26. Graedel, T.E.; Allwood, J.; Birat, J.-P.; Buchert, M.; Hagelüken, C.; Reck, B.K.; Sibley, S.F.; Sonnemann, G. What Do We Know About Metal Recycling Rates? *J. Ind. Ecol.* **2011**, *15*, 355–366. [CrossRef]

27. Harper, G.; Sommerville, R.; Kendrick, E.; Driscoll, L.; Slater, P.; Stolkin, R.; Walton, A.; Christensen, P.; Heidrich, O.; Lambert, S.; et al. Recycling Lithium-Ion Batteries from Electric Vehicles. *Nature* **2019**, *575*, 75–86. [CrossRef] [PubMed]

28. Wang, H.; Friedrich, B. Development of a Highly Efficient Hydrometallurgical Recycling Process for Automotive Li–Ion Batteries. *J. Sustain. Metall.* **2015**, *1*, 168–178. [CrossRef]

29. Georgi-Maschler, T.; Friedrich, B.; Weyhe, R.; Heegn, H.; Rutz, M. Development of a Recycling Process for Li-Ion Batteries. *J. Power Sources* **2012**, *207*, 173–182. [CrossRef]

30. Bale, C.W.; Bélisle, E.; Chartrand, P.; Decterov, S.A.; Eriksson, G.; Gheribi, A.E.; Hack, K.; Jung, I.-H.; Kang, Y.-B.; Melançon, J.; et al. FactSage Thermochemical Software and Databases, 2010–2016. *Calphad* **2016**, *54*, 35–53. [CrossRef]

31. Plewa, J.; Skrzypek, J. Kinetics of the Reduction of Copper Oxide with Carbon Monoxide. *Chem. Eng. Sci.* **1989**, *44*, 2817–2824. [CrossRef]

32. Vest, M.; Georgi-Maschler, T.; Friedrich, B.; Weyhe, R. Rückgewinnung von Wertmetallen aus Batterieschrott. *Chem. Ing. Tech.* **2010**, *82*, 1985–1990. [CrossRef]

33. Jones, R.T.; Erwee, M.W. Simulation of Ferro-Alloy Smelting in DC Arc Furnaces Using Pyrosim and FactSage. *Calphad* **2016**, *55*, 20–25. [CrossRef]

34. Sommerfeld, M.; Friedmann, D.; Kuhn, T.; Friedrich, B. "Zero-Waste": A Sustainable Approach on Pyrometallurgical Processing of Manganese Nodule Slags. *Minerals* **2018**, *8*, 544. [CrossRef]

35. Keber, S.; Brückner, L.; Elwert, T.; Kuhn, T. Concept for a Hydrometallurgical Processing of a Copper-Cobalt-Nickel Alloy Made from Manganese Nodules. *Chem. Ing. Tech.* **2020**, *92*, 379–386. [CrossRef]

36. Xiao, S.; Ren, G.; Xie, M.; Pan, B.; Fan, Y.; Wang, F.; Xia, X. Recovery of Valuable Metals from Spent Lithium-Ion Batteries by Smelting Reduction Process Based on MnO-SiO_2-Al_2O_3 Slag System. *J. Sustain. Metall.* **2017**, *3*, 703–710. [CrossRef]

37. Gražulis, S.; Chateigner, D.; Downs, R.T.; Yokochi, A.F.T.; Quirós, M.; Lutterotti, L.; Manakova, E.; Butkus, J.; Moeck, P.; Le Bail, A. Crystallography Open Database—An Open-Access Collection of Crystal Structures. *J. Appl. Crystallogr.* **2009**, *42*, 726–729. [CrossRef] [PubMed]

38. Downs, R.T.; Hall-Walace, M. The American Mineralogist Crystal Structure Database. *Am. Mineral.* **2003**, *88*, 247–250.

39. Gražulis, S.; Daškevič, A.; Merkys, A.; Chateigner, D.; Lutterotti, L.; Quirós, M.; Serebryanaya, N.R.; Moeck, P.; Downs, R.T.; Le Bail, A. Crystallography Open Database (COD): An Open-Access Collection of Crystal Structures and Platform for World-Wide Collaboration. *Nucleic Acids Res.* **2012**, *40*, D420–D427. [CrossRef]

40. Gražulis, S.; Merkys, A.; Vaitkus, A.; Okulič-Kazarinas, M. Computing Stoichiometric Molecular Composition from Crystal Structures. *J. Appl. Crystallogr.* **2015**, *48*, 85–91. [CrossRef]

41. Merkys, A.; Vaitkus, A.; Butkus, J.; Okulič-Kazarinas, M.; Kairys, V.; Gražulis, S. COD:CIF:Parser: An Error-Correcting CIF Parser for the Perl language. *J. Appl. Crystallogr.* **2016**, *49*, 292–301. [CrossRef]

42. Quirós, M.; Gražulis, S.; Girdzijauskaitė, S.; Merkys, A.; Vaitkus, A. Using SMILES Strings for the Description of Chemical Connectivity in the Crystallography Open Database. *J. Cheminf.* **2018**, *10*, 23. [CrossRef]

43. Li, C.-T.; Peacor, D.R. The Crystal Structure of $LiAlSi_2O_6$-II ("β Spodumene"). *Z. Kristallogr. Cryst. Mater.* **1968**, *126*, 46–65. [CrossRef]

44. Xu, H.; Heaney, P.J.; Beall, G.H. Phase Transitions Induced by Solid Solution in Stuffed Derivatives of Quartz: A Powder Synchrotron XRD Study of the $LiAlSiO_4$-SiO_2 Join. *Am. Mineral.* **2000**, *85*, 971–979. [CrossRef]

45. Nakagawa, K.; Izumitani, T. Metastable Phase Separation and Crystallization of Li_2O-Al_2O_3-SiO_2 Glasses: Determination of Miscibility Gap from the Lattice Parameters of Precipitated β-Quartz Solid Solution. *J. Non Cryst. Solids* **1972**, *7*, 168–180. [CrossRef]

46. Alekseeva, I.; Dymshits, O.; Ermakov, V.; Zhilin, A.; Petrov, V.; Tsenter, M. Raman Spectroscopy Quantifying the Composition of Stuffed β-Quartz Derivative Phases in Lithium Aluminosilicate Glass-Ceramics. *J. Non Cryst. Solids* **2008**, *354*, 4932–4939. [CrossRef]

47. Görlich, E.; Proniewicz, L.M. Laser Raman Spectroscopy Studies of Beta-Eucryptite Crystallization from Glass. *J. Mol. Struct.* **1982**, *79*, 247–250. [CrossRef]

48. Jochum, T.; Reimanis, I.E.; Lance, M.J.; Fuller, E.R. In Situ Raman Indentation of β-Eucryptite: Characterization of the Pressure-Induced Phase Transformation. *J. Am. Ceram. Soc.* **2009**, *92*, 857–863. [CrossRef]

49. Sharma, S.K.; Simons, B. Raman Study of Crystalline Polymorphs and Glasses of Spodumene Composition Quenched from Various Pressures. *Am. Mineral.* **1981**, *66*, 118–126.

50. Sprengard, R.; Binder, K.; Brändle, M.; Fotheringham, U.; Sauer, J.; Pannhorst, W. On the Interpretation of the Experimental Raman Spectrum of β-Eucryptite $LiAlSiO_4$ from Atomistic Computer Modeling. *J. Non Cryst. Solids* **2000**, *274*, 264–270. [CrossRef]

51. Zhang, M.; Xu, H.; Salje, E.K.H.; Heaney, P.J. Vibrational Spectroscopy of Beta-Eucryptite (LiAlSiO$_4$): Optical Phonons and Phase Transition(s). *Phys. Chem. Miner.* **2003**, *30*, 457–462. [CrossRef]

52. Ruismäki, R.; Rinne, T.; Dańczak, A.; Taskinen, P.; Serna-Guerrero, R.; Jokilaakso, A. Integrating Flotation and Pyrometallurgy for Recovering Graphite and Valuable Metals from Battery Scrap. *Metals* **2020**, *10*, 680. [CrossRef]

53. Elwert, T.; Goldmann, D.; Schirmer, T.; Strauß, K. Phase Composition of High Lithium Slags from the Recycling of Lithium Ion Batteries. *World Metall.* **2012**, *65*, 163–171.

54. Li, N.; Guo, J.; Chang, Z.; Dang, H.; Zhao, X.; Ali, S.; Li, W.; Zhou, H.; Sun, C. Aqueous Leaching of Lithium from Simulated Pyrometallurgical Slag by Sodium Sulfate Roasting. *RSC Adv.* **2019**, *9*, 23908–23915. [CrossRef]

55. Peltosaari, O.; Tanskanen, P.; Heikkinen, E.-P.; Fabritius, T. α→γ→β-Phase Transformation of Spodumene with Hybrid Microwave and Conventional Furnaces. *Miner. Eng.* **2015**, *82*, 54–60. [CrossRef]

56. Salakjani, N.K.; Singh, P.; Nikoloski, A.N. Acid Roasting of Spodumene: Microwave vs. Conventional Heating. *Miner. Eng.* **2019**, *138*, 161–167. [CrossRef]

57. Rosales, G.D.; Resentera, A.C.J.; Gonzalez, J.A.; Wuilloud, R.G.; Rodriguez, M.H. Efficient Extraction of Lithium from β-Spodumene by Direct Roasting with NaF and Leaching. *Chem. Eng. Res. Des.* **2019**, *150*, 320–326. [CrossRef]

58. Setoudeh, N.; Nosrati, A.; Welham, N.J. Phase Changes in Mechanically Activated Cpodumene-Na$_2$SO$_4$ Mixtures after Isothermal Heating. *Miner. Eng.* **2020**, *155*, 106455. [CrossRef]

59. Gasafi, E.; Pardemann, R. Processing of Spodumene Concentrates in Fluidized-Bed Systems. *Miner. Eng.* **2020**, *148*, 106205. [CrossRef]

60. Kallitsis, E.; Korre, A.; Kelsall, G.; Kupfersberger, M.; Nie, Z. Environmental Life Cycle Assessment of the Production in China of Lithium-Ion Batteries with Nickel-Cobalt-Manganese Cathodes Utilising Novel Electrode Chemistries. *J. Clean. Prod.* **2020**, *254*, 120067. [CrossRef]

61. Grudinsky, P.I.; Dyubanov, V.G.; Kozlov, P.A. Copper Smelter Dust Is a Promising Material for the Recovery of Nonferrous Metals by the Waelz Process. *Inorg. Mater. Appl. Res.* **2019**, *10*, 496–501. [CrossRef]

62. Murakami, Y.; Matsuzaki, Y.; Murakami, K.; Hiratani, S.; Shibayama, A.; Inoue, R. Recovery Rates of Used Rechargeable Lithium-Ion Battery Constituent Elements in Heat Treatment. *Metall. Mater. Trans. B* **2020**, *51*, 1355–1362. [CrossRef]

63. Halli, P.; Hamuyuni, J.; Revitzer, H.; Lundström, M. Selection of Leaching Media for Metal Dissolution from Electric Arc Furnace Dust. *J. Clean. Prod.* **2017**, *164*, 265–276. [CrossRef]

64. Aromaa, J.; Kekki, A.; Stefanova, A.; Makkonen, H.; Forsén, O. New Hydrometallurgical Approaches for Stainless Steel Dust Treatment. *Miner. Process. Extr. Metall.* **2016**, *125*, 242–252. [CrossRef]

Review

Industrial Recycling of Lithium-Ion Batteries—A Critical Review of Metallurgical Process Routes

Lisa Brückner [1,*], Julia Frank [2] and Tobias Elwert [3]

[1] Department of Mineral and Waste Processing, Institute of Mineral and Waste Processing, Waste Disposal and Geomechanics, Clausthal University of Technology, Walther-Nernst-Str. 9, 38678 Clausthal-Zellerfeld, Germany

[2] Northvolt Zwei GmbH & CO. KG, Frankfurter Straße 4, 38122 Braunschweig, Germany; julia.frank@zwei.northvolt.com

[3] New Energy Vehicle Research Centre, Qingdao University, Ningxia Road No. 308, Qingdao 266071, China; elwert@elwert-consulting.com

* Correspondence: lisa.brueckner@tu-clausthal.de; Tel.: +49-532-372-2633

Received: 20 July 2020; Accepted: 14 August 2020; Published: 18 August 2020

Abstract: Research for the recycling of lithium-ion batteries (LIBs) started about 15 years ago. In recent years, several processes have been realized in small-scale industrial plants in Europe, which can be classified into two major process routes. The first one combines pyrometallurgy with subsequent hydrometallurgy, while the second one combines mechanical processing, often after thermal pre-treatment, with metallurgical processing. Both process routes have a series of advantages and disadvantages with respect to legislative and health, safety and environmental requirements, possible recovery rates of the components, process robustness, and economic factors. This review critically discusses the current status of development, focusing on the metallurgical processing of LIB modules and cells. Although the main metallurgical process routes are defined, some issues remain unsolved. Most process routes achieve high yields for the valuable metals cobalt, copper, and nickel. In comparison, lithium is only recovered in few processes and with a lower yield, albeit a high economic value. The recovery of the low value components graphite, manganese, and electrolyte solvents is technically feasible but economically challenging. The handling of organic and halogenic components causes technical difficulties and high costs in all process routes. Therefore, further improvements need to be achieved to close the LIB loop before high amounts of LIB scrap return.

Keywords: lithium-ion battery; recycling; lithium; cobalt; nickel; manganese; graphite; mechanical processing; pyrometallurgy; thermal treatment; pyrolysis; hydrometallurgy

---

1. Introduction

Renewable energy sources have the potential to end the era of fossil fuels. Electrochemical storage systems, especially lithium-ion batteries (LIBs), are an important technology for the success of this transition. This is due to their sum of positive properties such as their high energy density and low self-discharge [1]. On the production site, the growing application of LIBs leads to high investments in research and development with a focus on questions such as optimized cell production or optimized active materials. These developments are accompanied by a critical discussion regarding the security of the necessary raw material supply, the environmental and social impact of the raw material production, and the responsible recycling of end-of-life batteries [2–5]. Beside base metals such as iron (Fe), aluminum (Al), copper (Cu), manganese (Mn), and nickel (Ni), LIBs also contain minor metals such as cobalt (Co) and lithium (Li) as well as graphite. The production of some of these raw materials can be connected to severe ecological and social impacts. Examples are the appearance of child labor in artisanal Co mining and the influence of Li production on the water balance in desert areas [6,7]. Furthermore, in the case of some raw materials, a small number of producers dominate the markets. This is especially true for Co and graphite [8]. In view of these developments, the recycling of LIBs is a key factor to manage the transition toward renewable energies in a sustainable way. However, LIB recycling is a challenging task due to their complex material composition and their electrical and chemical energy content leading to various health, safety, and environmental (HSE) risks [9].

In order to ensure a closed loop for LIBs (see Figure 1), extensive research activities started about 15 years ago. Several efforts led to the development of different recycling processes, which have been realized in a few pioneering industrial plants. All these process routes are characterized by long and complex process chains and use a combination of mechanical and/or thermal and/or pyrometallurgical and hydrometallurgical unit operations [10].

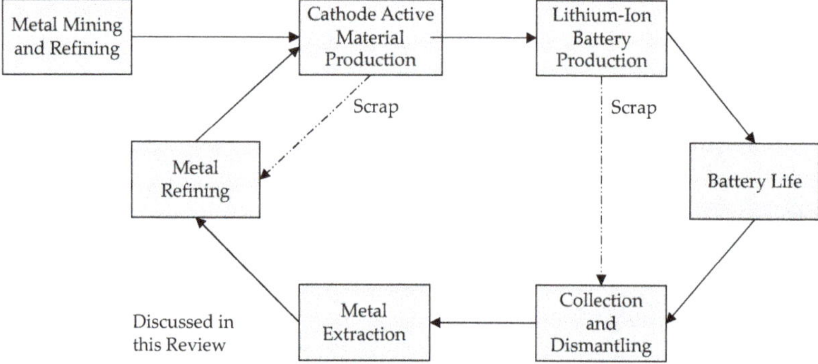

Figure 1. Closed loop for battery materials.

This paper reviews and discusses the current status of industrial metallurgical processes from a European perspective. In contrast to existing reviews [11–17], which often focus on research at a lower technology readiness level and specific process technologies, this review emphasizes the complex dependencies between legal framework, economic conditions, and technical boundaries of industrial process routes. The focus here lays on the metal extraction and metal refining from battery cells/modules. Pre-treatment (discharging and dismantling of battery systems) is not covered in this review. For further details on this subject, see [10,11]. Products of metal refining are metal salts, compounds, or the respective metals, e.g., cathodes that can be used in different industries depending on the final quality.

In the following, after a background on current and future LIBs as well as legislation is given in Section 2, the currently pursued industrial recycling routes are presented in Section 3. In Section 4, the recycling processes are critically discussed with respect to legislation, recovery rate, process robustness, economics, and HSE. Section 5 summarizes the main conclusions and gives an outlook regarding the upcoming developments.

2. Background

The LIB technology has reached a wide acceptance since its introduction and is applied in a growing number of applications. First, it was widely applied in batteries for mobile phones and laptops, followed by pedelecs and power tools. Nowadays, LIBs capture the markets for stationary storage systems and electric vehicles (xEVs) [1]. Table 1 summarizes the characteristics of LIBs in different applications.

Table 1. Characteristics of lithium-ion batteries (LIBs) used in battery electric vehicles (BEVs), plug-in hybrid electric vehicles (PHEVs), electric bikes (pedelecs), and mobile phones [1,11,16,18].

Characteristic	BEV	PHEV	Pedelec	Mobile Phone
Voltage U [V]	355–800	351	22.2–36	3.7
Capacity C [Ah]	60–117	26–34	8–10	0.7–1.2
Energy E [Wh]	21,000–93,000	9000–12,000	189–288	2.4–4.1
Mass m [kg]	235–680	80–135	1.3–4	0.021–0.038

LIBs are configured in cells, modules, and systems. Battery modules and especially systems need peripheral units such as a temperature and a battery management system. Depending on the field of application, the design of cells and modules varies considerably. Whereas applications with a smaller battery size often use cylindrical cells due to their dimensions, prismatic cells are primarily used for bigger battery systems, e.g., traction batteries. Pouch cells with an Al composite foil as casing instead of a rigid Al or steel casing are used among a wide range of applications in order to increase the energy density. In addition, battery systems without module levels are currently under development [1,19–21].

Despite the wide range of applications and different designs for cells, modules, and systems, the chemical composition of LIBs is similar. A typical composition of a battery system is given in Figure 2. Here, cells form 56% of the battery system [22]. In newer LIB systems, an increase of the cell fraction, especially when using pouch cells, is observed. The cells consist of five main components: the positive and negative electrodes, the ion-permeable separator, the electrolyte, and a cell casing. For a detailed explanation concerning the functionality and production of battery cells, modules, and systems, see [1,20].

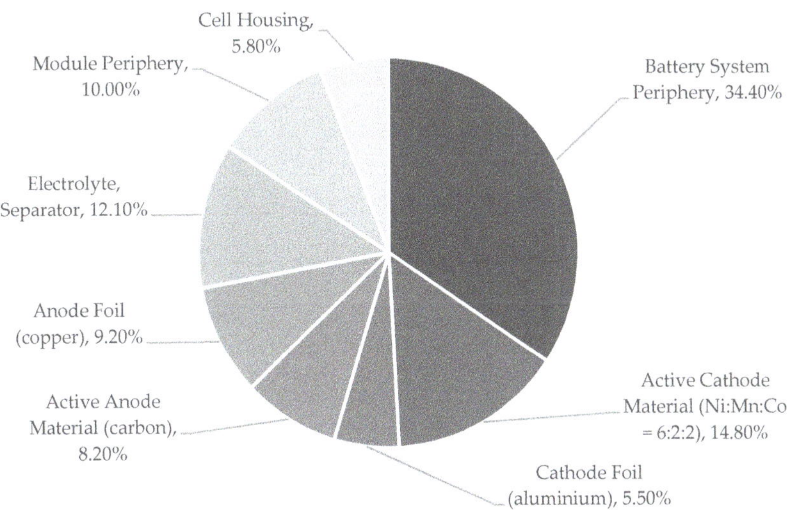

Figure 2. Typical composition of a generic traction battery system based on data from [22].

Table 2 gives concentrations of the metals of interest and graphite for recycling in a typical xEV battery module and their origin.

Table 2. Fraction of specific metals and graphite in a generic battery module with an NMC ($LiNi_xMn_yCo_zO_2$) chemistry of 6:2:2 [10].

Elemental	Fraction [%]	Origin
Al	25.2	Cell and module case, cathode current collector
C	12.5	Anode active material
Co, Mn, Ni	13.6	Cathode active material with Co (2.7%), Mn (2.7%), and Ni (8.2%)
Cu	14.0	Cables, anode current collector
Li	1.5	Cathode active material, conducting salt

2.1. Current Lithium-Ion Battery Composition

In this section, the main battery cell components, cathode, anode, electrolyte, and separator of current LIBs are presented, since these components have the biggest influence on the recycling processes.

2.1.1. Positive Electrode (Cathode)

The positive electrode consists of the active material, which is coated on the current collector foil, usually Al. In most cases, as binder, polyvinylidene fluoride (PVDF) with the addition of carbons as conducting agents is used. The active material must be able to deintercalate Li ions and oxidize the transition metals for charge balance [1]. There are three different types of cathode active materials: layered oxides ($LiMO_2$ with M = Co, Ni, Mn, Al), spinels (LiM_2O_4 with M = Mn, Ni), and phosphates ($LiMPO_4$ with M = Fe, Mn, Co, Ni). Commercially available active materials are presented in Table 3.

Table 3. Summary of the properties of the main commercially available cathode active materials [1,23–25].

Cathode Active Material	Reversible Capacity [Ah/Kg]	Specific Energy [Wh/Kg]	Advantage	Disadvantage
LCO ($LiCoO_2$)	150	624	specific energy	safety, stability and costs
NMC ($LiNi_xMn_yCo_zO_2$)	160–200	592–740	reversible capacity	capacity fade
NCA ($LiNi_{0.8}Co_{0.15}Al_{0.05}O_2$)	200	740	stability	safety, costs
LMO ($LiMn_2O_4$)	120	410	costs	stability
LFP ($LiFePO_4$)	160	544	costs, safety	specific energy

In most markets, layered oxides are the foremost commercially used cathode active materials, especially in traction batteries [25]. For optimized specific and reversible capacity compared to NMC 1:1:1 and a higher thermal stability compared to NCA, the Ni-content of NMC is currently maximized from 1:1:1 to Ni-rich compositions up to 8:1:1 [25–27]. In contrast, particularly in China, $LiFePO_4$ (LFP)-based batteries have been widely used in buses and xEVs, which means that in China, significant return flows of this battery chemistry have to be expected [10].

2.1.2. Negative Electrode (Anode)

On the anodic side of the battery, the current collector foil consists of Cu and the active material is graphite in most cases. Typically, as binder, styrene–butadiene rubber (SBR) in combination with the polymeric thickener carboxymethyl cellulose (CMC) and carbons as conducting agents is used [28]. Consumer applications usually use natural graphite because of the satisfactory properties at an acceptable price. For high-energy or high-power applications, artificial graphite is used [1,25]. To combine the benefits of both, a combination is possible as well. A limited but increasing number of cell producers include small fractions of about 5% silicon (Si) or SiO_2 to increase cell energy in their anode active material [25,29]. Furthermore, lithium titanate (LTO) is applied as anode active material in high-power applications. In this case, Al is used as a current collector foil instead of Cu [30].

Artificial and natural graphite have a market share of 89% while amorphous carbon has a share of 7%, which is in total a 96% market share of solely carbon-based anodic materials. C/Si_x composites and LTO share the remaining 4% market share equally [25].

2.1.3. Electrolyte and Separator

This section shortly summarizes the typically used electrolytes and separators. Electrolytes consist of the conducting salt, solvents, and additives. They need to provide high and stable conductivity and a manageable safety level. As electrolyte solvents carbonates, esters as well as ethers are used commercially. State of the art are mixtures of cyclic carbonates such as ethylene carbonate (EC) or propylene carbonate (PC) with open chained carbonates such as dimethyl carbonate (DMC), ethyl methyl carbonate (EMC), and/or diethyl carbonate (DEC). Usually, these mixtures strongly depend on the application of the battery [1]. Currently, new solvent components, that could for example substitute EC, are evaluated [29].

The conducting salt in all commercially available batteries is lithium hexafluorophosphate ($LiPF_6$). Other possible conducting salts are lithium bis(trifluormethyl)sulfonylimid (LiTFSI) and its derivates (e.g., lithium bis(fluorosulfonyl) imide (LiFSI)) or lithium [tris(pentafluorethyl)-trifluorphosphate] (LiFAP), lithium 4,5-dicyano-2-trifluoromethyl-imidazolide (LiTDI), and lithium bis(oxalate)borate (LiBOB) [1,29,31]. LiFSI and LiTDI are already commercially available [32].

Separators divide the space between the electrodes and are only permeable for ions. Four different separator types are existent: microporous membranes, ceramic-coated separators, non-woven mats, and solid inorganic and polymeric electrolytes. Polyolefin-based membranes currently dominate the market for separators [1,25]. Currently, coated separators (e.g., with PVDF or ceramics) experience increased application [26,32].

2.2. Future Lithium-Ion Battery Composition

For effective planning of recycling plants, knowledge about possible developments in the LIB technology is crucial. Consequently, this section presents the main possible future chemical compositions of LIB cells.

2.2.1. Positive Electrode (Cathode)

A short-term goal is to further improve NMC materials. Especially, the formation of spherical NMC particles that have a Ni-rich core for high capacity and a Mn-rich shell for stability seems interesting [27]. In addition, currently, a growing interest in LFP chemistries can be observed due to their low raw material costs and intrinsic safety [21]. Li- and Mn-rich oxides are discussed in research because of their high theoretical specific capacity and low costs. However, they are considered a medium- to long-term development because of the need for a different electrolyte system [27,33,34] also mentioned possible positive effects of doping or coating active material particles, which can also be considered a long-term optimization approach. Doping can promote high specific capacity and good long-term stability, which are two properties that usually do not appear together. Possible dopants are cations such as Ag^+, Al^{3+}, Cr^{3+}, Fe^{3+}, Mg^{2+}, Mo^{6+}, Ru^{2+}, Ti^{4+} and Zr^{4+}, as well as anions such as F^-. Coatings such as Al_2O_3, $AlPO_4$, AlF_3, $PrPO_4$, TiO_2, and V_2O_5 are also discussed, especially for protection measures of the particles against components of the electrolyte [27,33].

2.2.2. Negative Electrode (Anode)

Anode active materials with larger Si fractions of up to 40% or the use of tin (Sn) over Si are in discussion [35]. Li metal is a promising anode active material in the medium term, especially for all-solid-state batteries [1,25,29].

2.2.3. Electrolyte and Separator

The application of standard electrolyte systems with $LiPF_6$ will continue in the upcoming years. According to [25], even high voltage cathode materials are suitable for that system. Solid electrolytes that combine the functions of the separator and the electrolyte may gain importance in the long term. It is still unclear if the advantages of solid electrolytes are strong enough compared to already commercially functioning separators, since major technological breakthroughs are pending [25].

2.3. Legislative Framework

The recycling of batteries in Europe is regulated in the Batteries Directive (2006/66/EG) [36]. It came into force in 2006 and is implemented in national laws by each member of the EU. In terms of LIB recycling, the producer carries an extended responsibility and must bear the cost of collecting, treating, and recycling. Furthermore, the directive requires a minimum recycling efficiency. All collected and identifiable spent batteries must be treated and recycled according to the state of the art. For LIBs, a recycling efficiency of at least 50% by weight must be achieved. Since 2006, the directive has been amended several times, most recently in 2013 [37].

The Batteries Directive is currently under major revision, and the presentation of a draft version of the new directive is expected in the fourth quarter of 2020 [38]. Although details have not been published yet, increasing collection targets for consumer batteries, the introduction of element-specific recycling quotas for Li, Co, and Ni, more specific regulations for traction batteries, as well as uniform regulations for the calculation of recycling quotas are expected by most stakeholders. In China, similar regulations, especially regarding element-specific recycling targets, are in force [39,40].

3. Process Chains for the Recycling of Lithium-Ion Batteries

Section 2 described the current composition of LIBs and possible future developments. From a recycling point of view, the main challenges derive, on the one hand, from the complex material composition including halogenic and organic compounds as well as the high energy content and, on the other hand, from the requirements regarding high recycling rates, HSE, and economics. This results in comparatively long and complex process chains in comparison to earlier battery generations. In the last years, several different recycling processes have been developed, which can be classified into two general process routes, as shown in Figure 3. The first group of processes combines pyrometallurgy with hydrometallurgy and the second group consists of mechanical treatment prior to metallurgy.

In case of a pyrometallurgical treatment, a Co-, Cu-, and Ni-containing alloy (metallic phase) or matte (sulfidic phase), an Al-, Mn- and Li-containing slag (oxidic phase), and a fly ash are produced. Alloy/matte and slag can be treated by hydrometallurgy to recover the individual metals. The fly ash is usually used as an outlet for undesirable elements such as fluorine and hence, it is landfilled.

In the second group of processes, LIBs are treated mechanically, often after a thermal treatment step. Typical products of the mechanical process are ferrous and non-ferrous metal concentrates including Al and Cu concentrates as well as a fraction containing the active electrode materials, which is called black mass. The black mass can be either fed in pyrometallurgy or treated directly in hydrometallurgy. Depending on the overall process design, the black mass requires a thermal treatment prior to hydrometallurgy to remove the organic components and to concentrate the metal content. In hydrometallurgy, Co, Li, Mn, Ni, and, if applicable, graphite can be recovered.

In the following, both process routes are described in further detail, see Sections 3.2 and 3.3. For those not familiar with the applied unit operations, Section 3.1 gives a short introduction for a better understanding.

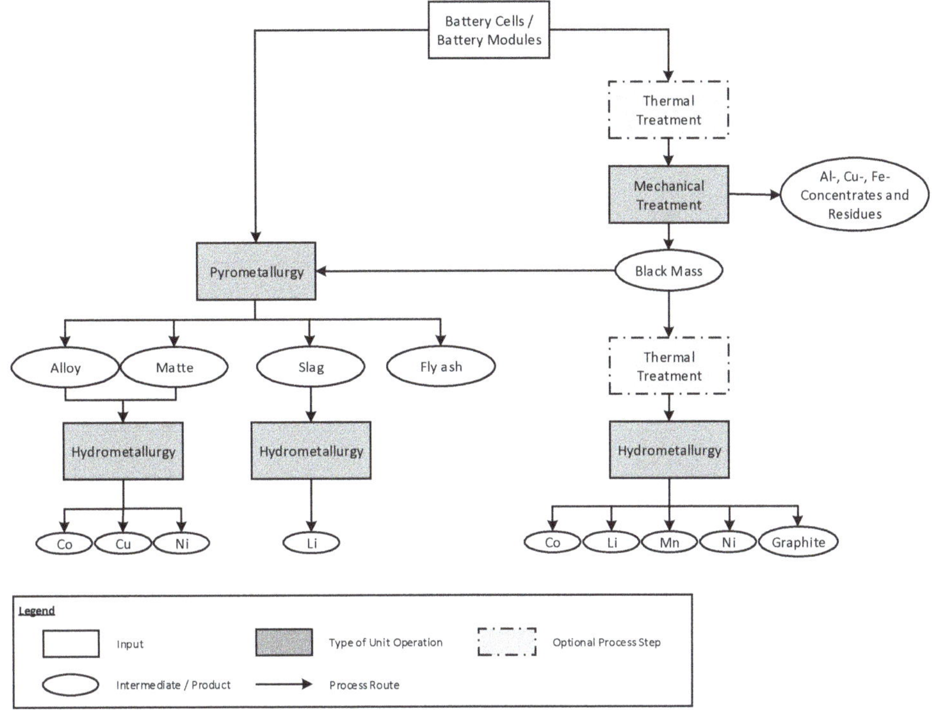

Figure 3. Possible process routes in the recycling of lithium-ion batteries.

3.1. Background on Processes Applied in Lithium-Ion Battery Recycling

Mechanical processes typically start with comminution aiming at liberation of the materials/components. Afterwards, the materials/components are sorted by their physical properties such as particle size, form, density, and electric and magnetic properties. Usually, in mechanical processing, concentrates for further metallurgical treatment are produced [41–43].

Pyrometallurgy includes high-temperature processes such as roasting or smelting for winning and refining metals. Roasting is the term for processes consisting of a gas–solid reaction, e.g., oxidizing roasting, with the goal of purifying the ore or secondary raw material. Smelting means reactions that are aiming at extracting a metal from an ore or secondary raw material. Smelting uses heat and a chemical reducing agent to decompose the ore/secondary raw material, driving off other elements as gases or slag and leaving the metal base behind. The reducing agent is commonly a source of carbon [41,44]. Pyrometallurgical processes have been used for a long time in history. Pyrometallurgy has different advantages such as high reaction rates, small plant size for a given throughput, and a high overall efficiency. On the downside, these processes often only produce intermediates that require further hydrometallurgical refining, require extensive off-gas treatment, and are uneconomic for low-grade concentrates [41].

In contrast, hydrometallurgy is based on aqueous chemistry, typically at low temperatures. Hydrometallurgical processes include three major process steps. The first one is leaching, which describes the dissolution of the metals, in most cases with the help of an acid, base, or salt. The second step is purification, which separates the metals via selective chemical reactions. This includes solid–liquid reactions, e.g., ion exchange, precipitation, and liquid–liquid reactions, e.g., solvent extraction. In the last step, the metals of interest need to be recovered from solution as a solid product, i.e., a metal, a metal salt, or a compound, by crystallization, ionic precipitation, reduction with gas, electrochemical reduction, or electrolytic reduction. Hydrometallurgical unit operations often occur as refining steps at the end of a process chain because of their ability to produce high-quality products. Their ability to produce high-quality products is opposed by comparatively large plant sizes for a given throughput, the need for waste water treatment, and a costlier residue management in case of sludges in comparison to slags [41,45,46].

3.2. Pyrometallurgy with Subsequent Hydrometallurgical Treatment

The first industrially implemented process routes for LIB recycling combine pyrometallurgical unit operations with hydrometallurgical unit operations. The processes can be divided into the co-processing of LIBs in existing primary or secondary Co, Cu, and Ni smelters and into dedicated plants specifically designed for LIB recycling. In the following, examples for the different process routes are given.

3.2.1. Co-Processing in Primary and Secondary Co-, Cu-, Ni-Smelters

Due to the high fluorine, Al, Li, and organic content, LIBs are a difficult feed for conventional Co, Cu, and Ni smelters with respect to corrosion, slag properties, energy, and mass balance. Fluorine and Li can severely attack the refractories, and the first one is also an issue in the off-gas treatment. Al increases the viscosity of the slag and is therefore only acceptable to a certain extent, which depends on the applied slag system and operating temperature of the furnace. Furthermore, in addition to the organic and graphite content, its participation in redox reactions adds a high amount of energy to the system, which also needs to be considered. Nevertheless, some smelters accept a certain amount of LIBs as feed material. Examples are Nickelhütte Aue GmbH (Aue, Germany) and Glencore Sudbury INO (Greater Sudbury, ON, Canada) [47,48]. The first smelter is a comparatively small smelter with an annual smelting capacity of 20,000 t for secondary raw materials such as Co-, Cu-, and Ni-bearing spent catalysts, electroplating sludges, filter dusts, and ashes [49]. The second one is capable of producing 95,000 t of Ni, Cu, and Co in matte annually, primarily from sulfidic ore concentrates. In this case, the batteries are calcined before smelting presumably in order to reduce the organic, graphite, and fluorine content and to avoid explosions of battery cells within the smelter. Afterwards, the calcined batteries are fed into an electric furnace together with calcine from the primary concentrates. Before granulation, the Fe content is reduced to about 2% by converting [48].

Both plants concentrate Co, Cu, and Ni in a matte, which is a typical intermediate in these processes. The Al, Li, and Mn content of LIBs is mainly transferred to the slag phase. As a result of the low LIB fraction in the feed, Li is highly diluted in the slag and therefore difficult to recover. The organic content and graphite are utilized as fuel and reductants. Halogens are captured in the off-gas treatment [48,49]. The matte is further refined by hydrometallurgy.

For the matte treatment, three main routes exist, which are based on oxygen–sulfuric acid, chlorine, and ammonia–air leaching [50]. The first two options are presented exemplary in the following, since they are currently applied by Nickelhütte Aue GmbH (oxygen–sulfuric acid leaching [49]) and Glencore Nikkelverk AS (Kristiansand, Norway) (chlorine leaching [50]).

Matte Processing at Nickelhütte Aue GmbH (Aue, Germany)

Nickelhütte Aue GmbH operates a comparatively small hydrometallurgical plant and produces approximately 3900 t Ni per year and smaller amounts of Co and Cu [49]. A simplified flowsheet is given in Figure 4.

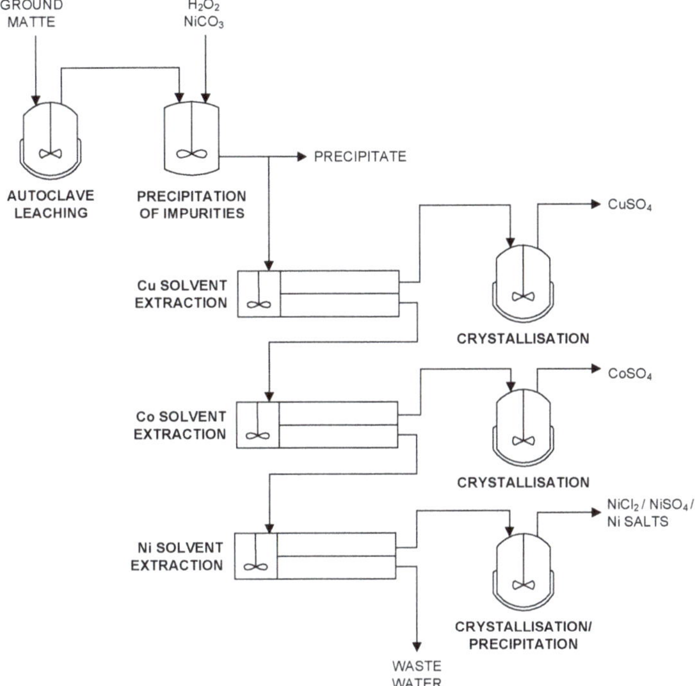

Figure 4. Simplified flowsheet of Nickelhütte Aue GmbH based on [49].

Matte processing starts with comminution followed by pressure oxidation leaching at 6 to 8 bar. Afterwards, impurities are removed prior to solvent extraction. For example, Fe is precipitated as goethite (FeOOH) by using H_2O_2 as an oxidizing agent and basic nickel carbonate for pH adjustment. Leaching and precipitation residues are recirculated into the smelter. Co, Cu, and Ni are separated and purified by several solvent extraction circuits. Depending on the process configuration, cobalt sulfate, nickel sulfate, nickel carbonate, nickel chloride, and copper sulfate are produced in electroplating grade or similar [47,49].

Matte processing at Glencore Nikkelverk AS (Kristiansand, Norway)

The Nikkelverk refinery [50,51] produces Cu, Co, and Ni electrodes, H_2SO_4 as well as by-products such as platinum group metals (PGMs). It treats matte from the Sudbury smelter (Canada) and other sources and has an annual capacity of 4700 t Co, 39,000 t Cu, 92,000 t Ni, and 115,000 t H_2SO_4. Table 4 gives the general composition of the matte. A simplified flow sheet of the complex process is shown in Figure 5.

Table 4. Composition of mattes entering Glencore Nikkelverk AS [50].

Component	Composition of Converter Matte [wt %]
Ni	38–54
Cu	18–36
Fe	2.5
S	22–23
Co	0.9–2.2

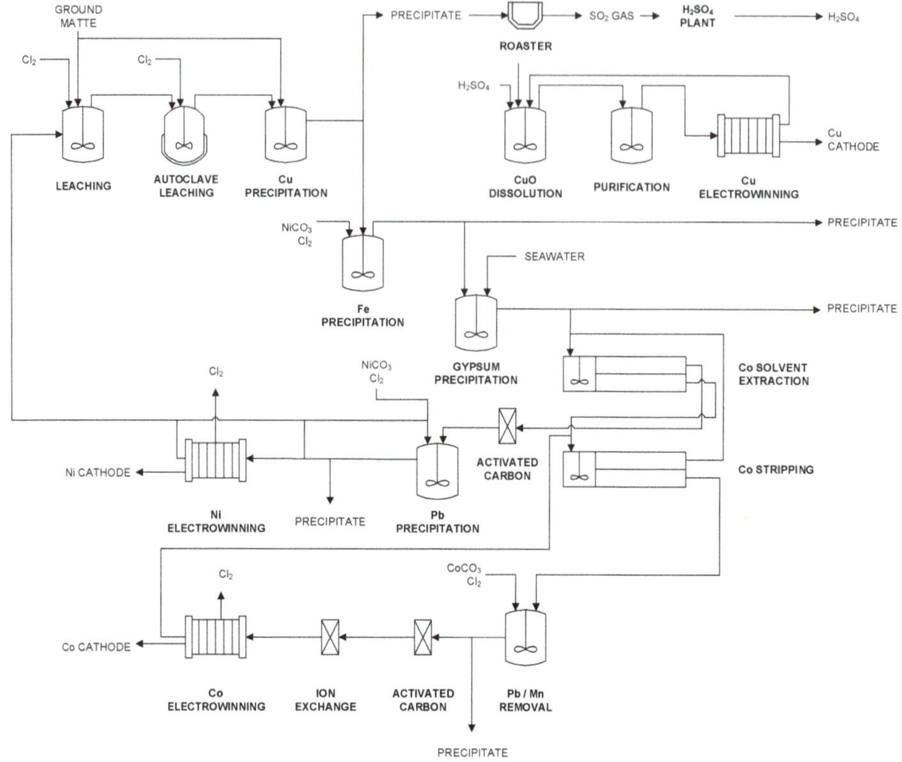

Figure 5. Simplified flowsheet of Glencore Nikkelverk AS refinery adapted from [50,51].

In the Nikkelverk process, the ground matte is pulped with solution from Ni electrowinning and fed to a multistage leaching operation. In the first stages, chlorine gas is injected. At this stage, the principal reactions are the dissolution of Ni_3S_2 and Cu_2S to yield Ni and Cu in solution and elemental sulfur in the residue. The reactions are exothermic and maintained at the boiling point. Afterwards, the slurry is transferred to autoclave leaching, which is operated at 150 °C. These conditions favor the following metathesis reaction

$$NiS(s) + 2\,CuCl(aq) \xrightarrow{150\,°C} NiCl_2(aq) + Cu_2S(s) \tag{1}$$

to separate Cu from Ni and Co. A further reduction of the Cu concentration to less than 0.5 g/L is achieved by precipitation using fresh matte as precipitant. The occurring chemical reactions are similar to Equation (1).

Subsequently, the slurry is washed and filtered. The residue, which contains about 50% Cu, 5% Ni, 40% S, 1% Co, and 1% Fe is transferred to the Cu roaster. The solution, which contains about 220 g/L Ni, 11 g/L Co, 7 g/L Fe, and 0.5 g/L Cu, is pumped to Fe removal.

Fe is precipitated from the solution as $Fe(OH)_3$ by using chlorine to oxidize Fe^{2+} to Fe^{3+} and $NiCO_3$ to raise the pH. After filtration, cooling, and gypsum removal, Co is extracted from the solution by solvent extraction. Before Co is recovered by electrowinning from the strip solution, the solution is purified by ion exchange and activated carbon to remove Zn and trace amounts of Cu as well as entrained organics.

The extraction raffinate, which contains 220 g/L Ni, 0.25 g/L lead (Pb), and 0.2 g/L Mn, is pumped through carbon columns to remove entrained organics from solvent extraction and is further purified by the precipitation of undesired metals such as Pb and Mn with $NiCO_3$ and chlorine gas. The purified solution is diluted to 65 g/L and fed to Ni electrowinning. The solution from electrowinning contains about 54 g/L Ni and is recycled.

The residue from Cu precipitation is roasted in a fluidized bed reactor to produce a calcine of CuO and SO_2 gas, which is sent to a double-adsorption H_2SO_4 plant after cleaning. The calcine is mixed with solution from Cu electrowinning and fresh acid to yield Cu into solution. After purification of the leach solution, Cu is recovered by electrowinning. A bleed is withdrawn from this stream and returned to chlorine leaching to remove Ni from the Cu circuit. PGMs are recovered from the residue of Cu leaching. Further details are given in [50,51].

3.2.2. Dedicated Processes

In contrast to co-processing, dedicated pyrometallurgical processes for LIB recycling are specifically developed for the treatment of LIBs. The processes enable the enrichment of Li in the slag, and the furnaces are designed to handle the highly corrosive feed material. Furthermore, the off-gas treatment is dimensioned to capture the high amounts of halogenic and organic compounds. Nevertheless, also in dedicated processes, LIBs are often co-processed with other feed materials in order to meet a suitable energy and mass balance as well as to enable sufficient plant utilization in this emerging market.

One of the most prominent examples is Umicore SA (Belgium), which operates a semi-industrial furnace at Hoboken (Belgium) with a capacity of 7000 t/year [52]. Umicore announced an increase of its recycling capacity by a factor of approximately 10 by the mid-2020s [53]. During the last 15 years, Umicore has published several patents regarding LIB recycling. Although little is known about the exact current direction, the history of patents indicates various changes regarding process design as well as furnace technology. In principal, all process options produce one or two alloys containing Co, Cu, and Ni, a Li-bearing slag, and a fly ash phase. Regarding process design, two principal flow sheets were published in the patents, which are shown in Figure 6.

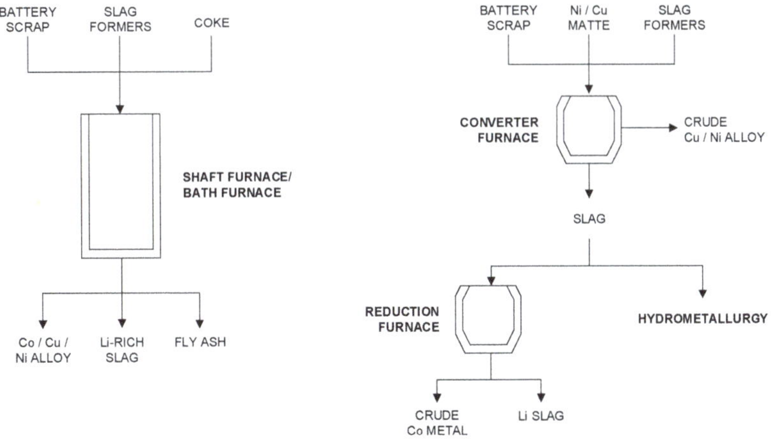

Figure 6. Simplified flowsheets of the two different processes published by Umicore SA based on [54–57].

According to the patent history, the first approach was to produce a Co/Cu/Ni alloy and a Li-bearing slag using a single-furnace process (Figure 6, left). The first patents, e.g., [54,55], describe the use of a shaft furnace, in which Co/Ni-bearing batteries (LIB and NiMH batteries) and other materials are fed together with slag formers, mainly CaO and SiO_2, and coke as a reductant. A shaft furnace was chosen in order to take advantage of the temperature gradient over the packed bed. The temperature of the batteries is slowly increased by the rising counter current gas generated in the smelting and reducing zone. Therefore, the electrolyte is slowly evaporated in the upper part of the bed (pre-heating zone), which lowers the risk of battery explosions. The furnace is operated at a comparatively high bath temperature of about 1450 °C, which is necessary to keep the viscosity of the Al-rich slag system sufficiently low.

According to [56], the drawbacks of the process are a very high coke consumption, which is necessary to carry out the reduction and to keep the packed bed sufficiently porous. Furthermore, the occurrence of size segregation in the shaft leads to an increased pressure drop over the packed bed, and large quantities of fines are carried over with gases, resulting in problems at the bag house. In order to solve these problems, [56] suggests the use of a bath smelting process instead of a shaft furnace. Due to the rapid heating of the battery cells in the molten bath, the batteries are susceptible to violent explosions, which can damage the lining. Therefore, and as protection against chemical attack, the bath smelter is equipped with freeze lining at the bath level. The operating temperature is again about 1450 °C, and SiO_2 and CaO are added to the feed in order to flux the Al.

The latest patent family indicates a change in direction (Figure 6, right). Instead of a fully dedicated process for Co/Ni-containing batteries, [57] suggests the processing of Co-bearing batteries and their scraps in a converter furnace together with Cu or Cu/Ni matte in order to produce a crude Cu/Ni alloy and a Co- and Li-bearing slag. Co can be recovered from the slag either by hydrometallurgy or by deep reduction, producing a crude Co metal phase and a Li-bearing slag.

The advantages of this approach are the ability to treat variable amounts of battery scrap in large-scale installations, an increased flexibility to meet a suitable energy and mass balance, as well as the pyrometallurgical separation of Co from Ni, which probably allows the hydrometallurgical refining of Co in existing plants designed for the refining of cobalt hydroxide produced from the African Cu/Co belt.

Processing of the Alloy

The further hydrometallurgical processing of the Co-, Cu-, and Ni-containing alloy is conducted in plants similar to those described in Section 3.2.1. Umicore produces a variety of Co and Ni products, including battery grade chemicals and cathode materials for LIBs [52,58].

Processing of the Slag

The Li content of the slags is comparable to the Li content of spodumene concentrates, which are beside Li-containing brines the major commercial source for Li [59]. In addition, the chemical composition shows similarities; see Table 5. Besides the listed oxides, minor amounts of Fe and Mn oxides occur. Therefore, an economic extraction might be feasible in comparison to slags with lower Li concentrations from co-processing plants.

Table 5. Composition of Li-rich slag and spodumene concentrate [59–63].

Component	Li-Rich Slag [wt%]	Spodumene Concentrate [wt %]
Li_2O	8–10	ca. 7
Al_2O_3	38–65	24.5–29
CaO	<55	0.1–0.5
SiO_2	<45	60–65

The extraction of Li from different slag compositions was investigated in the research project Lithium-Ion Battery Initiative (LiBRi) [64] as well as by [58]. A general flow sheet of the developed process is given in Figure 7.

In a first step, the slag is milled to obtain a micrometer-sized powder. Then, this powder is leached in H_2SO_4 or HCl at 80 °C. Good Li yields are obtained with both media at a free acid concentration of approximately 10 g/L. In order to ensure good filtration, the leach solution is neutralized to obtain a solution at pH 5. Al, the main impurity in solution, is removed under these conditions.

Several neutralization agents can be used. In a sulfate medium, CaO has the advantage that $CaSO_4$ is formed, which reduces the salt load of the solution. In a chloride medium, NaOH is preferred over CaO because Ca is more difficult to separate from Li. Na_2CO_3 can be used as a neutralization agent in both media but could decrease the Li yield of the process if it is already used in this stage of the flowsheet. After the removal of Al by filtration, it is advantageous to increase the Li concentration in solution by using the filtrate again in the next leach operation a few times until the Li concentration is high enough for the next steps. When the desired Li concentration is reached, the leach solution can be essentially purified of other metals by precipitation between pH 9 and pH 10 with Na_2CO_3. This is an efficient way to remove Ca and Mg by forming their carbonate salts. Ca and Mg carbonate salts are less soluble than Li_2CO_3. Finally, pure Li_2CO_3 is precipitated at high temperature (between 80 and 100 °C) and pH > 10. The solubility of Li_2CO_3 decreases at higher temperature and higher Na_2CO_3 content. Moreover, the Li yield in this step increases at higher initial Li concentration. Therefore, the yield can be improved if water is removed from the solution prior to precipitation, e.g., by evaporation [58,64].

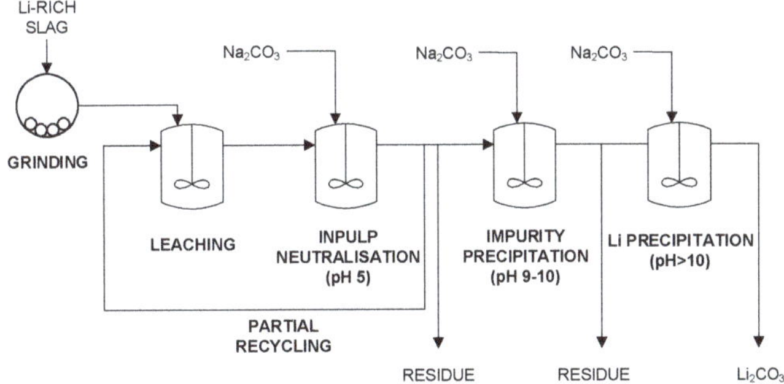

Figure 7. Simplified flowsheet for the production of Li_2CO_3 adapted from [58].

Investigations by [64] showed that silica management is of paramount importance. Under certain conditions, the dissolution of silica leads to gel formation, resulting in severe filtration problems and low yields. Overall Li yields under favorable conditions were reported to be between 60% and 70%.

Another critical issue is residue management. Due to the low Li content of the slags, the amount of residue from the leaching and precipitation of impurities is high. The disposal of such residues is restricted in Europe and is associated with high costs.

A possible solution to this problem is the application of the slags as a neutralization agent instead of conventional Ca-based agents in spodumene processing as described in [65]. Furthermore, this bypasses the need to establish a specific process for comparatively low amounts of slag. Spodumene is a pyroxene mineral consisting of lithium aluminium inosilicate ($LiAl(SiO_3)_2$). A simplified spodumene processing flow sheet is given in Figure 8.

After spodumene ores have been mined, concentrated, and comminuted, the finely divided material is submitted to a first high-temperature treatment step during which α-spodumene is converted into β-spodumene. Following the phase transformation, the material is mixed with H_2SO_4 and submitted to a roasting step that aims to liberate the Li from the mineral. This step is performed at 250–300 °C with an excess of acid with respect to Li.

The roasted material is subsequently mixed with water, upon which the Li_2SO_4 dissolves, together with the free H_2SO_4. Next, a conventional neutralizing agent such as $CaCO_3$, CaO, or $Ca(OH)_2$ is added to neutralize the free acid and to precipitate a number of impurities.

Typically, the neutralization step is performed at a pH of 5 to 6 to remove impurities such as Al, Si, and Fe from the solution. A solid–liquid separation step is applied to separate the crude Li_2SO_4 solution from the residue that mostly contains aluminum silicates, gypsum, and precipitated impurities.

Then, further purification steps are applied for the removal of Ca, Mg, and other impurities before Li precipitation as carbonate or hydroxide.

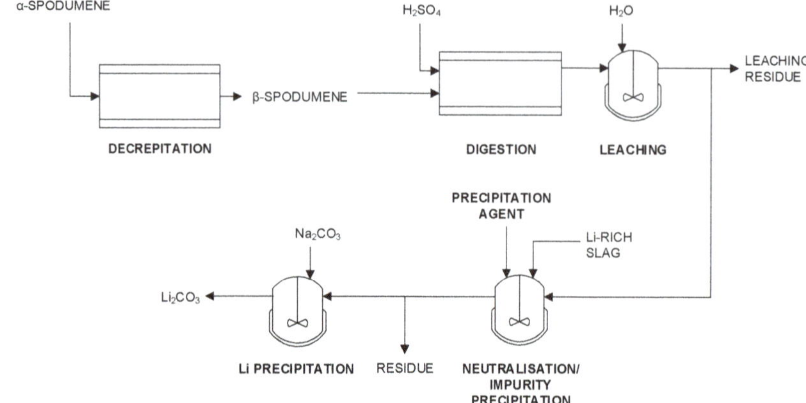

Figure 8. Simplified flowsheet of spodumene processing with partial replacement of Ca-based neutralizing agent based on [59,65].

In the described process, the Li-bearing metallurgical slag is used to substitute part of the conventional neutralizing agent. In this neutralization step, most of the Li in the slag is released and supplements the Li liberated from the spodumene. To ensure the optimum release of the Li from the slag, it is preferred to neutralize with Li-bearing slag up to a pH of less than 4. Above, it is recommended to proceed with a conventional neutralization agent to reach a pH between 5 and 7. So far, no information has been published about the industrial implementation. According to information published in the patent, overall Li yields around 90% from the slag are achievable, which is comparable to Li yields from spodumene concentrates [59].

3.3. Mechanical Processing with Subsequent Metallurgical Treatment of Black Mass

Apart from the process chains presented in Section 3.2, LIBs can be processed using a combination of mechanical treatment with pyrometallurgical and/or hydrometallurgical treatment of a certain fraction, which is called black mass. Specialized hydrometallurgical treatment of the black mass can either lead to intermediate products that can be fed into the hydrometallurgical processes described in Section 3.2 or directly to high-grade products. In the following, examples for the different process routes are given.

3.3.1. Mechanical Treatment

In most cases, LIBs undergo a thermal treatment, e.g., pyrolysis, before mechanical treatment in order to reduce the energy content in a controlled way, to eliminate the organic components and to reduce the halogenic content [19,66–68]. This process usually takes place at around 500 °C and is limited by the melting point of Al (660 °C) [69,70]. Afterwards, the pyrolyzed batteries can be treated mechanically without fire hazards.

Mechanical treatment starts with comminution. Afterwards, the crushed material is sorted by its physical properties using unit operations such as sieving, sifting, magnetic, and eddy current separation. Common fractions are an Al/Cu foil fraction (conducting foils), coarse non-ferrous metals (Al from casings, Cu), coarse ferrous metals (casings, screws), and a fine fraction called black mass (active electrode materials) [10,22,71].

Alternatively, a few companies follow different approaches without thermal treatment in order to avoid the complex thermal pre-treatment, which requires comparatively high throughputs to be economic. In this case, specific safety measures are mandatory to prevent explosion and ignition during the mechanical treatment [72]. One possibility is crushing under inert atmosphere, e.g., N_2, CO_2, or Ar, and a subsequent removal/recovery of the volatile components, e.g., by vacuum distillation [73,74] or drying at moderate temperatures. However, possible drawbacks are high costs for volatile organic component (VOC) abatement, low black mass yields, and high levels of cross-contamination with black mass and organic components. The latter is mainly because of an insufficient detachment of the active materials from the foils resulting from the strong bonding, which cannot be destroyed by purely mechanical processes [70,75].

Another possibility is comminution in a solution, e.g., in a slightly alkaline medium [76,77]. These processes have to deal with organic wastewater pollution, corrosion, wet product fractions, and also low black mass yields.

With the exception of black mass, all product fractions can be fed into established industrial recycling processes. Black mass is a relatively new intermediate product on the market and the most valuable fraction of the mechanical processing due to the Co and Ni content. It contains mainly the active electrode materials, i.e., graphite and Li transition metal compounds containing Co, Ni, and Mn. Further components are the fluorine containing conducting salt or its degradation products and several impurities such as Cu, Al, and Fe. Concentrations ranges of major black mass components produced from layered oxide chemistries are given in Table 6.

Table 6. Concentration ranges of major black mass components produced from layered oxides chemistries (see Section 2), adapted from [69,78,79]. $LiPF_6$: lithium hexafluorophosphate, PVDF: polyvinylidene fluoride.

Elements	Content [wt %]	Origin	Appearance
Al	1–5	conducting foil, NCA	metallic, oxidic
Co	3–33	LCO, NMC, NCA	oxidic
Cu	1–3	conducting foil	metallic
Fe	0.1–0.3	casing, screws, etc.	metallic
Li	3.5–4	LCO, NMC, NCA, $LiPF_6$	oxidic
Mn	3–11	NMC	oxidic
Ni	11–26	NMC, NCA	oxidic
Graphite	ca. 35	anode	-
F	2–4	$LiPF_6$, PVDF	-
P	0.5–1	$LiPF_6$	-

In case of LFP, Fe und P can be significantly higher than shown in Table 6. Without thermal treatment, the concentrations will be lower due to dilution by the organic content (solvent residues, binder and separator residue).

3.3.2. Metallurgical Processing of Black Mass

The black mass can either be fed into pyrometallurgical routes, described in the previous Section 3.2, or directly treated by hydrometallurgical methods. Both approaches are pursued industrially. Due to its unique composition and highly variable chemistry, the black mass does not fit into most available metallurgical processes. Furthermore, besides Co and Ni, the recycling of most other elements is under research and development, and an industrial implementation still pending.

Compared to batteries, the black mass fits better in most pyrometallurgical processes due to the significantly reduced Al and organic content and higher Co and Ni concentrations. Nevertheless, the fluorine and Li content lead to corrosion problems, and the efficient recovery of Li is still an issue.

In hydrometallurgy, a distinction can be made between two alternatives. The first one is the production of intermediates by leaching and precipitation with the aim of producing a Co/Ni and a separate Li product for further processing in existing refineries. The second one is the direct production of high-grade products by more complex processes. Compared to pyrometallurgy, the hydrometallurgical treatment of black mass can enable the recovery of more materials, i.e., Mn, graphite (and Li). However, it faces challenges regarding fluorine, Mn control, an efficient recovery of

Li, and the production of a marketable graphite product. Additionally, the processes are sensitive to organics because of the resulting contamination of process water and possible interference with the organic phase of solvent extraction processes. Therefore, typically only black mass from processes, which include a thermal treatment, is used as feed material.

Production of Intermediates

Due to the limited amount of black mass in Europe, currently, no specialized processes to produce high-grade products from black mass are in operation. Instead, the production of intermediates in pilot plants or treatment in existing variable plants can be observed.

These processes are typically based on leaching and precipitation and focus on the separation of Co and Ni from Mn, as most Co and Ni refineries can tolerate only limited amounts of Mn [80]. If the black mass still contains graphite after roasting, graphite can be recovered after leaching [12]. In addition, the production of Mn and Li intermediates is possible but only carried out in a few cases due to economic reasons. Possible flowsheets based on common precipitation processes are presented in Figure 9.

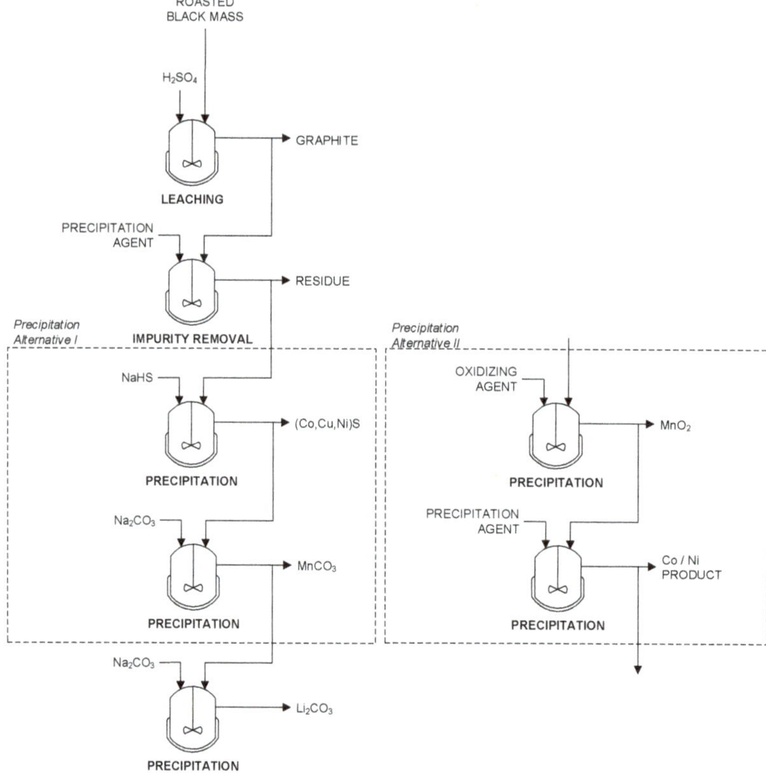

Figure 9. Simplified flowsheet for the hydrometallurgical route for the production of intermediates based on [12,80].

The leaching of active electrode material can be done with various acids [12]. In most cases, H_2SO_4 at elevated temperature assisted by H_2O_2 is used. Due to the fluorine content of most black masses, the formation of hydrofluoric acid takes place. This issue is discussed in the next section. Impurities such as Al and Fe can be precipitated as hydroxides. The separation of Co and Ni from Mn is a common challenge in the primary production of non-ferrous metals. In case of high Mn content, sulfide precipitation of Co and Ni is often performed utilizing the high selectivity of the reaction. Afterwards, Mn can be precipitated for example as carbonate (Figure 9, left) [66,80].

Alternatively, a selective precipitation of Mn as MnO_2 might be applied after leaching (Figure 9, right). Typically, the reaction requires a strong oxidizing agent such as ozone, Caro's acid, or hypochlorite. However, Co losses in the Mn precipitate can be very high: up to 25% according to the literature [80]. According to [80], better results can be achieved using SO_2/air mixtures. Co losses below 0.5% were reported. Co and Ni can then be precipitated, e.g., as carbonates. For the precipitation of Li, Li_2CO_3 is the preferred compound in case of intermediates [59].

Direct Production of High-Grade Products

Currently, the production of high-grade products from black mass and similar production waste only takes place in Asia due to the sufficient amount of available feed material, especially production wastes, as most battery producers are located in Asia [71,81]. Nevertheless, several European companies work on comparable processes, as European battery cell production is expected to increase significantly within the next years [82]. Most companies follow a similar process structure given by the chemistry of the elements and commercially available extractants for solvent extraction. A typical example with possible variants is shown in Figure 10.

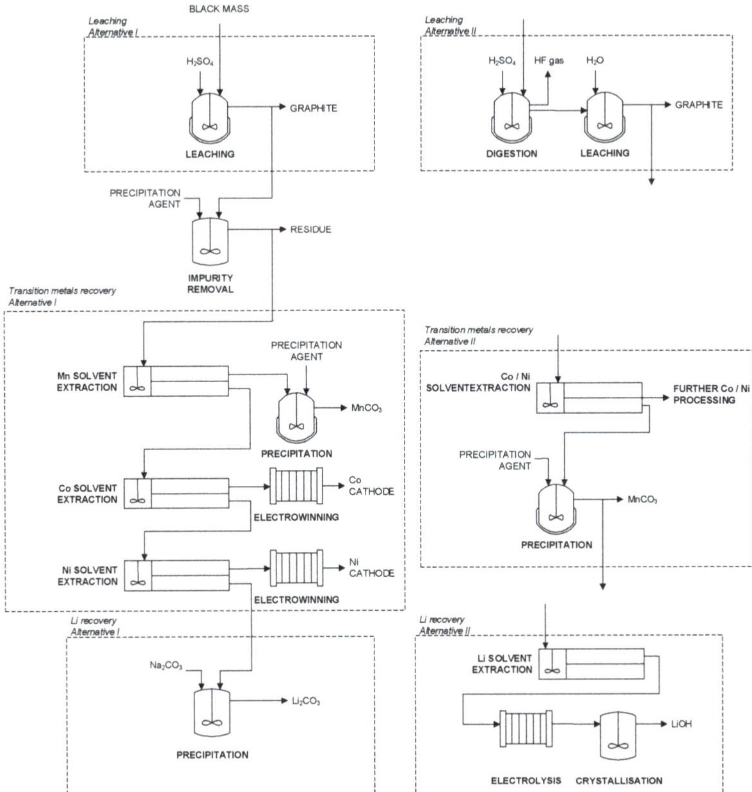

Figure 10. Simplified flowsheet of the process at JX Nippon Mining & Metals Corporation (left) based on [66] and process alternatives for leaching, transition metal recovery, and Li recovery (right) based on [83,84].

In comparison to the production of intermediates, these processes typically include solvent extraction to separate and purify the metals of interest. Therefore, the necessary investment and process complexity are high, but the processes enable a much higher product quality and added value.

For the production of high-grade products, Figure 10 shows simplified flowsheets. The left side shows the proposed process of JX Nippon Mining & Metals Corp. (Tokyo, Japan). It consists of leaching, precipitation, solvent extraction, and electrowinning. The metals are leached in H_2SO_4. In the following, impurities such as Al and Fe are removed via precipitation. Then, the purified leach liquor is fed into solvent extraction circuits to recover each transition metal individually. First, Mn is extracted with di-(2-ethylhexyl)phosphoric acid (DEHPA) and precipitated as manganese carbonate. After that, Co extraction is performed with a dialkyl phosphinic acid and Ni extraction is performed with a carboxylic acid. Co and Ni are recovered by electrowinning as pure metals. At last, Li is precipitated as Li_2CO_3 [66].

One concern when processing black mass is the formation of hydrofluoric acid during leaching due to the significant amount of fluorine in the black mass; see Table 6. The hydrofluoric acid requires measures concerning HSE. Moreover, it can cause corrosion and act as a complexing agent for certain metals such as Al, which leads to different chemical behaviors of these elements. However, so far, little has been published about the exact behavior of fluorine in such processes and possible measures. [66] suggests addressing this problem by precipitation, presumably as calcium fluoride (CaF). However, according to [85], the precipitation of fluoride as CaF is pH dependent and only takes place to a certain extent below pH 7. Furthermore, the co-precipitation of gypsum can be expected, and another cation is added to the system, leading to a higher impurity profile. The German battery recycling company Duesenfeld GmbH developed and patented an alternative approach, in which the black mass is digested with concentrated H_2SO_4 at elevated temperatures. In this process step, the fluorine is removed as hydrogen fluoride via the gas phase, and the metals are converted to water-soluble sulfates. The gas phase is scrubbed to remove hydrogen fluoride, and the dry digestion product is subsequently leached with water [83]. The advantages of an early-stage fluorine removal are opposed by higher requirements regarding reactor design and materials of construction. Another option to handle the fluorine content might be selective washing of the black mass under mildly alkaline conditions. The removal of halogens by selective washing is industrial practice, e.g., for Waelz oxide and fly ashes [86,87].

Another concern in the hydrometallurgical processing of black mass is the handling of Mn. From a circular economy point of view, the recovery of Mn is desirable, but it is challenging from an economic point of view. In the left flowsheet, the solvent extraction of Mn is a necessary purification step prior to the extraction of Co and Ni for chemical reasons. However, the revenues of the Mn product do not cover the costs of the solvent extraction process, which is therefore subsidized by the Co and Ni revenues. An alternative is the use of a dialkyl dithiophosphinic acid such as bis(2,4,4-trimethylpentyl)dithiophosphinic acid as extractant [80,88]. This configuration enables a selective Co and Ni extraction in the presence of Mn. After the extraction, Mn can be recovered by precipitation. However, the extractant is difficult to handle, as it is highly sensitive to oxidation and metal poisoning [89]. Therefore, its industrial use is limited.

In Figure 10, Co and Ni are won from the strip solution by electrowinning. Alternatively, the direct crystallisation of metal salts is possible. However, the starting materials for high-purity metal salts are often metal cathodes, as the electrowinning process provides an additional refining step [90].

In the left flowsheet, Li is directly recovered by precipitation as a carbonate. Due to the low Li concentration and an unfavorable Na:Li ratio, which results from the use of caustic soda as a neutralizing agent in solvent extraction, low yields and product quality can be expected. A possible solution to this problem might be the extraction of Li, using the newly developed phosphorus-based extractant Cyanex 936P (Solvay SA, Brussels, Belgium), which shows a high selectivity for Li over other alkaline metals [91]. From the strip solution, the Li can be precipitated as carbonate or directly converted into a LiOH product using electrolysis and crystallization [84,92]. This technology has been developed to extract Li from brines, and its application in recycling has not been proven so far. Due to the discussed difficulties, an earlier extraction of Li from the black mass is of high interest and is addressed in various research and development projects. Examples are the selective carbonation of Li [93,94] and the application of ion-selective membranes [95,96].

So far, graphite is not recovered in industrial processes, although it is often present in the black mass and can principally be recovered after leaching. Currently, several companies and research groups are working on the material recycling of graphite. Prerequisites for a graphite recovery are appropriate temperatures/atmosphere during the thermal treatment to prevent the loss of graphite and a sufficient removal of impurities to meet the specifications of established graphite products [12].

4. Discussion

In the following, the processes presented in Section 3 are going to be critically discussed from a European perspective based on the categories legislation, recovery rate, robustness, economics, and HSE. Compared to Asia, the battery production capacity in Europe is still small, but it is characterized by high growth rates. Hence, currently, only low amounts of production scrap are available. Due to projected production capacities of up to 2000 GWh in 2029 worldwide, thereof 500 GWh in Europe, sharply increasing quantities of production scraps can be expected within the next years. As a result of the complex production process, 5% to 10% of the production capacity end up as production scrap [97,98]. Within the next years, production scrap will be the main feed for LIB recycling plants. The available amount of end-of-life batteries is also currently low and dominated by consumer batteries (8200 t in 2020 according to [99]). The return flow of traction batteries is slowly increasing due to their long lifespan and is estimated to reach about 50,000 t/a in 2025 in Europe [100].

Consequently, the recycling in Europe currently takes place in small plants or co-processing plants. Therefore, the costs for recycling are still high, but they are expected to decrease significantly when higher recycling capacities are installed.

4.1. Legislation

In Europe, the legislative framework for the LIB recycling is defined by the Batteries Directive (2006/66/EG) from 2006. Due to the limited application of LIBs in the early 2000s, the recycling of LIBs from xEVs is not specifically addressed. Therefore, the Batteries Directive is currently under revision. From a processing point of view, an increased mass specific recovery rate as well as the introduction of specific recovery rates for individual metals would have the biggest influence. The current required recovery rate of 50 wt % is easily achieved in case of battery systems due to a relatively high fraction of peripheral materials such as the Al casing; see Figure 2.

In general, pyrometallurgy with subsequent hydrometallurgy will be more affected by increasing requirements regarding mass specific recovery rates due to their focus on Co, Cu, and Ni and the loss of metallic Al, Mn, non-metallic components, and Li in some cases (see Section 4.2.1). In contrast, a mechanical treatment enables a higher material recycling rate. Therefore, new process combinations, e.g., the separation of Al and Fe casing prior to pyrometallurgy, might be necessary in the future to balance the individual advantages and disadvantages.

One further legislative framework, which might influence the recycling of LIBs, is the European Union Emission Trading Scheme. Currently, the influence is low but it is expected to grow with increasing CO_2 prices. However, it is unclear which price level will be necessary for a significant steering effect, which processes will have a competitive advantage, and how the CO_2 price will influence the competitiveness of the European LIB recycling industry.

4.2. Recovery Rate

As presented in Section 2, batteries contain various metallic and non-metallic components. Currently, due to economic and thermodynamic reasons, many processes focus on the high recovery rates of Co, Cu, and Ni. Li recovery is still a challenge due to thermodynamic reasons despite its economic value and political interest. There is less focus on Al and Mn, but the first one is partly recovered in mechanical processes. So far, non-metallic components such as graphite and the solvents of the electrolyte are not recovered with a few exceptions. Depending on future legislations, the recovery of non-metallic components might be of interest to comply with required mass specific recovery rates, especially in case of battery chemistries with low Co and Ni contents such as LFP.

4.2.1. Pyrometallurgical with Hydrometallurgical Processing

Pyrometallurgical processes achieve high yields for Co, Cu, and Ni (>95%) [55,101]. Li can only be recovered in dedicated processes, in which a concentration of Li in the slag is realized. Consequently, co-processing plants have major disadvantages if a specific Li recovery rate is required by law. Al and non-metallic components are utilized as reductant and fuel and therefore substitute primary energy sources. Mn mainly reports to the slag, which can be used as a construction material or dumped.

In hydrometallurgy, metal losses for Co, Cu, and Ni are very low (<5%), which results in high overall recovery rates for these process routes. For Li recovery from the slag, no industrial process data are available. The literature indicates possible recovery rates for Li from the slag of around 90% [65]. However, the overall Li recovery is presumably lower as Li is partially fumed in pyrometallurgy according to [56].

4.2.2. Mechanical with Metallurgical Processing

Mechanical processes with subsequent hydrometallurgy enable the recovery of more elements and materials. Typically, mechanical processes produce different ferrous and non-ferrous metal concentrates. A high black mass yield is of paramount importance for economic reasons and in order to minimize the cross-contamination of other product fractions with carcinogenic Co/Ni-containing dusts. According to [75], processes with thermal pre-treatment of LIBs prior to mechanical processing reach black mass yields of up to 95%, whereas the yield in processes without thermal pre-treatment is significantly lower.

Whereas Al and Cu concentrates can be fed into established recycling processes, the black mass is a new type of concentrate, which requires at least the adaptation of existing processes due to the unique combination of metals and the fluorine content.

If the black mass is fed into pyrometallurgy, treatment follows the above (see Section 4.2.1) described processes with its respective advantages and disadvantages. In contrast, black mass processing in hydrometallurgy offers the possibility to recover more materials, especially graphite, Mn, and Li. However, the recovery of graphite and Mn are not established industrially due to economic reasons. Li recovery takes place in some plants, especially in Asia [102]. Industrial recovery rates of Li are not known but presumably lower than those of Ni or Co.

4.3. Robustness

High process robustness is a prerequisite for the long-term economic success of large-scale installations. As a result of the constant development of LIBs, the processes must be able to deal with changes in LIB design, chemistry, and size. Furthermore, especially in the case of consumer batteries, recycling processes need to be robust toward missorting. Another aspect concerning the process robustness is the required pre-treatment, especially discharging, as some batteries cannot be discharged for safety reasons, e.g., damaged batteries, or economic reasons, e.g., consumer batteries.

4.3.1. Pyrometallurgical with Hydrometallurgical Processing

Pyrometallurgical processes are generally robust. They are usually able to treat all kinds of LIB scraps within certain limitations, coming from LIB production as well as end-of-life. Discharging is not mandatory and only pursued by some battery recyclers in case of battery systems to enable a safe dismantling. Exceptions are LIB chemistries with low Co and Ni content, as these processes are primarily designed for the recovery of Co, Ni, and Cu and battery modules, which exceed a certain size and weight limit, depending on the applied furnace technology. In this case, a mechanical pre-treatment is necessary in order to achieve a size reduction, which allows feeding without damaging the furnace. Most critical elements are safely removed via the gas phase—for example, halogens or volatile toxic heavy metals such as mercury (Hg) or cadmium (Cd) from missorted batteries. Other metals of low economic value and high affinity to oxygen such as Mn or Ti are transferred to the slag phase. Nevertheless, the halogenic and Li content are challenging with respect to corrosion. Due to the high energy content, the share of LIBs in the feed is limited. However, mixing different feed materials allows the production of a homogeneous alloy or matte phase for the subsequent hydrometallurgical processing. Due to the above-mentioned pyrometallurgical separation, changes in LIB chemistry are less likely to affect the hydrometallurgical treatment.

4.3.2. Mechanical with Metallurgical Processing

Mechanical processes combined with metallurgical processes are generally less robust. In these kinds of process routes, the higher recovery rates, especially of non-metallic materials, often lead to a high process sensitivity. Therefore, processes with a thermal treatment prior to mechanical treatment are predominant.

If current LIBs undergo thermal treatment prior to mechanical processing, sufficient yields and product qualities are achieved. Future LIB generations might require process adaptations. Without thermal treatment, the organic content leads to a more complex mechanical treatment with respect to the risks of fire and explosion. In the case of dry processing, discharging of the LIBs is mandatory. Furthermore, product yield and quality might suffer from organic contamination and insufficient detachment of the electrode coatings. In both cases, the missorting of batteries leads to the contamination of products, which is especially critical in case of Cd, Hg, and Pb.

The direct treatment of the black mass in hydrometallurgy is more challenging than the treatment of an alloy or a matte. The reasons are higher levels of contaminants and a less homogenous feed material. Organic contaminants are typically removed via thermal treatment as organic components can interfere with solvent extraction and require additional wastewater treatment.

In contrast to pyrometallurgical processing, mechanical processing achieves a lower separation of elements, leading to higher separation effort in hydrometallurgy. Especially, fluorine, Mn, and trace elements lead to additional process steps to achieve high product qualities. Heterogenic feed material makes process control more challenging, especially in solvent extraction and with respect to impurity control. Due to the changing chemistry of batteries and various doping elements (see Section 2), there is an elevated risk for the enrichment of contaminants in the solvent extraction circuits.

4.4. Ecomonics

Currently, in Europe, most LIBs have a negative market value due to new and complex processes, high research and development expenses, and small plants. In the upcoming years, the market value of LIBs is expected to increase in case of Co- and Ni-containing chemistries because of the growing number of competitors operating larger plants. LFP batteries will presumably continue to have a negative market value.

Possible revenues mainly result from Co, Ni, Cu, and Li while other materials are of minor importance; see Table 7.

Figure 11 shows exemplarily possible revenues from one ton of black mass assuming a NMC 6:2:2 chemistry and 100% yield of each element based on the prices given in Table 7. For graphite, a low-quality product was assumed. It can be clearly seen that Co, Ni, and Li dominate the revenues under current market conditions. Other elements and components are only of minor importance or even generate costs, which is the case with fluorine and organics.

Table 7. Prices of materials based on an average 03/2019–02/2020 [103,104].

Material	Price Average [US$/t]	Quality
Al	1773	high grade primary
graphite (industrial use)	300–500	amorphous (<106 µm), 94–97%
graphite (battery applications)	2500–3000	large flakes (150–300 µm), >99%
Co	38,034	electrolytic, 99.8%
Cu	5965	grade A
Li_2CO_3	11,900	min. 99–99.5%
Mn	1776	electrolytic, 99.7%
Ni	14,085	primary 99.8%

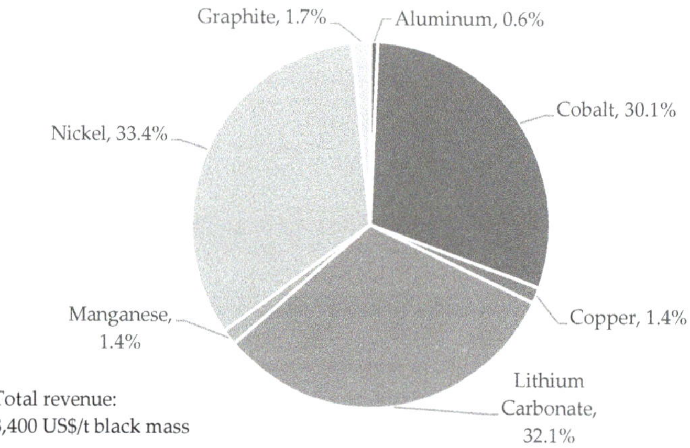

Figure 11. Theoretical revenue distribution of black mass with NMC 622.

4.4.1. Pyrometallurgical with Hydrometallurgical Processing

Generally, large-scale pyrometallurgical processes are highly cost efficient to treat complex secondary raw materials. In case of LIBs, the available quantities are presently low. Therefore, co-processing has a larger market share than the treatment in dedicated plants. In the future, this is expected to change because dedicated plants are able to enrich Li in the slag in addition to Co, Ni, and Cu in the alloy/matte and are better adapted to the specific challenges. Nevertheless, the economic Ni and Co dependency of such processes is high. Under current market conditions, the use of carbon and organics as reductants and fuel as well as the separation of Mn from the other transition metals seems to have economic advantages.

In the subsequent hydrometallurgical treatment of the intermediates, Co, Ni, and Cu can be recovered in established plants. At present, Li recovery from the slag is not established industrially but it is expected to be introduced with increasing available amounts. The costs for Li recovery are unknown, but based on the described processes, it can be assumed that the costs will be in the range of spodumene processing or even lower. According to [105], the processing costs from spodumene concentrates are around 4500 US$/t lithium carbonate equivalent (LCE).

4.4.2. Mechanical with Metallurgical Processing

Compared to the process described above, a higher number of materials can be recovered in mechanical processes in principle. However, it is unclear whether this translates to economic advantages under future market conditions.

In most cases, a cost-intensive thermal removal of organics takes place before comminution. Compared to smelting, the metals are concentrated but not separated.

The main advantages of this approach are the (partial) recovery of Al and Fe, which leads to additional revenues. In addition, the separation of Cu from the transition metals in the black mass might have economic advantages. However, Mn reports to the black mass and is currently considered as an impurity in hydrometallurgy due to its low value (see Table 7) and expensive separation from Co and Ni. Solvent and graphite recovery have been demonstrated at the pilot scale, but so far, there are no established markets for these products, and the large-scale economic viability is unknown.

The black mass is still a comparatively new intermediate product, which requires dedicated hydrometallurgical treatment due to its complex composition and to recover all components. The installation of these plants requires high investments and sufficient feed material. Currently, such plants are only available in Asia due to the high amounts of production scraps in this region [102].

4.5. Health, Safety, and Environment

In most countries, scrap LIBs are classified as hazardous waste. The main reasons are the electrical and chemical energy content, the carcinogenic Co- and Ni-containing cathode active material, as well as the toxic, corrosive, and hazardous conducting salt [9]. All of these hazards need to be handled in LIB recycling.

Regarding life-cycle assessment, available studies indicate positive results of LIB recycling in comparison to the primary production in most impact categories [106]. However, most of the results are based on laboratory or pilot-scale process data and need verification at an industrial scale. Moreover, comparative studies would be of interest but are difficult to generate due to the data sensitivity of industrial processes, complex process routes, and co-processing with other primary or secondary materials.

4.5.1. Pyrometallurgical with Hydrometallurgical Processing

As pyrometallurgical processes are high-temperature processes, the high energy content of LIBs does not pose specific risks from an HSE point of view. Halogens and Co- and Ni-containing dusts are handled by an extensive off-gas treatment.

In the further treatment of the alloy/matte, dust formation needs to be addressed during comminution. In the design of hydrometallurgical plants, the high water hazard class of Co and Ni salts has to be considered.

4.5.2. Mechanical with Metallurgical Processing

In most cases, thermal deactivation is used prior to mechanical processing in order to remove the high energy and organic content of LIBs in a controlled way to avoid fire and explosion during further processing. If thermal deactivation is not applied, special measures during the mechanical processing of the LIBs are necessary, see Section 3.3, including the capture of VOCs.

During mechanical processing, dust control is of paramount importance to handle the carcinogenic dusts. In addition, cross-contamination of Ni- and Co-containing black mass in other fractions can be an issue.

Fluorine is concentrated in the black mass and therefore transferred to the metallurgical treatment, where it needs to be addressed. Otherwise, for the hydrometallurgical treatment, the same measures apply as described above (see Section 4.5.1).

5. Conclusions and Outlook

In the industrial recycling of LIBs, there are two main process routes pursued. The first plants are in operation with typical annual capacities of a few thousand tons. Within the next decade, recycling capacities are expected to increase significantly to meet the growing demand, which will probably lead to decreasing processing costs as already observed in Asia [102].

Although the principal process routes are already defined, there are still open questions regarding the next generation of plants.

Currently, battery modules are often directly fed into the pyrometallurgical routes. However, a mechanical conditioning might be beneficial to reduce the Al and Fe content of the feed material and to simplify the feeding process by a size reduction of the modules. Al and Fe can be recycled, and their reduction is beneficial to the energy and mass balance of most pyrometallurgical processes.

The optimal processing of the black mass is still subject to ongoing discussions. Both process routes have their challenges and opportunities with respect to fluorine control and the handling of graphite, Li, and Mn. The further development will be strongly influenced by economic and regulatory incentives.

Efficient Li recovery is still a challenge in all process routes, and further research and development is required. Besides Co and Ni, Li is one of the main value carriers, and a solution for efficient Li recovery could be a competitive advantage, especially in case of battery chemistries with low Co and/or Ni contents.

The recovery of non-metallic components is in an early stage, and the technical and economic feasibility is still uncertain. Challenges derive from inhomogeneous feed, product quality requirements, and currently low revenues.

A widely ignored problem is the recycling of LFP batteries. Although their market share is comparatively low in Europe, they are present on the market and consequently appear in recycling. The current industrial practice is to co-process them with layered oxide chemistries to a limited extend (<20%). However, if their share increases and/or with more stringent legislative requirements, the introduction of specific processes might become necessary despite their low value.

Due to the complex process chains, most companies do not cover all process steps. Instead, the formation of consortia is observed. Typically, pre-treatment, metallurgy, refining, and cathode material production are operated by different companies. At present, mainly activities of European companies are observed in Europe. However, Asian companies might enter the market as currently seen in LIB production. Due to the experience advantage of some companies, they have to be considered as serious competitors.

Especially in the metallurgical processing, co-processing is pursued at various stages to take advantage of large-scale installations and to homogenize the feed material. A further advantage of using established process routes is an easier entry of recycling material into the loop, as no additional certification of the products is necessary. However, co-processing in pyrometallurgy without Li-enrichment in the slag phase might lose in importance with increasing significance of Li recovery. Furthermore, more dedicated installations can be expected with the growing availability of LIB scrap.

In conclusion, within the last years, significant progress has been achieved in industrial LIB recycling in Europe. Nevertheless, the developments are very dynamic, and further progress in many areas can be expected in the upcoming years.

Author Contributions: The authors conceived and wrote the article together. All authors have read and agreed to the published version of the manuscript.

Funding: This research received no external funding.

Acknowledgments: The authors acknowledge support by Open Access Publishing Fund of Clausthal University of Technology.

Conflicts of Interest: The authors declare no conflict of interest.

References

1.   Korthauer, R. *Handbuch Lithium-Ionen-Batterien*; Imprint: Springer Vieweg: Berlin/Heidelberg, Germany, 2013; ISBN 978-3-642-30652-5.
2.   Petroff, A. Carmakers and Big Tech Struggle to Keep Batteries Free from Child Labor. Available online: https://money.cnn.com/2018/05/01/technology/cobalt-congo-child-labor-car-smartphone-batteries/index.html (accessed on 22 May 2020).

3. Frankel, T.C. The Cobalt Pipeline: Tracing the path from deadly hand-dug mines in Congo to consumers' phones and laptops. *The Washington Post [Online]*. 30 September 2016. Available online: https://www.washingtonpost.com/graphics/business/batteries/congo-cobalt-mining-for-lithium-ion-battery/ (accessed on 22 May 2020).

4. Mau, K. Dreckige Rohstoffe für saubere Autos. *Zeit Online [Online]*. 11 December 2019. Available online: https://www.zeit.de/mobilitaet/2019-11/elektroautos-kobalt-lithium-batterie-akkus-rohstoffe-umweltschutz/komplettansicht (accessed on 22 May 2020).

5. Draper, R. This metal is powering today's technology—At what price?: As demand soars for powerful batteries, Bolivia dreams of striking it rich by tapping its huge lithium deposit. But will its people benefit? *National Geographic Magazine [Online]*. Available online: https://www.nationalgeographic.com/magazine/2019/02/lithium-is-fueling-technology-today-at-what-cost/ (accessed on 22 May 2020).

6. CBS News. CBS News finds children mining cobalt for batteries in the Congo. *CBC Interactive Corporation [Online]*. 5 March 2018. Available online: https://www.cbsnews.com/news/cobalt-children-mining-democratic-republic-congo-cbs-news-investigation/ (accessed on 22 May 2020).

7. Katwala, A. The spiralling environmental cost of our lithium battery addiction: As the world scrambles to replace fossil fuels with clean energy, the environmental impact of finding all the lithium required could become a major issue in its own right. *Wired on Energy [Online]*. 5 August 2018. Available online: https://www.wired.co.uk/article/lithium-batteries-environment-impact/ (accessed on 22 March 2020).

8. Olivetti, E.A.; Ceder, G.; Gaustad, G.G.; Fu, X. Lithium-Ion Battery Supply Chain Considerations: Analysis of Potential Bottlenecks in Critical Metals. *Joule* **2017**, *1*, 229–243. [CrossRef]

9. Lisbona, D.; Snee, T. A review of hazards associated with primary lithium and lithium-ion batteries. *Process. Saf. Environ. Prot.* **2011**, *89*, 434–442. [CrossRef]

10. Elwert, T.; Römer, F.; Schneider, K.; Hua, Q.; Buchert, M. Recycling of Batteries from Electric Vehicles. In *Behaviour of Lithium-Ion Batteries in Electric Vehicles: Battery Health, Performance, Safety, and Cost*; Pistoia, G., Liaw, B., Eds.; Springer: Cham, Switzerland, 2018; pp. 289–321. ISBN 978-3-319-69950-9.

11. Harper, G.; Sommerville, R.; Kendrick, E.; Driscoll, L.; Slater, P.; Stolkin, R.; Walton, A.; Christensen, P.; Heidrich, O.; Lambert, S.; et al. Recycling lithium-ion batteries from electric vehicles. *Nature* **2019**, *575*, 75–86. [CrossRef] [PubMed]

12. Larouche, F.; Tedjar, F.; Amouzegar, K.; Houlachi, G.; Bouchard, P.; Demopoulos, G.P.; Zaghib, K. Progress and Status of Hydrometallurgical and Direct Recycling of Li-Ion Batteries and Beyond. *Materials (Basel)* **2020**, *13*, 801. [CrossRef]

13. Li, L.; Zhang, X.; Li, M.; Chen, R.; Wu, F.; Amine, K.; Lu, J. The Recycling of Spent Lithium-Ion Batteries: A Review of Current Processes and Technologies. *Electrochem. Energ. Rev.* **2018**, *1*, 461–482. [CrossRef]

14. Pinegar, H.; Smith, Y.R. Recycling of End-of-Life Lithium Ion Batteries, Part I: Commercial Processes. *J. Sustain. Metall.* **2019**, *5*, 402–416. [CrossRef]

15. Pinegar, H.; Smith, Y.R. Recycling of End-of-Life Lithium-Ion Batteries, Part II: Laboratory-Scale Research Developments in Mechanical, Thermal, and Leaching Treatments. *J. Sustain. Metall.* **2020**, *6*, 142–160. [CrossRef]

16. Werner, D.; Peuker, U.A.; Mütze, T. Recycling Chain for Spent Lithium-Ion Batteries. *Metals* **2020**, *10*, 316. [CrossRef]

17. Velázquez-Martínez; Valio; Santasalo-Aarnio; Reuter; Serna-Guerrero. A Critical Review of Lithium-Ion Battery Recycling Processes from a Circular Economy Perspective. *Batteries* **2019**, *5*, 68. [CrossRef]

18. Zeng, X.; Li, J.; Singh, N. Recycling of Spent Lithium-Ion Battery: A Critical Review. *Crit. Rev. Environ. Sci. Technol.* **2014**, *44*, 1129–1165. [CrossRef]

19. Träger, T.; Friedrich, B.; Weyhe, R. Recovery Concept of Value Metals from Automotive Lithium-Ion Batteries. *Chem. Ing. Tech.* **2015**, *87*, 1550–1557. [CrossRef]

20. Julien, C.; Mauger, A.; Vijh, A.; Zaghib, K. Lithium Batteries. In *Lithium Batteries*; Julien, C., Mauger, A., Vijh, A., Zaghib, K., Eds.; Springer International Publishing: Cham, Switzerland, 2016; pp. 29–68. ISBN 978-3-319-19107-2.

21. Lazuen, J. Batteries: The true drivers behind LFP demand – new safety standards, costs, IP rights, ESG & simplified battery pack designs. *Roskill News [Online]*. 25 June 2020. Available online: https://roskill.com/news/batteries-the-true-drivers-behind-lfp-demand-new-safety-standards-costs-ip-rights-esg-simplified-battery-pack-designs/ (accessed on 8 July 2020).

22. Kwade, A.; Diekmann, J. *Recycling of Lithium-Ion Batteries: The LithoRec Way*; Springer International Publishing: Cham, Switzerland, 2018; ISBN 9783319705712.

23. Nitta, N.; Wu, F.; Lee, J.T.; Yushin, G. Li-ion battery materials: Present and future. *Mater. Today* **2015**, *18*, 252–264. [CrossRef]

24. Omenya, F.; Chernova, N.A.; Zhou, H.; Siu, C.; Whittingham, M.S. Comparative Study Nickel Rich Layered Oxides: NMC 622, NMC 811 and NCA Cathode Materials for Lithium Ion Battery. *Meet. Abstr.* **2018**, *MA2018-01*, 531.

25. Schmuch, R.; Wagner, R.; Hörpel, G.; Placke, T.; Winter, M. Performance and cost of materials for lithium-based rechargeable automotive batteries. *Nat. Energy* **2018**, *3*, 267–278. [CrossRef]

26. LG Chem Europe GmbH. Website 2019. Available online: https://www.lgchem.com/product/PD00000066 (accessed on 28 November 2019).

27. Susai, F.A.; Sclar, H.; Shilina, Y.; Penki, T.R.; Raman, R.; Maddukuri, S.; Maiti, S.; Halalay, I.C.; Luski, S.; Markovsky, B.; et al. Horizons for Li-Ion Batteries Relevant to Electro-Mobility: High-Specific-Energy Cathodes and Chemically Active Separators. *Adv. Mater.* **2018**, *30*, e1801348. [CrossRef] [PubMed]

28. Costa, C.M.; Lanceros-Méndez, S. *Printed Batteries: Materials, Technologies and Applications*; John Wiley & Sons: Hoboken, NJ, USA, 2018; ISBN 978-111-928-7889.

29. Blomgren, G.E. The Development and Future of Lithium Ion Batteries. *J. Electrochem. Soc.* **2017**, *164*, A5019–A5025. [CrossRef]

30. Sandhya, C.P.; John, B.; Gouri, C. Lithium titanate as anode material for lithium-ion cells: A review. *Ionics* **2014**, *20*, 601–620. [CrossRef]

31. Berhaut, C.L.; Porion, P.; Timperman, L.; Schmidt, G.; Lemordant, D.; Anouti, M. LiTDI as electrolyte salt for Li-ion batteries: Transport properties in EC/DMC. *Electrochim. Acta* **2015**, *180*, 778–787. [CrossRef]

32. Arkema Group. *PVDF Electrode Binders & Separator Coatings*. 2019. Available online: https://www.extrememberials-arkema.com/en/markets-and-applications/renewable-energy/lithium-ion-battery/?_ga=2.35991904.1158937935.1575031365-829431959.1575030587 (accessed on 29 November 2019).

33. Nayak, P.K.; Erickson, E.M.; Schipper, F.; Penki, T.R.; Munichandraiah, N.; Adelhelm, P.; Sclar, H.; Amalraj, F.; Markovsky, B.; Aurbach, D. Review on Challenges and Recent Advances in the Electrochemical Performance of High Capacity Li- and Mn-Rich Cathode Materials for Li-Ion Batteries. *Adv. Energy Mater.* **2018**, *8*, 1702397. [CrossRef]

34. Zhan, C.; Wu, T.; Lu, J.; Amine, K. Dissolution, migration, and deposition of transition metal ions in Li-ion batteries exemplified by Mn-based cathodes—A critical review. *Energy Environ. Sci.* **2018**, *11*, 243–257. [CrossRef]

35. Otero, M.; Heim, C.; Leiva, E.P.M.; Wagner, N.; Friedrich, A. Design-Considerations regarding Silicon/Graphite and Tin/Graphite Composite Electrodes for Lithium-Ion Batteries. *Sci. Rep.* **2018**, *8*, 15851. [CrossRef] [PubMed]

36. Europäische Union. Richtlinie 2006/66/EG des Europäischen Parlaments und des Rates: L 266/1. 2006. Available online: http://gww.de/wp-content/uploads/Batterierichtlinie.pdf (accessed on 4 July 2018).

37. European Commission. Batterie & Accumulators Waste Streams. *Legislation*. Available online: https://ec.europa.eu/environment/waste/batteries/legislation.htm (accessed on 23 April 2020).

38. EUWID Europäischer Wirtschaftsdienst. EU Battery Directive to be revised in Q4 2020. Available online: https://www.euwid-recycling.com/news/policy/single/Artikel/eu-battery-directive-to-be-revised-in-q4-2020.html (accessed on 22 May 2020).

39. Industry Standard Conditions for Comprehensive Utilization of Waste Power Batteries of New Energy veHicles (2019 Edition). Available online: http://www.miit.gov.cn/n1146295/n1652858/n1652930/n4509607/c7595282/content.html (accessed on 2 January 2020).

40. Interim Measures for Management of Industry Standard Announcement for Comprehensive Utilization of Waste Power Batteries of New Energy Vehicles (2019 Edition). Available online: http://www.miit.gov.cn/n1146295/n1652858/n1652930/n4509607/c7595282/content.html (accessed on 2 January 2020).

41. Hiskey, B. *Metallurgy, Survey. Kirk-Othmer Encyclopedia of Chemical Technology*; John Wiley & Sons, Inc: Hoboken, NJ, USA, 2000; ISBN 0471238961.

42. Schubert, H. *Aufbereitung Fester Mineralischer Rohstoffe*; Dt. Verl. für Grundstoffindustrie: Leipzig, Germany, 1989; ISBN 3-342-00289-1.

43. Schubert, H. *Sortierprozesse: Dichtesortierung, Sortierung in Magnetfeldern, Sortierung in Elektrischen Feldern, Flotation, Klauben, Sortierung nach Mechanischen und nach Thermischen Eigenschaften*; Dt. Verl. für Grundstoffindustrie: Stuttgart, Germany, 1996; ISBN 3342005556.

44. Habashi, F. *Principles of Extractive Metallurgy: Volume 1: General Principles*; Gordon and Breach Science Publishers Inc.: New York, NY, USA, 1969; ISBN 978-0677017709.

45. Habashi, F. *A Textbook of Hydrometallurgy*; Librairie des Presses de l'Université Laval: Québec, QC, Canada, 1993; ISBN 2980324701.

46. Gupta, C.K.; Mukherjee, T.K. *Hydrometallurgy in Extraction Processes, Volume I*; Routledge: Boca Raton, FL, USA, 1990; ISBN 9780849368042.

47. Nickelhuette Aue GmbH. Website. 2019. Available online: http://www.nickelhuette-aue.de/ (accessed on 20 August 2019).

48. Glencore. Sudbury Integrated Nickel Operations Website. 2019. Available online: https://www.sudburyino.ca/en/Pages/home.aspx (accessed on 18 December 2019).

49. Neumann, M.; Kuhnert, C. State-of-the-Art Recycling of Ni, Co, Cu and V. In *Production and Recycling of Non-Ferrous Metals: Saving Resources for a Sustainable Future: European Metallurgical Conference EMC 2017: June 25-28, Leipzig, Germany: Proceedings of EMC 2017*; GDMB Verlag GmbH: Clausthal-Zellerfeld, Germany, 2017; pp. 115–122. ISBN 978-3-940276-72-8.

50. Crundwell, F.K.; Moats, M.S.; Ramachandran, V.; Robinson, T.G.; Davenport, W.G. Hydrometallurgical Production of High-Purity Nickel and Cobalt. In *Extractive Metallurgy of Nickel, Cobalt and Platinum Group Metals*; Elsevier: Amsterdam, The Netherlands, 2011; pp. 281–299. ISBN 9780080968094.

51. Nikkelverk. Nikkelverk A Glencore Company Website. 2019. Available online: https://www.nikkelverk.no/en/about-us/production/Pages/flowchart.aspx (accessed on 12 December 2019).

52. Umicore AG & Co. KG. Umicore Website. 2019. Available online: https://www.umicore.de/de/industries/automotive/umicore-battery-recycling/ (accessed on 24 May 2019).

53. Manthey, N. Umicore to Ramp up Recycling of Electric Car Batteries. 2018. Available online: https://www.electrive.com/2018/08/07/umicore-to-ramp-up-recycling-of-electric-car-batteries/ (accessed on 7 August 2020).

54. Cheret, D.; Santen, S. Battery Recycling. Patent US000007169206A2, 30 January 2007.

55. Cheret, D.; Santén, S. Battery Recycling. Patent EP000001589121A1, 31 December 2018.

56. Verscheure, K.; Campforts, M.; van Camp, M. Process for the Valorization of Metals from Li-Ion Batteries. U.S. Patent 8,840,702, 23 September 2014.

57. Suetens, T.; van Hoerebeek, D. Process for Recycling Cobalt-Bearing Materials. Patent WO002011035915A1, 26 April 2018.

58. Verhaeghe, F.; Goubin, F.; Yazicioglu, B.; Schurmans, M.; Thijs, B.; Haesebroek, G.; Tytgat, J.; van Camp, M. Valorisation of battery recycling slags. In *Proceedings of the Second International Slag Valorisation Symposium*; The Transitition To Sustainable Materials Management, Katholieke Universiteit Leuven, 18.-20.04.2011; Jones, P.T., Ed.; Umicore: Brussels, Belgium, 2011.

59. Wietelmann, U.; Steinbild, M. Lithium and Lithium Compounds. In *Ullmann's Encyclopedia of Industrial Chemistry*; Wiley: Chichester, UK, 2010; ISBN 9783527306732.

60. Treffer, F. *Entwicklung eines Realisierbaren Recyclingkonzeptes für die Hochleistungsbatterien zukünftiger Elektrofahrzeuge-Lithium-Ionen Batterierecycling Initiative-LiBRi: Verbundprojekt; Gemeinsamer Abschlussbericht des Konsortiums; Berichtszeitraum: 01.09.2009 bis 30.09.2011*; Technische Informationsbibliothek u. Universitätsbibliothek: Hanover, Germany; Umicore AG & Co.KG: Hanau, Germany, 2011.

61. Quix, M.; van Horebeek, D.; Suetens, T. Lithium-rich Metallurgical Slag. Patent WO002017121663A1, 20 July 2017.

62. Albermarle. Spodumene Concentrate, SC 6.8: Technical Data Sheet. Available online: https://www.albemarle.com/storage/components/T402211.PDF (accessed on 26 January 2017).

63. Albermarle. Spodumene Concentrate, SC 7.2 premium: Technical Data Sheet. Available online: https://www.albemarle.com/storage/components/T402209.PDF (accessed on 26 January 2017).

64. Elwert, T.; Goldmann, D.; Schirmer, T.; Strauß, K. Recycling von Li-Ionen-Traktionsbatterien: Das Projekt LiBRi. In *Recycling und Rohstoffe*; Thomé-Kozmiensky, K.J., Goldmann, D., Eds.; TK-Verl.: Neuruppin, Germany, 2012; pp. 679–690. ISBN 9783935317818.

65. Oosterhof, H.; Dupont, D.; Drouard, W. Process for the Recovery of Lithium. *Patent WO201818 4876A1*, 2018.

66. Haga, Y.; Saito, K.; Hatano, K. Waste Lithium-Ion Battery Recycling in JX Nippon Mining & Metals Corporation. In *Materials Processing Fundamentals 2018*; Lambotte, G., Lee, J., Allanore, A., Wagstaff, S., Eds.; Springer International Publishing: Cham, Switzerland, 2018; pp. 143–147. ISBN 978-3-319-72130-9.
67. Nottez, E. RecLionBat Project: Lithium-ION Battery Recycling; Viviez. Available online: http://ec.europa.eu/environment/life/project/Projects/index.cfm?fuseaction=home.showFile&rep=file&fil=LIFE05_ENV_F_000080_LAYMAN.pdf (accessed on 25 October 2019).
68. Arnberger, A. Recycling von Lithium-Ionen-Batterien. Available online: https://www.ffg.at/sites/default/files/allgemeine_downloads/thematische%20programme/Produktion/6_arnberger_recycling_von_lithium-ionen-batterien.pdf (accessed on 8 July 2020).
69. Wang, H.; Friedrich, B. Development of a Highly Efficient Hydrometallurgical Recycling Process for Automotive Li–Ion Batteries. *J. Sustain. Metall.* **2015**, *1*, 168–178. [CrossRef]
70. Zhang, G.; He, Y.; Feng, Y.; Wang, H.; Zhang, T.; Xie, W.; Zhu, X. Enhancement in liberation of electrode materials derived from spent lithium-ion battery by pyrolysis. *J. Clean. Prod.* **2018**, *199*, 62–68. [CrossRef]
71. Vezzini, A. *Manufacturers, Materials and Recycling Technologies. Lithium-Ion Batteries*; Elsevier: Amsterdam, The Netherlands, 2014; pp. 529–551. ISBN 9780444595133.
72. Diekmann, J.; Hanisch, C.; Froböse, L.; Schälicke, G.; Loellhoeffel, T.; Fölster, A.-S.; Kwade, A. Ecological Recycling of Lithium-Ion Batteries from Electric Vehicles with Focus on Mechanical Processes. *J. Electrochem. Soc.* **2017**, *164*, A6184–A6191. [CrossRef]
73. Hanisch, C. Recycling Method for Treating Used Batteries, in Particular Rechargeable Batteries, and Battery Processing Installation. Patent WO002018073101A1, 26 April 2018.
74. Hanisch, C. Recycling Method for Treating Used Batteries, in Particular Rechargeable Batteries, and Battery Processing Installation. Patent EP000003312922A1, 25 April 2018.
75. Sojka, R.T. Sichere Aufbereitung von Lithium-Ionen-basierten Batterien durch thermische Konditionierung. In *Recycling und Sekundärrohstoffe, Band 13*; Thomé-Kozmiensky, E., Holm, O., Friedrich, B., Goldmann, D., Eds.; Thomé-Kozmiensky Verlag GmbH: Nietwerder, Germany, 2020; pp. 506–523. ISBN 978-3-944310-51-0.
76. Retriev Technologies Inc. Retriev Website. Available online: https://www.retrievtech.com/ (accessed on 15 June 2020).
77. PROMESA GmbH & Co. KG. PROMESA Website. Available online: http://promesa-tec.de/ (accessed on 15 June 2020).
78. Vest, M.; Georgi-Maschler, T.; Friedrich, B.; Weyhe, R. Rückgewinnung von Wertmetallen aus Batterieschrott. *Chemie Ingenieur Technik-CIT* **2010**, *82*, 1985–1990. [CrossRef]
79. Kwade, A.; Diekmann, J.; Hanisch, C.; Spengler, T.; Thies, C.; Herrmann, C.; Dröder, K.; Cerdas, J.F.; Gerbers, R.; Scholl, S.; et al. *Recycling von Lithium-Ionen-Batterien-LithoRec II: Abschlussberichte der Beteiligten Verbundpartner*; Technische Universität Braunschweig: Braunschweig, Germany, 2016.
80. Ferron, C.J. The Control of Mangenese in Acidic Leach Liquors, with Special Emphasis to Laterite Leach Liquors. 2002. Available online: https://pdfs.semanticscholar.org/2224/690b72a8adc61c531ab5e87ef2d4c6e7c795.pdf (accessed on 1 April 2020).
81. Rapier, R. Why China Is Dominating Lithium-Ion Battery Production. 2019. Available online: https://www.forbes.com/sites/rrapier/2019/08/04/why-china-is-dominating-lithium-ion-battery-production/#6f1f4a893786 (accessed on 1 April 2020).
82. Infinity Lithium Industry News. March News-European Li-ion Battery Supply Chain. Available online: https://www.linkedin.com/feed/update/urn:li:activity:6650314043599003649/ (accessed on 15 June 2020).
83. Hanisch, C.; Elwert, T.; Brückner, L. Verfahren zum Verwerten von Lithium-Batterien. Patent WO002019149698A1, 29 January 2019.
84. Lipp, J. Lithium Solvent Extraction (LiSX TM) Process Evaluation using Tenova Pulsed Columns (TPC). In Proceedings of the ISEC 2017-The 21st International Solvent Extraction Conference, Miyazaki City, Japan, 5–9 November 2017.
85. Dietrich, G. *Hartinger Handbuch Abwasser- und Recyclingtechnik*; 3.; Vollständig Überarbeitete Auflage; Hanser: Munich, Germany, 2017; ISBN 9783446449015.
86. Schwab, B.; Ruh, A.; Manthey, J.; Drosik, M. *Zinc. Ullmann's Encyclopedia of Industrial Chemistry*; Wiley: Chichester, UK, 2010; pp. 1–25. ISBN 9783527306732.
87. Thomé-Kozmiensky, K.J.; Amsoneit, N.; Baerns, M.; Majunke, F. Waste, 6. Treatment. In *Ullmann's Encyclopedia of Industrial Chemistry*; Wiley: Chichester, UK, 2010; p. 484. ISBN 9783527306732.

88. Solvay AS. CYANEX®301: Product Information. Available online: https://www.solvay.com/en/product/cyanex-301#product-documents (accessed on 2 April 2020).

89. Peek, E.; Åkre, T.; Asselin, E. Technical and business considerations of cobalt hydrometallurgy. *JOM* **2009**, *61*, 43–53. [CrossRef]

90. Lascelles, K.; Morgan, L.G.; Nicholls, D.; Beyersmann, D.; Institute, N. *Nickel Compounds. Ullmann's Encyclopedia of Industrial Chemistry*; Wiley-VCH Verlag GmbH & Co. KGaA: Weinheim, Germany, 2000; pp. 1–17. ISBN 9783527306732.

91. Solvay SA. Solvay Website: CYANEX®936P Product Information. Available online: https://www.solvay.com/en/product/cyanex-936p (accessed on 15 June 2020).

92. Tenova S.p.A. Tenova Website: Lithium Processing. Available online: https://www.tenova.com/product/lithium-processing/ (accessed on 15 June 2020).

93. Bertau, M.; Martin, G. Integrated Direct Carbonation Process for Lithium Recovery from Primary and Secondary Resources. *MSF* **2019**, *959*, 69–73. [CrossRef]

94. Sabarny, P.; Peters, L.; Sommerfeld, M.; Stallmeister, C.; Schier, C.; Friedrich, B. Early-Stage Lithium Recovery (ESLR) for Enhancing Efficiency in Battery Recycling. 2020. [CrossRef]

95. Zhang, Y.; Wang, L.; Sun, W.; Hu, Y.; Tang, H. Membrane technologies for Li+/Mg2+ separation from salt-lake brines and seawater: A comprehensive review. *J. Ind. Eng. Chem.* **2020**, *81*, 7–23. [CrossRef]

96. Park, S.H.; Kim, J.H.; Moon, S.J.; Jung, J.T.; Wang, H.H.; Ali, A.; Quist-Jensen, C.A.; Macedonio, F.; Drioli, E.; Lee, Y.M. Lithium recovery from artificial brine using energy-efficient membrane distillation and nanofiltration. *J. Membr. Sci.* **2020**, *598*, 117683. [CrossRef]

97. Zenn, R. About 500 GWh/a Li-Ion Battery Cell Production Capacity Announced in Europe. Available online: https://www.linkedin.com/posts/roland-zenn_gigafactories-battery-europe-activity-6673867930679218176-Hebm (accessed on 30 June 2020).

98. Green Car Congress Website. Roskill Forecasts Li-Ion Battery Demand to Increase more than Ten-Fold by 2029 to >1,800GWh. Available online: https://www.greencarcongress.com/2020/06/20200609-roskill.html (accessed on 30 June 2020).

99. Weyhe, R.; Yang, X. Investigations about Lithium-Ion Battery Market Evolution and Future Potential of Secondary Raw Material from Recycling. 2018. Available online: https://accurec.de/wp-content/uploads/2018/04/0-2Market-Research_YXF_3.0.pdf (accessed on 13 July 2020).

100. Reimer, M.V.; Schenk-Mathes, H.Y. Systemdynamische Modellierung und Prognose der Rückflüsse von Lithium-Ionen-Batterien in der Circular Economy. In *Forschungsfeldkolloquium 2020*; Langefeld, O., Mrotzek-Blöß, A., Eds.; Papierflieger: San Francisco, CA, USA, 2020; pp. 45–55. ISBN 978-3-86948-767-0.

101. Industry Expert. Lithium-Ion Battery Recycling in Asia. Oral Communication. 2019.

102. Jara, A.D.; Betemariam, A.; Woldetinsae, G.; Kim, J.Y. Purification, application and current market trend of natural graphite: A review. *Int. J. Min. Sci. Technol.* **2019**, *29*, 671–689. [CrossRef]

103. Deutsche Rohstoffagentur (DERA) in der Bundesanstalt für Geowissenschaften und Rohstoffe. DERA Preismonitor. February 2020. Available online: https://www.bgr.bund.de/DE/Themen/Min_rohstoffe/Produkte/Preisliste/pm_20_02.pdf?__blob=publicationFile (accessed on 24 April 2020).

104. Reuter, B.; Hendrich, A.; Hengstler, J.; Kupferschmid, S.; Schwenk, M. Rohstoffe für innovative Fahrzeugtechnologien-Herausforderungen und Lösungsansätze. 2019. Available online: https://www.e-mobilbw.de/fileadmin/media/e-mobilbw/Publikationen/Studien/Material-Studie_e-mobilBW.pdf (accessed on 24 April 2020).

105. Buchert, M.; Sutter, J. Aktualisierte Ökobilanzen zum Recyclingverfahren EcoBatRec für Lihtium-Ionen-Batterien (Stand 09/206). 2016. Available online: https://www.erneuerbar-mobil.de/sites/default/files/2017-01/EcoBatRec-LCA-Update%202016.pdf (accessed on 30 June 2020).

106. Buchert, M.; Sutter, J. Aktualisierte Ökobilanzen zum Recyclingverfahren LithoRec II für Lithium-Ionen-Batterien (Stand 09/2016). 2016. Available online: https://www.erneuerbar-mobil.de/sites/default/files/2017-01/LithoRec%20II-LCA-Update%202016.pdf (accessed on 30 June 2020).

Article

Selective Precipitation of Metal Oxalates from Lithium Ion Battery Leach Solutions

Eva Gerold *, Stefan Luidold and Helmut Antrekowitsch

Chair of Nonferrous Metallurgy of Montanuniversitaet Leoben, 8700 Leoben, Austria;
stefan.luidold@unileoben.ac.at (S.L.); helmut.antrekowitsch@unileoben.ac.at (H.A.)
* Correspondence: eva.gerold@unileoben.ac.at; Tel.: +43-3842-402-5207

Received: 23 September 2020; Accepted: 25 October 2020; Published: 29 October 2020

Abstract: The separation of cobalt and nickel from sulfatic leach liquors of spent lithium-ion batteries is described in this paper. In addition to the base metals (e.g., cobalt and nickel), components such as manganese and lithium are also present in such leach liquors. The co-precipitation of these contaminants can be prevented during leach liquor processing by selective precipitation. For the recovery of a cobalt-nickel mixed material, oxalic acid serves as a suitable reagent. For the optimization of the precipitation retention time and yield, the dependence of the oxalic acid addition must be taken into account. In addition to efficiency, attention must also be given to the purity of the product. After this procedure, further processing of the products by calcination into oxides leads to better marketability. A series of experiments confirms the suitability of oxalic acid for precipitation of cobalt and nickel as a mixed oxalate from sulfatic liquors and also suggests a possible route for further processing of the products with increased marketability. The impurities in the resulting oxides are below 3%, whereby a sufficiently high purity of the mixed oxide can be achieved.

Keywords: precipitation; lithium-ion battery; oxalic acid; mixed oxalate

---

1. Introduction

Lithium-ion batteries (LIBs) have been available on the market since the early 1990s [1]. Technological innovations driven by various branches of industry have led to a large number and variety of different electronic devices worldwide [2]. These developments have greatly stimulated the production and consumption of LIBs [3]. Due to their desirable characteristics such as reduced size and weight, high cell voltage, low self-discharge rates and high energy density, LIBs are increasingly replacing other types of batteries (e.g., Ni-MH or Ni-Cd batteries) [1,4]. Nevertheless, it must be considered that electronic waste is the fastest growing solid waste problem worldwide, including LIBs for electronic devices and vehicles [5]. For this reason, the recycling of lithium-ion batteries must be addressed, not only from an environmental point of view, but also for its economic benefits due to the increasing price of cobalt [6]. The main valuable metals in LIBs (cobalt, nickel and lithium) were evaluated by the European Union in terms of criticality. Critical raw materials are highlighted and located within their criticality zone of the graph exhibited in Figure 1 [7,8]. For example, nickel has a very high economic importance as an alloying element in advanced stainless steels [9,10], but a significant low supply risk. In contrast, the supply risk for cobalt is considerably higher. Critical raw materials are highlighted and located within the criticality zone of the graph. [7,8]

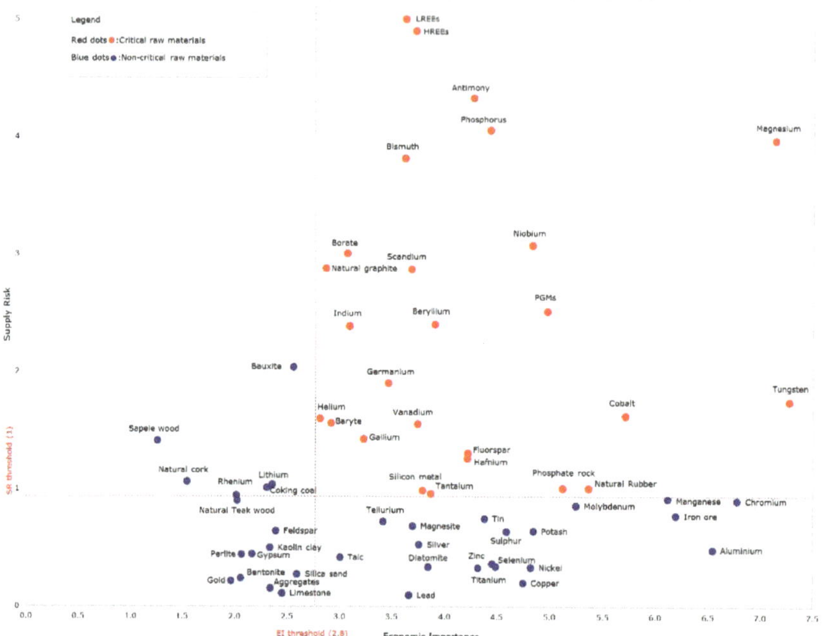

Figure 1. Economic importance and supply risk of raw materials in the European Union 2020 [7].

The analysis of the worldwide supply indicates that cobalt is mined in 19 countries but, as demonstrated in Figure 2, the Democratic Republic of Congo (DRC) represents the most important global supplier with a share of 64% (based on a five-year average between 2010 and 2014). In Europe, cobalt mining takes place exclusively in New Caledonia (France) and Finland, corresponding to 2% and 1% of the global market, respectively. Although this provides a partial independence and a sufficient supply of cobalt as a raw material to the European Union, recycling mechanisms for cobalt are yet to be addressed [7].

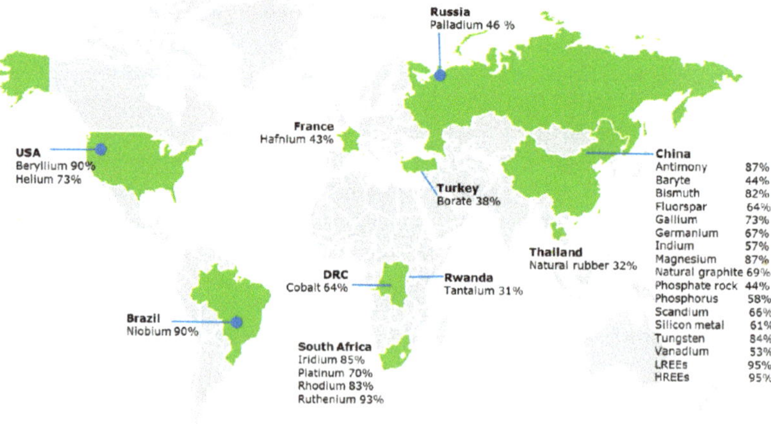

Figure 2. Countries accounting for largest share of the global supply of critical raw materials [7].

Spent LIBs usually contain significant amounts of inorganic compounds such as heavy metals as well as organic substances and therefore are classified as hazardous wastes, resulting in stricter worldwide regulations regarding their disposal [2,4,5]. The anode of LIBs contains a copper foil coated with graphite (see Figure 3). The cathode consists of an aluminum foil covered with an active material (e.g., lithium cobalt oxide). Due to the presence of valuable metals in LIBs mainly in the active material, the development of efficient recycling technologies is of paramount importance for industry. In addition, the invention of new cathode materials aimed at increasing the efficiency of LIBs will possibly lead to more complicated waste streams and new challenges in the field of recycling. With advanced compositions of active materials, the separation of the contained valuable metals may be even harder, reinforcing the need for the development of innovative recycling technologies [3,5].

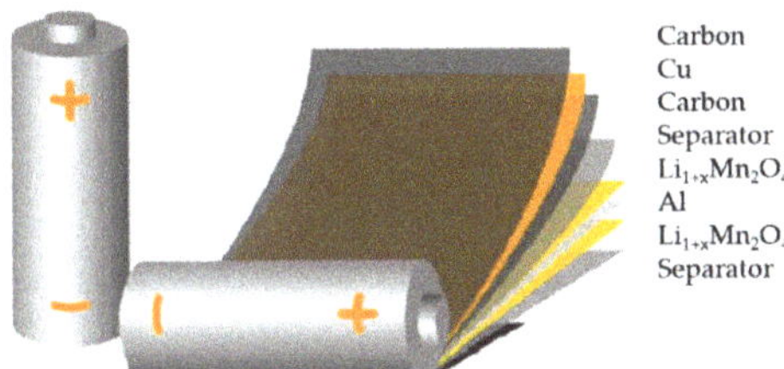

Figure 3. Schematic drawing showing the components of a lithium-ion battery [11].

In principle, pyro- and hydrometallurgical processes are already used for the recycling of LIBs. The pyrometallurgical operations have been used by several companies (such as Inmetco, Umicore and Xstrata), but are currently discouraged due to some disadvantages such as the thermal treatment of binder and organic electrolytes, which is expensive due to high energy requirement, emission of hazardous gases and dust as well as loss of critical metals (e.g., lithium) in the slag [1,5,6].

Hydrometallurgical processes often comprise several stages to allow a clean separation of individual valuable metals. In general, they include dismantling, physical separation, crushing, acid leaching with or without additives as well as numerous separation and purification steps. The segregation of cobalt and nickel is not simple due to their physico-chemical similarities. One attractive possibility to separate these metal ions from acidic leachates comprises solvent extraction, since Co(II) has a higher tendency to form stable complexes than nickel. However, as this process is complicated and laborious, the research area has been extended to precipitation combinations. Several papers mentioned the use of different precipitants in multi-step processes [2,5,12–14].

The recovery of cobalt and nickel can be accomplished in several ways. Nickel can be selectively precipitated by adding dimethylglyoxime reagent (DMG, $C_2H_8N_2O_2$) to the leach liquor. After the dissolution of the filtered precipitate with hydrochloric acid, the DMG can be regenerated and reused as a precipitant, while nickel is recovered as $NiCl_2$ in the filtrate. This process step depends not only on the temperature but also on the set pH-value. The pH of the acidic leaching solutions has to be adjusted by adding a base (e.g., NaOH) in order to achieve the highest efficiency of the precipitation. The return of dimethylglyoxime to the process cycle is associated with losses (up to approx. 40%), since the chemical cannot be completely regenerated. Although high-grade mixed nickel-cobalt precipitates are generated in various processes and are expected to dominate the feed materials used in the next decade for the production of battery-grade nickel and cobalt sulphates [15], due to the high cost of the dimethylglyoxime as a precipitation agent, this method in particular only finds application on a small scale [1,3,5,16].

For the recovery of cobalt, selective precipitation as well as solvent extraction are used. Solvent extraction is used to remove cobalt from liquors that also contain nickel due to reagent costs. This leaves behind a liquor containing nickel [17]. In the course of the precipitation, oxalic acid or ammonium oxalate are applied as reagents. The efficiency of this process highly depends on temperature, with the best results observed around a temperature of 50–55 °C. The pH also plays a role, although contradictory data are found in the literature. The resulting oxalate can be processed into an oxide via a calcination process. Alternatively, ion exchange can be used for the separation of cobalt from the leach solution, in order to subsequently precipitate nickel as an oxalate [1–5,14].

Furthermore, lithium and manganese are often dissolved in these leach solutions, allowing these metals to be removed through several different process steps. In order to extract manganese from solution, either precipitation with potassium permanganate ($KMnO_4$) can be implemented or co-extraction via oxalate precipitation can be performed. Lithium can be removed either as a carbonate or phosphate with the addition of precipitants, usually in the last stage of the process [3,5,18].

This work is aimed at testing a hydrometallurgical process to recover nickel and cobalt from sulfuric acid leachates of spent lithium-ion batteries followed by selective precipitation via oxalic acid ($C_2H_2O_4$). The recycling process proposed in this work is then optimized by identifying and adapting relevant processing parameters.

2. Materials and Methods

2.1. Composition of the Input Material

The used active material from lithium-ion batteries was obtained from NMC cells with low levels of other metal impurities. The main components were cobalt, nickel, manganese and lithium, along with small residual amounts of the copper and aluminum foils. The composition was determined by inductively coupled plasma mass spectrometry (ICP-MS, Agilent 8800, Santa Clara, CA, United States) with an upstream peroxide digestion, as shown in Table 1. The carbon contained was pre-determined by an analysis of the total carbon content, and the ICP-MS data were adjusted accordingly on the basis of these results.

Table 1. Chemical composition of the material used in this work.

Element	C	Al	Co	Fe	Li	Mg	Mn	Ni	Si	Cu
(wt%)	38.7	3.4	15.0	0.4	4.0	0.1	0.8	17.0	<1.0	3.7

2.2. Experimental Procedure

The first stage of work aimed for the optimization of the leaching process. In this study, the main parameters of this method were investigated (using the experiment setup shown in Figure 4). At a given concentration of the acid (1–2 mol/L), the optimal parameter combination comprises 80 °C, 100 g/L solids, magnetic stirrer speed (500 min^{-1}) and 4 h of leaching time [19,20]. The acid concentration was found within this low range to result in higher selectivity of the leaching. These specific parameters prevented copper from being dissolved and, as a result, this contaminant can be separated in the first step [21].

Figure 4. Setup of the leaching process under optimized conditions in the laboratory (double walled reaction vessel with a reaction volume of one liter and a thermostat to control the chosen process temperature).

After leaching, selective precipitation of the valuable metals cobalt and nickel was conducted to generate a high quality product. Since almost no appropriate thermodynamic data are available in the literature for these concentrated metal-containing solutions and their precipitates such as mixed oxalates, the equilibrium concentration of the metals could not be calculated. In addition, the interactions between the metals in solution have not been determined. For this purpose, an optimization of the relevant parameters of precipitation with oxalic acid takes place in the context of this work. The yield of a pure cobalt-nickel mixed oxalate from the leaching solution was targeted. The parameters optimized were the stoichiometric factor of oxalic acid addition, the adjusted pH value and the retention time during the precipitation (see Table 2). The pH was adjusted before the start of the precipitation process and thus before the addition of the precipitant at room temperature by adding sodium hydroxide solution. The measurement was carried out using a pH meter (InLab Science, Mettler-Toledo, Vienna). The stoichiometric amount of precipitant was calculated based on the concentration of the valuable metals in the leaching solution, which was determined via ICP-MS. The temperature was set at 55 °C for these experiments based on previous work [4,21]. This approach should result in identification of the main influencing factors and a detection of the dependencies on each parameter. After conducting the experiments, the filtrates were analyzed by ICP-MS, while the solids were characterized by SEM/EDS (Scanning Electron Microscopy with Energy Dispersive Spectroscopy). The results were evaluated with the help of a statistical experimental design software (MODDE 12.1, Goettingen, Germany). At this stage of research, the test parameters were screened using a full factorial experimental design to enable evaluation of a linear model that considers the interaction of the factors.

Table 2. Parameters of the executed experiments.

Experiment	Stoichiometric Addition of Oxalic Acid	pH	Retention Time (h)
F1	1.5×	0	4
F2	2×	0	4
F3	1.5×	1	4
F4	2×	1	4
F5	1.5×	2	4
F6	2×	2	4
F7	1.5×	0	8
F8	2×	0	8
F9	1.5×	1	8
F10	2×	1	8
F11	1.5	2	8
F12	2	2	8

3. Results

In the course of the precipitation test work, several parameters (duration of precipitation, pH value and the added amount of precipitant) were varied and their overall influence on the process was evaluated. Table 3 shows the concentrations of metals contained in the leaching solution. These concentrations were determined by ICP-MS and thus enable the calculation of the corresponding yields of recovered fractions after the precipitation process.

Table 3. Concentrations of the metals in the leaching liquor, which was used for the subsequent precipitation tests.

Li (g/L)	Al (g/L)	Mn (g/L)	Fe (g/L)	Mg (g/L)	Co (g/L)	Ni (g/L)
4.1	2.1	0.7	0.3	0.1	12.7	13.3

3.1. Influence of the Retention Time

For this series of experiments, the holding periods after addition of the precipitant were set to between four and eight hours based on a preliminary test. During the tests, the precipitation solution was stirred uniformly by using a magnetic stir plate at a speed of 500 min^{-1} to ensure efficient mixing. The experiments F1–F6 represent those with a retention time of four hours. In comparison, the experiments F7–F12 describe tests with an 8 h retention time. Figure 5 shows the composition of the filtrates obtained through precipitation with oxalic acid after the stated holding periods. In order to make it possible to use the obtained product as a recycled battery material, the concentration of impurities in the precipitate must be low, due to the high quality demands of battery manufacturers. Therefore, all impurity elements, such as magnesium, should be located in the filtrate after this precipitation stage.

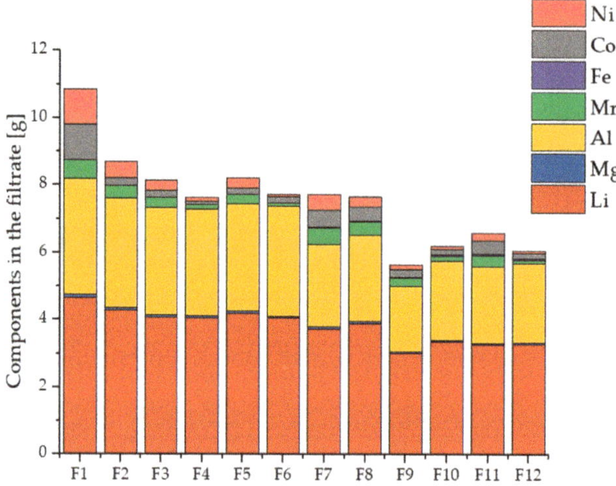

Figure 5. Compositions of the filtrate after precipitation with oxalic acid.

For the evaluation of the experiments, special emphasis was given to the residual cobalt and nickel contents in the filtrates. The content of valuable metals in the solutions should be as low as possible, while the precipitate should contain low levels of impurities. Higher contents in the filtrate indicate poor precipitation and thus lead to a lower yield in the corresponding filter cake. The results for the valuable metals cobalt and nickel are indicated separately in Figure 6. Experiment F1 shows very high residual cobalt and nickel contents compared to the other tests. Since there were no other changes to the parameters for F2, only the amount of precipitant added can be decisive. However, since the further experiments (e.g., F7) show that the used 1.5× stoichiometric amount of oxalic acid sufficed for an efficient precipitation, experiment F1 was excluded from the following considerations due to the result of a statistical evaluation (the corresponding measured value is outside four times the standard deviation).

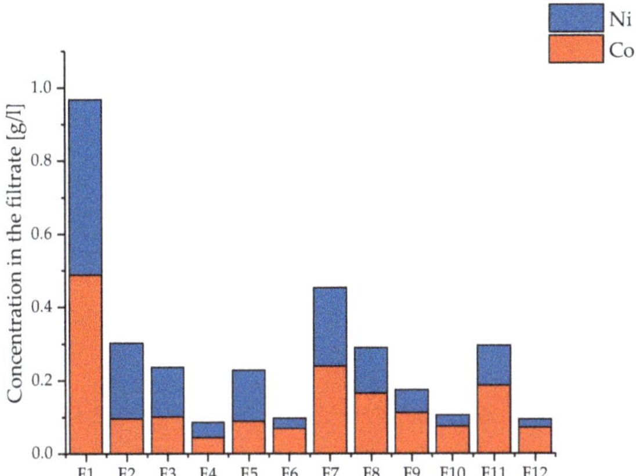

Figure 6. Concentration of cobalt and nickel in the filtrate after precipitation with oxalic acid.

The comparison of the four-hour tests with the corresponding eight-hour tests (with otherwise the same parameters) shows that an extension of the precipitation time is of minor advantage. The residual contents are in a very low range and in all cases considered, the results of the four-hour experiments are the same or better than that with longer test periods except for the comparison of experiments F1 and F7. In general, the residual cobalt contents after eight hours exhibit higher values than after four hours in contrast to the behavior of nickel. This can be traced back to the redissolution of Co (see Figure 7). In this diagram, the solid lines indicate the upper and lower confidence intervals, while the dashed line reflects the predicted values according to the model used for this evaluation. To create these diagrams, a full factorial model was used and its factors were fitted statistically to the experimental results. This resulted in the following model equation for cobalt and nickel, where x indicates the stoichiometric factor (-) of precipitant, pH (-) is the pH value, and t (h) is the precipitation time of the process.

$$c_{Co} = 0.303 - 0.754 * x_{Oxalic\ acid} - 0.348 * pH + 0.028 * t + 0.162 * x_{Oxalic\ acid} * pH$$

$$c_{Ni} = 0.330 - 0.443 * x_{Oxalic\ acid} - 0.073 * pH - 0.025 * t$$

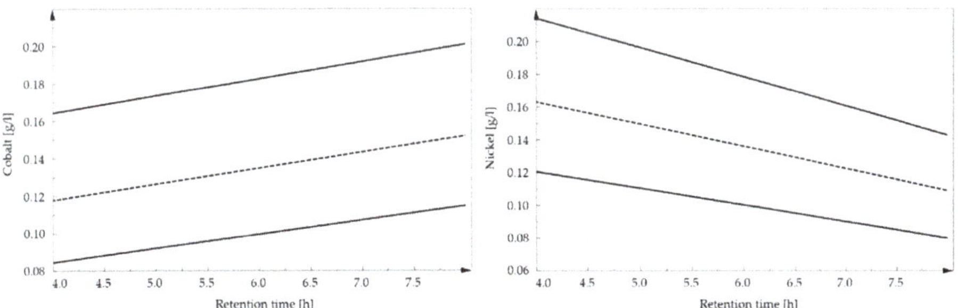

Figure 7. Behavior of soluble cobalt and nickel as a function of retention time (solid lines indicate the upper and lower confidence intervals, while the dashed line reflects the predicted values according to the model).

These tendencies could not have been foreseen because, as mentioned earlier, the compounds of these two valuable metals often show similar behavior. This behavior can be explained by the mutual influence of the components in the concentrated solution in dependence of the concentrations of cobalt and nickel regarding their solubility (salt effect). Much of the data available in the literature relate exclusively to dilute solutions of single metals, and interactions between different metals are often neglected. Therefore, further investigations must be carried out for concentrated solutions, not only of pure substances but also mixtures containing two or more metals and their behavior.

This effect, shown in Figure 7, is less pronounced at higher pH values than at lower ones. The same applies to the dependence on the amount of precipitant added. In order to expand this model further, additional tests were carried out, up to a pH value of 4, which was set before the precipitation process. The analysis of the residual dissolved concentrations in the filtrate yielded the results shown in Figure 8. From an economic perspective, use of a precipitation period of four hours instead of eight is sufficient for the efficient precipitation of cobalt and nickel with oxalic acid from the enriched solutions with recovery yields above 95%. However, tests to further reduce the holding period should be carried out in order to define the critical precipitation time more precisely. In the following diagrams, the residual concentrations of cobalt and nickel in the filtrate are plotted in g/L.

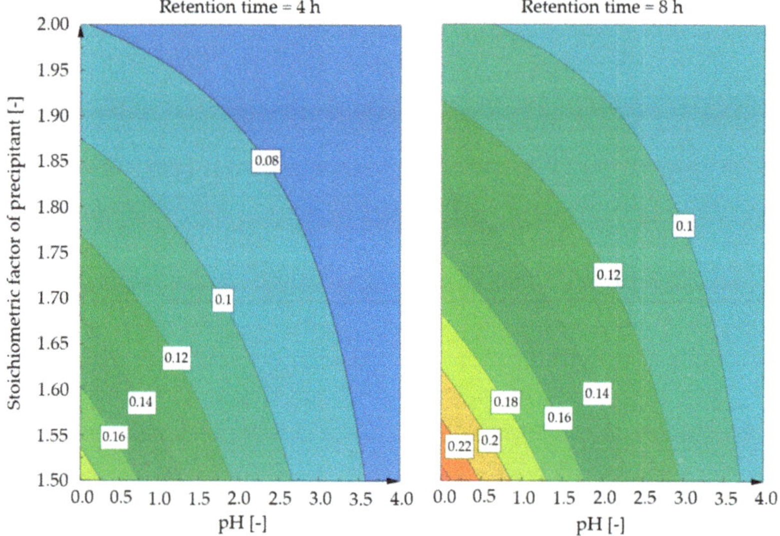

Figure 8. Statistical evaluation of the residual dissolved concentrations (g/L) of cobalt for 4 and 8 h retention time, respectively.

The strong influence of the pH value can be seen in Figure 8, since at high acid concentrations, twice as much cobalt remains in the residual solution as at pH values above 3.5. Furthermore, the excess of precipitation agent only plays an increasingly significant role as the pH value decreases, irrespective of the retention time.

Dependencies are different for nickel because of linear relations without interaction parameters. As shown in Figure 9, twice the residual nickel concentrations are obtained with a reduction from eight to four hours with a small stoichiometric addition of precipitant. This effect is significantly reduced at higher oxalic acid levels or higher pH values. In the case of nickel, minimal residual levels occur for a precipitation time of eight hours, and at a pH value of 4, no nickel could be detected in the residual solution.

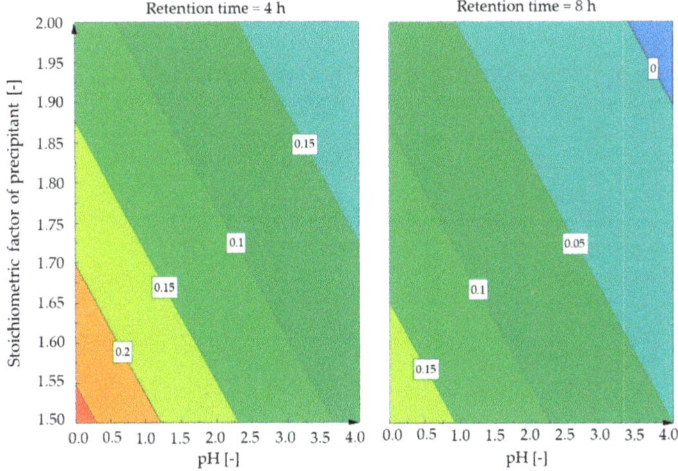

Figure 9. Statistical evaluation of the residual dissolved concentrations (g/L) of nickel for 4 and 8 h retention time, respectively.

3.2. Influence of the pH Value

An examination regarding the higher-priced valuable cobalt shows that at elevated pH values, lower residual contents are expected in the solution. From a pH value of 3–3.5, there is no further improvement in the precipitation efficiency. Furthermore, a direct dependency on the amount of precipitant added can no longer be recognized above this pH value, since even a smaller amount of oxalic acid leads to the maximum possible precipitation yield. It was also shown that no selective precipitation of cobalt or nickel is possible over the entire pH range tested, but that a mixed oxalate was always obtained. The comparison of the minimum and maximum pH value indicates a doubling of the cobalt content in the residual solution at low pH values and low amounts of precipitant. This effect can no longer be observed with an efficient oxalic acid supply. In order to visualize the combined dependency on the pH value and stoichiometrical factor of oxalic acid, the trends are shown in Figure 10.

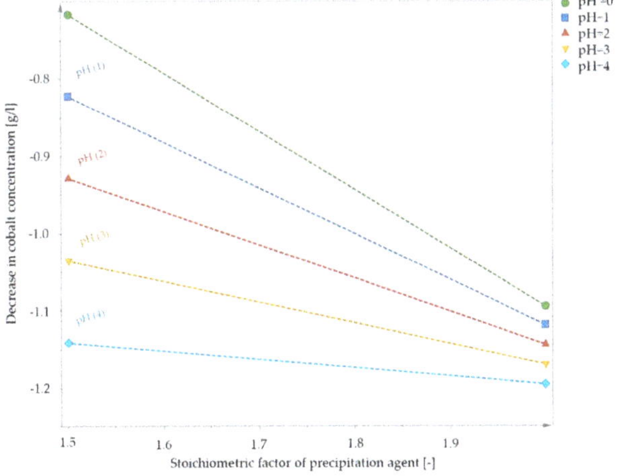

Figure 10. Influence of the pH on the difference of cobalt concentration between the stoichiometric addition of 1.5 and 2.0.

In general, for both cobalt and nickel, it turns out that a higher stoichiometric factor for oxalic acid leads to a lower influence of the pH value. This fact is presented in Figure 11. The areas marked indicate the parameter combination with the highest precipitation efficiency. However, the parametric region for an efficient recovery of cobalt (shown in blue) significantly exceeds that for nickel.

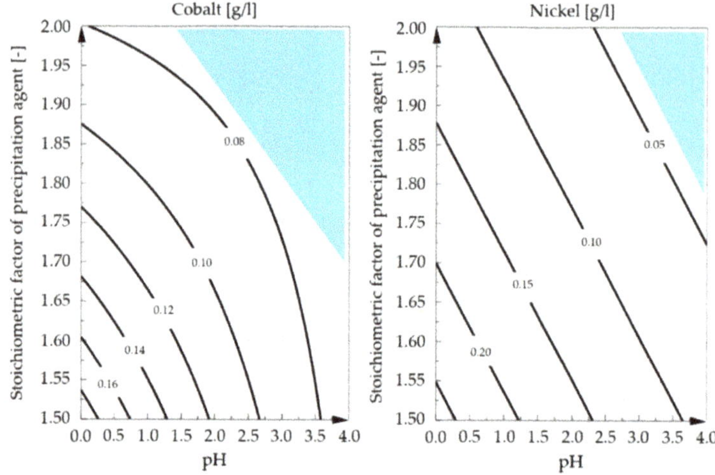

Figure 11. Influence of pH on the residual dissolved concentrations of cobalt and nickel.

3.3. Influence of the Added Amount of Precipitant

To ensure the effectiveness of the precipitation, it is important that sufficient precipitation reagent is dosed to generate a high yield of precipitated valuable metals. In contrast, economic and resource-saving approaches have to be considered. For this reason, the stoichiometric factor of precipitant that enables an improvement of the process performance was evaluated. The tested range was set between 1.5 and 2 times the theoretically required stoichiometric amount of oxalic acid. All experiments showed that an addition of the lower value of precipitant sufficed to precipitate the majority of valuable metals. The addition of higher amounts led to only a slight improvement of the yields and therefore makes little sense economically and environmentally. Nevertheless, it has previously been discussed that the dependencies of precipitation on pH value and duration were more pronounced with smaller amounts of oxalic acid added than with higher ones. Figure 12 shows the influence of oxalic acid on the levels of cobalt and nickel in the residual solution. In this figure, the solid lines indicate the upper and lower confidence intervals, while the dashed line reflects the predicted values according to the model. In future investigations, a reduction in the amount added should be tested in order to determine the minimum needed amount of precipitant.

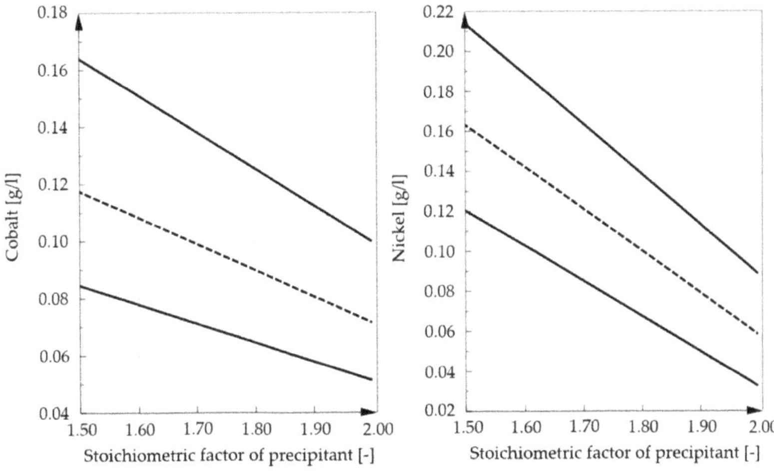

Figure 12. Dependence of residual cobalt and nickel concentrations on the stoichiometric amount of precipitation agent added for a fixed pH of 2 and a precipitation time of four hours (solid lines indicate the upper and lower confidence intervals, while the dashed line reflects the predicted values according to the model).

3.4. Composition of the Obtained Product

The precipitate obtained (mixed cobalt-nickel oxalate) was separated from the filtrate by means of vacuum filtration and rinsed with 200 mL of boiling, deionized water in order to remove impurities such as sulfates. After drying for 24 h at 105 °C in a drying oven, a semi-quantitative analysis was carried out by scanning electron microscopy and energy dispersive X-ray spectroscopy (Jeol Ltd., Tokyo, Japan) (SEM/EDS). Table 4 indicates the corresponding metal contents normalized to 100% for selected elements, given that lithium cannot be detected using EDS by the available device at our institute. It can be observed that the levels of impurities are low. In addition, no copper could be detected in the precipitate, which is a relevant feature for further processing of the product and later use in various industries (e.g., stainless steel industry). On the basis of the data obtained from the software of the EDS measurement, no significant amount of copper could be detected, which could be clearly separated from the background noise of the measurement. Since this analysis method only outputs reproducible values to a limited extent, because of the detection limits of EDS and also the possibly existing inhomogeneity in the sample, especially in the case of low contents, ICP-AES analysis is planned for future studies to specifically analyze minor elements such as Cu, Al, Mn and Fe.

Table 4. Metal contents of the obtained mixed cobalt-nickel oxalates.

No.	Oxalic Acid	pH	Retention Time (h)	Co (wt%)	Ni (wt%)	Cu (wt%)	Al (wt%)	Mn (wt%)	Fe (wt%)
1	1.5×	0	4	45.7	50.5	0.0	1.3	1.0	1.5
2	2×	0	4	45.2	51.2	0.0	1.1	1.5	1.1
3	1.5×	1	4	46.2	50.5	0.0	1.0	1.4	0.9
4	2×	1	4	46.0	50.3	0.0	0.9	1.8	1.0
5	1.5×	2	4	46.8	49.4	0.0	1.1	1.4	1.3
6	2×	2	4	46.7	50.5	0.0	0.8	2.1	0.0
7	1.5×	0	8	45.1	51.7	0.0	1.2	1.2	0.8
8	2×	0	8	44.8	51.4	0.0	1.2	1.6	1.0
9	1.5×	1	8	46.9	49.6	0.0	1.0	1.4	1.1
10	2×	1	8	46.3	50.7	0.0	1.0	2.1	0.0

An examination of the data in Table 4 indicates that the sum of the impurities reaches a maximum of 3.8%. These amounts of contaminants correspond to a sufficient high quality of the recycled product, especially as an intermediate that can be used in several industries (e.g., stainless steel industry), as well as a precursor for generation of battery quality materials. In addition, it can be seen that with higher quantities of precipitants, slightly increased contents of impurities in the precipitate also occur due to the more aggressive conditions during the precipitation process. In order to make it possible to use it as a recycled battery material, the quality still needs improvements. In order to separate the existing impurities and obtain a high-quality product, further research is necessary. The precipitate obtained is shown in Figure 13. In addition to the overview picture, the distribution images of the valuable metals and their impurities are also shown.

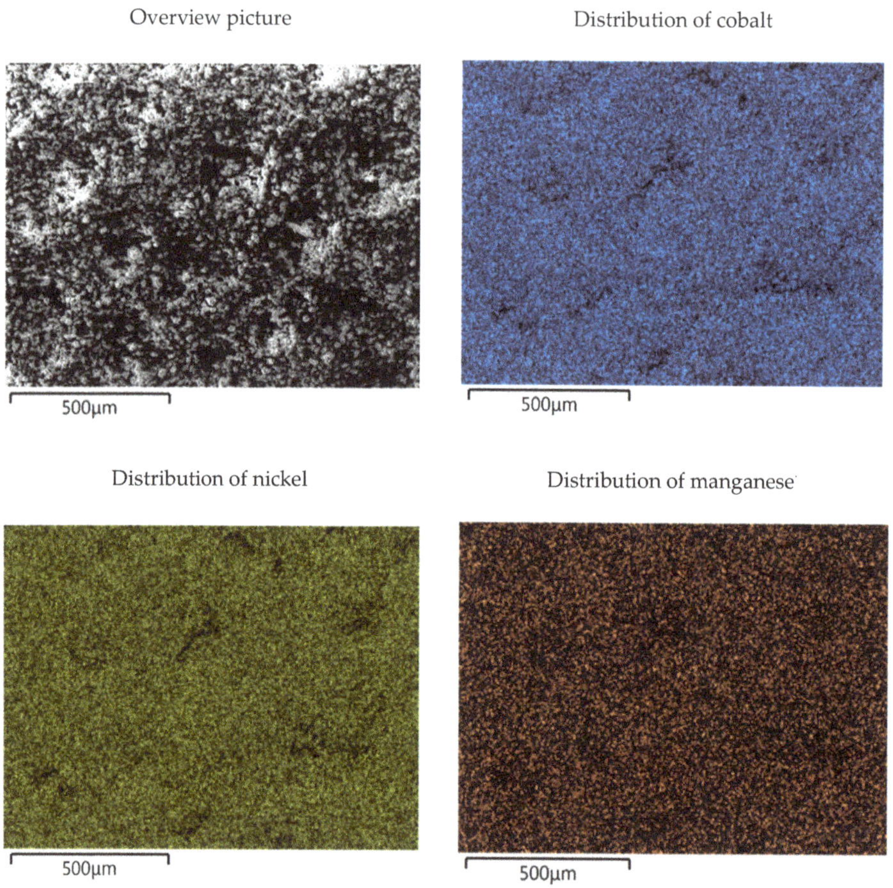

Figure 13. SEM/EDS images of the obtained precipitate.

Figure 13 illustrates that a homogeneous distribution of the metals such as cobalt and nickel and thus a homogeneous mixed product was achieved. Essentially, no segregation of the individual oxalates and thus no major inhomogeneity can be recognized at this magnification. This statement can also be made for manganese and iron, with aluminum accumulations being found in certain areas. However, reference should be made to the generally very low concentration of aluminum in the product. The precipitate can be used as a high-quality raw material for various cobalt and nickel processing industries, with further processing into oxide or alloy being possible [21].

4. Discussion

The investigations performed in this work demonstrated that various parameters can influence the precipitation of metals when oxalic acid is used in the recycling processes. Small amounts of oxalic acid were observed to induce significant dependencies of precipitation efficiency on pH and holding period. These tendencies were much less pronounced for higher amounts of precipitant addition. These descriptions apply in principle to both valuable metals (cobalt and nickel), although competing behavior of the two metals could be determined during the evaluation of the test data, since, for example, precipitation times of more than four hours lead to a decrease in the nickel concentration in the solution, while the concentration of cobalt increased again slowly over time and a redissolution process occurred. Since the interdependencies between nickel and cobalt precipitation are difficult to describe, they are represented diagrammatically in Figure 14.

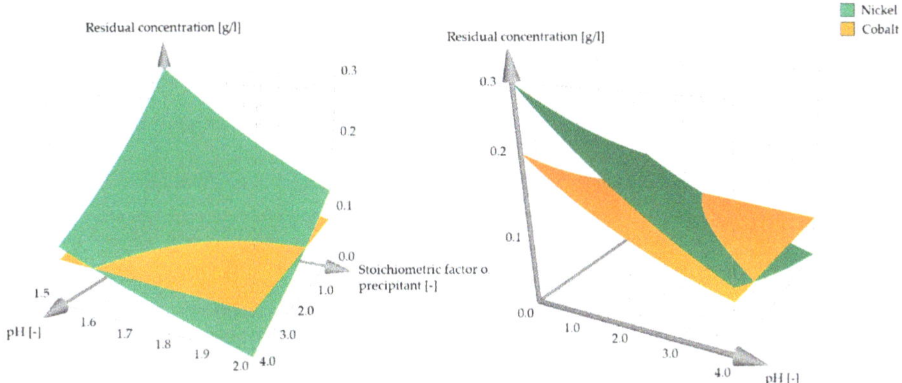

Figure 14. Surface plots for cobalt and nickel precipitation as oxalates and their dependency on the pH value and the stoichiometric factor of oxalic acid addition.

5. Conclusions and Outlook

Oxalic acid is an effective precipitant for the simultaneous recovery of cobalt and nickel from leaching solutions generated from the recycling of lithium-ion batteries. However, it was found that the effectiveness of this organic precipitant has various dependencies on different parameters, which must be taken into account during the precipitation process

In general, smaller amounts of precipitation reagent results in significantly stronger dependencies on pH and duration of precipitation. As redissolution of cobalt oxalate occurs, the holding time must be limited to a maximum of four hours in order to minimize losses.

Based on the promising results so far, this holding time could be further reduced. The pH value plays an essential role, since the adaptation is often difficult due to buffer effects.

It was shown that an increase in the pH of the leach solutions was necessary, with acceptable results achieved at a pH of 1.5–2. Using this optimized combination of parameters, low levels of cobalt and nickel in the residual solutions and homogenous, sufficient pure products can be obtained.

By determining the competing behaviors of cobalt and nickel depending on the duration of the precipitation process with oxalic acid, further studies of this behavior should be carried out in order to study a potentially new separation mechanism for cobalt and nickel.

Author Contributions: E.G. developed the theoretical formalism, performed the analytic calculations and performed thermodynamical simulations. Both S.L. and H.A. contributed to the final version of the manuscript and supervised the project. All authors have read and agreed to the published version of the manuscript.

Funding: This research was funded by the Austrian Federal Ministry of Science, Research and Economy and the National Foundation for Research, Technology and Development.

Acknowledgments: The financial support by the Austrian Federal Ministry of Science, Research and Economy and the National Foundation for Research, Technology and Development and the support from the industry by A. Arnberger is gratefully acknowledged.

Conflicts of Interest: The authors declare no conflict of interest.

References

1. Chen, X.; Zhou, T.; Kong, J.; Fang, H.; Chen, Y. Separation and recovery of metal values from leach liquor of waste lithium nickel cobalt manganese oxide based cathodes. *Sep. Purif. Technol.* **2015**, *141*, 76–83. [CrossRef]

2. Fernandes, A.; Afonso, J.C.; Dutra, A.J.B. Separation of nickel(II), cobalt (II) and lanthanides from spent Ni-MH batteries by hydrochloric acid leaching, solvent extraction and precipitation. *Hydrometallurgy* **2013**, *133*, 37–43. [CrossRef]

3. Chen, X.; Chen, Y.; Zhou, T.; Liu, D.; Hu, H.; Fan, S. Hydrometallurgical recovery of metal values from sulphuric acid leaching liquor of spent lithium-ion batteries. *Waste Manag.* **2015**, *38*, 349–356. [CrossRef] [PubMed]

4. Kang, J.; Sohn, J.; Chang, H.; Senanayake, G.; Shin, S.M. Preparation of cobalt oxide from concentrated material of spent lithium ion batteries by hydrometallurgical method. *Adv. Powder Technol.* **2010**, *21*, 175–179. [CrossRef]

5. Sattar, R.; Ilyas, S.; Bhatti, H.N.; Ghaffar, A. Resource recovery of critically-rare metals by hydrometallurgical recycling of spent lithium ion batteries. *Sep. Purif. Technol.* **2019**, *209*, 725–733. [CrossRef]

6. Golmohammadzadeh, R.; Rashchi, F.; Vahidi, E. Recovery of lithium and cobalt from spent lithium-ion batteries using organic acids: Process optimization and kinetic aspects. *Waste Manag.* **2017**, *64*, 244–254. [CrossRef]

7. European Commission. *Study on the Review of the List of Critical Raw Materials: Final Report*; Publications Office of the European Union: Luxembourg, 2020.

8. Community Research and Development Information Service (CORDIS). Economic Importance and Supply Risk of Raw Materials in the European Union. Available online: https://cordis.europa.eu/docs/results/h2020/660/660885_PS/crit-metals.jpg (accessed on 27 October 2020).

9. Arpe, H.-J.; Ullmann, F. (Eds.) *Part A6-Ullmann's Encyclopedia of Industrial Chemistry*, 5th ed.; Wiley-VCH Verlag GmbH &. Co. KGaA: Weinheim, Germany, 1991; pp. 293–300.

10. Arpe, H.-J.; Ullmann, F. (Eds.) *Part A18-Ullmann's Encyclopedia of Industrial Chemistry*, 5th ed.; Wiley-VCH Verlag GmbH &. Co. KGaA: Weinheim, Germany, 1986; pp. 213–215.

11. Rahimzei, E.; Sann, K.; Vogel, M.; VDE Verband der Elektrotechnik. Kompendium: Li-Ionen-Batterien. 2015. Available online: https://www.dke.de/resource/blob/933404/3d80f2d93602ef58c6e28ade9be093cf/kompendium-li-ionen-batterien-data.pdf (accessed on 27 October 2020).

12. Takacova, Z.; Dzuro, V.; Havlik, T. Cobalt Precipitation from Leachate Originated from Leaching of Spent Li-ion Batteries Active Mass – Characterization of Inputs, Intermediates and Outputs. *World Metall.* **2017**, *70*, 336c343.

13. Ferreira, D.A.; Prados, L.M.; Majuste, D.; Mansur, M.B. Hydrometallurgical separation of aluminium, cobalt, copper and lithium from spent Li-ion batteries. *J. Power Sources* **2009**, 238–246. [CrossRef]

14. Korthauer, R. (Ed.) *Handbuch Lithium-Ionen-Batterien*; Springer: Berlin/Heidelberg, Germany, 2013.

15. Rath, M.; Behera, L.P.; Dash, B.; Sheik, A.R.; Sanjay, K. Recovery of dimethylglyoxime (DMG) from Ni-DMG complexes. *Hydrometallurgy* **2018**, *176*, 229–234. [CrossRef]

16. Habashi, F. *Handbook of Extractive Metallurgy*; Wiley-VCH: Weinheim, Germany, 1997.

17. Godlewska, L.; Grohol, M. *Study on the Review of the List of Critical Raw Materials: Critical Raw Materials Factsheets*; Publications Office of the European Union: Luxembourg, 2020.

18. Luidold, S.; Honner, M.; Antrekowitsch, H. Leaching of metal-containing residues from the battery sector; Gerold, E. *World Metall.* **2019**, *5*, 267–273.

19. Gerold, E.; Luidold, S.; Scheiber, S.; Honner, M. Synergy effect at the recycling of metal containing waste from the waste industry. *Proc. Eur. Metall. Conf. EMC* **2019**, *2019*, 973–982.

20. Gerold, E.; Luidold, S.; Bartelme, C.; Antrekowitsch, H. Pyrometallurgische Aufarbeitung von Zwischenprodukten beim Batterierecycling. In *Proceedings, Recycling und Sekundärrohstoffe, Berlin (2–3 March 2020)*; Thomé-Kozmiensky Verlag GmbH: Neuruppin, Germany, 2020.

21. Chen, L.; Tang, X.; Zhang, Y.; Li, L.; Zeng, Z.; Zhang, Y. Process for the recovery of cobalt oxalate from spent lithium-ion batteries. *Hydrometallurgy* **2011**, *108*, 80–86. [CrossRef]

Publisher's Note: MDPI stays neutral with regard to jurisdictional claims in published maps and institutional affiliations.

Article

Recycling Potential of Lithium–Sulfur Batteries—A First Concept Using Thermal and Hydrometallurgical Methods

Lilian Schwich *, Paul Sabarny and Bernd Friedrich

IME, Institute for Process Metallurgy and Metal Recycling, RWTH Aachen University, 52056 Aachen, Germany; psabarny@ime-aachen.de (P.S.); bfriedrich@ime-aachen.de (B.F.)
* Correspondence: lschwich@ime-aachen.de; Tel.: +49-241-80-95194

Received: 13 October 2020; Accepted: 8 November 2020; Published: 13 November 2020

Abstract: High-energy battery systems are gaining attention in the frame of global demands for electronic devices and vehicle electrification. This context leads to higher demands in terms of battery system properties, such as cycle stability and energy density. Here, Lithium–Sulfur (Li–S) batteries comprise an alternative to conventional Li-Ion battery (LIB) systems and can be asserted to next-generation electric storage systems. They offer a promising solution for contemporary needs, especially for applications requiring a higher energy density. In a global environment with increasing sustainable economics and ambitions towards commodity recirculation, the establishing of new technologies should also be evaluated in terms of their recycling potential. In this sense, innovative recycling considers highly valuable metals but also mobilizes all technologically relevant materials for reaching a high Recycling Efficiency (RE). This study uses an approach in which the recycling of Li–S batteries is addressed. For this purpose, a holistic recycling process using both thermal and hydrometallurgical steps is suggested for a safe treatment in combination with a maximum possible recycling efficiency. According to the batteries' chemical composition, the containing elements are recovered separately, while a multi-step treatment is chosen. Hence, a thermal treatment in combination with a subsequent mechanical comminution separates a black mass powder containing all recoverable resources from the metal casing. The black mass is then treated further in an aqueous solution using different solid/liquid ratios: 1:20, 1:50, 1:55, and 1:100. Different basic and acidic leaching solutions are compared with one another: sulfuric acid (H_2SO_4), nitric acid (HNO_3), hydrochloric acid (HCl), and NaOH. For further precipitation steps, different additives for a pH adjustment are also contrasted: sodium hydroxide (NaOH) and potassium hydroxide (KOH). The results are evaluated by both purity and yield; chemical analysis is performed by ICP-OES (inductively coupled plasma optical emission spectrometry). The aim of this recycling process comprises a maximum yield for the main Li–S battery fractions: Li, S, C, and Al. The focal point for the evaluation comprises lithium yields, and up to 93% of lithium could be transferred to a solid lithium carbonate product.

Keywords: battery recycling; lithium–sulfur batteries; metallurgical recycling; metal recovery; recycling efficiency; lithium-ion batteries; circular economy

---

1. Introduction

Looking at today's society, it can be observed that the demand and desire of greenhouse gas saving and sustainable technologies is higher than ever before. As a result of this social rethinking, numerous branches of industry are being affected, including the automotive industry. This has led to a rapid change in drive technology from the combustion to electric engine [1–3]. One of the core technologies for the implementation of the electric engine, which has already established itself in the small electronics sector (smartphones, laptops, etc.), are lithium batteries [2,4]. In this context, sales of one million electric vehicles per year were already recorded worldwide in 2017, which represents an increase of 56% compared to the previous year, 2016 [5,6].

Since the 1990s, the lithium-ion battery (LIB) has been the most common form of lithium battery technology. However, despite the constant development and different forms (NMC, NCA, LMO, LFP, etc.), the most modern LIBs reach a physical limit at a specific gravimetric energy density of 350–400 Wh/kg [6–8]. Furthermore, critical raw materials such as graphite (C) or cobalt (Co), but also strategic raw materials such as lithium (Li) or nickel (Ni) are required for the production of these batteries [4,6,9,10]. Not only the incidence of the mentioned raw materials, but also the location of the mining areas and the associated import dependency are major challenges for European countries. Furthermore, the socially critical aspects of some raw materials should be considered (e.g., cobalt) [4,6].

Promising alternative future technologies to LIBs are lithium–air (L–A) batteries and lithium–sulfur batteries (LSBs). Both systems are currently in the development phase and will probably eliminate many of the disadvantages of LIBs [7,11]. Although the basic structure (electrodes, liquid electrolyte, etc.) of LSBs is similar to that of LIB, the two systems are fundamentally different in cell chemistry. The LSB has a pure lithium metal anode and a sulfur–carbon composite cathode. Carbon is indispensable, since sulfur is electrically non-conductive. The sulfur content within the cathode can vary between 50 and 70 wt.%. The remaining proportion is accounted for carbon and small quantities of the binder. In general, lithium bis(trifluoromethanesulfonyl)imide ($LiN(SO_2CF_3)_2$, LiTFSI) is used as the conducting salt [7]. The actual cell reaction takes place through the complex formation of several polysulfides (from S_8 over Li_2S_4 up to Li_2S). Therefore, an LSB is exactly a lithium polysulfide battery [8].

Due to the materials used, an LSB can theoretically achieve a specific energy density of 600 Wh/kg. On a cell level, 460 Wh/kg have already been reached [12]. A further advantage besides the high energy density is the use of toxicologically harmless, inexpensive, and readily available sulfur as an active material [7,8,13]. However, despite the numerous advantages of LSBs compared to LIBs, the cycle stability of LSBs currently represents a major challenge. The shuttle mechanism (a shuttle mechanism or effect is the cycle in which the cathodically dissolved polysulfides $[S_2]^{2-}$ diffuse unwantedly to the lithium anode, where they are reduced to lower polysulfides $[S_{n-x}]^{2-}$ and migrate back again. As a result, a part of the cathode reaction takes place at the anode, and the cell is continuously discharging [8]) leads to a continuous self-discharge of the cell. Due to these problems, LSBs currently achieve only a few hundred charging cycles [13–15].

In [7,16], a detailed elaboration on the chemical structure and redox reactions in Lithium-Sulfur Batteries is given. During charging the Li-ions diffuse from the carbon-sulfur cathode to the lithium anode and vice versa during discharging from the lithium anode to the carbon-sulfur cathode [7,16]. Here, the different polysulfides are related to the state of charge (SOH), namely for charging the sequence of polysulfide formation will be S_8, Li_2S_8, Li_2S_6, Li_2S_4, Li_2S_3, Li_2S_2 and finally Li_2S [16]. So, depending on the SOH the shares of Li_2S_8 and Li_2S can be different, showing also different properties in terms of solubility, e.g. in the electrolyte [16].

Regardless of which lithium battery technology will finally prevail, a spent lithium battery represents a versatile and important secondary raw material source. Based on the import dependency of almost all materials required for battery production, the importance of recycling is underlined once again. In this context, the recycling process must be highly efficient, environmentally compatible, and economical [6].

In the case of LIBs, there are already some proven recycling process routes that use a combination of mechanical, pyro- and/or hydrometallurgical processes [1]. Since LSBs are a new, innovative battery system, there are no significant approaches for a recycling process so far. However, EU-funded projects, such as HELIS (High Energy Lithium Sulphur cells and batteries) [17] or LISA (Lithium sulphur for SAfe road electrification) [18] aim to combine battery development and circular economy approaches, but no outcome regarding a recycling path has been published until now. In addition, there are already approaches to improve the design for recycling by using recyclable components, such as a Co_3Mo_3C-separator [19]. This paper presents for the first time a recycling concept for the lithium–sulfur battery with the aim of recovering all elements, considering the legal requirement of 50 wt.%. This legal frame is demanded in the EU battery directives 2006/66/EG [20] and 493/2012 [21], where the threshold of 50 wt.% is described as recycling efficiency (RE) of at least 50 wt.% based on a battery's cell level, as can be seen in Formula (1) [21]:

$$\text{RE } [\%] = \frac{\sum m_{output}}{m_{input}} \times 100 \text{ [mass . \%]} \tag{1}$$

To reach this target, a suitable recycling path has to be developed for any battery system. Since LSBs do have a metallic lithium anode, which is critical in terms of a high reactivity leading to exothermal oxidation and, hence, safety issues [22,23], a suitable pre-treatment before entering the metallurgical processing can be helpful. It is also crucial to work in an inert atmosphere to prevent atmosphere-related oxidation [24]. In addition, metallic lithium can ignite when heated beyond 180 °C in air [25], which can even occur due to mechanical strain, such as shredding [26], or when being in contact with moisture [27]. One form of thermal pre-treatments is a pyrolysis, where the cells are deactivated in the absence of oxygen at temperatures of maximum 600 °C [28]. In order to remove the containing organics, such as binders, an optimal temperature of 550 °C in air has been defined by Chen et al., using LCO cells [29]. Both pyrolysis and incineration, with some oxygen shares, are thermal pre-treatments for a safe cell deactivating and facilitating of further downstream recycling without an uncontrolled thermal runaway [30,31]. Another benefit comprises an eased detachment of substrate foils and active mass [32], and especially for hydrometallurgical processing, a thermal pre-treatment is suggested [33]. For a metallurgical recycling, both hydro- and pyrometallurgical processes are available [34]. Pyrometallurgy is established regarding the production of Co, Ni, and Cu alloys, hence, rather ignoble and valuable component recycling [35], whereas hydrometallurgy is also able to selectively separate rather ignoble components, such as graphite, aluminum, or lithium [35–37]. This is why this study considers hydrometallurgical treatments as suitable for the purpose of a circular economy approach. Although hydrometallurgy comprises generally slower kinetics [37], it leads to higher yields and lower energy consumption [36].

Based on LIBs, different strategies for optimal wet-chemical processing are investigated around the world. Within this variety of approaches, an overview on relevant literature is given as follows. Generally, besides physical separation methods in aqueous environments, such as flotation [38], chemical processing is mainly based on leaching, precipitation, solvent extraction, and ion exchanging [39]. Within acidic leaching, both organic or inorganic solvents can be used. For LIBs recycling approaches, studies have examined hydrochloric acid (HCl), sulfuric acid (H_2SO_4), phosphoric acid (H_3PO_4), and nitric acid (HNO_3) in terms of inorganic acids, and citric acid, malic acid, acetic acid, lactic acid, or trichloroacetic acid have been studied in terms of organic acids. For every solvent, different concentrations, leaching temperatures, or solid/liquid ratios have been reported [40]. Within these solvents, Zou et al. report a lithium leaching efficiency of 100% using 4-molar H_2SO_4 and adding of 30 wt.% H_2O_2, and a lithium yield of 80% recovered as lithium carbonate by using sodium carbonate (Na_2CO_3) [41]. A similar approach is presented by Wang et al. [42,43]. In this context, 2 molar H_2SO_4 and 4 molar HCl with or without addition of 50 g/L H_2O_2 were examined, leading to a lithium yield of maximum 64% when using H_2SO_4. Castillo et al. investigated 0.5 to 5-molar HNO_3, adjusted the pH value by NaOH, and obtained 100% leaching efficiency for lithium, when applying acidic concentrations between 1 and 2 moles [44]. Besides acidic leaching, alkaline leaching can also be applied, which is rarely investigated for LIBs [45]. Ferreira et al. have followed an approach starting with alkaline leaching in NaOH, since aluminum shows a better recyclability in basic environment, and then, the pH value was adjusted step-wise by using H_2SO_4 and H_2O_2. This set-up leads to a leaching efficiency of 100% for lithium [46]. The approach for the novel and innovative recycling process for lithium–sulfur batteries is based on the knowledge and experience gained from the hydrometallurgical recycling processes of lithium-ion batteries. Elements such as Co, Ni, or Cu are no longer inserted into an LSB, so the process scheme is adjusted. However, during charging and discharging, many different, intermediate polysulfides with Li and S are formed, making the recycling process challenging [47]. This study's target is a zero-waste recovery of the components C, S, Li, and Al.

Research Needs and Work Hypothesis

Since Li–S batteries are a promising alternative to conventional Li-ion batteries, investigating their recycling process options is crucial regarding the concepts of circular economy and waste minimization. The intrinsic materials' value of Li–S batteries is comparatively lower due to the dispensing of cobalt and nickel, while this work focuses on the recovery of lithium. Other material fractions, such as carbon and aluminum, will be separated, too, but no discussion on recovery yields is taking place at this early-stage of recycling considerations for Li–S batteries. The lithium contents in Li–S batteries are higher than in conventional Li-ion batteries, while lithium has a higher impact on the recycling efficiency than in Li-ion batteries.

2. Materials and Methods

In order to enable a high selectivity and treat the rather ignoble LSB components with a combination from thermal treatment, mechanical treatment and hydrometallurgical processing is presented in this study, as shown in the process flow chart in Figure 1.

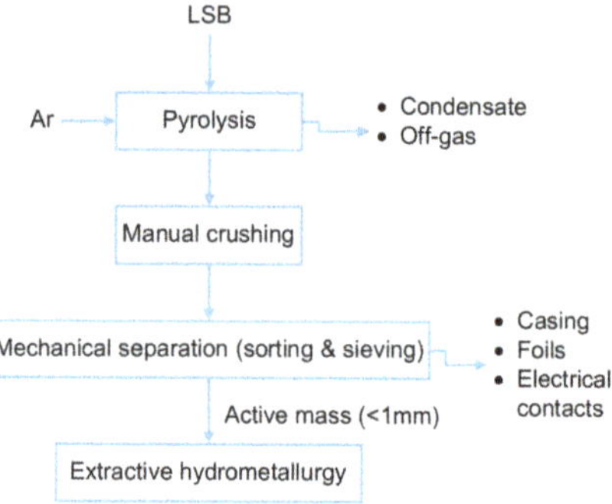

Figure 1. General process for a lithium–sulfur battery (LSB) recycling process based on extractive hydrometallurgy for elemental recovery.

LSB pouch cells are provided by Fraunhofer IWS (see Figure 2b), of whose every cell has a specific composition. The cells are pyrolyzed in a Thermostar resistance furnace (Thermostar, Aachen, Germany), which is flood with Argon to displace oxygen and, thus, avoids exothermal reactions with the environment. This incineration is to be prevented due to formations of $\{CO_2\}$, $\{SO_2\}$, and thus, active mass losses. A specifically constructed steel chamber with small holes ensures controlled off-gas release and hence prevents a sudden excess pressure. This chamber is placed in a closed and sealed steel reactor, which is then inserted in the furnace (see Figure 2a). The off-gas can leave the system at two exit points: Firstly, two scrubbers in a row containing deionized H_2O clean the main off-gas stream, neutralizing acidic gases. The residual permanent gas leaves the reactor to an off-gas cleaning system. Secondly, an FTIR (Fourier transform infrared spectroscopy) analyzer (Gasmet Technologies Oy, Helsinki, Finnland) drains a defined volume flow from the off-gas for identifying gaseous phases. The pyrolysis temperature comprises 500 °C with a holding time of 1 h, which is continuously measured within the furnace and between the excess pressure chamber and steel reactor (see Figure 2a).

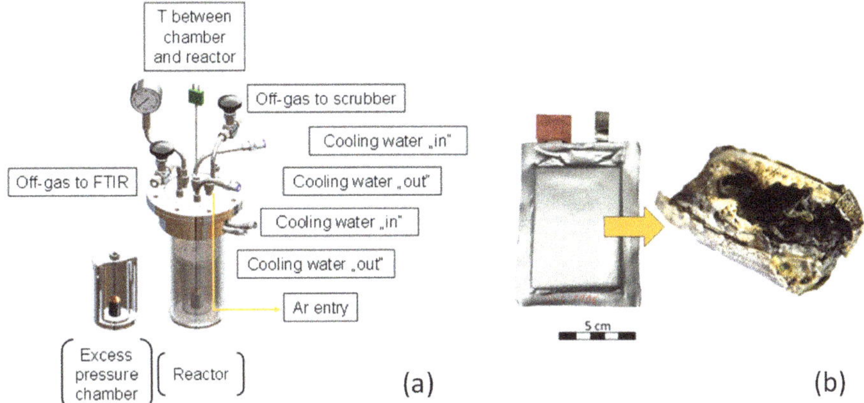

Figure 2. (**a**) Schematic pyrolysis set-up with relevant entries into and exits for gaseous and liquid components. (**b**) LSB cell from Fraunhofer IWS before and after pyrolysis.

As Figure 3 shows, the pyrolysis treatment leads to a cell opening temperature mean value of 90 °C, made visible by an abrupt rise in different gaseous phases, such as propene, methane, or formaldehyde.

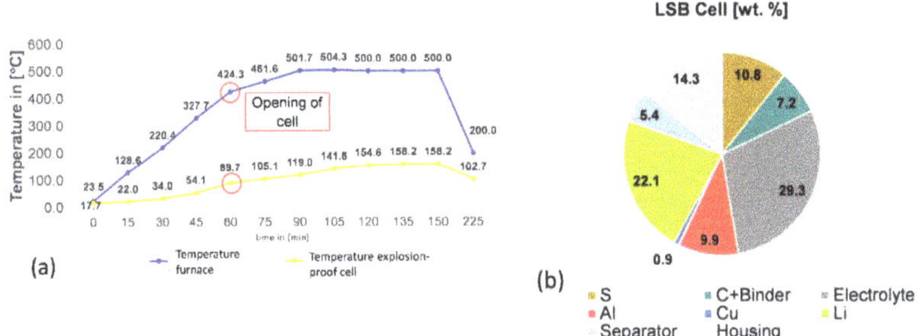

Figure 3. (**a**) Exemplarily pyrolysis temperature profile, where the furnace-measured temperature is higher than the temperature reached between the excess pressure chamber and the furnace temperature. (**b**) Exemplarily composition of an LSB 1 cell, whose chemical composition changes due to pyrolysis in terms of removing up to 5.4 wt.% from the separator, 29.3 wt.% from the electrolyte, and maximum 7.2 wt.% from binder.

Moreover, pyrolysis results in an averaged weight loss of 27.6%, due to the evaporation of volatile components stemming from electrolyte, separator, and binder. The next step comprises the manual separating of casing from the active mass and substrate foils in a glovebox. The downstream comminution by grinding in mortar is conducted, also in a glove box. The material still releases gaseous compounds, which can be toxic, such as H_2S. Furthermore, this treatment manner prevents metallic lithium from the anode from oxidizing. Subsequently, after separation and grinding, sieving is performed to extract the active mass of <1 mm, whose exemplarily composition can be seen in Table 1. It has to be pointed out that heterogeneities persist both within the batteries due to the current research regarding their cell design, and also within the samples taken for analyzing active mass. The averaged, extracted active mass of a cell comprises 14.6 g. Therefore, calculating yields by extractive hydrometallurgy does not take the chemical analysis as reference but instead sums up the first filter cake and the first solution to get information about the real composition.

Table 1. Chemical composition of exemplarily cell "LSB 2", in particular its active mass (<1 mm).

Al	Fe	Li	F	S	C
3.15 wt.%	0.04 wt.%	15.8 wt.%	4 wt.%	10.6 wt.%	15 wt.%

The chemical compositions within this study is measured by ICP-OES (Spectro, Kleve, Germany) for the present metals and by combustion method in the case of S and C, and combustion ion chromatography (CIC) in the case of F.

The hydrometallurgical process applied in this study in can be described as follows: The first step comprises a leaching with a subsequent filtration of the C-product, a precipitation including a subsequent filtration of copper sulfide, a pH adjustment for Al-precipitation, and a carbonation with a precipitation and a subsequent Li_2CO_3 filtration. The leaching step is conducted by different solvents, HCl, HNO_3, H_2SO_4, and NaOH. Here, carbon stemming from both cathode material and from pyrolysis soot is insoluble and, thus, can be filtrated. During leaching, 20 mL of H_2O_2 are added. Precipitation 1 makes use of the ongoing reaction between copper and sulfur. Cu has a high affinity to S, forming Cu_2S. During the leaching process S-containing gases are liberated and are transferred into a $CuSO_4$ solution. Here, copper matte is created, which is a precursor for copper production in the established Cu-production path, and then filtrated. The filtrate after the first filtration (C-filter cake) is colorless liquid free from solid particles (filtrate). The next step, precipitation 2, works by adjusting the pH using NaOH, KOH, or HNO_3, hence, basic solvents, to precipitate aluminum selectively. Some F-contaminations in the active mass due to abrasion of the steel pyrolysis reactor can be removed along with Al, since the Eh-pH diagram of Fe and Al show a possible precipitation in the same pH range [48].

$$Al^{3+} + 3OH^- \leftrightarrow Al(OH)_3 \tag{2}$$

The process described can be visualized as follows (see Figure 4):

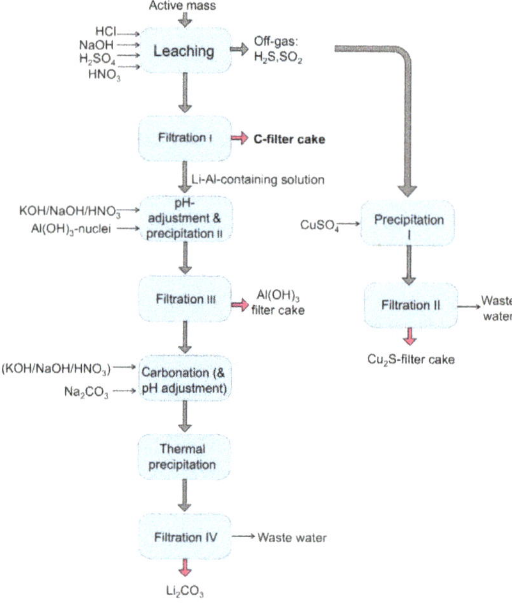

Figure 4. Proposed process flow chart for a hydrometallurgical treatment of lithium–sulfur batteries.

Therefore, the pH value was increased to a value of 3, followed up by an addition of H_2O_2 to assure a dissolution of Al. Afterwards, the pH value was increased another time in an area to 5, where the best precipitation of Al and Fe is reported. Subsequently, the addition of $Al(OH)_3$ nuclei lead to a turbidity, representing the on-going precipitation. Before precipitation 2, 20 mL of H_2O_2 are added to the solution. The step "filtration 3" separates the generated precipitate. Precipitation 3 makes use of the Li carbonates' property to have a lower solubility product at higher temperatures. Therefore, the amount of water can be reduced and therefore the formed Li_2CO_3 can precipitate. This behavior is promoted by the reduced liquid, in other words, approaching the solubility product. Thus, the last process step comprises a temperature increase to 100 °C, depending on the present pH value, followed by a slight pH adjustment in order to reach a neutral/basic area. As a side effect of the temperature increase, the solubility product of the formed Li_2CO_3 is reduced from 13.2 to 7.2 g/L [49]. In this case, the pH value must be until ~7. Then, the addition of Na_2CO_3 leads to a pH increase in the area of ~9–10. This is also beneficial, because other studies have calculated a formation and thus precipitation of lithium carbonate in alkaline areas [50]. Boiling and subsequent filtrating, namely filtration 4, of the solution enables lithium recovery as Li_2CO_3.

It is important to highlight the fact that different reactions occur, depending on the lithium sulfide compound as educt: During charging and discharging, different polysulfide phases arise, for example S_8, Li_2S_8, or Li_2S. Hence, the input phases have a direct impact on the ability to dissolve in aquatic media.

Figure 5 displays the reactor used for the first process sequence, namely the leaching step. An entirely sealed glass reactor fully transfers the arising gaseous phases into the $CuSO_4$ solution. This is crucial from a circular economy perspective, hence, for obtaining a maximum S recovery as Cu_2S, but also in terms of canalizing toxic off-gas products. The transfer from leaching reactor to the scrubbers is promoted by an N_2 carrier gas, which is led into the reactor area. During the leaching, both H_2S and SO_2 can be detected at the outlet of the second scrubber bottle. This shows an incomplete reaction between gas and product, but during leaching, the color of the $CuSO_4$ solution changed from blue into dark green, resulting in a solid precipitate. In virtue of the small input mass amount, the formed precipitate mass shows a few milligrams.

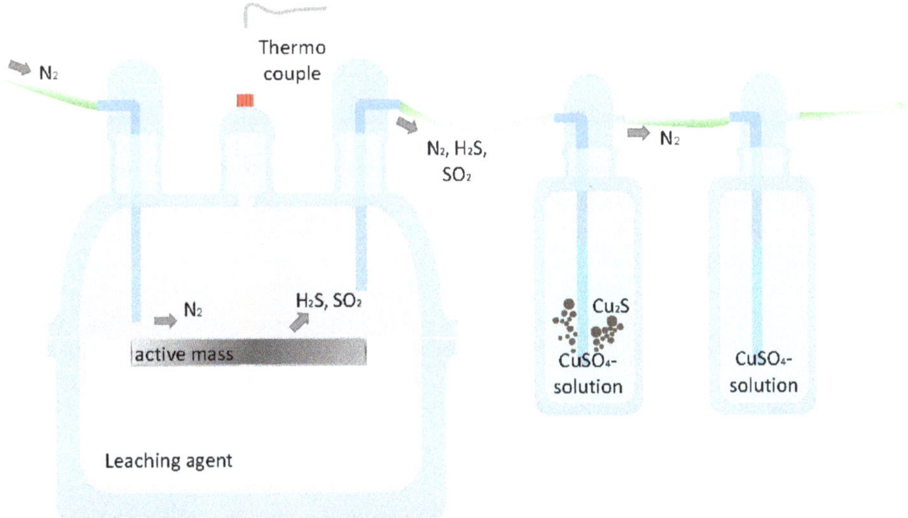

Figure 5. Schematic illustration of the applied leaching step set-up: Li–S black mass is inserted into a sealed three-neck flask and the add-on reactors for S recovery are represented by gas washing bottles (both borosilicate glass).

Table 2 displays an overview of the parameters examined within the conducted trials, resulting in 37 parameter combinations, where each set-up was performed one time within the pre-trial series (indicated as "VV1-29") to perform a screening. The best-case scenarios were repeated within the main trials (indicated as EV1-3 and VD1-5).

Table 2. Overview of the parameters used in this study. The leaching agent columns match the leachate concentration columns. The pH additive columns match the pH additive concentrations. Other columns are not to read as matching parameter specifications, e.g., HNO_3 as leaching agent can also comprise trials with a 120 min leaching time.

Parameter	Parameter Specifications			
Leaching agent	HNO_3	H_2SO_4	HCl	NaOH
Leaching agent concentration	2-, 4-, and 8-molar	2- and 4-molar	4-molar	4-, and 8-molar
Leaching time [min]	60		120	
Solid/liquid (S/L) ratio (g/mL)	1:20 (13 g/250 mL)	1:30 (13.3 g/400 mL)	1:50 (10 g/500 mL or 5 g/250 mL)	1:55 (4.5 g/250 mL) 1:100 (2.5 g/250 mL)
Leaching temperature [°C]	60 °C	23.15 °C (room temperature)	100 °C	
pH-additive	NaOH		KOH	
pH-additive concentration	4-, 8-, 14-molar		2-, 4-, 8-molar	

3. Results and Discussion

For the following discussion, the focus is based on lithium yields, since lithium is the key driver for the recycling of LSBs, especially by value. Lithium values in filter cakes and solutions were detected by ICP-OES in a certified laboratory.

As already reported above, input analyses from the active mass show less accuracy due to the heterogenous chemical composition of the black mass/active mass. Bringing these aspects together, the lithium yields were calculated by summing up the lithium mass in every filter cake (fc) (C-fc, $Al(OH)_3$ fc and Li_2CO_3 fc), and the mass of lithium in the residual filtrate after filtration IV($Li_{filtrate, IV}$). An alternative calculation would be summing up the lithium mass in the carbon filter cake (C -fc) and the lithium mass in the first filtrate (filtrate $_I$), but since this calculation leads to the same results, the first presented option was chosen. Finally, the calculation of lithium yields was performed as follows:

$$\eta_{Li} = \frac{Li_{Li_2CO_{3-fc}} \, [g]}{Li_{total} \, [g]} \tag{3}$$

$$with \; Li_{total} \, [g] = \sum \left(Li_{C-fc} + Li_{Al-fc} + Li_{Li_2CO_{3-fc}} + Li_{filtrate \; IV} \right) \tag{4}$$

In order to visualize the distribution of lithium, exemplary Sankey diagrams reveal lithium distribution within the different filter cakes in Figures 6 and 7. This behavior is explainable by the high reactivity of lithium, entering the leaching step also in metallic form stemming from the anode.

"VV5" shows the highest lithium losses in the C filter cake. This can be attributed to the room temperature leaching. "VD4" represents the lowest lithium losses within the Al filter cake. This can be explained by the amount of H_2O_2 added in the process step "pH adjustment and precipitation II". In "VV5", 15 mL H_2O_2 was added at once; whereas in "VV19", 10 mL H_2O_2 was added in two 5 mL steps; in "VV28", 10 mL H_2O_2 was added at once; and in "VD4", 30 mL of H_2O_2 was added in three 10 mL steps.

When comparing the leaching agents HCl, NaOH, and H_2SO_4 in terms of lithium distribution, only "VV10" shows low lithium losses within the residual solution. With regard to the impurity evaluation of lithium carbonate, it can be seen that "VV7" shows high shares of Cl, and "VV29" shows high shares of S and K. Thus, the formation of more stable phases suppresses the precipitation of lithium carbonate. The highest lithium losses in the C filter cake occur when using NaOH. This can be due to an incomplete dilution of lithium phases within the active mass in alkaline areas.

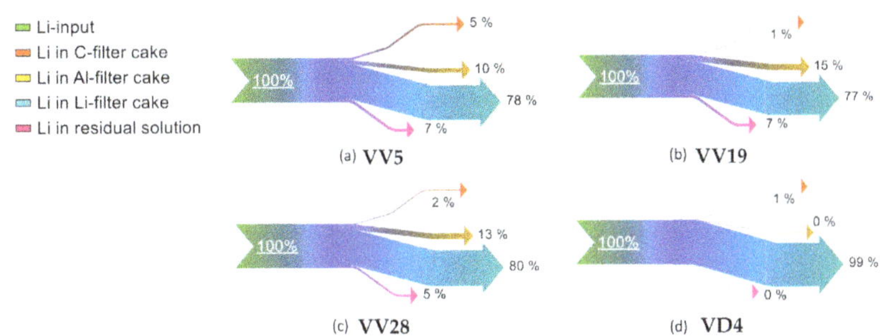

Figure 6. Sankey diagrams of the best HNO_3 trials for the target metal lithium. (**a**) 2-molar HNO_3, solid/liquid ratio 1:100, leaching for 60 min at 23.15 °C, pH-additive 8-molar NaOH, using of $Al(OH)_3$ nuclei and adding of Na_2CO_3 (**b**) 4-molar HNO_3, solid/liquid ratio 1:55, leaching for 60 min at 60 °C, pH-additive 14-molar NaOH, using of $Al(OH)_3$ nuclei and adding of Na_2CO_3 (**c**) 2-molar HNO_3, solid/liquid ratio 1:50, leaching for 60 min at 60 °C, pH-additive 8-molar KOH, using of $Al(OH)_3$ nuclei and adding of Na_2CO_3 (**d**) 2-molar HNO_3, solid/liquid ratio 1:50, leaching for 60 min at 60 °C, pH-additive 2-molar KOH, using of $Al(OH)_3$ nuclei and adding of Na_2CO_3.

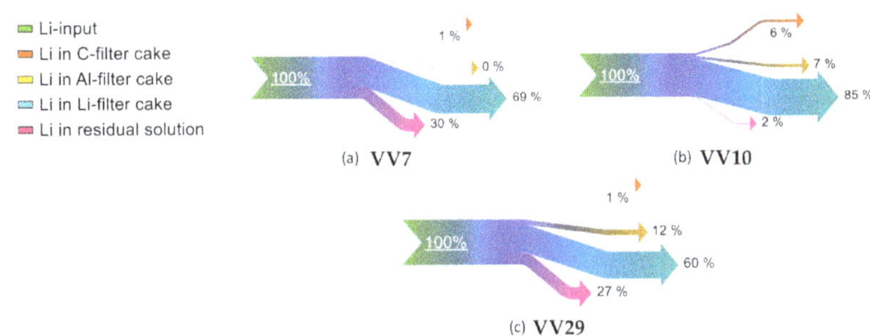

Figure 7. Sankey diagrams of the best-case (**a**) HCl-leaching agent (VV7), (**b**) NaOH-leaching agent (VV10), and (**c**) H_2SO_4-leaching agent (VV29).

A qualitative X-ray powder diffractometry (XRD) evaluation shows the following main lithium phases within the pyrolyzed active mass (see Table 3). However, it should be noted that within the samples, deviations can occur due to the state-of-charge and the post mortem cells history. This will consequently lead to different cell-internal reactions and different phases within the active mass.

Table 3. Main Li phases detected by XRD in an exemplarily active mass stemming from post-mortem Li–S cells.

Detected Li Phases:	LiOH	Li_2CO_3	LiF	Li_2SO_4

Table 4 gives an overview on parameter combinations using H_2SO_4. It can be seen that 2- and 4-molar sulfuric acid was paired with NaOH and KOH.

Table 4. H_2SO_4 parameter combinations applied in the pre-trials. S/L represents the solid/liquid ratio applied, hence, black mass per leaching liquid.

Trial	Leaching Agent Conc.	S/L Ratio [g/mL]	Leaching Time and Temperature	pH Additive and Concentration	Adding Al(OH)$_3$	Adding Na$_2$CO$_3$
VV1	2-molar H_2SO_4	1:100	60 min at 100 °C	8-molar NaOH	yes	yes
VV4	2-molar H_2SO_4	1:100	60 min at 23.15 °C	4/8-molar NaOH	yes	yes
VV11	2-molar H_2SO_4	1:55	120 min at 60 °C	4-molar NaOH	yes	yes
VV12	2-molar H_2SO_4	1:55	120 min at 60 °C	4/8-molar NaOH	yes	yes
VV15	2-molar H_2SO_4	1:55	60 min at 60 °C	14-molar NaOH	no	yes
VV16	2-molar H_2SO_4	1:55	60 min at 60 °C	14-molar NaOH	no	yes
VV20	4-molar H_2SO_4	1:55	60 min at 60 °C	4/8-molar NaOH	yes	yes
VV21	4-molar H_2SO_4	1:55	60 min at 60 °C	8/14-molar NaOH	yes	yes
VV22	2-molar H_2SO_4	1:50	60 min at 60 °C	8-molar NaOH	yes	no
VV25	2-molar H_2SO_4	1:50	60 min at 60 °C	8-molar KOH	yes	no
VV29	2-molar H_2SO_4	1:50	60 min at 60 °C	8-molar KOH	yes	yes

Figure 8 displays the yields of H_2SO_4 in terms of the pre-trials. It can be seen that the lithium yield has not crossed the 50% threshold in all trials, except "VV29".

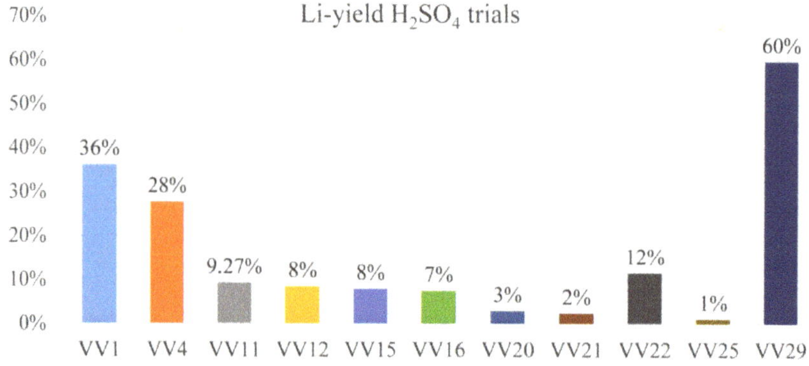

Figure 8. Overview on the lithium yields reached in the pre-trials by using H_2SO_4 as solvent.

"VV29" differs from the other H_2SO_4-trials in terms of the pH additive used: in this case, KOH was used. In the other H_2SO_4-trials, lithium losses can be asserted to high contents either in the C filter cake but especially in the Al filter cake. However, most of the lithium remains in the residual solution and is therefore irrecoverable if the solution after the Li filter cake filtration will not be circulated or the filtration is not optimized. This is theoretically feasible to avoid losses but was not conducted here to keep the trial procedure constant among all trials.

In contrast to that, the lithium yield by using HNO_3 shows in most cases a higher yield than 50%, as can be seen in Figure 9. Table 5 displays the matching parameters used for the HNO_3 pre-trials.

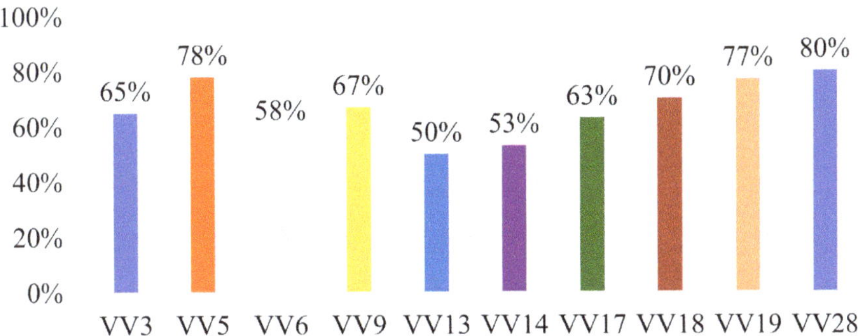

Figure 9. Overview on the lithium yields reached in the pre-trials by using HNO_3 as solvent.

Table 5. HNO_3 parameter combinations applied in the pre-trials.

Trial	Leaching Agent Conc.	S/L Ratio [g/mL]	Leaching Time and Temperature	pH Additive and Concentration	Adding Al(OH)$_3$	Adding Na$_2$CO$_3$
VV3	2-molar HNO_3	1:100	60 min at 100 °C	4/8-molar NaOH	yes	yes
VV5	2-molar HNO_3	1:100	60 min at 23.15 °C	8-molar NaOH	yes	yes
VV6	2-molar HNO_3	1:100	120 min at 60 °C	4/8-molar NaOH	yes	yes
VV9	2-molar HNO_3	1:50	120 min at 60 °C	4-molar NaOH	yes	yes
VV13	2-molar HNO_3	1:55	60 min at 60 °C	4/14-molar NaOH	yes	yes
VV14	2-molar HNO_3	1:55	60 min at 60 °C	4/14-molar NaOH	no	yes
VV17	2-molar HNO_3	1:55	60 min at 60 °C	8/14-molar NaOH	no	yes
VV18	4-molar HNO_3	1:55	60 min at 60 °C	8/14-molar NaOH	yes	yes
VV19	4-molar HNO_3	1:55	60 min at 60 °C	14-molar NaOH	yes	yes
VV28	2-molar HNO_3	1:50	60 min at 60 °C	8-molar KOH	yes	yes

In the case of using NaOH as leaching agent, trials "VV10", "VD1", and "VD2" were performed. Table 6 shows the parameter combinations tested.

Table 6. NaOH parameter combinations applied in the pre-trials.

Trial	Leaching Agent Conc.	S/L Ratio [g/mL]	Leaching Time and Temperature	pH Additive and Concentration	Adding Al(OH)$_3$	Adding Na$_2$CO$_3$
VV10	4-molar NaOH	1:50	60 min at 60 °C	4-molar HNO3	yes	yes
VD1	8-molar NaOH	1:29	60 min at 60 °C	8-molar HNO3	yes	yes
VD2	8-molar NaOH	1:50	60 min at 60 °C	4/8-molar HNO3	yes	yes

The NaOH trials show poor lithium yields, except "VV10", as can be seen by Figure 10a. The HNO$_3$ trials show the best consistency in yields, independent from the parameters chosen, as can be seen in Figure 10b. "VD3" shows low lithium results, which can be explained by only use of an 8-molar acid. This concentration can, hence, be classified as an inadequate parameter.

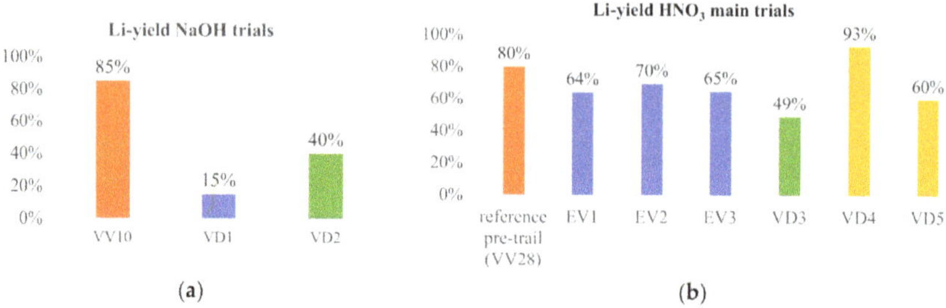

Figure 10. (**a**) Overview of the lithium yields reached in the pre-trial (VV10) and in the main trials (VD1 and VD2) by using NaOH as solvent. (**b**) Overview of the lithium yields reached in the best-case pre-trial (VV28) and the main trials (EV1-3, VD3-5) by using HNO$_3$ as solvent.

In terms of lithium yields, it can be concluded that "VV5", "VV19", and "VV28" represent the best outcome with the leaching agent HNO$_3$ for the pre-trials, and "VD4" the best outcome for the main trials. In case of H$_2$SO$_4$, it is "VV29", and in case of NaOH, it is "VV10". "VV10" is different from "VD1" and "VD2", since a leaching concentration of 4 mol/L was chosen, instead of 8 mol/L in "VD1" and "VD2". In addition, the solid/liquid ratio shows an impact when comparing "VD1" and "VD2": a solid/liquid ratio of 1:50 reflects higher lithium yields. The HCl trial shows a comparatively low lithium yield of 69%, in combination with high Cl contaminations within the lithium carbonate filter cake. Hence, this solvent was not used repeatedly and is therefore not represented in the bar charts. However, the parameter specifications for the best yields in Table 7 give an overview on successful combinations.

Table 7. Summary of parameter combinations with the highest lithium yields of each leaching agent. Since the yields of pre-trials VV5, VV19, and VV28 show similar lithium yields, all three are highlighted. Every parameter combination is represented once.

Trial	Leaching Agent and Concentration	Solid/Liquid Ratio [g/mL]	Leaching Time and Temperature	pH Additive andConcentration	Addition ofAl(OH)$_3$ Nuclei
VV5	2-molar HNO$_3$	1:100	60 min at RT	8-molar NaOH	yes
VV7	4-molar HCl	1:100	60 min at RT	8-molar NaOH	yes
VV10	4-molar NaOH	1:50	60 min at 60 °C	4-molar HNO3	yes
VV19	4-molar HNO$_3$	1:55	60 min at 60 °C	14-molar NaOH	yes
VV28	2-molar HNO$_3$	1:50	60 min at 60 °C	8-molar KOH	yes
VV29	2-molar H$_2$SO$_4$	1:50	60 min at 60 °C	8-molar KOH	yes
VD4	2-molar HNO$_3$	1:50	60 min at 60 °C	2-molar KOH	yes

Hence, it can be derived that the following parameters lead to enhanced Li yields:

- HNO$_3$ and NaOH reach the highest lithium yields. H$_2$SO$_4$ and HCl yields have not been satisfying when comparing all leaching agents in terms of lithium yields.
- All best-case scenarios in show best results when leaching for 60 instead of 120 min.
- Addition of H$_2$O$_2$ is beneficial.

- Addition of Al(OH)$_3$ as nuclei is suggested.

Since not only reached yields are decisive, arising lithium filter cake impurities are also to be focused upon. Hence, Table 8 focuses on the chemical composition of several lithium carbonate filter cakes.

Table 8. Impurities occurring within the best-case scenarios of each solvent yields based on ICP-OES analysis.

Trial	K Impurities [wt.%]	S Impurities [wt.%]	Na Impurities [wt.%]	Cl Impurities [wt.%]	F Impurities [wt.%]	Al Impurities [wt.%]	Sum of Impurities [wt.%]
VV5	n/a	n/a	25.6	n/a	0.2	n/a	25.8
VV7	n/a	n/a	10.8	17.9	n/a	0.2	28.9
VV10	n/a	0.2	14.4	n/a	n/a	n/a	14.6
VV19	n/a	n/a	9.4	n/a	n/a	n/a	9.4
VV28	4.9	n/a	0.7	n/a	n/a	n/a	5.8
VV29	33.9	15	1.7	n/a	n/a	1.6	52.2
VD4	5.4	13.5	0.9	n/a	0.6	0.2	20.6

By evaluating the impurities, HNO$_3$ and NaOH also show better results, especially trial "VD4". Hence, another preferred parameter can be derived from the impurities' evaluation:

- KOH is slightly more suitable. "VV29" is excluded from this evaluation, since the leaching agent H$_2$SO$_4$ has not been suitable in yields, as well. Although "VV29" (H$_2$SO$_4$ leaching agent) represents an unsatisfactory result compared to other experiments, the highest Li recovery could also be determined in this experiment by using KOH.
- According to Figures 10 and 11, lithium shows the highest shares within the C filter cake among the HNO$_3$-trials when leaching at room temperature ("VV5"). Additionally, the sum of impurities is also higher in "VV5", according to Table 8. Thus, a leaching temperature at 60 °C is preferred. In addition, the yields of aluminum within the Al filter cakes were higher when applying leaching at 60 °C.

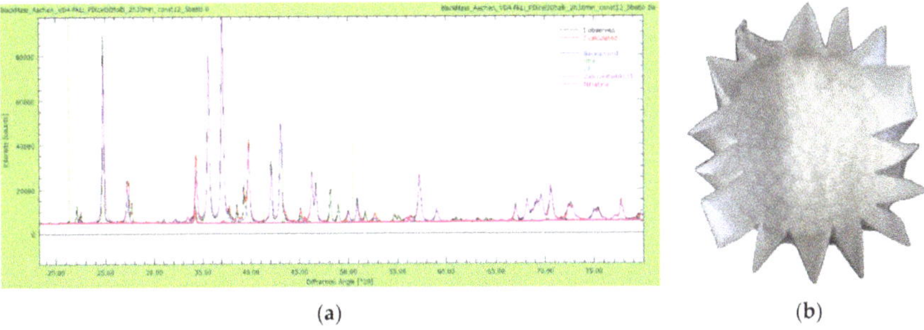

(a)   (b)

Figure 11. XRD analysis of VD4's lithium filter cake (**a**). Main phases detected comprise Li$_2$CO$_3$, and in smaller shares LiF, KNO$_3$, and NaNO$_3$. On the right side (**b**), a visualization of extracted lithium carbonate can be found.

Due to the pre-trial results, the main trials "ED1-3" and "VD4-5" were performed with different process parameters based on the combination of KOH and NaOH. Moreover, "VV10" showed a high lithium yield. Hence, more trials with NaOH were performed within the main trials. Table 9 gives an overview on the used parameters.

Table 9. Parameter combinations of the main trials.

Trial	Leaching Agent and Concentration	Solid/Liquid Ratio [g/mL]	Leaching Time and Temperature	pH Additive and Concentration	Addition of Al(OH)$_3$ Nuclei
ED1-ED3	4-molar HNO$_3$	1:50	60 min at 60 °C	8-molar KOH	yes
VD1	8-molar NaOH	1:30	60 min at 60 °C	8-molar HNO$_3$	yes
VD2	8-molar NaOH	1:50	60 min at 60 °C	4-molar HNO$_3$	yes
VD3	8-molar HNO$_3$	1:20	60 min at 60 °C	4-molar KOH	yes
VD4 and VD5	2-molar HNO$_3$	1:50	60 min at 60 °C	2-molar KOH	yes

"VD1" shows low lithium yields (15%), which can be attributed to a higher solid/liquid ratio than both "VV10" and "VD2". In addition, the direct comparison between a 4-molar NaOH (VV29) and an 8-molar NaOH ("VD2") shows significantly lower yields for the 8-molar NaOH, so the parameter combination of a lower solid/liquid ratio and a lower molality ("VV10") is the preferential combination. "VD3" has not reached high lithium yields due to its solid/liquid ratio of 1:20, the influence of the parameter combination of an 8-molar HNO$_3$ and the use of a 4-molar HNO3 cannot be detected at this point. The comparatively low lithium yields of "VD5" can be asserted to a high aluminum-containing input fraction. Only one third of the lithium in "VD4's" input material is present in the input material of "VD5", thus lithium losses in other filter cakes according to Figures 8 and 9 have a bigger impact. As already reported above, the material used in this study shows a high degree of heterogeneity and especially within the trial set-ups of using a few grams per trial, a deviation within the composition can potentially change the system behavior.

From Figure 10a and Table 9, it can be derived that the trials VD1 and VD2, whose concentration was an 8-molar NaOH, could not lead to high yields. The pH additive was selected as follows: VD1 with an 8-molar HNO$_3$ solution and VD2 with a 4-molar HNO$_3$ solution. Thus, it can be concluded that the optimal parameter combination for NaOH consists of a 4-molar NaOH solution with a 4-molar HNO$_3$ pH additive.

Since it is crucial for a further refining of the lithium filter cakes to gain knowledge on the prevalent phases, XRD analyses provide insight into the presence of lithium carbonate and impurities on a phase level. This is exemplarily shown by the XRD analysis of VD4's lithium filter cake (see Figure 11a). Here, mainly Li$_2$CO$_3$ but also LiF, KNO$_3$, and NaNO$_3$ are present. The impurities containing KNO$_3$ can be asserted to the use of KOH as pH additive, and the impurities containing NaNO$_3$ can be asserted to the addition of Na$_2$CO$_3$ for forming Li$_2$CO$_3$.

VD4 shows the highest lithium yields, while the chemical composition of the other filter cakes, namely C filter cake, CuS filter cake, Al filter cake, and lithium carbonate filter cake are to be found in Table 10. VD4 is chosen exemplarily for efficiency reasons. Here, "fc" indicates the filter cakes, and no oxygen and hydrogen can be detected by ICP-OES, while the sum of detected elements is not 100%.

Table 10. Chemical composition of VD4 filter cakes in [wt.%]. The label "fc" indicates filter cakes.

Filter Cake	Li	Al	Fe	K	Na	C	S	F	Cu
					[wt.%]				
C–fc	0.7	n/a	n/a	n/a	n/a	60.7	14.7	0.6	n/a
CuS–fc	n/a	n/a	n/a	n/a	n/a	n/a	26.7	0.01	72.9
Al–fc	0.4	38.3	n/a	0.6	0.2	n/a	n/a	2.4	n/a
Li–fc	17.6	0.2	n/a	5.4	0.9	13.5	0.12	0.6	n/a

The majority of C enriches within the C filter cake that the principle of S recovery as copper sulfide was successful, and Al mostly precipitates in the Al filter cake. Interestingly, also F is mainly leaving the aqueous system along with the Al filter cake. This is explainable by the formation of Al–F-containing phases, such as AlF_3, but needs to be proven by XRD analysis. Besides the high S-shares, which have already been discussed before, the set-up of VD4 shows a proof-of-concept of this study's scope.

According to the elemental distribution of the obtained filter cakes in Table 10, the filter cakes obtained from HNO_3 leaching with KOH as pH additive are visualized in Table 11. In the C fc, metallic Al flakes are still visible, since metallic Al indicates a low leaching efficiency for metallic Al. The color of the CuS fc together with the chemical composition in Table 10 suggests the formation of CuS or Cu_2S. Some $CuSO_4$ shares are also possible but further XRD analysis would be needed to investigate this. The Al fc color suggests the formation of $Al(OH)_3$ and AlF_3.

Table 11. Pictures of the obtained filter cakes apart from Li–fc (see Figure 11b) by leaching with HNO_3.

C fc	CuS fc	Al fc

Finally, a process for lithium recycling from lithium–sulfur batteries was proven: A combination of pyrolysis, manual extraction of black mass, and subsequently, leaching of black mass in HNO_3 have shown lithium yields of 93%, with a Li_2CO_3 purity of 92.78%. H_2SO_4, HCl, and NaOH, which were validated as suitable leaching agent for Li-ion battery active mass, show poor results for lithium–sulfur batteries.

4. Outlook

For future research, several aspects are to be investigated further: A repetition of the set-up in trials "VV10", "VV28", and "VD4" for statistical validation will be one focus. These set-ups are indicated in Table 11 as "3×", representing three repetitions. Besides these repetition trials, different set-ups in terms of detailed HNO_3 investigation will be added. These parameter combinations are indicated in Table 11 as "1×", representing one trial. The combination of 4-molar HNO_3 with 8-molar KOH is neglected due to the comparatively low yields in "ED1-3". Although the combination of 8-molar HNO_3 and 2-molar KOH as pH additive was investigated in "VD3", it will be tested again with a more promising solid/liquid (S/L) ratio. An overview on the trials to be performed as next steps is given in Table 12.

The most successful set-up in terms of Table 11 will be performed with the battery fraction > 1 mm, as well, since in Li–S batteries, lithium distributes between the fine fraction (black mass) and the coarse fraction (>1 mm) by almost 50:50.

Moreover, another carrier gas could be used instead of N_2 in order to avoid a possible formation of Li_3N. Ar is a suitable replacement in this case. In addition, the $CuSO_4$ solution in the gas washing bottles could be replenished with a basic additive, improving the scrubbing effect. This then leads to a further reduction in gaseous emissions, such as HF and H_2S. On the other hand, another chemical additive is then required. With the set-up described in this study, hence without basic additive, (see Figure 5), 5 ppm of H_2S and 3 ppm of HF were measured after the second washing bottle. In addition, the C fc could be treated to remove the remaining sulfur. On the other hand, a refurbishing of the cathode for second use in LSBs is thinkable. For the detection of C-within solutions, a Total Organic Carbon (TOC) analyzer will be used.

Table 12. Parameter set-ups to be investigated for extracting the best-case scenario for treating Li–S black mass by the hydrometallurgical process suggested in this study.

Trial	Leaching Agent and Concentration	Solid/Liquid Ratio [g/mL]	Leaching Time and Temperature	pH Additive and Concentration	Addition of Al(OH)$_3$ Nuclei
3× VV10	4-molar NaOH	1:50	60 min at 60 °C	4-molar HNO$_3$	yes
3× VD4	2-molar HNO$_3$	1:50	60 min at 60 °C	2-molar KOH	yes
1× VD6	2-molar HNO$_3$	1:50	60 min at 60 °C	4-molar KOH	yes
3× VV25	2-molar HNO$_3$	1:50	60 min at 60 °C	8-molar KOH	yes
1× VD7	4-molar HNO$_3$	1:50	60 min at 60 °C	2-molar KOH	yes
1× VD8	4-molar HNO$_3$	1:50	60 min at 60 °C	4-molar KOH	yes
1× VD9	8-molar HNO$_3$	1:50	60 min at 60 °C	2-molar KOH	yes
1× VD10	8-molar HNO3	1:50	60 min at 60 °C	4-molar KOH	yes
1× VD11	8-molar HNO3	1:50	60 min at 60 °C	8-molar KOH	yes

Moreover, waste water should be avoided by recirculating the residual filtrate after Li$_2$CO$_3$-filtration. Besides waste reduction, it also implies the benefit to minimize lithium losses.

Another strategy for an early-stage lithium recovery (ESLR) has shown good results for conventional Li-ion batteries and will also be investigated for Li–S batteries. This can be realized by using CO$_2$ for lithium carbonation before using acidic media or also as a substitute for Na$_2$CO$_3$.

For a better understanding of on-going reactions within the hydrometallurgical processing, equipment for online measurement of both ionic concentrations and potential is going to be applied. Thus, the area of Eh-pH diagrams can be asserted more effective, and especially the lithium precipitation can be designed more accurately: Na- and K-ion concentration can be detected to filtrate solution before the solubility product of both ions is reached, avoiding impurities within the Li filter cake. For better post-processing, XRD analyses of the C filter cake, the CuS filter cake, and the Al filter cake are to be evaluated, as well.

5. Conclusions

In this study, a recycling approach for Li–sulfur batteries is examined for the first time. A proof-of-concept of the set-up presented in Figure 4 is shown. Especially when applying NaOH and HNO$_3$ leaching, high lithium yields are reached. In the case of "VV5" 78%, in the case of "VV28" 80%, and in the case of "VD4" 93% of lithium in the input material could be transferred to a lithium carbonate filter cake. However, this proof-of-concept study shows every experimental set-up only once. "ED1-3" and "VD4-5" were repeated three times or twice, respectively. This is why a statistical validation will be performed in the future. Moreover, the filter cakes' purities are to be improved for a second use of the generated products. Hence, a hydrometallurgical purification step should be added. Finally, since the intrinsic value of lithium–sulfur batteries on a commodity level is lower in comparison to lithium-ion batteries, the use of organic acids will be used to recycle more cost efficiently.

Author Contributions: Conceptualization, L.S. and P.S. and B.F.; methodology, L.S.; validation, L.S., P.S. and B.F.; formal analysis, L.S. and P.S.; investigation, L.S. and P.S.; resources, L.S. and P.S.; data curation, L.S. and P.S.; writing—original draft preparation, L.S. and P.S.; writing—review and editing, L.S.; visualization, L.S.; supervision, B.F. All authors have read and agreed to the published version of the manuscript.

Funding: This research received no external funding.

Acknowledgments: This study has been enabled by the post-mortem Li–S cells provided by Fraunhofer Institute for Material and Beam Technology IWS in Dresden. In addition, the cooperation with Helmholtz Institute Freiberg for Resource Technology (HZDR) and the Institute of Crystallography at RWTH Aachen has supported the evaluation of XRD filter cakes.

Conflicts of Interest: The authors declare no conflict of interest.

References

1. Al Rashdan, M.; Al Zubi, M.; Al Okour, M. Effect of Using New Technology Vehicles on the World's Environment and Petroleum Resources. *J. Ecol. Eng.* **2019**, *20*, 16–24. [CrossRef]
2. Kawamoto, R.; Mochizuki, H.; Moriguchi, Y.; Nakano, T.; Motohashi, M.; Sakai, Y.; Inaba, A. Estimation of CO_2 Emissions of Internal Combustion Engine Vehicle and Battery Electric Vehicle Using LCA. *Sustainability* **2019**, *11*, 2690. [CrossRef]
3. Thackeray, M.M.; Wolverton, C.; Isaacs, E.D. Electrical energy storage for transportation—Approaching the limits of, and going beyond, lithium-ion batteries. *Energy Environ. Sci.* **2012**, *5*, 7854. [CrossRef]
4. Helbig, C.; Bradshaw, A.M.; Wietschel, L.; Thorenz, A.; Tuma, A. Supply risks associated with lithium-ion battery materials. *J. Clean. Prod.* **2018**, *172*, 274–286. [CrossRef]
5. IEA: Global EV Outlook 2018: Organisation for Economic Cooperation and Development (OECD)/International Energy Agency (IEA). Available online: https://www.iea.org/reports/global-ev-outlook-2018 (accessed on 11 November 2020).
6. Harper, G.; Sommerville, R.; Kendrick, E.; Driscoll, L.; Slater, P.; Stolkin, R.; Walton, A.; Christensen, P.; Heidrich, O.; Lambert, S.; et al. Recycling lithium-ion batteries from electric vehicles. *Nature* **2019**, *575*, 75–86. [CrossRef]
7. Korthauer, R. *Handbuch Lithium-Ionen-Batterien*; Springer: Berlin/Heidelberg, Germany, 2013.
8. Kurzweil, P.; Dietlmeier, O.K. *Elektrochemische Speicher. Superkondensatoren, Batterien, Elektrolyse-Wasserstoff, Rechtliche Grundlagen*; Springer: Wiesbaden, Germany, 2015.
9. British Geological Survey; Bureau de Recherches Géologiques et Minières; Directorate-General for Internal Market; Industry, Entrepreneurship and SMEs (European Commission); Deloitte Sustainability; TNO. *Study on the Review of the List of Critical Raw Materials*; Final Report; Publications Office of the European Union: Luxembourg, 11 September 2017.
10. Regett, A.; Fischhaber, S. Reduction of Critical Resources Consumption through Second Life Applications of Lithium Ion Trection Batteries. In Proceedings of the 10th Internationale Energiewirtschaftstagung (IEWT), Wien, Austria, 15–17 February 2017.
11. Bruce, P.G.; Hardwick, L.J.; Abraham, K.M. Lithium-air and lithium-sulfur batteries. *MRS Bull.* **2011**, *36*, 506–512. [CrossRef]
12. Dörfler, S.; Althues, H.; Härtel, P.; Abendroth, T.; Schumm, B.; Kaskel, S. Challenges and Key Parameters of Lithium-Sulfur Batteries on Pouch Cell Level. *Joule* **2020**, *4*, 539–554. [CrossRef]
13. Li, T.; Bai, X.; Gulzar, U.; Bai, Y.-J.; Capiglia, C.; Deng, W.; Zhou, X.; Liu, Z.; Feng, Z.; Zaccaria, R.P. A Comprehensive Understanding of Lithium–Sulfur Battery Technology. *Adv. Funct. Mater.* **2019**, *29*, 1901730. [CrossRef]
14. He, Y.; Qiao, Y.; Chang, Z.; Cao, X.; Jia, M.; He, P.; Zhou, H. Developing A "Polysulfide-Phobic" Strategy to Restrain Shuttle Effect in Lithium-Sulfur Batteries. *Angew. Chem. Int. Ed. Engl.* **2019**, *58*, 11774–11778. [CrossRef]
15. Liu, J.; Chen, H.; Chen, W.; Zhang, Y.; Zheng, Y. New Insight into the "Shuttle Mechanism" of Rechargeable Lithium-Sulfur Batteries. *ChemElectroChem* **2019**, *6*, 2782–2787. [CrossRef]
16. Frey, M. Neue Energiespeichermaterialien für Lithium-Schwefel-Batterien. Ph.D. Thesis, University of Stuttgart, Stuttgart, Germany, 16 October 2017.
17. HELIS Project Website HELIS. High Energy Lithium Sulphur Cells and Batteries. European Union's Horizon 2020 Research and Innovation Programme. 2015. Available online: https://www.helis-project.eu/ (accessed on 15 April 2020).
18. LISA Project Website LISA. Lithium Sulphur for Safe Road Electrification. European Union's Horizon 2020 Research and Innovation Programme. 2019. Available online: https://www.lisaproject.eu/ (accessed on 1 April 2020).
19. Zhang, Z.; Wang, J.-N.; Shao, A.-H.; Xiong, D.-G.; Liu, J.-W.; Lao, C.-Y.; Xi, K.; Lu, S.-Y.; Jiang, Q.; Yu, J.; et al. Recyclable cobalt-molybdenum bimetallic carbide modified separator boosts the polysulfide adsorption-catalysis of lithium sulfur battery. *Sci. China Mater.* **2020**. [CrossRef]
20. Directive 2006/66/EC of the European Parliament and of the Council of 6 September 2006 on Batteries and Accumulators and Waste Batteries and Accumulators and Repealing Directive 91/157/EEC. Available online: https://eur-lex.europa.eu/eli/dir/2006/66/oj (accessed on 11 November 2020).

21. Commission Regulation (EU) No 493/2012 of 11 June 2012 Laying down, Pursuant to Directive 2006/66/EC of the European Parliament and of the Council, Detailed Rules Regarding the Calculation of Recycling Efficiencies of the Recycling Processes of Waste Batteries and Accumulators. Available online: https://eur-lex.europa.eu/legal-content/EN/TXT/?uri=CELEX%3A32012R0493 (accessed on 11 November 2020).

22. Placke, T.; Kloepsch, R.; Dühnen, S.; Winter, M. Lithium ion, lithium metal, and alternative rechargeable battery technologies: The odyssey for high energy density. *J. Solid State Electrochem.* **2017**, *21*, 1939–1964. [CrossRef]

23. Yamaki, J.; Sakurai, Y.; Saito, K.; Hayashi, K. Safety evaluation of rechargeable cells with lithium metal anodes and amorphous V2O5 cathodes. *J. Appl. Electrochem.* **1997**, *28*, 135–140. [CrossRef]

24. Heimes, H.; Kampker, A.; Vom Hemdt, A.; Schön, C.; Michaelis, S.; Rahimzei, E. *Produktion einer All-Solid-State-Batteriezelle*, 1st ed.; PEM der RWTH Aachen University: Aachen, Germany, 2018.

25. Kamienski, C.W.; McDonald, D.P.; Stark, M.W.; Papcun, J.R. Lithium and Lithium Compounds. In *Encyclopedia of Chemical Technology*; Kirk, R.E., Othmer, D.F., Eds.; Wiley: New York, NY, USA, 2003; Volume 6, p. 484.

26. Hanisch, C.; Diekmann, J.; Stieger, A.; Haselrieder, W.; Kwade, A. Recycling of Lithium-Ion Batteries. In *Handbook of Clean Energy Systems*; Yan, J., Ed.; John Wiley & Sons, Ltd.: Chichester, UK, 2015; Volume 130, pp. 1–24.

27. Xu, J.; Thomas, H.R.; Francis, R.W.; Lum, K.R.; Wang, J.; Liang, B. A review of processes and technologies for the recycling of lithium-ion secondary batteries. *J. Power Sources* **2008**, *177*, 512–527. [CrossRef]

28. Martens, H.; Goldmann, D. *Recyclingtechnik. Fachbuch für Lehre und Praxis*, 2nd ed.; Springer: Wiesbaden, Germany, 2016.

29. Chen, Y.; Liu, N.; Hu, F.; Ye, L.; Xi, Y.; Yang, S. Thermal treatment and ammoniacal leaching for the recovery of valuable metals from spent lithium-ion batteries. *Waste Manag.* **2018**, *75*, 469–476. [CrossRef] [PubMed]

30. Diekmann, J.; Sander, S.; Sellin, G.; Petermann, M.; Kwade, A. Crushing of Battery Modules and Cells. In *Recycling of Lithium-Ion Batteries*; Kwade, A., Diekmann, J., Eds.; The LithoRec Way; Springer International Publishing: Cham, Switzerland, 2018; pp. 127–138.

31. Sojka, R.T. Sichere Aufbereitung von Lithium-basierten Batterien durch thermische Konditionierung. Safe treatment of Lithium-based Batteries through Thermal Conditioning. In *Recycling und Sekundärrohstoffe, Band 13*; Thomé-Kozmiensky, E., Holm, O., Friedrich, B., Goldmann, D., Eds.; Thomé-Kozmiensky Verlag GmbH: Nietwerder, Germany, 2020; pp. 506–523.

32. Werner, D.; Peuker, U.A.; Mütze, T. Recycling Chain for Spent Lithium-Ion Batteries. *Metals* **2020**, *10*, 316. [CrossRef]

33. Chen, Y.; Liu, N.; Jie, Y.; Hu, F.; Li, Y.; Wilson, B.P.; Xi, Y.; Lai, Y.; Yang, S. Toxicity Identification and Evolution Mechanism of Thermolysis-Driven Gas Emissions from Cathodes of Spent Lithium-Ion Batteries. *ACS Sustain. Chem. Eng.* **2019**, *7*, 18228–18235. [CrossRef]

34. Zheng, X.; Zhu, Z.; Lin, X.; Zhang, Y.; He, Y.; Cao, H.; Sun, Z. A Mini-Review on Metal Recycling from Spent Lithium Ion Batteries. *Engineering* **2018**, *4*, 361–370. [CrossRef]

35. Pinegar, H.; Smith, Y.R. Recycling of End-of-Life Lithium Ion Batteries, Part I: Commercial Processes. *J. Sustain. Metall.* **2019**, *5*, 402–416. [CrossRef]

36. Lombardo, G.; Ebin, B.; Foreman, M.R.S.J.; Steenari, B.-M.; Petranikova, M. Incineration of EV Lithium-ion batteries as a pretreatment for recycling—Determination of the potential formation of hazardous by-products and effects on metal compounds. *J. Hazard. Mater.* **2020**, *393*, 122372. [CrossRef]

37. Becker, J.; Beverungen, D.; Winter, M.; Menne, S. *Umwidmung und Weiterverwendung von Traktionsbatterien*; Springer: Wiesbaden, Germany, 2019.

38. He, Y.; Zhang, T.; Wang, F.; Zhang, G.; Zhang, W.; Wang, J. Recovery of LiCoO2 and graphite from spent lithium-ion batteries by Fenton reagent-assisted flotation. *J. Clean. Prod.* **2017**, *143*, 319–325. [CrossRef]

39. Ekberg, C.; Petranikova, M. Lithium Batteries Recycling. In *Lithium Process Chemistry. Resources, Extraction, Batteries, and Recycling*; Chagnes, A., Światowska, J., Eds.; Elsevier: Amsterdam, The Netherlands, 2015; pp. 233–267.

40. Or, T.; Gourley, S.W.D.; Kaliyappan, K.; Yu, A.; Chen, Z. Recycling of mixed cathode lithium-ion batteries for electric vehicles: Current status and future outlook. *Carbon Energy* **2020**, *2*, 6–43. [CrossRef]

41. Zou, H.; Gratz, E.; Apelian, D.; Wang, Y. A novel method to recycle mixed cathode materials for lithium ion batteries. *Green Chem.* **2013**, *15*, 1183. [CrossRef]

42. Wang, H. *Development of a High Efficient Hydrometallurgical Recycling Process for Automotive Li-Ion Batteries*; Dissertations of IME, Vol. 26; Shaker: Aachen, Germany, 2015; ISBN 978-3-8440-3294-9.

43. Wang, H.; Friedrich, B. Development of a Highly Efficient Hydrometallurgical Recycling Process for Automotive Li–Ion Batteries. *J. Sustain. Metall.* **2015**, *1*, 168–178. [CrossRef]

44. Castillo, S. Advances in the recovering of spent lithium battery compounds. *J. Power Sources* **2002**, *112*, 247–254. [CrossRef]

45. An, L. (Ed.) *Recycling of Spent Lithium-Ion Batteries*; Springer International Publishing: Cham, Switzerland, 2019.

46. Ferreira, D.A.; Prados, L.M.Z.; Majuste, D.; Mansur, M.B. Hydrometallurgical separation of aluminium, cobalt, copper and lithium from spent Li-ion batteries. *J. Power Sources* **2009**, *187*, 238–246. [CrossRef]

47. Yan, J.; Liu, X.; Li, B. Capacity Fade Analysis of Sulfur Cathodes in Lithium-Sulfur Batteries. *Adv. Sci* **2016**, *3*, 1600101. [CrossRef]

48. Gupta, C.K. *Chemical Metallurgy: Principles and Practice*, 1st ed.; Wiley-VCH: Weinheim, Germany, 2003.

49. Haynes, W.M.; Lide, D.R.; Bruno, T.J. *CRC Handbook of Chemistry and Physics*; A Ready-Reference Book of Chemical and Physical Data; CRC Press: Boca Raton, FL, USA; London, UK; New York, NY, USA, 2017.

50. Choubey, P.K.; Chung, K.-S.; Kim, M.-S.; Lee, J.-C.; Srivastava, R.R. Advance review on the exploitation of the prominent energy-storage element Lithium. Part II: From sea water and spent lithium ion batteries (LIBs). *Miner. Eng.* **2017**, *110*, 104–121. [CrossRef]

Publisher's Note: MDPI stays neutral with regard to jurisdictional claims in published maps and institutional affiliations.

Article

Recycling Strategies for Ceramic All-Solid-State Batteries—Part I: Study on Possible Treatments in Contrast to Li-Ion Battery Recycling

Lilian Schwich [1,*], Michael Küpers [2], Martin Finsterbusch [2,3], Andrea Schreiber [4], Dina Fattakhova-Rohlfing [2,3,5], Olivier Guillon [2,3,6] and Bernd Friedrich [1]

[1] IME, Institute for Process Metallurgy and Metal Recycling, RWTH Aachen University, Intzestraße 3, 52056 Aachen, Germany; bfriedrich@ime-aachen.de

[2] Forschungszentrum Jülich GmbH, Institute of Energy and Climate Research (IEK-1): Materials Synthesis and Processing, Wilhelm-Johnen-Straße, 52425 Julich, Germany; m.kuepers@fz-juelich.de (M.K.); m.finsterbusch@fz-juelich.de (M.F.); d.fattakhova@fz-juelich.de (D.F.-R.); o.guillon@fz-juelich.de (O.G.)

[3] IEK-12, Forschungszentrum Jülich GmbH, 52425 Julich, Germany

[4] IEK-STE Forschungszentrum Jülich GmbH, 52425 Julich, Germany; a.schreiber@fz-juelich.de

[5] Faculty of Engineering and Center for Nanointegration Duisburg-Essen (CENIDE), Universität Duisburg-Essen, Lotharstraße 1, 47057 Duisburg, Germany

[6] Jülich Aachen Research Alliance, JARA-Energy, 52425 Julich, Germany

* Correspondence: lschwich@ime-aachen.de; Tel.: +49-241-8095194

Received: 23 October 2020; Accepted: 14 November 2020; Published: 17 November 2020

Abstract: In the coming years, the demand for safe electrical energy storage devices with high energy density will increase drastically due to the electrification of the transportation sector and the need for stationary storage for renewable energies. Advanced battery concepts like all-solid-state batteries (ASBs) are considered one of the most promising candidates for future energy storage technologies. They offer several advantages over conventional Lithium-Ion Batteries (LIBs), especially with regard to stability, safety, and energy density. Hardly any recycling studies have been conducted, yet, but such examinations will play an important role when considering raw materials supply, sustainability of battery systems, CO_2 footprint, and general strive towards a circular economy. Although different methods for recycling LIBs are already available, the transferability to ASBs is not straightforward due to differences in used materials and fabrication technologies, even if the chemistry does not change (e.g., Li-intercalation cathodes). Challenges in terms of the ceramic nature of the cell components and thus the necessity for specific recycling strategies are investigated here for the first time. As a major result, a recycling route based on inert shredding, a subsequent thermal treatment, and a sorting step is suggested, and transferring the extracted black mass to a dedicated hydrometallurgical recycling process is proposed. The hydrometallurgical approach is split into two scenarios differing in terms of solubility of the ASB-battery components. Hence, developing a full recycling concept is reached by this study, which will be experimentally examined in future research.

Keywords: battery recycling; all-solid-state batteries; metallurgical recycling; metal recovery; recycling efficiency

---

1. Introduction

Generally, continued operation of batteries after their typical end of life (80% of nominal capacity), often referred to as "second life", has both environmental and economic benefits. However, due to required testing protocols and safety as well as reliability issues, this second life exploitation is challenging [1]. At the end of the first or the second life, normal recycling needs to be exploited to lower environmental impact of battery fabrication and move further towards a circular economy. The recycling of LIBs is already established on the industrial scale using specific process routes [2]. The following section provides an overview on possible recycling paths. However, many more processes are investigated in both industry and research, so this elaboration is an overview and not full literature review. Moreover, in the following section, industry and research activities are not differentiated, as the focus will be presenting options for Li-Ion battery recycling and hence, their possible application to ASBs. The most direct way to reuse battery materials is by reconditioning the active materials for direct fabrication of new cells [2]. If this is not possible (e.g., due to cell performance), elemental recycling strategies based on the combination of two different process modules are in place: Pre-treatments and metal extraction processes [3].

Within pre-treatments, discharging is crucial to safely process the batteries further. This can be realized by immersing the used battery into a salt solution [4] or by a thermal treatment, e.g., a pyrolysis, as a deactivation step [5]. During this thermal pre-treatment, organics and binders are decomposed and evaporated, leaving the reactor via the off-gas stream [6]. The applied thermal treatment can either occur by generating an inert gas or vacuum atmosphere, thus in an oxygen-free environment (pyrolysis), or in oxygen-containing environment (incineration/thermolysis) [4,7]. After completing the thermal treatment, a mechanical treatment consisting of comminution and separation is possible. The separation can be realized by making use of physical properties or a sieving operation [8]. An alternative sequence is the deactivation of the batteries in a salt solution, a mechanical treatment, and a subsequent thermal treatment [9–12]. Here, before the thermal treatment, the batteries are either shredded and then sorted by means of physical properties [10], or directly manually dismantled to extract specifically the electrode foils with the attached active mass [9,12]. A third method combines inert/cryogenic shredding and a subsequent thermal treatment [13]. This brief overview gives the three main approaches regarding pre-treatment steps. There are differences in applied atmospheres and temperatures, but in most cases, a better removability of black mass, especially on the cathode side, from substrate foils is reported when applying a thermal treatment. This can be explained considering the following: Within battery design, a good adhesion between substrate foils and active materials is crucial, which is realized by applying a strong binder. The binder compounds can be cracked and removed in the thermal treatment. Moreover, the removal of the binder and other organics is a suitable tool for easing the downstream hydrometallurgical treatments, since the organic compounds are hardly soluble in leaching steps [7,12]. On the other hand, a challenge in direct shredding and in thermal pre-treatment is the need for extensive off-gas cleaning [4,14].

Within the metal extraction, there are different chemical approaches, such as hydrometallurgy and pyrometallurgy [15]. Hydrometallurgical processes offer many different solvents for leaching and also target phases of the battery components. They can be classified by the type of leaching media: Mineral acids, alkalis, and organic acids. For instance, processes based on mineral acids, such as sulfuric acid (H_2SO_4) or hydrochloric acid (HCl), or organic acids, such as ascorbic acid together with hydrogen peroxide (H_2O_2), have reached satisfying yields [16]. Altogether, hydrometallurgy requires less energy, and due to its selectivity high purities can be obtained. Nonetheless, mechanical and/or thermal pre-treatments are essential [14].

Pyrometallurgy generally is a high efficient concentration operation and comprises of the smelting of batteries transferring noble metals to an alloy, which then will be purified by hydrometallurgical refining steps [14,17]. Rather ignoble, and depending on the operation temperatures, partly volatile metals will be transferred to a slag phase or a flue dust [18]. This path is industrially widely applied [18] due to its high robustness and productivity [19]. In addition, pyrometallurgy is robust regarding the input stream's heterogeneity, which is why battery scrap can be treated along with persisting primary production lines of metals like cobalt (Co) [20]. Since lithium (Li) cannot be recovered by pyrometallurgical methods as metal, a downstream slag purification by means of hydrometallurgy is also a subject of research [21].

In conclusion, different process modules can be combined to design the most effective recycling process for each cell design and, vice versa, cell can be designed to promote effective recycling.

The main scrap volumes of LIBs recycling are presently based on consumer batteries, and partly also on Hybrid Electric Vehicles/ Electric Vehicles (HEV/EV) traction batteries [22]. The EV scrap volumes that are available for recycling are comparatively low, but if this status-quo changes in the future towards an increased use of electrification in the mobility sector, some recycling paths will turn out to be more viable if they benefit from scale effects [2]. Today's industrial recycling paths focus mainly on the valuable elements; however, LIBs are complex and materials from different applications or generations are often accompanied by changing compositions [22,23]. Since the second life application rates are currently very low in industrial processes [24], a lot of research activity is pursued targeting elemental materials' recycling. Public-funded German projects such as MERCATOR or InnoRec tend to follow innovative recycling steps in order to mobilize Li, and even in the case of MERCATOR, the critical raw material graphite will be reused for non-battery applications [25,26]. Methods for mobilizing Li before it is deposited in a slag or flue dust are already under fundamental investigation. These investigations are relevant since extracting Li from a slag is energy and resource intensive, and the current Li price does not assure economic viability for this add-on process [20]. Early-stage recovery methods for transforming of Li phases in the battery active mass are proposed, making use of supercritical CO_2 or by thermally activation [27–30]. Innovative methods for a holistic recycling based on graphite pursue the flotation technology [31,32]. Another important research topic is the recovery of the electrolyte, which is challenging due to its high reactivity [33] and the ecotoxicity of prevailing F-compounds [34]. Phosphorous (P), present in almost all LIBs, is another critical topic in recycling. P needs to be removed in pyrometallurgical nickel (Ni) production since it affects the properties of specific Ni alloys [35]. In conclusion, there is still a strong need for research in recycling of conventional batteries [36], especially based on recycling efficiency and added value. Moreover, research on non-chemical recycling optimization is needed, for example in the field of collection and scrap logistics [2].

ASBs are regarded as promising future batteries, as they have advantages like enhanced stability, safety, and energy density over conventional LIBs [37]. This is mainly due to the solid electrolyte's beneficial properties such as a better thermal stability, non-flammability, resistance against overcharging, and long cycle stability [38].

Different classes of solid-state electrolyte materials like polymers, sulfides, and oxides are under investigation. Due to easier processing, polymer based ASBs are the closest system in terms of market introduction. Sulfides are also easy to process and show the highest Li-ion conductivity, increasing the power density greatly, but their difficult synthesis and chemical instability towards water/air impede the large-scale fabrication. Considering safety aspects, oxide-ceramic materials stand above all other Li electrolytes due to their chemical, thermal, and oxidation stability. They are non-flammable, non-toxic, and can be handled in air. The chemical stability of garnet-based $Li_7La_3Zr_2O_{12}$ electrolytes towards Li allows the direct use of metallic Li, making this material one of the most promising electrolytes for ceramic all solid-state batteries [39]. However, expensive dopant elements like lanthanum (La) and tantalum (Ta) as well as the required high temperature processing steps are the biggest hurdles for large scale market introduction. Additionally, a first analysis of the resource availability have shown that with a market share of 10% in traction and stationary applications, Li, La and zirconium (Zr) can be classified as critical [40]. However, there is no general consensus on the assessment of resource availability and criticality of raw materials. Different studies evaluate the criticality of materials (especially Li) in different ways and with different results [41–49]. For example, Helbig et al. published a study to assess the supply risks associated with 10 elements used in different LIBs [45]. Li and Co have the highest supply risk scores. The high score for Li mainly emerges from a lack of end-of-life recycling and the high future technology demand. The high supply risk score of Co, in contrast, results from the by-product dependence and the high risk from political instability. Aluminium (Al) shows the lowest supply risk score followed by titanium (Ti), copper (Cu), iron, Ni, graphite, manganese (Mn), and P. Schultz and Kuckshinrichs [50] analyzed the need for Li for electrochemical energy storages. By analyzing data from known Li sources, the authors conclude that there is no major risk in terms of exhausting world reserves, especially if a market for Li recycling is introduced. However, with possible strongly rising demand in mind, the authors see serious potential for risks on the supply side, which may result in temporary shortage situations and rising price levels at the Li world market. The European Commission [48] identified 26 raw materials and material groups as critical. This includes Co, La, phosphate rock, and Ta, which are used for ASBs. Li, Mn, Ni, Ti, Al, and Cu are considered non-critical according to the EU critical material list with Li and Mn being close to the threshold.

Moreover, a consideration of possible recycling processes was not part of the study by Troy et al. [40]. Generally, no recycling concepts for ceramic ASBs are in place yet. In an indirect way, Piana et al. investigated the reusability of industrial waste products for the synthesis of a sodium (Na)-on ASB electrolyte [51], and Wang et al. extracted end-of-life $LiMn_2O_4$ cathodes for resynthesizing a Li-Ion ASB electrolyte [52]. This shows that only rudimentary knowledge regarding a circular economy of ASBs exist. Therefore, this study aims to generate the first ceramic ASB recycling concept. In particular, a theoretical approach is chosen, taking the presented LIB treatment methods as a starting point, then evaluating to what extent specific tools can be translated to ASB recycling.

2. Materials and Methods

2.1. Cell Concept

In terms of industrial and technological value, the cell concept with the lowest cost and highest energy density for oxide-ceramic based ASBs is a flat cell, which is housed in a pouch bag. The design of the cell itself (see Figure 1) is rather simple and consists of five different layers: (1) Anode, (2) Separator, (3) Cathode, (4), and (5) Current Collectors.

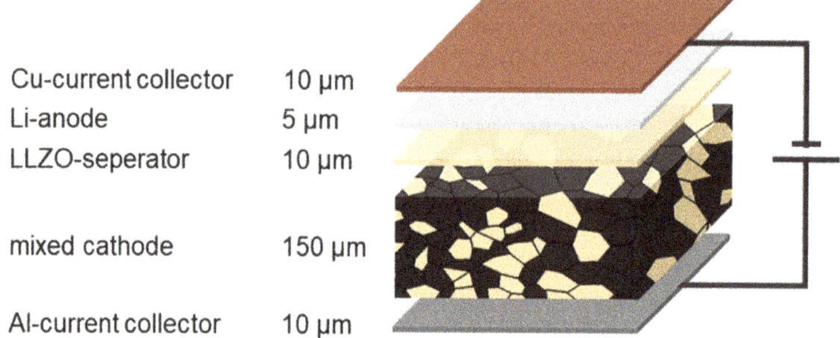

Figure 1. Pouch cell design of a ceramic ASB: A thick mixed cathode is separated by a thin LLZ-layer from the Li-metal anode. The currents are collected by a thin Cu or Al foil.

(1) Anode: The electrochemical stability of some oxide-ceramic electrolytes towards metallic Li allows for the direct use of Li metal as an anode material. Li metal shows a very high theoretical capacity and the lowest electrochemical potential, making it the most promising anode material in Li based batteries in terms of energy density. Theoretically, all active Li ions are located in the cathode material. Nevertheless, a thin layer of metallic Li on the anode side is necessary to compensate for irreversible Li losses and a homogeneous Li plating during cycling. Thus, we used a thin layer of 5 µm thickness in our cell design, knowing that Li free anode concepts are also heavily researched at the moment.

(2) The separator prevents electrical short circuiting due to the direct contact between anode and cathode material, while it allows Li-ions to pass between the two electrodes. In ASBs, the separator can be made from the solid electrolyte itself, therefore having the same high ionic conductivity while

the electronic conductivity is low. However, it needs to be chemically stable towards the anode material (Li metal) on the one hand and towards the cathode material (e.g., LCO, NMC) on the other hand. So far, the only materials that combines all these properties are garnet-based compounds like $Li_7La_3Zr_2O_{12}$ (LLZ). Ta substitution ($Li_{6.4}La_3Zr_{1.6}Ta_{0.4}O_{12}$) stabilizes a cubic garnet structure and enhances the conductivity of the material up to 1 ms/cm [53]. Since the separator is an inactive part of the battery, it should be as thin as possible while still showing sufficient electronic and mechanical properties. A dense layer of 10 µm LLZ is used as a compromise between total resistance, sufficient mechanical properties, and realistic processing. In the future, thinner separators and interlayer to improve the contact resistance are of course realistic.

(3) Cathode: In a conventional LIB, the cathode consists of a rather porous structure of active material, conductive additives, and polymer binders in which the liquid electrolyte can be infiltrated and achieves a good interface contact. As cathode active materials (CAM) for ASBs, the same materials which are already used in LIB can be incorporated. To achieve a good surface contact between the two solid phases, the cathode material needs to be co-sintered together with the electrolyte. One promising material combination for oxide-based ASBs is the combination of $LiCoO_2$ (LCO) and LLZ. The mixture of these materials is chemically stable up to 1085 °C [54], which allows a co-sintering of the cathode. Batteries of this material mixture are already realized on lab scale [1,12,39,55]. However, to obtain high-energy density cells, the amount of CAM in the mixed electrode should be as high as possible. In our concept, we consider a CAM: solid electrolyte ratio of 2:1 as reasonable, allowing for percolation of both phases. The thickness of the cathode is set to 150 µm to achieve areal capacities of approx. 4 mAh/cm^2, which is a common value for conventional LIBs. By substituting Co in LCO by Ni and Mn the capacity of the cathode material can be increased. Therefore, we also consider the use of $LiNi_{0.8}Mn_{0.1}Co_{0.1}O_2$ (NMC811) in an ASB although this material shows a lower chemical stability at elevated temperatures towards LLZ then LCO [56]. However, stabilizing coatings or lowering the sintering temperatures by advanced processing technologies could allow for usage of NMC811 as cathode material for LLZ bases ASBs in the future. To further increase energy density, another highly conductive ceramic electrolyte should be considered: NASICON structure based $Li_{1.5}Al_{0.5}Ti_{1.5}(PO_4)_4$ (LATP). It shows a comparable or even higher total Li-ion conductivity (~1 ms/cm) than LLZ, while the density is lower and the raw materials needed for synthesis are less critical and cheaper [57]. However, the major drawback is its lower chemical stability, especially towards metallic Li. Without any stabilizing coating, it can therefore not be used as a separator material. To combine both advantages of the individual materials, we also investigate a cell concept with LLZ as separator and LATP instead of LLZ in the mixed cathode [58].

(4) Current collector anode side: the current collector on the anode side needs to fulfill only two requirements: high electronic conductivity and chemical stability towards metallic Li. One material that fulfills these requirements is Cu. Since the ceramic cathode and separator material construct a mechanically stable backbone, a rather thin layer of Cu (10 µm) is sufficient.

(5) Current collector cathode side: Since the cathode side is chemical less active then the anode side, the requirements for the current collector on the cathode side are reduced to only high electronic conductivity. To keep the cell as light as possible, we use a thin Al foil (10 µm) on this side.

Due to the mechanical stability of the cell stemming from the ceramic cell itself, the thicknesses of the Cu and Al current collectors could be decreased further (even only the metallic Li could be used as current collector on the anode side). However, since this will depend mainly on the specific production process and design of the cell stacks (e.g., bi-polar vs. parallel), all possible variations cannot be investigated within this paper.

In total, we consider four different cell designs of promising as future ceramic ASBs, all containing a LLZ separator, Cu and Al current collectors, but different cathode composite materials. The energy density of these four cell designs were calculated and are listed in Table 1.

The LATP-based cells (2.1 and 2.2) generally show higher energy densities, stemming from its lower density compared to LLZ. NMC811 with its higher capacity than LCO results in the highest total energy densities.

Table 1. Capacities of the four different cell designs.

Cell Number	Cathode Materials	Energy Density (Wh/kg)
1.1	LLZ + LCO	309
1.2	LLZ + NMC	406
2.1	LATP + LCO	352
2.2	LATP + NMC	463

2.2. Assembling Process

One crucial step during the assembling of a working cell is to ensure a good interface conductivity between the separator and the cathode material. This can be achieved by a co-sintering process, where the cathode and the separator will be baked together at elevated temperatures. The current collectors, on the other hand, will be attached rather loose on the electrodes and mainly kept in place by the packaging of the cell. The cell housing itself will most likely be adapted from LIBs in pouch cell format, since it has the lowest impact on the energy density and enables the use of the same foil, tabs, and ultrasonic welding to obtain cell stacks and batteries. After the sintering process, cathode and separator will be chemically bonded, making a mechanical separation impossible. However, due to the brittle nature of ceramics, crack formation is still a major concern. The Li anode will be either pressed on the separator as a foil or, more likely, will be evaporated on the separator material. Cycling and the ductility of Li metal will make it hard to mechanically separate the anode from the rest of the cell.

2.3. Material Demand

The market for batteries will increase drastically within the next years. With politics pushing worldwide towards an electrification of the automobile sector, the demand for batteries will multiply within the next years, from 500 GWh in 2017 to several TWh in 2050 [59]. Considering our four cell concepts, we calculated the amount of raw material that would be necessary to produce 1 TWh of oxide-based ASB.

To discuss possible bottlenecks in supply of the elements for the assumed 1 TWh ASB application, we present the material requirements in comparison to the current world production as a first rough estimate (Table 2).

The material compositions (weight percent) of all four ASB cells are shown in Figure 2. Around one-third of the cell consists of transition metals from the cathode side containing Co, Ni, and Mn. The electrolyte metals in the LLZ cells (1.1 and 1.2) make around one quarter of the complete cell, with the main part being the rare earth element La. 10 wt.% is Cu, while Li makes up around 6 wt.%. The total demand of the materials that are necessary for a battery production of 1 TWh is listed in Table 2. The right side of the table shows the share (%) on current world production for 1 TWh. All four cell types require a critical demand of Li, La, and Ta. In addition, the Co and zirconia demand can be seen as critical for cell 1.1, 1.2, and 2.1. The criticality can be defused when using Co pure cathode material NMC as it is in cell 1.2 and 2.2, since Ni and Mn can be seen less critical. Looking at the LLZ-based batteries, the demand of La, Zr and especially Ta are in a range that goes far beyond the current world production. Including LATP into the mixed cathode (Cell 2.1 and 2.2) lowers the criticality for La, Zr, and Ta, but being still in a critical range for Ta and La. This shows that a good recyclability of the battery cells is inevitable, if this type of batteries should take a reasonable market share. However, any future developments in the raw material production market are not taken into account yet and can change the results noticeably. It is very likely that as demand increases, new mines will start production and others will maximize their output [40].

The following table shows battery components and their annual production, and relates those numbers to the material demand when producing the cells presented in this study (see Figure 2 and Table 2).

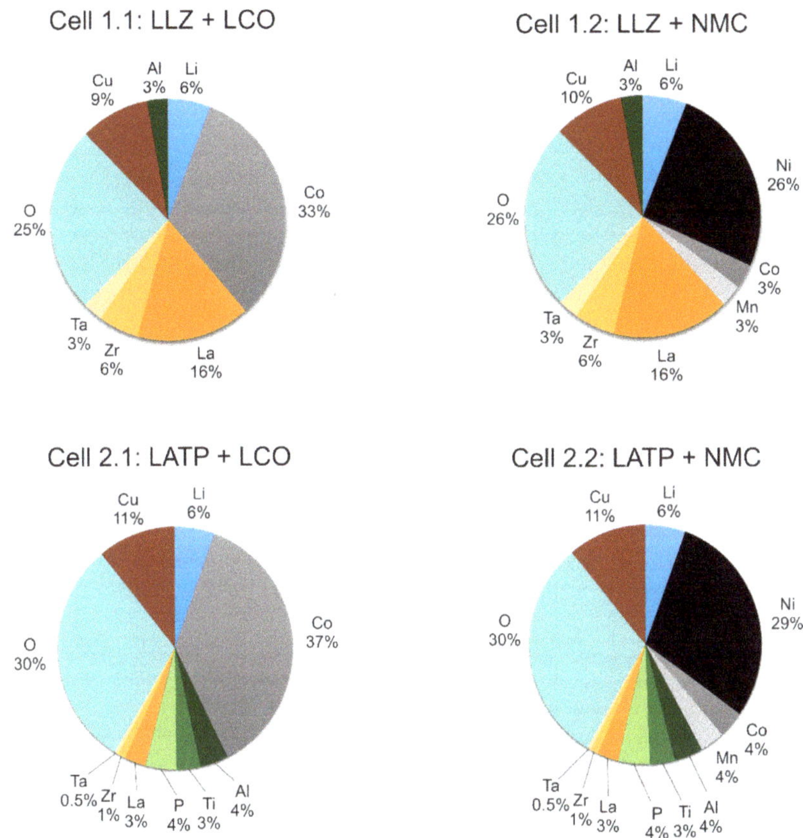

Figure 2. Material compositions of the four different cell designs.

Table 2. Material demand required for 1 TWh ASB and share of demand on current world production.

	World Production (T)	Material Demand for 1 TWh ASB in 10^5 T				Share (%) on Current World Production in 2030			
		Cell 1.1	Cell 1.2	Cell 2.1	Cell 2.2	Cell 1.1	Cell 1.2	Cell 2.1	Cell 2.2
LI	77,000 [60]	1.93	1.47	1.57	1.19	251	191	204	155
NI	2,700,000 [60]	–	6.35	–	6.35	–	24	–	24
CO	140,000 [60]	10.56	0.80	10.56	0.80	755	57	754	57
MN	15,414,509 [61] 53,000,000 [a] [62]	–	0.74	–	0.74	–	0.5	–	0.5
AL	64,000,000 [60]	0.93	0.72	1.11	0.86	0.1	0.1	0.2	0.1
TI	4,394,500 [62]	–	–	0.95	0.73	–	–	2.2	1.7
P	36,650,402 [61]	–	–	1.23	0.95	–	–	0.34	0.26
LA	56,700 [b] [60]	5.11	3.95	0.85	0.66	901	697	150	116
ZR	112,471 [c] [62]	1.79	1.39	0.30	0.23	159	123	27	20
TA	1800 [60]	0.89	0.69	0.15	0.11	4928	3816	833	611
CU	20,000,000 [60]	3.07	2.38	3.07	2.38	1.5	1.2	1.5	1.2

[a] Mn ore; [b] La production accounts for 27% of the total world rare earth production of 210,000 T [63]; [c] based on 1,256,000 T $ZrSiO_4$, (zircon), assumption: 18% of the total zircon amount is used for zirconia production (ZrO_2).

3. Recycling Approach

A battery recycling process should be tailored to the battery components used and relevant elements targeted for recovery. As indicated before, many different ASB systems are promoted currently in laboratories all over the world. Moreover, within the three superordinate systems (polymers, sulfides, and oxides), various components are being tested during the cell design phase aiming at a high-performance solid-state battery. Therefore, exemplarily battery compositions have to be selected for generating first ASBs recycling considerations. In this study, recycling approaches for the two systems LLZ + NMC and LATP + NMC are designed to match the different available systems. These approaches are then compared in terms of their sustainability.

Due to the different nature of both composition and physical form of ASBs, the recycling concepts also differ quite a lot from those for conventional LIBs: Most integral components of the cell are chemically rather ignoble, and besides, do not show a high vapor pressure. ASBs do not contain combustible components, and thus, do not contribute to exothermic reactions. Hence, pyrometallurgy-based recycling steps are not the tool of choice for establishing a suitable recycling process, such avoiding the generation of only a large volume of slags. Moreover, the ASB composition according to Figure 2 shows that the battery system contains even more elements in comparison to conventional LIBs, leading to a more complex recycling chain. Regarding ecotoxicity, ASBs have beneficial properties for performing a recycling process. They do not contain fluorine or phosphorous compounds, which is why the amount of hazardous gas phases will be reduced when applying a thermal treatment. According to the state-of-the-art of LIBs recycling (see Introduction), the occurrence of hazardous off-gases is one main drawback of thermal pre-treatments. Recycling is generally eased due to the absence of organic compounds, like binders. They are removed by sintering in the battery production.

In order to select a suitable recycling process, Figure 3 sums up the Introduction chapter. Here, the reference level comprises of charged battery cells, assuming that they have already been extracted from their modules in case of EV-battery packs. Even though a discharge of whole modules is possible, for reasons of clarity, this option is neglected in this case.

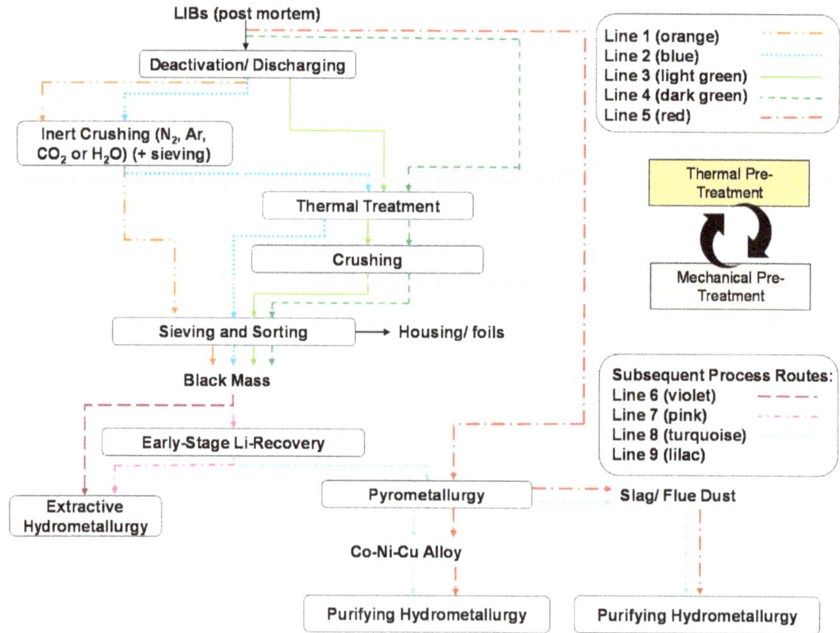

Figure 3. Options for selecting a recycling method based on the available recycling paths according to the state-of-the-art.

As described above, Line 5 (red path, pyrometallurgy) is not seen as optimal treatment. Besides, since ASBs are beneficial for safety and risk for fire incidents, the selection of pre-treatments is more flexible in terms of direct inert or aqueous shredding (Lines 1 and 2). The pre-treatment or LIBs generally requires a discharge step to protect the comminuting unit operations from fire incidents, a thermal treatment with discharging (Line 3), or a direct thermal treatment without discharging (Lines 4 and 5). In Line 4, a thermal treatment can by bypassed due to a robust system being able to handle thermal runaways and moreover comprising a waste heat recovery system. In Line 5, a thermal treatment can by bypassed due to the high temperature requirements in smelters, where thermal runaways do not represent a processing challenge. For ASBs, either Line 1 or 2 is preferred since the solid-state battery components are sintered and thus are expected to be mechanically and chemically stable. Inert shredding in H_2O can be beneficial for the ASBs treated in this study, since their anodes do not consist of metallic lithium, which is highly reactive and thus could go up in flames. Hence, high costs for vacuum and inert gas shredding could be avoided by aqueous shredding. Since no binders are applied, a thermal treatment is not necessarily required for ASBs. It has been, however, reported in the literature that an eased component liberation is reached when applying a thermal treatment. Thus, whether Line 1 or 2 is beneficial is to be examined in Part II of this study. Whether the sintered battery components can be separated from one another can only be figured out by conducting practical experiments. Because of the comparatively high Li contents (~3 wt.% Li in LIBs [64]) in comparison to 6 wt.% Li in the ASBs (see Figure 2), a treatment for an Early-Stage Li-Recovery (ESLR) would be more viable, and thus, seems a good option for the selected concept. Finally, extractive hydrometallurgy is highly selective and can separate battery components like Mn, La, and Zr element-wise, which is why Line 7 (violet) or Line 8 (pink) are chosen for a subsequent process route. Whether ESLR is a suitable tool for ASBs is to be investigated experimentally in Part II of this study.

According to this discussion and based on know-how regarding hydrometallurgical extraction processes for recycling LIBs, as can be seen in [8,18,65], the resulting flow chart for LLZ + NMC batteries can be seen in Figure 4.

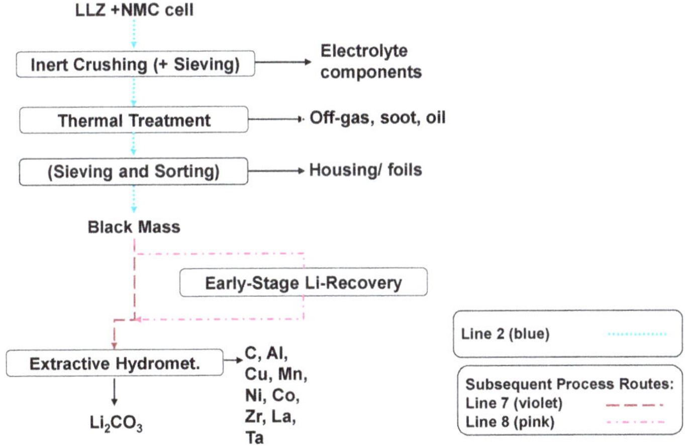

Figure 4. Experimental plan for a suitable LLZ + NMC recycling. C represents both graphite, active charcoal, and soot.

It has to be further examined whether a shredding process together with a component separation by means of sieving will lead to satisfying results. Eventually, a further comminution step by a ball mill is required to grind the oxidic fractions and liberate them from the substrate foils and casing.

Moreover, research will be necessary considering the necessity of a thermal treatment. The ESLR, represented by a thermal or CO_2-driven phase transformation of Li compounds into water soluble Li-carbonate, and the subsequent dissolution of Li phases in H_2O as a neutral leach is one option. Here, Li, which will be dissolved, can be separated from the residual active mass, which is mainly insoluble in water. The CO_2-driven option for the ESLR consists of a phase transformation by means of supercritical CO_2 in an autoclave reactor filled with water or by gaseous CO_2 in an aqueous medium. When using an autoclave reactor, an excess pressure of 73.8 bar [66] is needed to reach the supercritical state.

The above-mentioned garnet structure (LLZ) is a mineral compound, whose dissolution conditions are not experimentally proven yet. According to own studies, elevated temperatures and strong acids/bases are required to dissolve sintered LLZ. On this account, two scenarios are to be discussed within this study. They discuss the two border cases, in which either the whole LLZ structure is chemically dissolved (scenario 1) or only Li is dissolved from the structure, whereas the main garnet structure remains insoluble (scenario 2). In these scenarios, the underlying solvent is not further specified. However, it can be predicted that the solvent in scenario 1 will tend to be stronger than the used solvent in scenario 2, since the Li^+/H^+ exchange reaction is well investigated for LLT [67–69]. Both scenarios are presented in Table 3.

Table 3. Two hydrometallurgical recycling scenarios discussed within this study.

Criteria of Scenario	Scenario 1	Scenario 2
Characteristics	LLZ is fully dissolved	Only Li is dissolved, the other LLZ components remain solid
Leaching conditions	High temperature/aggressive leaching	Moderate leaching

The extractive hydrometallurgy according to Wang [18,65] can be modified adding specific precipitation steps for La, Zr, and Ta. Depending on the scenario, the multistep hydrometallurgy is to be designed differently. In scenario 1, the elements Zr, Ta, and La are integrated in the leaching and precipitation sequences according to their behavior in aqueous solutions. In scenario 2, they are extracted as a concentrate. Here, no dissolution is taking place and cross-contaminations by similar precipitation areas, e.g., Co, is neglected. The following elaboration treats the behavior of Zr, Ta, and La in aqueous systems. For this purpose, both StabCal simulations and literature-based properties are considered. The StabCal simulations show the precipitation behavior in the case of ionic dissolution of Zr, Ta, and La. This means that no information on the degree of dissolution is given. Moreover, the oxidation states of the input material cannot be specified by the StabCal simulations since the level of consideration is the ionic dissolution. The concentration applied refers to the chemical composition given, and the StabCal database used is indicated in the image captions. However, it has to be noted that no specification on leaching medium, pH-additive and further components in the system can be detailed. When realizing the presented simulations experimentally, the parameters leaching medium, pH-additive, and further components have an influence on the precipitate phase and thus on the pH-value of precipitation. Experimental validation by titration is a suitable tool to validate the calculations. Titration is therefore going to be performed in the frame of this publication's Part II. The literature-based information gives an insight on the degree of dissolution, and besides, on the precipitation behavior. Thus, it is complementary to the simulation data. However, the data presented are also specific studies and might not be transferrable to any system. The data derived from literature provide a first orientation to forecast the hydrometallurgical behavior and thus construct a suitable recycling path. Experimental validation within Part II is going to evaluate the approaches presented in this study.

Figure 5 shows a thermochemical modelling by StabCal in terms of the behavior of the element Zr when being brought into aqueous solution. It can be seen that the Zr-ions are starting to precipitate from the leaching liquor at a pH-value of ~4.5 as ZrO_2.

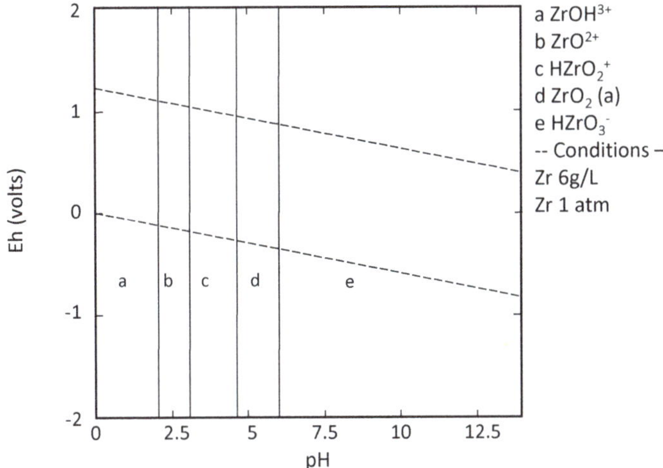

Figure 5. EpH diagram of Zr dependent of pH calculated by StabCal. Database used: Helgeson (SupCrt).

In the literature, this phenomenon has been discussed, as well: Ma et al. built a phase diagram and calculated a precipitation of ZrO_2 from the leaching liquor at a pH-value of ~2.5 (see Figure 6) [70]. The results show slight deviations but general accordance with the StabCal simulation in Figure 5, especially taking into account the high degree of simplification discussed above. Moreover, Ma et al. treated an eudialyte concentrate in H_2SO_4, and adjusted the pH-value by sodium carbonate (Na_2CO_3). Thus, deviations by further components in the system are likely to occur.

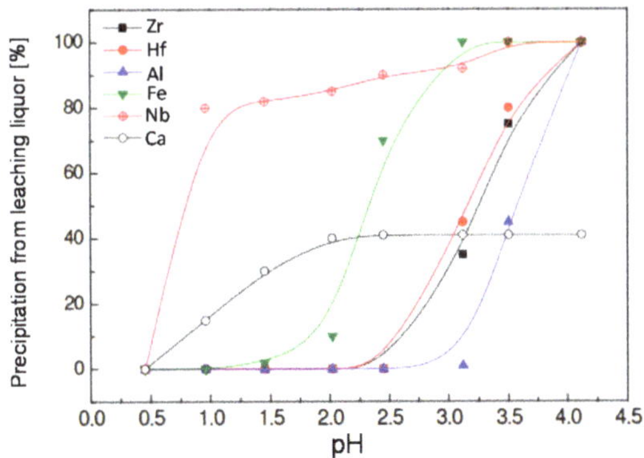

Figure 6. Influence of pH-value in terms of precipitation of Zr stemming from eudialyte from a H_2SO_4 solution, based on [70].

In general, ZrO_2 can be leached in an acidic medium, as reported by Ferreira et al. in case of nitric acid (HNO_3). They have shown a Zr dissolution of 95% with a leaching time of 4 h, an operating temperature of 70 °C, and a molality of 12.0 mol/L [71].

Figure 7 shows a thermochemical modelling by StabCal in terms of the behavior of the element La when being brought into aqueous solution. From this calculation, a precipitation of La-ions in a pH-area of ~7.4 is predicted.

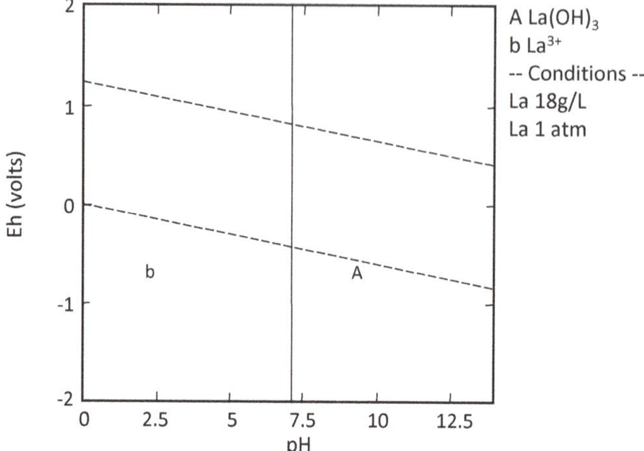

Figure 7. EpH diagram of La dependent on pH calculated by StabCal. Database used: LLNL (Cp(T)).

Literature on the dissolution of La-phases has shown a good accordance with the StabCal simulation presented. Orhanovic et al. reported a precipitation pH-value of 7.36–7.56 for La(OH)$_3$ [72]. Um et al. examined the dissolution of La$_2$O$_3$ in H$_2$SO$_4$, leading to a full leachability, see Figure 8 [73].

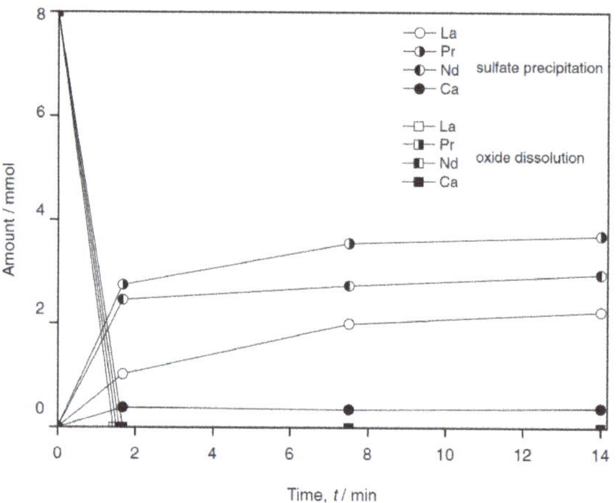

Figure 8. La$_2$O$_3$ dissolution and conversion into La$_2$(SO$_4$) [73].

Figure 9 shows thermochemical modelling by StabCal in terms of the behavior of the element Ta when being brought into the aqueous solution. From the simulation, no pH-dependent precipitation threshold can be identified.

However, a literature research gives more information on the dissolution and precipitation of Ta-compounds dependent on the pH-value. Chen et al. report a low Ta leaching efficiencies in HCl, H$_2$SO$_4$ and HNO$_3$ [74]. When applying hydrogen fluoride (HF) based pressure leaching, at 23 bar, 180 °C for 3 h, a leaching efficiency of 99% could be obtained [74]. Regarding Ta-recovery from the solution by solvent extraction, an extraction efficiency of 99.5% could be reached at pH = 1 (see Figure 10). Nevertheless, solvent extraction can extract ions selectively dependent on the solvent applied, and thus, the system behaviour cannot be transferred linearly to acidification/basification-driven precipitation (conventional hydrometallurgy by pH-adjustment). Hence, the extraction efficiency describes the dissolution within the solvent; research on basification-driven precipitation will be presented below.

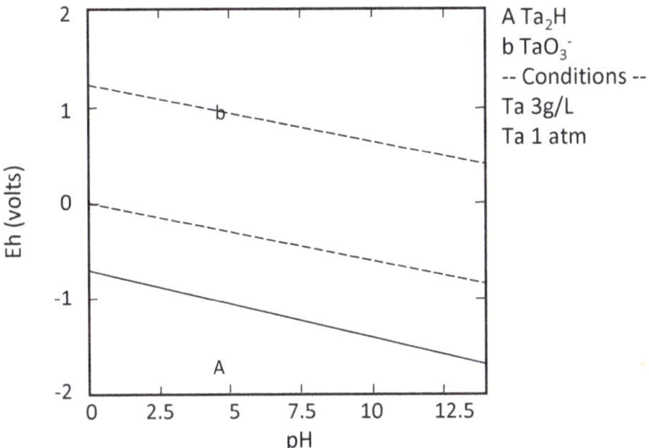

Figure 9. EpH diagram of Ta dependent of pH calculated by StabCal. Database used: HSC (Outotch).

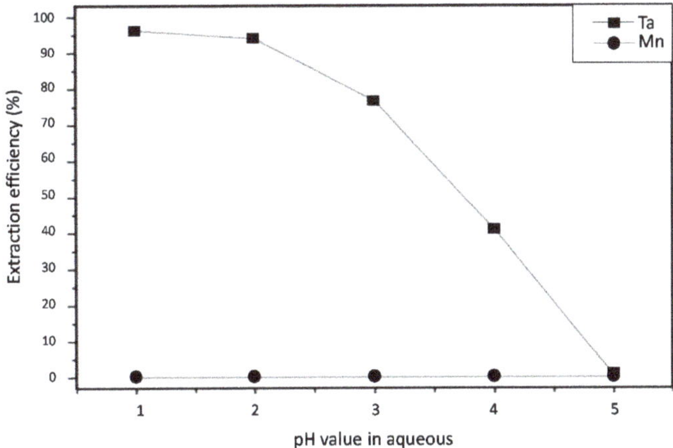

Figure 10. Influence of pH-value 1–5 on the extraction efficiency of Ta using Alamine 336 for solvent extraction, based on [74].

Clark and Brown confirm the dissolution behaviour detected by Chen et al. stating the capability of Ta to dissolve in alkaline solutions [75]. This is also supported by Deblonde et al. [76,77]. Here, alkaline leaching in sodium hydroxide (NaOH) or potassium hydroxide (KOH) shows a good hydrometallurgical alternative to fluorine-based leaching, which is critical in terms of toxic emissions [77]. Deblonde et al. report on a combination process consisting of alkaline leaching and pH-value adjustment to pH = 2–7 in order to precipitate Ta [77]. Thus, it can be concluded that Ta is likely to precipitate in an acidic area and might cause contaminations in the Cu-cementation step or the Al-Fe-hydroxide precipitation according to Wang [18]. The same option is valid for La and Zr, which tend to precipitate in acidic areas, too, as described above. When assuming a highly selective hydrometallurgical treatment, a precipitation into element-specific product will be realizable though.

The simulated properties were, hence, combined with a literature research in order to extract a recycling process in terms of multi-step hydrometallurgy. The process design of a multi-step hydrometallurgy can be applied to the two scenarios shown in Table 3. In combination with the behavior of the elements in the active mass, Zr, La, and Ta, and the conventional NMC-chemistry, one flow-chart for scenario 1 and one flow-chart for scenario 2 regarding a multi-step hydrometallurgy are presented in Figures 11 and 12. Since the oxidation states of the input material cannot be forecasted, neither the phase of the precipitates is specified in this elaboration.

Scenario 1

As already suggested in Table 3, scenario 1 represents the case of bringing all battery components into solution. Since LLZ implies garnets, an aggressive leaching medium has to be chosen here. Then, the specific elements can be precipitated according to an increased pH-value e.g., as hydroxides, sulfates, or oxides. Since the precipitation of metals can be estimated according to the elaborations from Figures 5–10, the following multi-step hydrometallurgy is suggested for scenario 1.

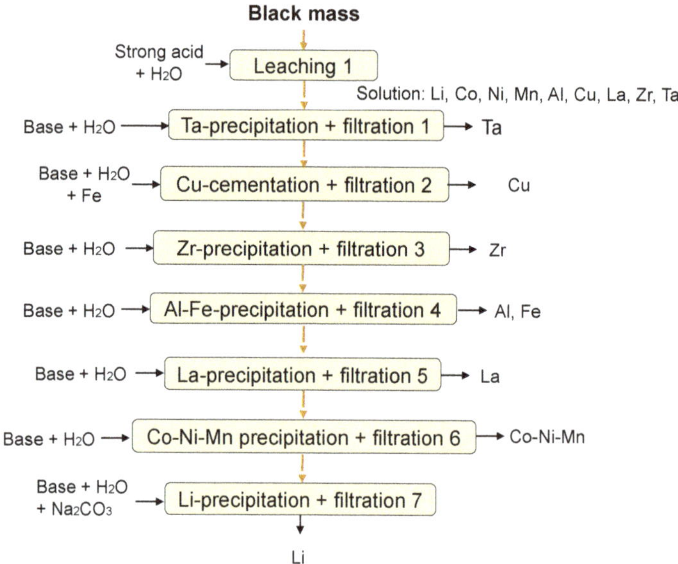

Figure 11. Developed recycling path for multi-step hydrometallurgy (Scenario 1: Full dissolution of garnet structure → aggressive leaching).

Scenario 2

In the case of moderate leaching, the garnet structure is not dissolved. However, depending on the acid chosen, the presented system behavior of scenario 2 can also occur although an aggressive leaching medium is chosen. In contrast, a milling step, as indicated in Figure 4 by the term "Crushing", could also enhance Li liberation. The Li^+/H^+ exchange is also realizable in aqueous or moderate leaching. Hence, a combination of milling and moderate leaching is to be experimentally evaluated in terms of leaching behaviour according to scenario 1. As described above, Ta would not be brought into solution, even if choosing a strong acid, except of HF. If La is not diluted, which can be the case for an unsuitable leaching medium as well, it will be filtered along with Ta. Subsequently, alkaline leaching could dissolve Ta and thus, enable a separation between La and Ta. Hence, all the garnet oxides can be selectively separated also by scenario 2.

Thus, the focal point of the experimental investigations will be a determination of suitable pre-treatments, both mechanically and thermally. Furthermore, the extractive hydrometallurgy aiming for a zero-waste and maximum yield, especially based on Co and Ni, is to be developed.

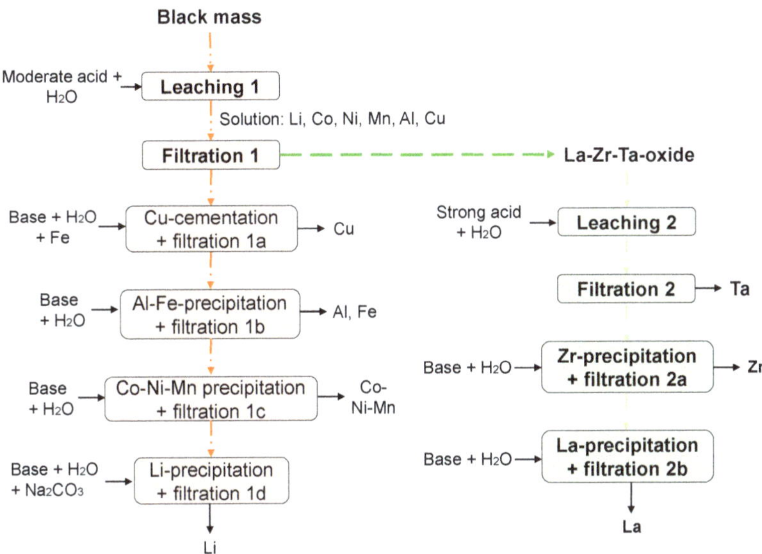

Figure 12. Developed recycling path for multi-step hydrometallurgy (Scenario 2: No dissolution of garnet structure → moderate leaching).

4. Conclusions

A technology's recyclability at its prototype development stage is a progressive and, according to the authors opinion, crucial approach for contributing to a both high-tech and highly sustainable world. Investigating a technology's recyclability at this stage generates an understanding of its future viability and, moreover, drawing up recommendations in terms of a "design for recycling". In this way, performance and sustainability are brought together, leading to the best technology concept in both economic and ecological value.

This study is the first attempt in approaching suitable recycling paths for oxide-based ASBs. Different options have been pointed out, focusing either on conventional LIB treatment steps or on innovative methods for a specifically tailored process.

According to our evaluation, the most promising pre-process is shown in Figure 4, and in combination with a detailed multi-step hydrometallurgy steps in Figure 12 leads to an optimal recycling yield. The dissolution of the garnet structure could require strong acids, which can lead to undesired environmental impacts and thus, should be avoided. Moreover, less cross-contaminations with conventional NMC-components and thus a higher recycling efficiency can be expected. If a thermal treatment is required, it should be tested experimentally, and if an ESLR is implied (see Figure 4), the final step regarding the Li carbonate recovery in Figure 12 can be foregone. However, in contrast to environmental and recycling efficiency considerations, the economic perspective is not taken into account here.

5. Outlook

Future research will focus on experimental implementation of the theoretical concepts drawn up within this study. Thus, mass balances and chemical analysis will examine the validity of the theoretical concept and thereby contribute to sustainability assessments of next-generation batteries. This research will be shown in "Recycling Concept for Ceramic All-Solid-State Batteries—Part II: Experimental validation for a LLZ + NMC-based System".

Author Contributions: Conceptualization, L.S. and M.K.; methodology, L.S., M.K., and A.S.; software, L.S.; validation, L.S., M.K., A.S., M.F., and B.F.; investigation, L.S., M.K., A.S., and M.F.; resources, L.S., M.K., A.S., and M.F.; data curation, L.S., M.K., A.S., and M.F.; writing—original draft preparation, L.S., M.K., A.S., and M.F.; writing—review and editing, L.S. and M.F.; visualization, L.S., M.K., and A.S.; supervision, M.F., D.F.-R., O.G., and B.F.; project administration, M.F.; funding acquisition, M.F., D.F.-R., and O.G. All authors have read and agreed to the published version of the manuscript.

Funding: This research was funded by the Helmholtz Association and the German Federal Ministry of Education and Research (BMBF) in the framework of the project "FestBatt-Oxide" (FKZ: 13XP0173A).

Conflicts of Interest: The authors declare no conflict of interest.

References

1. Olivetti, E.A.; Ceder, G.; Gaustad, G.G.; Fu, X. Lithium-ion battery supply chain considerations: Analysis of potential bottlenecks in critical metals. *Joule* **2017**, *1*, 229–243. [CrossRef]

2. Harper, G.; Sommerville, R.; Kendrick, E.; Driscoll, L.; Slater, P.; Stolkin, R.; Walton, A.; Christensen, P.; Heidrich, O.; Lambert, S.; et al. Recycling lithium-ion batteries from electric vehicles. *Nature* **2019**, *575*, 75–86. [CrossRef] [PubMed]

3. Zheng, X.; Zhu, Z.; Lin, X.; Zhang, Y.; He, Y.; Cao, H.; Sun, Z. A Mini-Review on Metal Recycling from Spent Lithium Ion Batteries. *Engineering* **2018**, *4*, 361–370. [CrossRef]

4. Zhao, S.; He, W.; Li, G. Recycling Technology and Principle of Spent Lithium-Ion Battery. In *Recycling of Spent Lithium-Ion Batteries*; An, L., Ed.; Springer International Publishing: Cham, Switzerland, 2019; pp. 1–26.

5. Träger, T.; Friedrich, B.; Weyhe, R. Recovery Concept of Value Metals from Automotive Lithium-Ion Batteries. *Chem. Ing. Tech.* **2015**, *87*, 1550–1557. [CrossRef]

6. Diaz, F.; Wang, Y.; Moorthy, T.; Friedrich, B. Degradation Mechanism of Nickel-Cobalt-Aluminum (NCA) Cathode Material from Spent Lithium-Ion Batteries in Microwave-Assisted Pyrolysis. *Metals* **2018**, *8*, 565. [CrossRef]

7. Ekberg, C.; Petranikova, M. Lithium Batteries Recycling. In *Lithium Process Chemistry. Resources, Extraction, Batteries, and Recycling*; Chagnes, A., Światowska, J., Eds.; Elsevier: Amsterdam, The Netherlands, 2015; pp. 233–267.

8. Wang, H.; Vest, M.; Friedrich, B. Hydrometallurgical processing of Li-Ion battery scrap from electric vehicles. In Proceedings of the EMC 2011, Düsseldorf, Germany, 26–29 June 2011; Harre, J., Ed.; GDMB: Clausthal-Zellerfeld, Germany, 2011.

9. Lombardo, G.; Ebin, B.; Foreman, M.R.S.J.; Steenari, B.-M.; Petranikova, M. Chemical Transformations in Li-Ion Battery Electrode Materials by Carbothermic Reduction. *ACS Sustain. Chem. Eng.* **2019**, *7*, 13668–13679. [CrossRef]

10. Zhang, G.; Du, Z.; He, Y.; Wang, H.; Xie, W.; Zhang, T. A Sustainable Process for the Recovery of Anode and Cathode Materials Derived from Spent Lithium-Ion Batteries. *Sustainability* **2019**, *11*, 2363. [CrossRef]

11. Meshram, P.; Pandey, B.D.; Mankhand, T.R. Recovery of valuable metals from cathodic active material of spent lithium ion batteries: Leaching and kinetic aspects. *Waste Manag.* **2015**, *45*, 306–313. [CrossRef]

12. Sun, L.; Qiu, K. Vacuum pyrolysis and hydrometallurgical process for the recovery of valuable metals from spent lithium-ion batteries. *J. Hazard. Mater.* **2011**, *194*, 378–384. [CrossRef]

13. Hanisch, C.; Loellhoeffel, T.; Diekmann, J.; Markley, K.J.; Haselrieder, W.; Kwade, A. Recycling of lithium-ion batteries: A novel method to separate coating and foil of electrodes. *J. Clean. Prod.* **2015**, *108*, 301–311. [CrossRef]

14. Or, T.; Gourley, S.W.D.; Kaliyappan, K.; Yu, A.; Chen, Z. Recycling of mixed cathode lithium-ion batteries for electric vehicles: Current status and future outlook. *Carbon Energy* **2020**, *2*, 6–43. [CrossRef]

15. Pagliaro, M.; Meneguzzo, F. Lithium battery reusing and recycling: A circular economy insight. *Heliyon* **2019**, *5*, e01866. [CrossRef] [PubMed]

16. Zheng, Y.; Song, W.; Mo, W.-T.; Zhou, L.; Liu, J.-W. Lithium fluoride recovery from cathode material of spent lithium-ion battery. *RSC Adv.* **2018**, *8*, 8990–8998. [CrossRef]

17. Chen, Y.; Liu, N.; Jie, Y.; Hu, F.; Li, Y.; Wilson, B.P.; Xi, Y.; Lai, Y.; Yang, S. Toxicity Identification and Evolution Mechanism of Thermolysis-Driven Gas Emissions from Cathodes of Spent Lithium-Ion Batteries. *ACS Sustain. Chem. Eng.* **2019**, *7*, 18228–18235. [CrossRef]

18. Wang, H.; Friedrich, B. Development of a Highly Efficient Hydrometallurgical Recycling Process for Automotive Li–Ion Batteries. *J. Sustain. Metall.* **2015**, *1*, 168–178. [CrossRef]

19. Yin, H.; Xing, P. Pyrometallurgical Routes Pyrometallurgical Routes for the Recycling of Spent Lithium-Ion Batteries. In *Recycling of Spent Lithium-Ion Batteries*; An, L., Ed.; Springer International Publishing: Cham, Switzerland, 2019; pp. 57–83.

20. Pinegar, H.; Smith, Y.R. Recycling of End-of-Life Lithium Ion Batteries, Part I: Commercial Processes. *J. Sustain. Metall.* **2019**, *5*, 402–416. [CrossRef]

21. Georgi-Maschler, T.; Friedrich, B.; Weyhe, R.; Heegn, H.; Rutz, M. Development of a recycling process for Li-ion batteries. *J. Power Sources* **2012**, *207*, 173–182. [CrossRef]

22. Heelan, J.; Gratz, E.; Zheng, Z.; Wang, Q.; Chen, M.; Apelian, D.; Wang, Y. Current and Prospective Li-Ion Battery Recycling and Recovery Processes. *JOM* **2016**, *68*, 2632–2638. [CrossRef]

23. Chagnes, A.; Pospiech, B. A brief review on hydrometallurgical technologies for recycling spent lithium-ion batteries. *J. Chem. Technol. Biotechnol.* **2013**, *88*, 1191–1199. [CrossRef]

24. Swain, B. Recovery and recycling of lithium: A review. *Sep. Purif. Technol.* **2017**, *172*, 388–403. [CrossRef]

25. Kwade, A. Project Website InnoRec. 2019. Available online: https://www.prozell-cluster.de/en/projects/innorec/ (accessed on 24 March 2020).

26. VDI/VDE Innovation + Technik GmbH. Project Website Mercator. 2019. Available online: https://www.erneuerbar-mobil.de/en/node/1232 (accessed on 16 November 2020).

27. Stallmeister, C.; Schwich, L.; Friedrich, B. Early-Stage Li-Removal -Vermeidung von Lithiumverlusten im Zuge der Thermischen und Chemischen Recyclingrouten von Batterien. In *Recycling und Sekundärrohstoffe, Band 13*; Thomé-Kozmiensky, E., Holm, O., Friedrich, B., Goldmann, D., Eds.; Thomé-Kozmiensky Verlag GmbH: Neuruppin, Germany, 2020; pp. 545–557.

28. Hu, J.; Zhang, J.; Li, H.; Chen, Y.; Wang, C. A promising approach for the recovery of high value-added metals from spent lithium-ion batteries. *J. Power Sources* **2017**, *351*, 192–199. [CrossRef]

29. Li, J.; Wang, G.; Xu, Z. Environmentally-friendly oxygen-free roasting/wet magnetic separation technology for in situ recycling cobalt, lithium carbonate and graphite from spent $LiCoO_2$/graphite lithium batteries. *J. Hazard. Mater.* **2016**, *302*, 97–104. [CrossRef] [PubMed]

30. Bertau, M.; Martin, G. Integrated Direct Carbonation Process for Lithium Recovery from Primary and Secondary Resources. *MSF* **2019**, *959*, 69–73. [CrossRef]

31. Vanderbruggen, A.; Rudolph, M. *Flotation of Spheroidized Graphite from Spent Lithium Ion Batteries*. 2020. Available online: https://www.researchgate.net/publication/338630299_Recovery_of_spheroidized_graphite_from_spent_lithium_ion_batteries_Talk_at_AABC_Europe (accessed on 16 November 2020).

32. He, Y.; Zhang, T.; Wang, F.; Zhang, G.; Zhang, W.; Wang, J. Recovery of $LiCoO_2$ and graphite from spent lithium-ion batteries by Fenton reagent-assisted flotation. *J. Clean. Prod.* **2017**, *143*, 319–325. [CrossRef]

33. Paulino, J.F.; Busnardo, N.G.; Afonso, J.C. Recovery of valuable elements from spent Li-batteries. *J. Hazard. Mater.* **2008**, *150*, 843–849. [CrossRef]

34. Hosseini-Bab-Anari, E.; Boschin, A.; Mandai, T.; Masu, H.; Moth-Poulsen, K.; Johansson, P. Fluorine-free salts for aqueous lithium-ion and sodium-ion battery electrolytes. *RSC Adv.* **2016**, *6*, 85194–85201. [CrossRef]

35. Davenport, W.G. *Extractive Metallurgy of Nickel, Cobalt, and Platinum Group Materials*; Elsevier: Amsterdam, The Netherlands, 2011.

36. Pinegar, H.; Smith, Y.R. Recycling of End-of-Life Lithium-Ion Batteries, Part II: Laboratory-Scale Research Developments in Mechanical, Thermal, and Leaching Treatments. *J. Sustain. Metall.* **2020**, *6*, 142–160. [CrossRef]

37. Janek, J.; Zeier, W.G. A solid future for battery development. *Nat. Energy* **2016**, *1*, 1167. [CrossRef]

38. Kurzweil, P.; Dietlmeier, O.K. *Elektrochemische Speicher: Superkondensatoren, Batterien, Elektrolyse-Wasserstoff, Rechtliche Grundlagen, 1. Aufl.*; Springer Fachmedien Wiesbaden: Wiesbaden, Germany, 2015.

39. Tsai, C.-L.; Ma, Q.; Dellen, C.; Lobe, S.; Vondahlen, F.; Windmüller, A.; Grüner, D.; Zheng, H.; Uhlenbruck, S.; Finsterbusch, M.; et al. A garnet structure-based all-solid-state Li battery without interface modification: Resolving incompatibility issues on positive electrodes. *Sustain. Energy Fuels* **2019**, *3*, 280–291. [CrossRef]

40. Troy, S.; Schreiber, A.; Reppert, T.; Gehrke, H.-G.; Finsterbusch, M.; Uhlenbruck, S.; Stenzel, P. Life Cycle Assessment and resource analysis of all-solid-state batteries. *Appl. Energy* **2016**, *169*, 757–767. [CrossRef]

41. Erdmann, L.; Graedel, T.E. Criticality of non-fuel minerals: A review of major approaches and analyses. *Environ. Sci. Technol.* **2011**, *45*, 7620–7630. [CrossRef] [PubMed]

42. Graedel, T.E.; Barr, R.; Chandler, C.; Chase, T.; Choi, J.; Christoffersen, L.; Friedlander, E.; Henly, C.; Jun, C.; Nassar, N.T.; et al. Methodology of metal criticality determination. *Environ. Sci. Technol.* **2012**, *46*, 1063–1070. [CrossRef]

43. Ambrose, H.; Kendall, A. Understanding the future of lithium: Part 1, resource model. *J. Ind. Ecol.* **2020**, *24*, 80–89. [CrossRef]

44. Simon, B.; Ziemann, S.; Weil, M. Criticality of metals for electrochemical energy storage systems—Development towards a technology specific indicator. *Metall. Res. Technol.* **2015**, *111*, 191–200. [CrossRef]

45. Helbig, C.; Bradshaw, A.M.; Wietschel, L.; Thorenz, A.; Tuma, A. Supply risks associated with lithium-ion battery materials. *J. Clean. Prod.* **2018**, *172*, 274–286. [CrossRef]

46. Buchholz, P.; Huy, D.; Sievers, H. *DERA-Rohstoffliste 2012, Angebotskonzentration bei Metallen und Industriemineralien—Potenzielle Preis-und Lieferrisiken*; Deutsche Rohstoffagentur (DERA) in der Bundesanstalt für Geowissenschaften und Rohstoffe: Berlin, Germany, 2012; Volume 10.

47. LITHOREC. Recycling von Lithium-Ionen-Batterien—LithoRec II. Abschlussberichte der Beteiligten Verbundpartner. 2012. Available online: https://www.erneuerbar-mobil.de/sites/default/files/2017-01/Abschlussbericht_LithoRec_II_20170116.pdf (accessed on 1 October 2020).

48. Deloitte Sustainability; British Geological Survey; Bureau de Recherches Géologiques et Minières; Netherlands Organisation for Applied Scientific Research. Study on the Review of the List of Critical Raw Materials. Criticality Assessments. 2017. Available online: https://op.europa.eu/en/publication-detail/-/publication/08fdab5f-9766-11e7-b92d-01aa75ed71a1/language-en (accessed on 16 November 2020).

49. Joint Research Centre. *The European Commission's In-House Science Service: Annual Report 2013*; Report EUR 26372 EN; Publications Office of the European Union: Luxembourg, 2014.

50. Schultz, D.; Kuckshinrichs, W. The need for lithium—An upcoming problem of electrochemical energy storages? *J. Energy Chall. Mech.* **2016**, *3*, 180–185.

51. Piana, G.; Ricciardi, M.; Bella, F.; Cucciniello, R.; Proto, A.; Gerbaldi, C. Poly(glycidyl ether)s recycling from industrial waste and feasibility study of reuse as electrolytes in sodium-based batteries. *Chem. Eng. J.* **2020**, *382*, 122934. [CrossRef]

52. Wang, Y.-Y.; Diao, W.-Y.; Fan, C.-Y.; Wu, X.-L.; Zhang, J.-P. Benign Recycling of Spent Batteries towards All-Solid-State Lithium Batteries. *Chemistry* **2019**, *25*, 8975–8981. [CrossRef]

53. Tsai, C.-L.; Roddatis, V.; Chandran, C.V.; Ma, Q.; Uhlenbruck, S.; Bram, M.; Heitjans, P.; Guillon, O. Li7La3Zr2O12 Interface Modification for Li Dendrite Prevention. *ACS Appl. Mater. Interfaces* **2016**, *8*, 10617–10626. [CrossRef]

54. Uhlenbruck, S.; Dornseiffer, J.; Lobe, S.; Dellen, C.; Tsai, C.-L.; Gotzen, B.; Sebold, D.; Finsterbusch, M.; Guillon, O. Cathode-electrolyte material interactions during manufacturing of inorganic solid-state lithium batteries. *J. Electroceram.* **2017**, *38*, 197–206. [CrossRef]

55. Park, K.; Yu, B.-C.; Jung, J.-W.; Li, Y.; Zhou, W.; Gao, H.; Son, S.; Goodenough, J.B. Electrochemical Nature of the Cathode Interface for a Solid-State Lithium-Ion Battery: Interface between $LiCoO_2$ and Garnet-$Li_7La_3Zr_2O_{12}$. *Chem. Mater.* **2016**, *28*, 8051–8059. [CrossRef]

56. Ren, Y.; Liu, T.; Shen, Y.; Lin, Y.; Nan, C.-W. Chemical compatibility between garnet-like solid state electrolyte Li6.75La3Zr1.75Ta0.25O12 and major commercial lithium battery cathode materials. *J. Mater.* **2016**, *2*, 256–264. [CrossRef]

57. DeWees, R.; Wang, H. Synthesis and Properties of NaSICON-type LATP and LAGP Solid Electrolytes. *ChemSusChem* **2019**, *12*, 3713–3725. [CrossRef] [PubMed]

58. Kato, T.; Iwasaki, S.; Ishii, Y.; Motoyama, M.; West, W.C.; Yamamoto, Y.; Iriyama, Y. Preparation of thick-film electrode-solid electrolyte composites on $Li_7La_3Zr_2O_{12}$ and their electrochemical properties. *J. Power Sources* **2016**, *303*, 65–72. [CrossRef]

59. Thielmann, A.; Neef, C.; Hettesheimer, T.; Döscher, H.; Wietschel, M.; Tübke, J. Energiespeicher-Roadmap. 2017. Available online: https://www.isi.fraunhofer.de/content/dam/isi/dokumente/cct/lib/Energiespeicher-Roadmap-Dezember-2017.pdf (accessed on 23 March 2020).

60. United States Geological Survey. *U.S. Geological Survey: Mineral Commodity Summaries*; United States Geological Survey: Reston, VA, USA, 2020.

61. Reichl, C.; Schatz, M.; Zsak, G. World-Mining-Data Welt-Bergbau-Daten. 2014. Available online: https://www.univie.ac.at/Mineralogie/docs/Weltbergbaudaten_2014.pdf (accessed on 1 October 2020).

62. Brown, T.J.; Idoine, N.E.; Wrighton, C.E.; Raycraft, E.R.; Hobbs, S.F.; Shaw, R.A.; Everett, P.; Kresse, C.; Deady, E.A.; Bide, T. *World Mineral Production 2014–2018*; British Geological Survey: Nottingham, UK, 2019.

63. Available online: https://www.statista.com/statistics/280060/share-of-global-rare-earth-supply-by-element-2015/ (accessed on 1 October 2020).

64. Peters, L.; Schier, C.; Friedrich, B. Lithium- und Kobalt-Rückgewinnung aus Elektrolichtbogenofenschlacken des Batterie-Recyclings. In *Mineralische Nebenprodukte und Abfälle. Aschen, Schlacken, Stäube und Baurestmassen*; Thiel, S., Thomé-Kozmiensky, E., Friedrich, B., Pretz, T., Quicker, P., Senk, D., Wotruba, H., Eds.; Thomé-Kozmiensky Verlag GmbH: Neuruppien, Germany, 2018; pp. 338–359.

65. Wang, H.; Friedmann, D.; Vest, M.; Friedrich, B. *Innovative Recycling of Li-Based Electric Vehicle Batteries, General Hydrometallurgy, General Pyrometallurgy/Vessel Integrity/Process Gas Treatment, Recycling. Posters, Authors Index, Keywords Index*; GDMB Verlag: Clausthal-Zellerfeld, Germany, 2013.

66. Lide, D.R. (Ed.) A ready-reference book of chemical and physical data. In *CRC Handbook of Chemistry and Physics*; CRC Press: Boca Raton, FL, USA, 2003.

67. Larraz, G.; Orera, A.; Sanjuán, M.L. Cubic phases of garnet-type Li7La3Zr2O12: The role of hydration. *J. Mater. Chem. A* **2013**, *1*, 11419. [CrossRef]

68. Liu, X.; Chen, Y.; Hood, Z.D.; Ma, C.; Yu, S.; Sharafi, A.; Wang, H.; An, K.; Sakamoto, J.; Siegel, D.J.; et al. Elucidating the mobility of H+ and Li+ ions in (Li 6.25–xHxAl0.25)La3Zr2O12 via correlative neutron and electron spectroscopy. *Energy Environ. Sci.* **2019**, *12*, 945–951. [CrossRef]

69. Ma, C.; Rangasamy, E.; Liang, C.; Sakamoto, J.; More, K.L.; Chi, M. Excellent stability of a lithium-ion-conducting solid electrolyte upon reversible Li(+)/H(+) exchange in aqueous solutions. *Angew. Chem. Int. Ed. Engl.* **2015**, *54*, 129–133. [CrossRef]

70. Ma, Y.; Stopic, S.; Wang, X.; Forsberg, K.; Friedrich, B. Basic Sulfate Precipitation of Zirconium from Sulfuric Acid Leach Solution. *Metals* **2020**, *10*, 1099. [CrossRef]

71. Ferreira, C.A.; Formiga, T.S.; Morais, C.A. Study of the Zircon Processing Aiming the Recovery of Zirconium and Silica. In Proceedings of the 24th International Mining Congress and Exhibition of Turkey, IMCET 2015, Antalya, Turkey, 14–17 April 2015; pp. 1351–1356.

72. Orhanovic, Z.; Pokric, B.; Füredi, H.; Branica, M. Precipitation and Hydrolysis of Metallic Ions. III. Studies on the Solubility of Yttrium and Some Rare Earth Hydroxides. *Croat. Chem. Acta* **1966**, *38*, 269–276.

73. Um, N.; Hirato, T. Dissolution Behavior of La_2O_3, Pr_2O_3, Nd_2O_3, CaO and Al_2O_3 in Sulfuric Acid Solutions and Study of Cerium Recovery from Rare Earth Polishing Powder Waste via Two-Stage Sulfuric Acid Leaching. *Mater. Trans.* **2013**, *54*, 713–719. [CrossRef]

74. Chen, W.-S.; Ho, H.-J.; Lin, K.-Y. Hydrometallurgical Process for Tantalum Recovery from Epoxy-Coated Solid Electrolyte Tantalum Capacitors. *Materials* **2019**, *12*, 1220. [CrossRef] [PubMed]

75. Clark, R.J.H.; Brown, D.; Bailar, J.C.; Emeléus, H.J.; Nyholm, R. *The Chemistry of Vanadium, Niobium and Tantalum: Pergamon Texts in Inorganic Chemistry*; Elsevier: Amsterdam, The Netherlands, 2013; Volume 20.

76. Deblonde, G.J.-P.; Chagnes, A.; Bélair, S.; Cote, G. Solubility of niobium(V) and tantalum(V) under mild alkaline conditions. *Hydrometallurgy* **2015**, *156*, 99–106. [CrossRef]

77. Deblonde, G.J.-P.; Chagnes, A.; Weigel, V.; Cote, G. Direct precipitation of niobium and tantalum from alkaline solutions using calcium-bearing reagents. *Hydrometallurgy* **2016**, *165*, 345–350. [CrossRef]

Publisher's Note: MDPI stays neutral with regard to jurisdictional claims in published maps and institutional affiliations.

Article

A Combined Pyro- and Hydrometallurgical Approach to Recycle Pyrolyzed Lithium-Ion Battery Black Mass Part 2: Lithium Recovery from Li Enriched Slag—Thermodynamic Study, Kinetic Study, and Dry Digestion

Jakub Klimko [1,*], Dušan Oráč [1], Andrea Miškufová [1], Claudia Vonderstein [2], Christian Dertmann [2], Marcus Sommerfeld [2], Bernd Friedrich [2] and Tomáš Havlík [1]

[1] Institute of Recycling Technologies, Faculty of Materials, Metallurgy and Recycling, Technical University of Košice, Letna 9, 04200 Košice, Slovakia; dusan.orac@tuke.sk (D.O.); andrea.miskufova@tuke.sk (A.M.); tomas.havlik@tuke.sk (T.H.)

[2] IME Process Metallurgy and Metal Recycling, Institute of RWTH Aachen University, Intzestraße 3, 52056 Aachen, Germany; cvonderstein@ime-aachen.de (C.V.); cdertmann@ime-aachen.de (C.D.); msommerfeld@ime-aachen.de (M.S.); bfriedrich@ime-aachen.de (B.F.)

* Correspondence: jakub.klimko@tuke.sk; Tel.: +421-905-507-327

Received: 28 October 2020; Accepted: 19 November 2020; Published: 23 November 2020

Abstract: Due to the increasing demand for battery raw materials, such as cobalt, nickel, manganese, and lithium, the extraction of these metals, not only from primary, but also from secondary sources, is becoming increasingly important. Spent lithium-ion batteries (LIBs) represent a potential source of raw materials. One possible approach for an optimized recovery of valuable metals from spent LIBs is a combined pyro- and hydrometallurgical process. The generation of mixed cobalt, nickel, and copper alloy and lithium slag as intermediate products in an electric arc furnace is investigated in part 1. Hydrometallurgical recovery of lithium from the Li slag is investigated in part 2 of this article. Kinetic study has shown that the leaching of slag in H_2SO_4 takes place according to the 3-dimensional diffusion model and the activation energy is 22–24 kJ/mol. Leaching of the silicon from slag is causing formation of gels, which complicates filtration and further recovery of lithium from solutions. The thermodynamic study presented in the work describes the reasons for the formation of gels and the possibilities of their prevention by SiO_2 precipitation. Based on these findings, the Li slag was treated by the dry digestion (DD) method followed by dissolution in water. The silicon leaching efficiency was significantly reduced from 50% in the direct leaching experiment to 5% in the DD experiment followed by dissolution, while the high leaching efficiency of lithium was maintained. The study takes into account the preparation of solutions for the future trouble-free acquisition of marketable products from solutions.

Keywords: lithium-ion battery; recycling; lithium; slag; hydrometallurgy; leaching; dry digestion; critical raw materials

---

1. Introduction

Lithium-ion batteries (LIBs) are currently considered as one of the most important energy storage systems, which is reflected in a wide range of applications, especially for portable devices [1–7]. Due to the extensive electrification expected in the field of electromobility, batteries will have another key role in the future ensuring the transition towards a climate-neutral economy [8]. In addition to the implementation of electromobility and their widespread use for portable applications, lithium-ion batteries are also indispensable as intermediate storage for the stabilization of decentralized power systems [2–5,9,10]. Compared to other battery types, LIBs have advantageous technical properties that substantiate their dominance as energy storage systems, including, e.g., high energy density and low self-discharge [10,11]. As a result of increasing applications of lithium-ion batteries, a significantly higher demand for battery containing critical or strategic raw materials, such as cobalt, lithium, and nickel, is to be expected. Those crucial metals are only available in limited quantities and are currently obtained mainly from primary sources [2]. Recycling is an essential aspect of closing the entire substance cycle of LIBs and securing the supply of raw materials for new battery production. To meet the increasing demand for strategic metals, the development of a raw material recycling economy, in addition to the expansion of mining capacities, is therefore unavoidable [5].

In the European Union, Directive 2006/66/EC applies to the recycling of LIBs. This directive requires the recycling of fifty percent of the average weight of used batteries, including spent LIBs [12]. The extensive recycling of battery components that exceed the recycling quota of fifty percent is of central importance to ensure the supply of materials for the new battery production and consequently the transition to a climate-neutral economy. This also requires consideration of battery components such as lithium.

In 2020, lithium was newly included among the European Union (EU)'s critical raw materials, which made its recovery, especially in the battery systems, where lithium is an indispensable cathode component, essential [13]. In the field of battery recycling, research projects have been carried out for several years dealing with both single and combined mechanical, pyrometallurgical, and hydrometallurgical processes as well as pyrolysis to recover battery components [5,10,11,14–24]. However, the focus is mainly set on more valuable metals, which is the reason why lithium as a component has not been sufficiently considered [25]. Overall, the recovery of lithium from active electrode mass has not been solved satisfactorily since the recovery is made more difficult by the ignoble character of the spent LIBs. Only about 20 tons of lithium were reported in 2016 as secondary raw material on a global scale [13,26]. In the EU, 4 tons were produced from secondary raw materials, representing an end-of-life recycling-input rate of less than 1% [13,27]. In pyrometallurgical processes, lithium is converted into slag, which is either used as construction material, undergoes further hydrometallurgical treatment, or can be sold, e.g., for the cement industry [28]. Hydrometallurgical processes allow lithium to be recovered from black mass, for example as lithium carbonate [29].

Within the scope of this work, a combined pyro- and hydrometallurgical process was designed, which enables a complete recovery of the valuable elements present in the active mass of spent LIBs. In a first process step presented in part 1 of this article, the melting of pellets with SiO_2 and CuO in the electric arc furnace was realized, where metal alloy and Li slag were produced. Production of the slag enabled the recovery of lithium, while more valuable components, such as cobalt, copper, and nickel, were enriched in a metal alloy [30].

In part 2 of this article, the optimal leaching conditions of Li slag in sulfuric acid were studied. The design of the leaching step was adapted for the subsequent trouble-free recovery of lithium from the obtained solution. The main complication of leaching of materials with a higher silicon content is the formation of silicon gels [31–35]. In addition to the thermodynamic and kinetic study of the slag leaching process, the work also describes the conditions of gel formation and proposes the leaching method, in which the formation of gels does not occur.

2. Materials and Methods

2.1. Sample Preparation and Characterization

Pyrolyzed lithium-ion (Li-ion) battery black mass used for this study was generated by Accurec Recycling GmbH and further processed in the Electric Arc furnace (EAF) at the IME Process Metallurgy and Metal Recycling, Institute of RWTH Aachen University, Germany with the SiO_2 used as flux and CuO used to react with excess carbon from pyrolyzed battery black mass. A detailed description of the trials is presented in part 1 [30]. The chemical composition of the input materials, intermediates, and solutions obtained after hydrometallurgical treatment was analyzed by atomic absorption spectrometry (AAS) using spectrAA20+ spectrometer (Varian, Autralia). The phase composition of materials was analyzed by X-ray diffraction phase analysis (XRD) using Philips X'Pert Pro Co-Kα (PANanalytical, Almelo, Netherlands) and identified by PANanalytical HighScore Plus software version 3.0.

Generated Li slag showed in Figure 1 was crushed in a jaw crusher, ground in a mechanical mortar, subjected to a two-stage manual magnetic separation using a ferrite magnet, and subjected to sieving. Magnetic separation aimed to reduce the content of metallic impurities such as cobalt, copper, nickel, and iron in the Li slag. The first stage of magnetic separation was performed after crushing (Figure 2a) and the second after grinding (Figure 2b). Particle size of +0 −0.5 mm was used in this study. The chemical composition of Li slag and slag after mechanical pre-treatment is shown in Table 1.

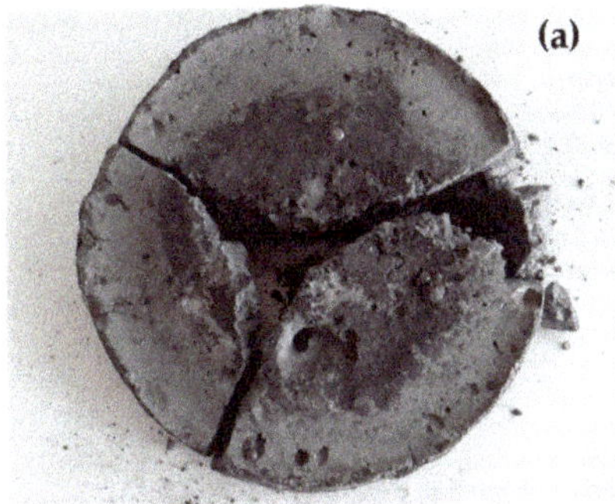

Figure 1. Li slag from EAF: (**a**) before mechanical pretreatment, (**b**) after mechanical pre-treatment consisting of crushing, grounding, magnetic separation, and sieving.

Figure 2. Metal phases obtained by magnetic separation: (**a**) after crushing (**b**) after grounding.

Table 1. Chemical composition of the slag before and after magnetic separation.

Sample	Li	Co	Cu	Al	Fe	Si	Ca	Ni	Mn
Li slag	6.80	1.17	1.53	16.52	0.51	48.62	1.16	0.15	0.65
De-metalized Li slag	6.96	0.00	0.11	16.40	0.20	51.10	1.26	0.01	0.88

The price and estimated value of Li slag elements in their marketable products are shown in Table 2. The theoretical values of the elements were calculated from the content of the element in Li slag and from the content of the element in the marketable product and their price. The most valuable element for its recovery from used Li slag is lithium marketed as Li_2CO_3. From one metric ton of Li slag, it is theoretically possible to obtain Li_2CO_3 with an estimated value of US $2500.27. The second most valuable element in the slag is aluminum with a theoretical value of US $367.99 per metric ton of Li slag. In addition to Li and Al, the study is focused on the leaching rates of silicon due to its high content in the Li slag and potential silica gel formations. Other elements are considered as impurities due to their low content.

Table 2. Estimated value of the elements in the Li slag, data from [36,37].

Elements	Content	Marketable Products	Price of Marketable Products (US $/t)	Estimated Value of Marketable Products in 1 t of Li Slag (US $)
Li	6.96%	Li_2CO_3 (Li = 18.79%)	6750	2500.27
Al	16.40%	Al	2243.86	367.99
Si	51.10%	SiO_2 (Si = 46.74%)	100–120	109–131
Mn	0.88%	Mn	1525.83	13.43
Cu	0.11%	Cu	7703.93	8.47
Ca	1.26%	CaO (Ca = 71.47%)	80	1.41–1.76
Ni	0.01%	Ni	17,935.94	1.79
Fe	0.20%	Fe	285	0.57
Co	0.00%	Co	32,985	-

XRD analysis of de-metalized Li slag is shown in Figure 3. The analysis shows that lithium is present in the slag as $LiAlSiO_4$ and Li_2SiO_3. Other phases containing SiO_2 were not identified in XRD diffraction pattern, which implies that excessive amount of SiO_2 in the slag is present in amorphous form $SiO_{2\ (am)}$.

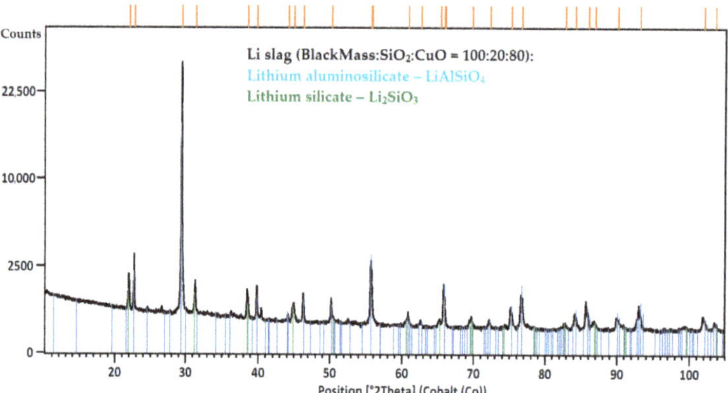

Figure 3. XRD analysis of the Li slag.

2.2. Thermodynamic Study of Lithium Slag Leaching

A thermodynamic analysis was performed using HSC software (Outotec, Espoo, Finland) [38]. Table 3 shows potential reactions of identified lithium phases with sulfuric acid and their $\Delta G°$ values at 20 °C and 80 °C. A negative value of standard Gibbs energy indicates that these reactions are spontaneous and thermodynamically feasible.

Figures 4–6 show E-pH diagram for Li-Al-S-Si-H_2O, Al-Li-S-Si-H_2O, and Si-Al-Li-S-H_2O systems respectively. Thermodynamic study confirmed the presence of lithium as Li^+ ion in acidic leaching system at both selected temperatures. At pH above 13 and at 80 °C LiOH*H_2O might precipitate. Aluminum and silicon are also thermodynamically stable in ionic form under pH five. At pH above five at 20 °C and pH above four at 80 °C Al_2O_3*SiO_2 precipitates. Reactions (1), (2), and (3) from Table 3 indicated three possible silicon phases (SiO_2, H_2SiO_3, and H_4SiO_4) present in the leaching system (formed by means of reaction of $LiAlSiO_4$ with H_2SO_4). According to the Eh-pH diagram, the most thermodynamically stable is H_4SiO_4.

Table 3. Potential reactions of slag phases and $\Delta G°$ values in the H_2SO_4 leaching system, data from [38].

Equation	Reaction	$\Delta G°_{293.15}$ [kJ]	$\Delta G°_{353.15}$ [kJ]
(1)	$2LiAlSiO_4 + 4H_2SO_{4(l)} = Li_2SO_{4(ia)} + Al_2(SO_4)_{3(ia)} + 2H_4SiO_{4(a)}$	−274.727	−274.783
(2)	$2LiAlSiO_4 + 4H_2SO_{4(l)} = Li_2SO_{4(ia)} + Al_2(SO_4)_{3(ia)} + 2H_2SiO_{3(a)} + 2H_2O$	−300.098	−272.695
(3)	$2LiAlSiO_4 + 4H_2SO_{4(l)} = Li_2SO_{4(ia)} + Al_2(SO_4)_{3(ia)} + 2SiO_2 + 4H_2O$	−302.950	−303.060
(4)	$2LiAlSi_2O_6 + 4H_2SO_{4(l)} + 4H_2O = Li_2SO_{4(ia)} + Al_2(SO_4)_{3(ia)} + 4H_4SiO_{4(a)}$	−216.581	−218.504
(5)	$Li_2SiO_3 + H_2SO_{4(l)} = Li_2SO_{4(ia)} + H_2SiO_{3(a)}$	−151.744	−141.622
(6)	$Li_2SiO_3 + H_2SO_{4(l)} + H_2O_{(l)} = Li_2SO_{4(ia)} + H_4SiO_{4(a)}$	−152.092	−142.666

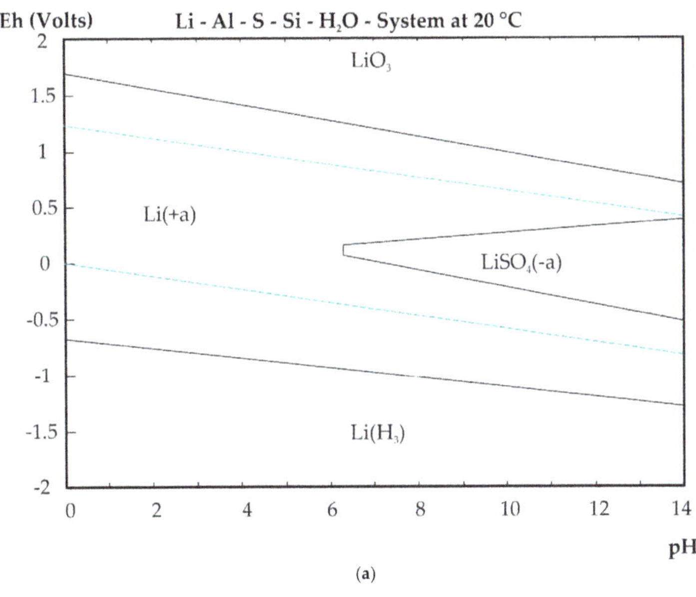

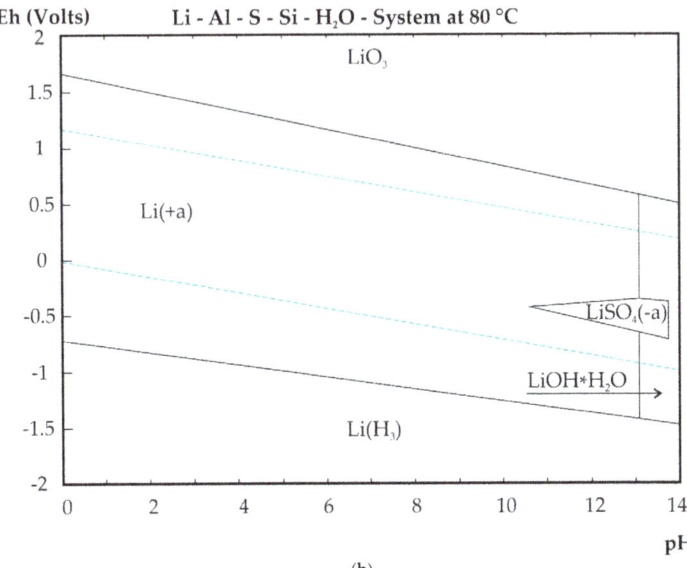

Figure 4. Eh-pH diagram for Li-Al-S-Si-H₂O system: (**a**) 20 °C, (**b**) 80 °C. Molarity: (Li) = 1 M, (Al) = 0.5 M, (S) = 0.5 M and (Si) = 0.05 M.

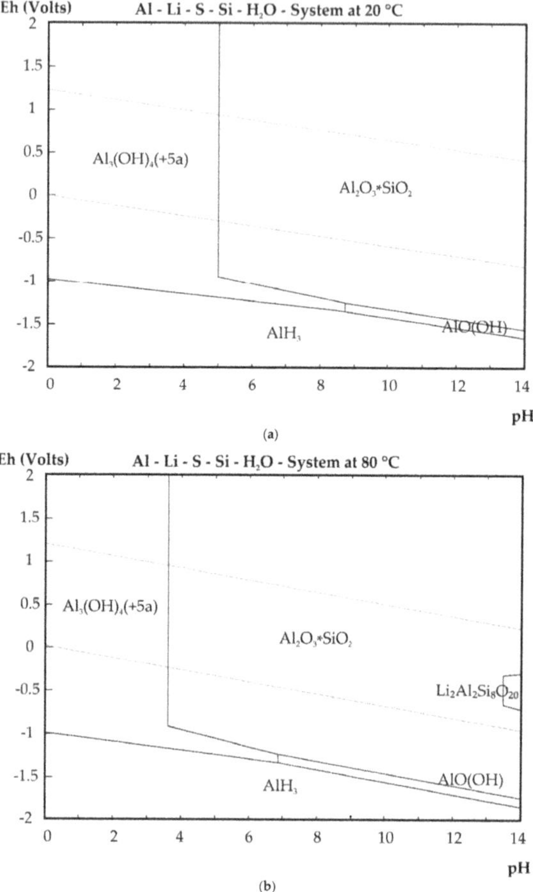

Figure 5. Eh-pH diagram for Al-Li-S-Si-H$_2$O system: (**a**) 20 °C; (**b**) 80 °C. Molarity: (Li) = 1 M, (Al) = 0.5 M, (S) = 0.5 M and (Si) = 0.05 M.

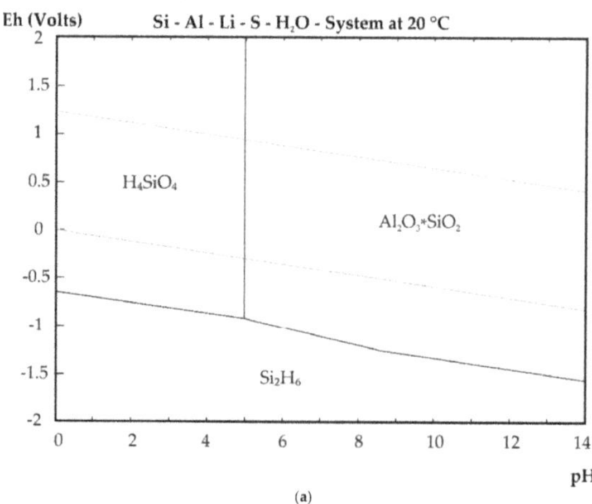

Figure 6. *Cont.*

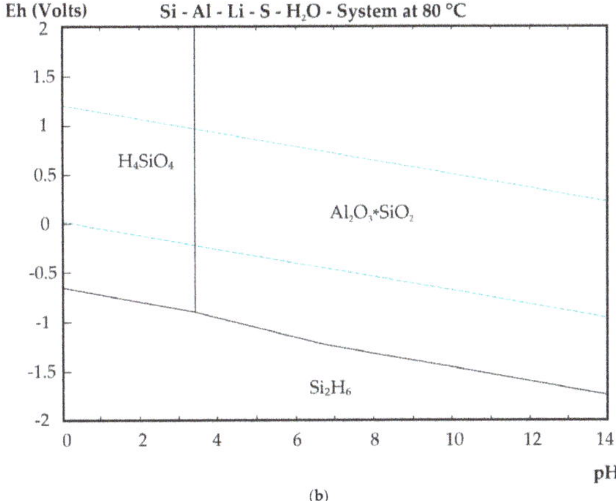

Figure 6. Eh-pH diagram for Si-Al-Li-S-H$_2$O system: (**a**) 20 °C, (**b**) 80 °C. Molarity: (Li) = 1 M, (Al) = 0.5 M, (S) = 0.5 M and (Si) = 0.05 M.

2.3. Methodology of Direct Li-Slag Leaching

The direct leaching experiments were carried out in an 800 mL glass reactor placed in a thermostatically controlled water bath. The experiments were performed at temperatures from 20 °C, 40 °C, 60 °C, and 80 °C using constant 400 rpm stirring speed. The aqueous solution of sulfuric acid at concentrations of 0.5 M and 1 M were used as leaching reagents. The pH of the solutions was measured with pH-meter (Inolab, WTW 3710, Germany). The volume of the leaching reagent was 500 mL. Ten grams of de-metalized Li slag samples were used for the experiments, which represents liquid to solid ratio L:S = 50. Solid Li slag samples for the experiments were obtained by manual quartering. The total duration of the experiment was 30 min with sampling time after 2, 5, 10, 15, 30 min.

2.4. Methodology of Dry Digestion Followed by Dissolution in Water

In addition to direct slag leaching, dry digestion (DD) experiments were performed to minimize gel formation. These experiments were carried out in 800 mL glass reactors. The experiments consisted of mixing a Li slag samples with concentrated (17.9 M) sulfuric acid and deionized water. The duration of the DD experiments was set to 60 min, after which the resulting dried mixture was dissolved in additional 500 mL of deionized water. The first series of experiments consisted of finding the optimal ratio of sample, acid, and water in DD step. The weight of Li slag samples was 10 g. The volumes of sulfuric acid used in this set of experiments was 5, 10, and 15 mL and volume of deionized water was 4, 12, and 24 mL. The experiments were also performed without the addition of water. In the second series of experiments, the constant ratio of Li slag sample, acid and water in DD step and volume of water in water dissolution step was maintained, while the total amount of the dried mixture was increased in the dissolution step. The prepared mixtures consisted of slag [g], acid [mL] and water [mL] in the ratios 10:10:24, 25:25:60, 50:50:120, and 100:100:240.

3. Results and Discussion

3.1. Direct Li Slag Leaching and Kinetic Study

The first hydrometallurgical experiment was direct leaching of Li slag in 500 mL of 0.5 M and 1 M H$_2$SO$_4$ solution at 20 °C, 40 °C, 60 °C, and 80 °C for 30 min. Figures 7–9 show the dependence of leaching efficiency over time at different temperatures and acid concentrations for lithium, aluminum, and silicon, respectively.

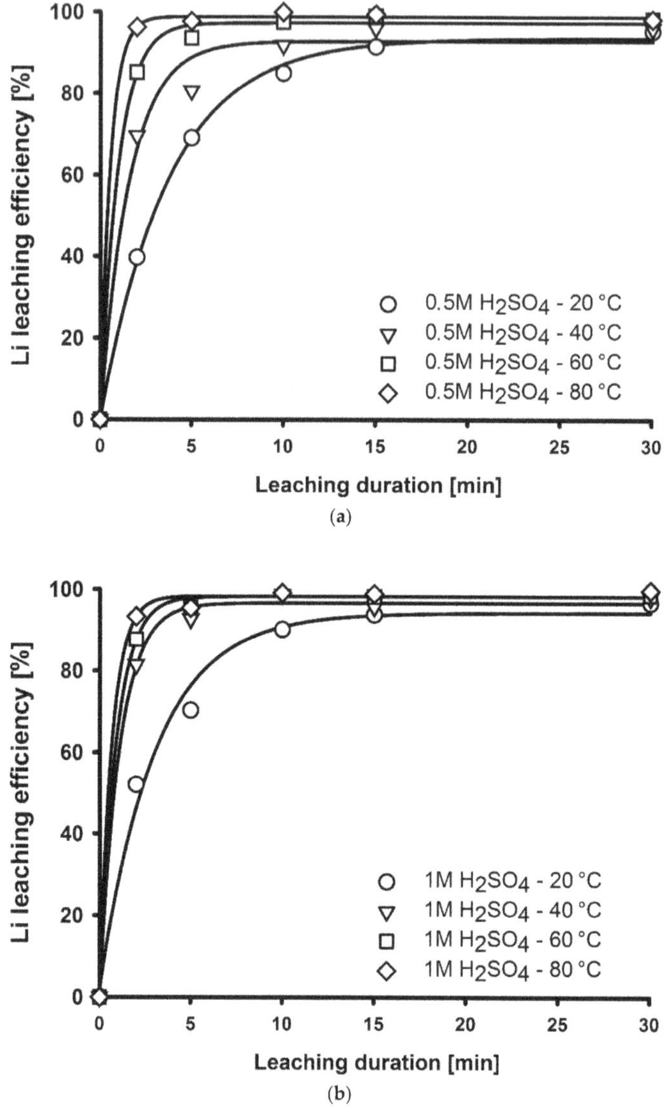

Figure 7. Lithium leaching efficiency at 20–80 °C in 0.5 M (**a**) and 1 M (**b**) H_2SO_4.

Previous analysis of chemical composition shows that the most valuable element for the recovery from Li slag is lithium itself. High lithium leaching efficiencies close to 100% were achieved after 30 min of leaching at 20 °C using both H_2SO_4 concentrations. As the leaching temperature increases, the time required to leach lithium gradually decreases to 15 min at 40 °C and 60 °C and to five minutes at 80 °C. An increase in Li leaching efficiency can be observed in the first minutes of the experiments when leaching results of 0.5 M and 1 M H_2SO_4 are compared. In addition to lithium, the leaching efficiencies of aluminum and silicon were also studied, as these three elements are present in the $LiAlSiO_4$ phase and their content in the slag is high. From the efficiency of aluminum leaching, it is possible to observe a slight effect of increasing the concentration of sulfuric acid, especially at lower temperatures, but more significant is the effect of temperature increase. Silicon is leached with an efficiency of up to 50%. This corresponds to stoichiometric calculations according to which approximately 50% of the silicon is present in the LiAlSiO$_4$ phase, which is leachable in H$_2$SO$_4$, and approximately 50% in the SiO$_2$ phase, which is not leachable in H$_2$SO$_4$. Thermodynamic study confirmed that leaching takes place according to Equations (1) and (6), in which orthosilicic acid (H$_4$SiO$_4$) is formed. A lower Si leaching efficiency was achieved at 20 °C. The highest Si leaching efficiencies were achieved at temperatures of 40 and 60 °C. As the leaching temperature is increased to 80 °C, a decrease in Si leaching efficiency was observed, which might be related to the fact that in addition to reaction (1), reaction (3) also takes place, in which solid SiO$_2$ and H$_2$O are formed instead of H$_4$SiO$_4$.

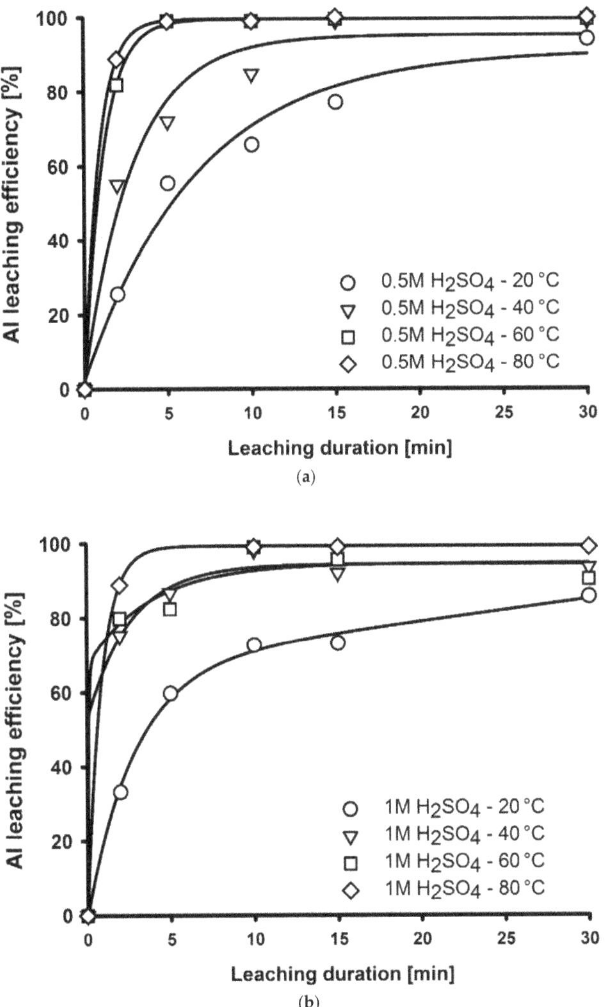

Figure 8. Aluminum leaching efficiency at 20–80 °C in 0.5 M (**a**) and 1 M (**b**) H$_2$SO$_4$.

The apparent activation energy E_a of lithium was determined from measured experiments. The first step was the calculation of lithium conversion degree α for 0.5 and 1 M H$_2$SO$_4$, according to Equation (7):

$$\alpha_{(Li)} = m_0 - m/m_0 \qquad (7)$$

where m_0 is the amount of Li in the solid sample in time t = 0 and m is the calculated amount of lithium in the solution at the specific time of the leaching. Different kinetic models were applied to evaluate the linear relationship between model function $f(\alpha)$ over time in the first 15 min of the leaching. The model is confirmed if a linear relationship between $f(\alpha)$ and t with a high coefficient of determination (R^2) is found.

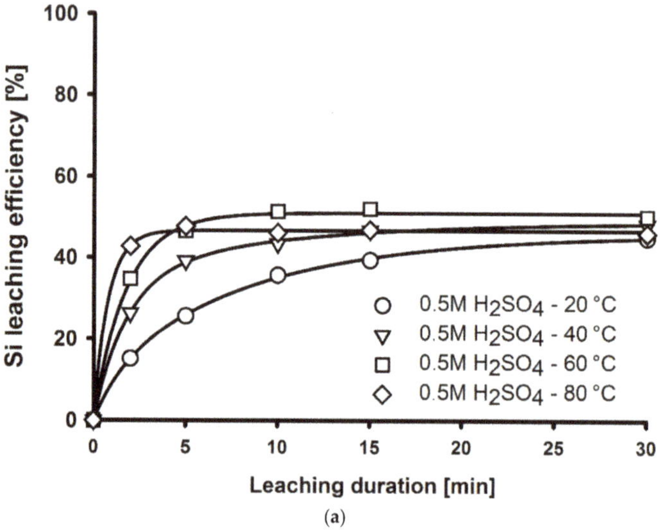

(a)

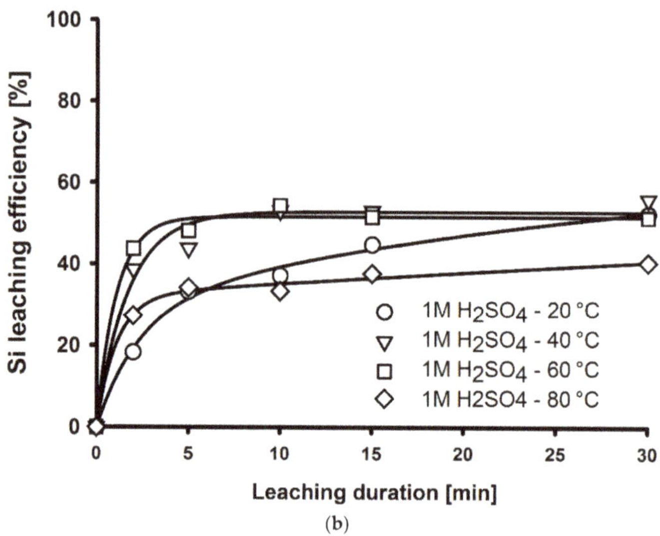

(b)

Figure 9. Silicon leaching efficiency at 20–80 °C in 0.5 M (**a**) and 1 M (**b**) H_2SO_4.

Table 4 shows the applied kinetic models [39] and their calculated values of R^2 from the 0.5 M H_2SO_4 leaching in the time interval from 0 to 15 min. Low R^2 values of the leaching at 80 °C indicated that none of these models could be applied to linearize the dissolution rate over time. This was caused by high dissolution rates in the first two minutes of the leaching. Therefore, only data obtained by leaching at 20 °C, 40 °C, and 60 °C, showed in Figure 10 were used for the E_a calculations. The R^2 values given in Table 4 confirmed the three-dimensional diffusion model. The time dependence of the function of the selected diffusion model and the value of R^2 are shown in Figure 11. Apparent rate constants were extracted from the slopes of the selected model function at different temperatures and their natural logarithm values were plotted against reciprocal temperature (Figure 12). Apparent activation energy was calculated from the slope according to the Arrhenius Equation (8) in the form:

$$ln\frac{k_2}{k_1} = \frac{-E_a}{R}\left(\frac{1}{T_2} - \frac{1}{T_1}\right) \tag{8}$$

Table 4. Kinetics models and linear relationship of model functions over time represented by coefficient of determination (R^2) value, kinetic models from [39].

Kinetic Models	Linear Relationship (R^2)			
	20 °C	**40 °C**	**60 °C**	**80 °C**
1. Linear α-t dependency				
No kinetic model applied : $f(\alpha) = \alpha = \frac{\%Me}{100}$	0.80	0.59	0.45	0.37
2. Deceleratory α–t curves				
2.1. Based on geometrical models				
Contracting area (cylindrical symmetry) : $f(\alpha) = 1-(1-\alpha)^{\frac{1}{2}}$	0.90	0.77	0.63	0.47
Contracting volume (spherical symmetry) : $f(\alpha) = 1-(1-R)^{\frac{1}{3}}$	0.93	0.83	0.72	0.53
2.2. Based on diffusion mechanism				
1D diffusion : $f(\alpha) = \alpha^2$	0.93	0.78	0.56	0.40
2D diffusion : $f(\alpha) = (1-\alpha)\cdot\ln(1-\alpha) + \alpha$	0.97	0.89	0.69	0.46
3D diffusion (cylindrical) : $f(\alpha) = \left[1 - (1-\alpha)^{\frac{1}{3}}\right]^2$	1.00	0.99	0.92	0.64
3D Ginstling–Brounstein diffusion (spherical) : $f(\alpha) = \left(1 - \frac{2\alpha}{3}\right) - (1-\alpha)^{\frac{2}{3}}$	0.98	0.94	0.78	0.53
2.3. Based on order of reaction				
First order : $f(\alpha) = -\ln(1-\alpha)$	0.98	0.93	0.89	0.59
Second – order chemical reaction : $f(\alpha) = (1-\alpha)^{-1}$	0.99	0.95	0.90	0.17
Third – order chemical reaction : $f(\alpha) = (1-\alpha)^{-2}$	0.86	0.80	0.73	0.11
3. Acceleratory α –t curves				
Power law : $f(\alpha) = \alpha^{\frac{1}{n}}$ ($n = 0.25$)	0.98	0.95	0.73	0.45
Power law : $f(\alpha) = \alpha^{\frac{1}{n}}$ ($n = 0.5$)	0.93	0.78	0.56	0.40
Power law : $f(\alpha) = \alpha^{\frac{1}{n}}$ ($n = 1$)	0.80	0.59	0.45	0.37
Power law : $f(\alpha) = \alpha^{\frac{1}{n}}$ ($n = 2$)	0.62	0.47	0.40	0.36
Exponential law : $f(\alpha) = \ln\alpha$	0.13	0.12	0.15	0.14
4. Sigmoidal α–t curves				
Avrami–Erofeev nucleation and growth ($n = 2$) : $f(\alpha) = [-\ln(1-\alpha)]^{\frac{1}{2}}$	0.83	0.73	0.68	0.54
Avrami–Erofeev nucleation and growth ($n = 3$) : $f(\alpha) = [-\ln(1-\alpha)]^{\frac{1}{3}}$	0.71	0.61	0.58	0.49
Avrami–Erofeev nucleation and growth ($n = 4$) : $f(\alpha) = [-\ln(1-\alpha)]^{\frac{1}{4}}$	0.63	0.55	0.52	0.46
Prout–Tompkins : $f(\alpha) = \ln\left[\frac{R}{1-\alpha}\right]$	0.97	0.95	0.90	0.17

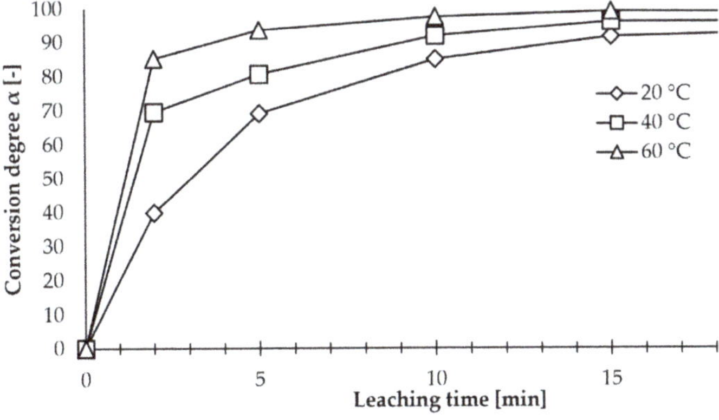

Figure 10. Input kinetic data of lithium leaching used for apparent activation energy calculations.

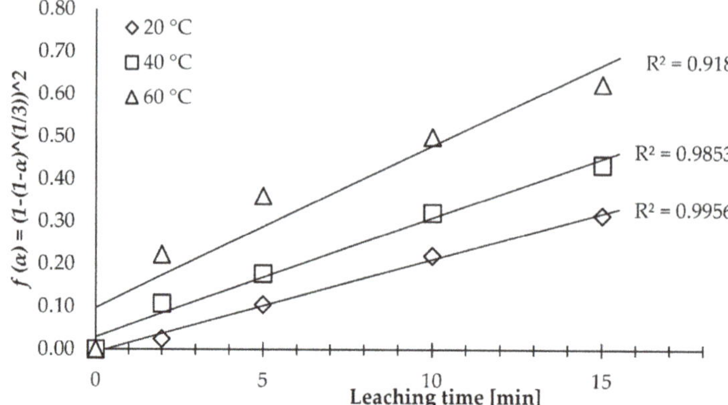

Figure 11. Kinetic data after linearization by cylindrical diffusion kinetic model.

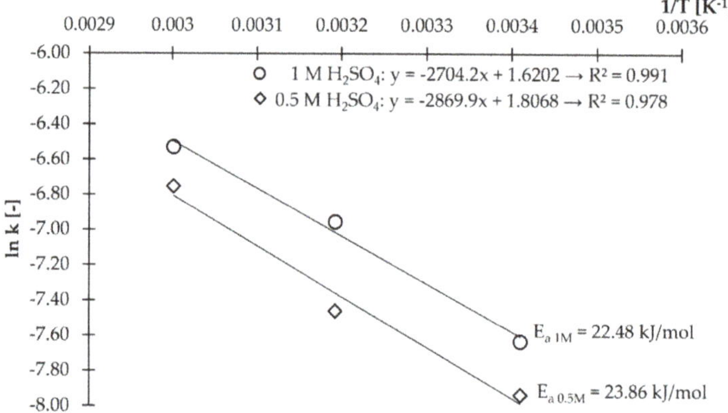

Figure 12. Apparent activation energy determination from the slope of Arrhenius plot at two different molarities of H_2SO_4.

The same procedure for the apparent activation energy calculation was used for the data obtained from the Li slag leaching in 1 M H_2SO_4.

Activation energy values under 21 kJ/mol indicate that the rate-determining step of the reactions is diffusion, in the range of E_a values from 21 to 35 kJ/mol it is a mixed mechanism and activation energy values above 35 kJ/mol indicate that the rate-determining step is a chemical reaction [40]. Apparent activation energy of lithium leaching in 0.5 M H_2SO_4 is 23.86 kJ/mol (R^2 = 0.978) and in 1 M H_2SO_4 is 22.48 kJ/mol (R^2 = 0.991). Takáčová et al. [41] studied direct leaching of black mass fromLIBs, where lithium was present in $LiCoO_2$. The apparent activation energy of lithium leaching from LCO black mass was in the interval from 16 to 19 kJ/mol. The comparison of apparent activation energies shows that, unlike black mass leaching, where the rate-determining step is diffusion only, in the case of Li slag leaching it is a mixed mechanism, where both diffusion and the rate of chemical reaction affect the overall course of the reaction. To maintain high leaching efficiencies of lithium from slag, it is necessary to leach at increased temperatures with sufficient stirring speed.

3.2. Investigation of the Silica Gel Formation

The results of Li slag leaching in H_2SO_4 confirmed that achieving high lithium leaching efficiency is possible, but the problem of conventional direct leaching is the high silicon leaching efficiency, which leads to the formation of gels. According to XRD phase analysis and stoichiometric calculations, 50% of the silicon in the Li slag is present in the Li_2SiO_3 and $LiAlSiO_4$ phases and the remaining silicon, which was not detected by the XRD method, should be present in the amorphous SiO_2 phase. The solubility of silicon oxide $SiO_{2\,(cr)}$ is 10 µg/mL and the solubility of amorphous $SiO_{2\,(am)}$ ranges from 100 µg/mL to 120 µg/mL [42]. However, the concentration of Si in the solutions after direct leaching of the Li slag reaches up to 6000 µg/mL. The low solubility of SiO_2 and high silicon concentration in solutions close to 50% confirmed the theory that silicon is leached from $LiAlSiO_4$ and Li_2SiO_3 phases only according to Equations (1) and (6) with high relative efficiency. Products of these reactions are lithium sulfates, aluminum sulfates, and orthosilicic acid (H_4SiO_4). A higher concentration of orthosilicic acid leads to oligomerization, according to Equations (9)–(11) [43].

$$H_4SiO_{4(a)} + H_4SiO_{4(a)} = H_6Si_2O_{7(a)} + H_2O_{(l)} \tag{9}$$

$$H_4SiO_{4(a)} + H_6Si_2O_{7(a)} = H_8Si_3O_{10(a)} + H_2O_{(l)} \tag{10}$$

$$H_4SiO_{4(a)} + H_8Si_3O_{10(a)} = H_{10}Si_4O_{12(a)} + H_2O_{(l)} \tag{11}$$

Chemically bound water is gradually separated from orthosilicic acid by oligomerization, causing formation of gels. Under some leaching conditions on a small laboratory scale, it is possible to filter the solution and leaching residue from each other before the gels are formed. The gel-like solution formed after filtration is shown in Figure 13a. Gel formation can significantly complicate the whole process of recycling Li slag on an industrial scale, especially if a continuous leaching process is used or if the gels are formed in a batch process before the solid residues are filtered off. The separation of solid residues entrapped in the gels (Figure 13b) is impossible without additional technological operations, which can affect economic and environmental aspects of the whole recycling process.

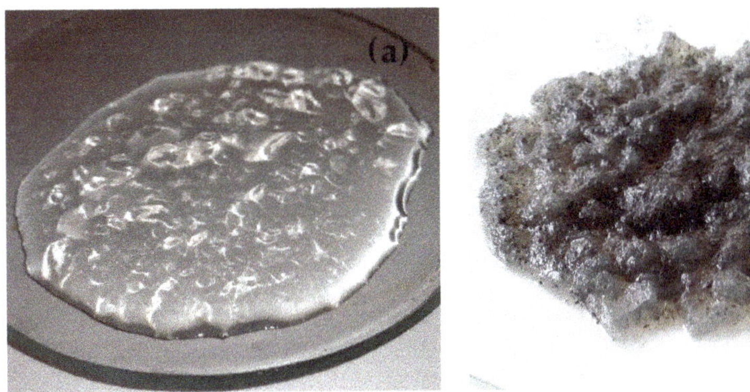

Figure 13. Gel aggregates present in the solutions after direct leaching of the Li slag: (**a**) formed after filtration of the leaching residue; (**b**) formed before filtration with entrapped leaching residues particles.

The presence of gel-like aggregates in solutions obtained by direct Li slag leaching with high silicon concentration (6000 µg/mL) and low pH (~0) was observed immediately after the leaching was completed. According to Icopini et al. [43], gel formation occurrence is dependent on pH and on the concentration of silicon in the solutions. Therefore, the gel formation was monitored for two months after the leaching experiments. With decreasing silicon concentration in the solution and increasing pH at the end of the experiments, gel formation slowed to a duration of four weeks. No gel presence was observed in the solutions with a Si concentration lower than 1000 µg/mL or solutions with a pH higher than 0.75. The addition of concentrated H_2SO_4 to a solution with a silicon concentration above the threshold limit of 1000 µg/mL resulted in the rapid gel formation (Figure 14). pH adjustment of gel-like solutions to values above the limit of 0.75 does not cause fast gels dissolution, which means that the mechanism of gel formation is not reversible and it is not possible to eliminate the problem of gelation by this method.

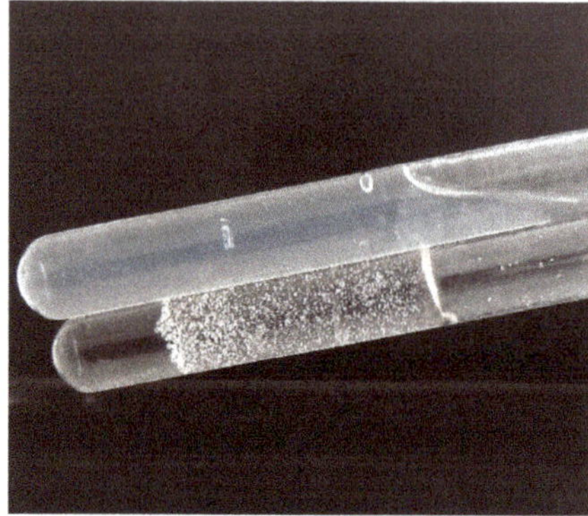

Figure 14. Solution before (**top**) and 30 min after addition of 17.9 M H_2SO_4 (**bottom**).

3.3. Suppression of Gel Formation—Thermodynamic Analysis

High concentration of silicon in the solutions confirmed, that leaching of the Li slag takes place according to Equations (1) and (6), where H_4SiO_4 is produced. The calculated apparent activation energy of this ongoing reaction is 22–24 kJ/mol. In addition to this reaction, it should be also possible to precipitate SiO_2 out of solution directly, or precipitate SiO_2 from already produced H_4SiO_4. Precipitation of SiO_2 could be a solution to gel-free leaching. Thermodynamic values for direct precipitation of SiO_2 are listed in Table 3. Reactions of SiO_2 precipitation from H_4SiO_4 and their $\Delta H°$ and $\Delta G°$ values are shown in Table 5.

Table 5. Thermodynamic data of reactions in which SiO_2 precipitates from H_4SiO_4.

Equation	Reaction	$\Delta H°_{293.15}$ (kJ)	$\Delta H°_{353.15}$ (kJ)	$\Delta G°_{293.15}$ (kJ)	$\Delta G°_{353.15}$ (kJ)
(12)	$H_4SiO_{4(a)} = SiO_2 + 2H_2O_{(l)}$	−13.642	−14.203	−14.112	−14.139
(13)	$H_4SiO_{4(a)} = SiO_2 + 2H_2O_{(g)}$	74.782	69.217	4.199	−9.698

Figure 15 shows a potential energy diagram of $LiAlSiO_4$ leaching according to the Equations (1), (3), (12) and (13). The diagram shows only the reaction path of the $LiAlSiO_4$ with the focus on silicon phase, but a similar principle can be applied for leaching of the Li_2SiO_3 phase as well.

Thermodynamic analysis shows, that reaction (3) is thermodynamically favorable, but high Si concentration in the solution confirms that the leaching takes place according to reaction (1). After leaching, it was not possible to observe a decrease in the concentration of silicon in the solution, from which it can be concluded that reactions (12) and (13) do not proceed despite the fact that their $\Delta G°$ values are negative. One of the possible reasons why SiO_2 does not precipitate out from solutions may be the fact, that E_a of reactions (3), (12) and (13) were not reached. Specific values of these activation energies cannot be determined experimentally due to the fact that these reactions do not take place during direct leaching. Therefore, these values are marked in Figure 15 as unknown with dashed lines. Part of the energy generated by the exothermic direct leaching, according to reaction (1), could be sufficient to overcome the activation energy of reaction (12) or even (13), but this direct leaching takes place at high S:L ratios equal 1:50 and the heat generated is thus consumed for heating of a large volume of leach solution. More energy must be supplied to the leaching system in next experiments to overcome E_a of the reactions in which SiO_2 precipitates out of the solution.

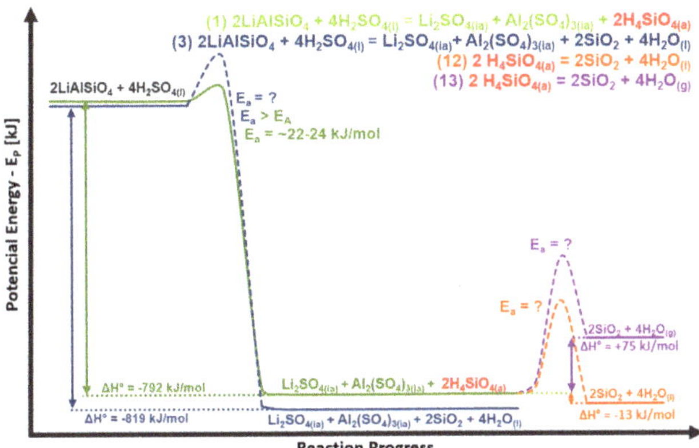

Figure 15. Potential energy diagram of LiAlSiO$_4$ phase leaching with the focus on the silicon phases.

3.4. Dry Digestion—the Optimal Ratio of Li Slag Sample, Acid, and Water

Thermodynamic study from chapter 3.3 confirmed that one of the solutions to prevent the formation of gels is to precipitate SiO$_2$. To achieve precipitation, it is necessary to overcome the unknown activation energies of reactions (3), (12), and (13), and therefore it is necessary to increase the energy to the system. It is also appropriate to reduce the volume of the leaching solution to prevent energy consumption for its heating. It was proposed to treat the slag by the dry digestion (DD) method in the experiments, which has been applied in the treatment of eudialyte concentrates [31–35] and iron-depleted red mud [35].

The DD principle consists in mixing a slag sample with concentrated acid and water and this method aimed to verify whether the heat generated from the exothermic acid dilution and the exothermic slag decomposition reaction may be sufficient to overcome the unknown E$_a$ needed to precipitate SiO$_2$ from solution according to theory showed in Figure 15. Small addition of water (4, 12, 24 mL) to slag and acid resulted in a strong exothermic reaction. Under these conditions, it is assumed that insoluble SiO$_2$ and soluble lithium and aluminum sulfates were formed according to reaction (3).

One-hour DD experiment was thus followed by dissolution of obtained solid mixtures in 500 mL of deionized water. Figure 16 shows lithium-leaching efficiencies. The addition of 5 mL of H$_2$SO$_4$ was not sufficient to achieve high Li leaching efficiency, but 10 mL H$_2$SO$_4$ resulted in efficiency close to 100%. No further increase in H$_2$SO$_4$ addition was necessary. Further increase of H$_2$SO$_4$ to 15 mL was neither necessary nor appropriate, due to the fact that lowering the pH may promote gel formation. The use of 10 mL H$_2$SO$_4$ seems to be suitable also in the case of aluminum leaching showed in Figure 17.

The silicon leaching efficiencies up to 50% were achieved in previous direct Li slag leaching (Figure 9), which corresponds to complete dissolution of silicon from the LiAlSiO$_4$ and Li$_2$SiO$_3$ phases. Figure 18 shows the silicon leaching efficiency in DD experiments followed by water dissolution. To maintain the high leaching efficiency of lithium, it is advisable to use 10 mL H$_2$SO$_4$. Reduction in the leaching efficiency of silicon from 50% to 30% was achieved, when 10 mL of H$_2$SO$_4$ without the addition of water was used. Heat was generated only from the decomposition of the LiAlSiO$_4$ and Li$_2$SiO$_3$ phases in these experiments, which was not sufficient to precipitate SiO$_2$ in significant amounts. When water was added during DD, which caused additional production of heat by exothermic dilution of sulfuric acid, a further reduction of silicon leaching efficiency to 15% using 4 mL of H$_2$O, 6% using 12 mL of H$_2$O and 4% using 24 mL of H$_2$O was achieved. From the leaching curves, follows increasing of Si leaching efficiency over time and therefore further reduction in the Si leaching efficiency could be possible by shortening of dissolution time of DD mixture. Maximum lithium leaching efficiency was achieved using 10 mL of H$_2$SO$_4$ and 24 mL of H$_2$O (ratio 10:10:24) after 10 min of dissolution. Figure 18d shows Si leaching efficiency in the 10th minute of dissolution. In a 10 min experiment at a ratio of

10:10:24, it was possible to achieve 99.12% (1492 µg/mL) and 92.36% (2792 µg/mL) dissolution of lithium and aluminum, respectively, and reduce the dissolution efficiency of silicon to 1.25% (134 µg/mL).

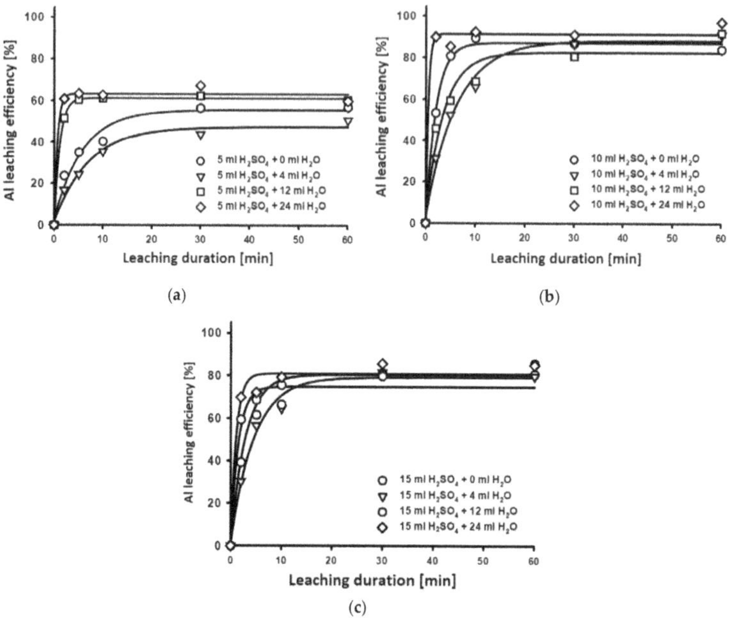

Figure 16. Lithium leaching efficiency in H_2O using (**a**) 5 mL of H_2SO_4, (**b**) 10 mL of H_2SO_4, (**c**) 15 mL of H_2SO_4.

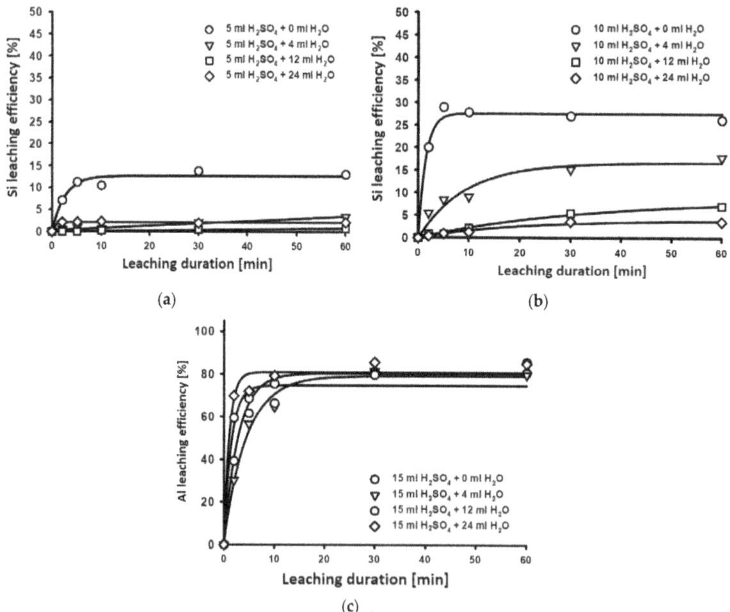

Figure 17. Aluminum leaching efficiency in H_2O using (**a**) 5 mL of H_2SO_4, (**b**) 10 mL of H_2SO_4, (**c**) 15 mL of H_2SO_4.

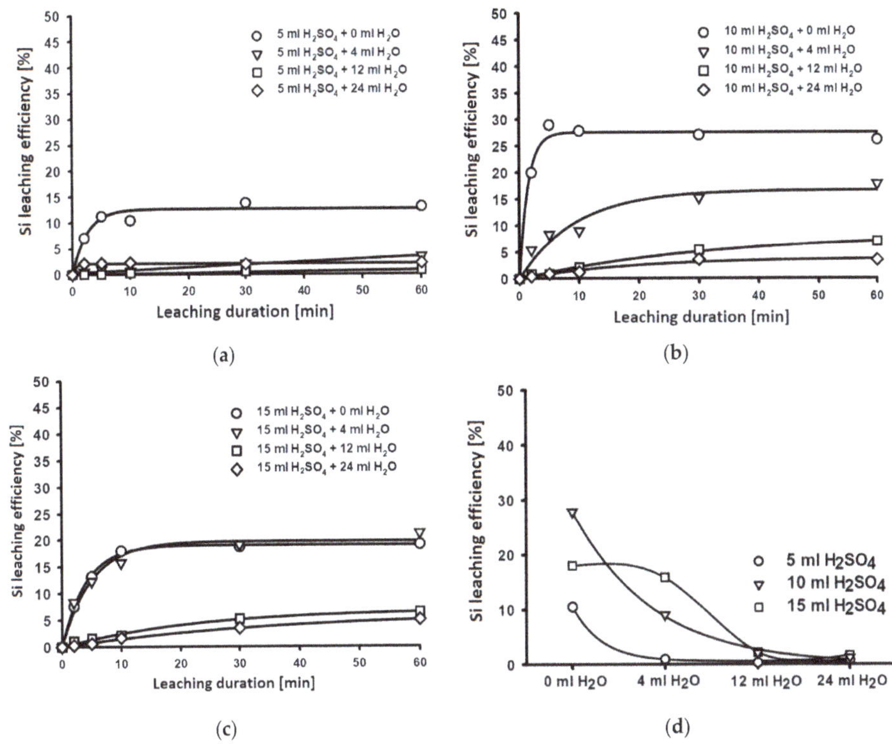

Figure 18. Silicon leaching efficiency in H_2O using (**a**) 5 mL of H_2SO_4, (**b**) 10 mL of H_2SO_4, (**c**) 15 mL of H_2SO_4, (**d**) silicon dissolution in 10th min of neutral leaching.

The new precipitated phase was observed at the bottom of the beaker after dissolution of the DD mixture (Figure 19). The precipitated phase was separated from the leaching residue and the samples were subjected to XRD analysis. Figure 20 shows an XRD analysis comparison of the input Li slag sample (black), the leaching residue obtained by dissolution of DD mixture at ratio 10:10:24 (red) and the precipitated phase (blue). The significant peak intensity decrease of the $LiAlSiO_4$ and Li_2SiO_3 phases in the XRD patterns indicates that the leaching residue and precipitate consists of unknown amorphous phases. Taking into account the theoretical analysis, the XRD, and the AAS analysis, it is concluded that the precipitated phase is amorphous SiO_2.

The results confirmed that energy generated in DD process was sufficient to overcome the activation energy of reactions (3), (6), (12), and (13). This resulted in the precipitation of insoluble SiO_2 from the mixture in DD step. Precipitation reduced the concentration of silicon in the solution below 1000 µg/mL and pH of the obtained solution was 1.322, which confirmed that both gel prevention conditions were met. Another advantage of dry digestion over direct leaching is the increase in the weight of the solid residue from 20.01% to 51.05%. DD method followed by dissolution makes the separation of the leaching residue from the solution possible. Residues can be thus landfilled or used in the construction industry [28].

3.5. Dry Digestion—Increase of Dry Digested Mixture Used per Constant Volume of Water in Dissolution Step

The optimal mixture ratio of 10:10:24 for DD was determined in chapter 3.4 The solution obtained by dissolution of dry digested mixture in 500 mL of deionized water contained 1492 µg/mL of lithium (1.18 g of Li_2SO_4/100 mL). One suitable option for future lithium recovery from solutions is the precipitation of Li_2CO_3. Precipitation of lithium carbonate (Li_2CO_3) will occur only if the concentration of lithium in the solution exceeds 2498 µg/mL ($K_{sp\ Li_2CO_3}$ = 1.33 g/100 mL; molar ratio of Li in Li_2CO_3 is 18.79%), but current concentration is lower. Therefore, it is necessary to increase lithium concentration in solution significantly.

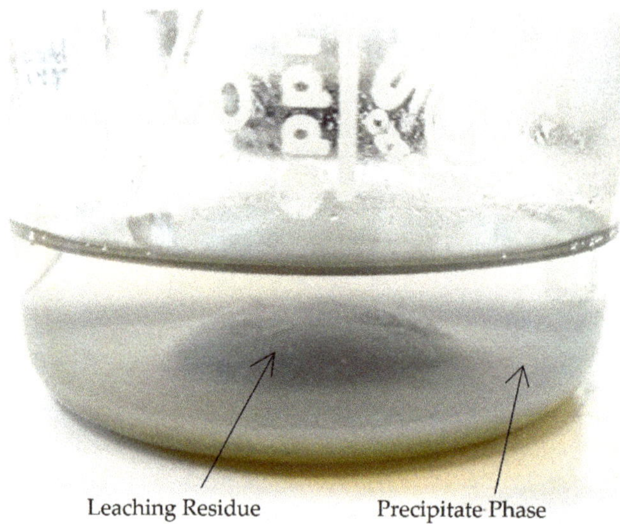

Figure 19. Leaching residue and precipitated phase after dry digestion followed by neutral leaching.

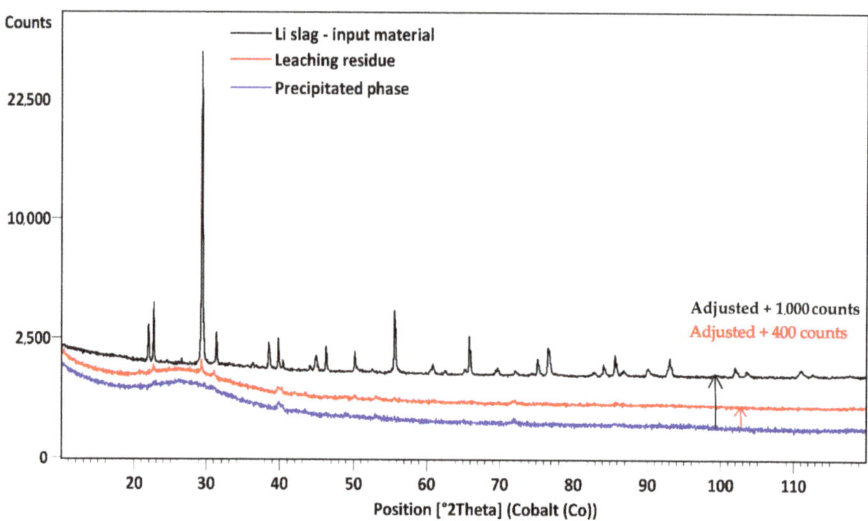

Figure 20. XRD analysis of input slag sample (black) leaching residue after leaching (red) and precipitated phase (blue).

Lithium concentration increase was experimentally verified in followed chapter by increasing the amount of dry digested mixture used in the next dissolution step. Mixtures of 10:10:24, 25:25:60, 50:50:120 and 100:100:240 were dissolved in constant 500 mL of water in the following experiments. The length of the experiments was extended to 4 h because silicon concentration was continuously increasing throughout the first 60 min of leaching in the previous dissolution experiments (Figure 18). Figures 21 and 22 shows the leaching efficiency of lithium and aluminum, respectively.

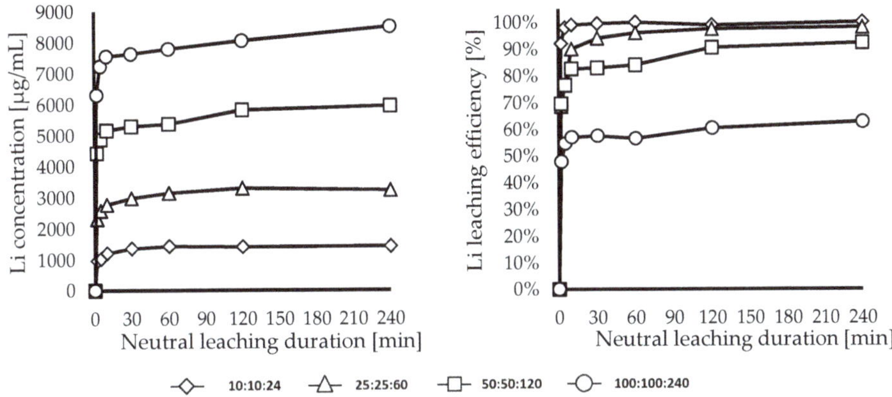

Figure 21. Lithium concentration and leaching efficiency.

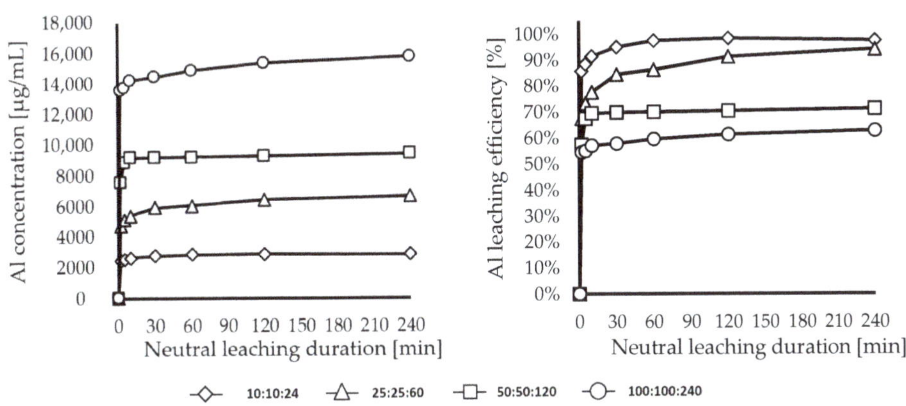

Figure 22. Aluminum concentration and leaching efficiency.

The results show that the concentration of lithium and aluminum increases with the increased amount of dry digested mixture, but the overall leaching efficiency of lithium and aluminum decreased from 99.98% and 97.46% in the trial 10:10:24 to 62.74% and 62.71% in the trial 100:100:240. However, such a significant reduction in efficiency should not occur due to the fact that the ratio Li slag: H_2SO_4: H_2O in DD step was maintained.

A possible reason for the reduction in leaching efficiency of lithium and aluminum could be in reaching of their maximum solubility. The maximum theoretical concentration of lithium in sulfuric acid is 43940 µg/mL ($K_{sp\ Li_2SO_4}$ = 34.8 g/100 mL; molar ratio of Li in Li_2SO_4 is 12.63%) and the maximum theoretical concentration of aluminum is 57409 µg/mL ($K_{sp\ Al_2(SO_4)_3}$ = 36.4 g/100 mL; molar ratio Al in $Al_2(SO_4)_3$ is 15.77%). The maximum concentration of lithium in this solution will be affected by the presence of aluminum and vice versa, since both of these elements are bound to the same SO_4^{2-} anions. However, even taking into account the mutually limited solubility, it can be stated that the maximum concentrations of lithium and aluminum were not exceeded in these experiments. Leaching residues obtained in the dissolution step were washed and analyzed by AAS for that reason. The results are shown in Table 6.

Table 6. Solid sample atomic absorption spectrometry (AAS) analysis and calculated dissolution rate of lithium and aluminum.

Sample (Slag:H_2SO_4:H_2O)	Leaching Residue Weight	Li		Al	
		Solid (AAS)	Solution (Calculated)	Solid (AAS)	Solution (Calculated)
Input sample	100.00%	100.00%	-	100.00%	-
10:10:24	54.16% (5.41 g)	0.07%	99.93%	0.36%	99.64%
25:25:60	55.32% (13.83 g)	0.07%	99.93%	0.57%	99.43%
50:50:120	45.03% (22.51 g)	0.34%	99.66%	1.59%	98.41%
100:100:240	49.96% (49.96 g)	2.76%	97.24%	3.81%	96.19%

The chemical analysis of leaching residues from experiments 10:10:24 and 25:25:60 confirmed the dissolution of lithium and aluminum above 99%. The leaching residue from the 50:50:120 experiment contained only 0.34% of Li and not 7.82%, as it was assumed according to Figure 21, where Li leaching efficiency was 92.17% only. The leaching residue from the 100:100:240 experiment also contained less lithium and aluminum than expected from the leaching results.

Due to the difference between the results of the solutions and the leaching residues, the experiments were repeated and the leaching residues were washed in a vacuum filtration funnel with a constant volume of 500 mL. Analysis of the washing solutions (Table 7) confirmed that the solution obtained by washing of solids from trial 50:50:120 contained 546 µg/mL of lithium, which represented remaining 7.8% of the lithium. The 100:100:240 wash solution contained only 25.39 of 37.5% remaining lithium, suggesting that washing such a large amount of the mixture should be performed with a larger volume of washing solution. The washing solution results indicate that lithium and aluminum sulfates are probably adsorbed on the surface of solid insoluble residues. In the DD experiments, 50:50:120 and 100:100:240, the surface area of the insoluble residues is so large that there is a significant reduction in the leaching efficiencies of these elements.

Table 7. AAS analysis of the solutions obtained by washing of the leaching residues by 500 mL of H_2O.

Trial (Slag:H_2SO_4:H_2O)	Li		Al		Si	
	(µg/mL)	(%)	(µg/mL)	(%)	(µg/mL)	(%)
10:10:24	16.07	1.15%	136	4.70%	83	0.81%
25:25:60	34.3	0.98%	553	7.65%	249	0.97%
50:50:120	546	7.80%	1548	10.71%	289	0.57%
100:100:240	3555	25.39%	6460	22.34%	311	0.30%

Figure 23 shows silicon concentration and leaching efficiency of experiments 10:10:24–100:100:240. The threshold value of the silicon concentration of 1000 µg/mL was exceeded at about 60 min of the leaching in the trial 10:10:24. In trials, where a larger volume of the mixture was dissolved, the threshold concentration was exceeded in the first minutes of leaching.

The analysis in Chapter 3.2 shows, that gels are formed in solutions with higher silicon concentrations only if the pH is lower than 0.75. Table 8 shows pH values of the solutions at the beginning and after the end of leaching. Gel presence could be observed only in the solution obtained in trial 100:100:240.

Table 8. pH values of solutions at the beginning and after the end of DD mixture water leaching.

Slag:H_2SO_4:H_2O	10:10:24	25:25:60	50:50:120	100:100:240
pH_0	0.466	0.097	−0.159	−0.383
pH_1	1.322	1.033	0.752	0.494

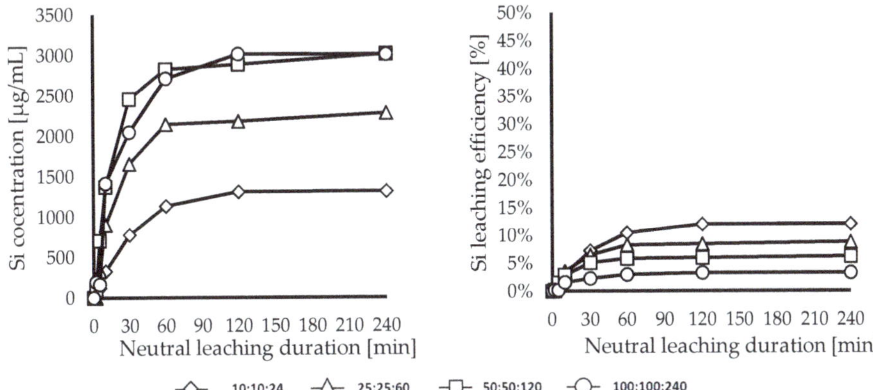

Figure 23. Silicon concentration and leaching efficiency.

The results of this chapter confirm that 92.17% Li leaching efficiency (5825 µg/mL) can be achieved by leaching a 50:50:120 mixture in the dissolving step and a suitable dissolution time is 1 h. The remaining 7.8% of Li can be recovered by quick washing of solids in a vacuum filtration funnel. To prevent lithium loses by washing of leaching residue, the reuse of wash water in dissolution step should be considered. No gel formation should be present under these conditions. Remaining washed solids (45 wt. % of input Li slag used in DD step) can thus be landfilled or used in various industries.

3.6. Further Lithium Concentration Increase and Precipitation of Products (Theoretical Analysis)

The results show that by dry digestion in a ratio of 50:50:120 and by subsequent leaching in water, it is possible to obtain a solution with 5950 µg/mL of Li and 9410 µg/mL of Al. The pH of the solutions under these conditions should be above 0.75 and therefore no gelation should be present. Precipitation of lithium carbonate (Li_2CO_3) from obtained solution with the lithium concentration of 5950 µg/mL should be possible, but the amount obtainable Li_2CO_3 would be relatively small, since Li_2CO_3 solubility is 1.33 g/100mL, which represent Li concentration of 2498 µg/mL. As much as 41.98% of Li (2498 µg/mL) would remain dissolved in the solution and only 58% (3452 µg/mL) would precipitate out of the solution. From the practical point of view, it is therefore appropriate to consider the multiple reuse of the solution for leaching of new 50:50:120 batches/mixtures. It should be possible to double (±12,000 µg/mL) or triple (±18,000 µg/mL) the amount of lithium in the solution. By treatment of this solution with tripled concentration of lithium (±18,000 µg/mL) by carbonates, the amount of dissolved lithium remaining in the solution would be reduced from 41.98% to 13.88% and therefore 86.18% (15,502 µg/mL) of the lithium would be precipitated as Li_2CO_3. The maximum concentration of lithium based on Li_2SO_4 solubility is 43,939.73 µg/mL, but the Li_2SO_4 solubility will be affected by leaching of aluminum. Multiple reuse of the same solution would also lower the pH of the solution below the critical limit, causing gels to form.

For this reasons, it seems appropriate to adjust pH of the solution to pH 4–6, at which all the aluminum precipitates out of solution according to the fraction diagrams (Figure 24a), while lithium should stay in ionic form (Figure 24b).

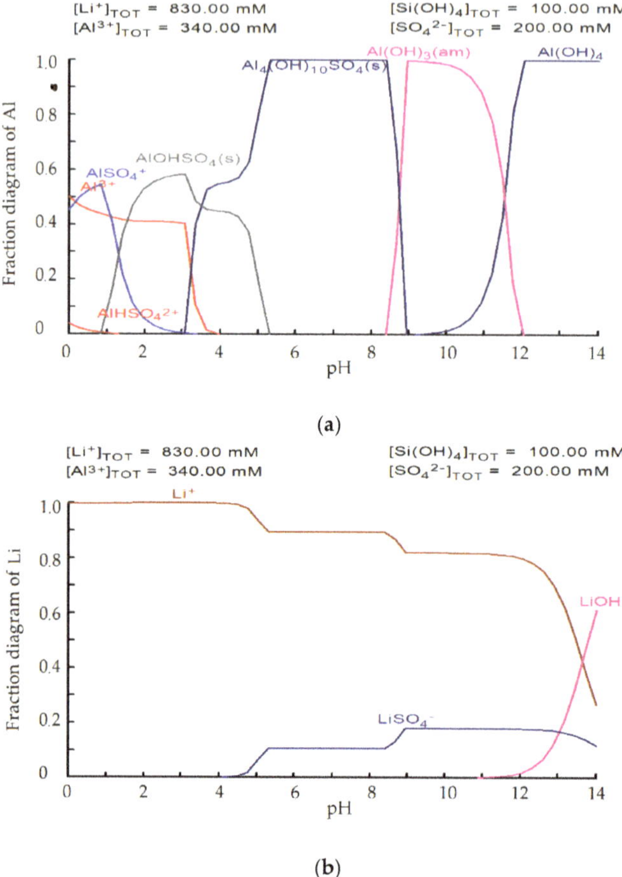

(a)

(b)

Figure 24. Fraction diagrams of precipitation: (**a**) aluminum; (**b**) lithium.

After pH adjustment and filtration of aluminum precipitates, refined solution should be ready for next leaching of fresh DD 50:50:120 according to Figure 25.

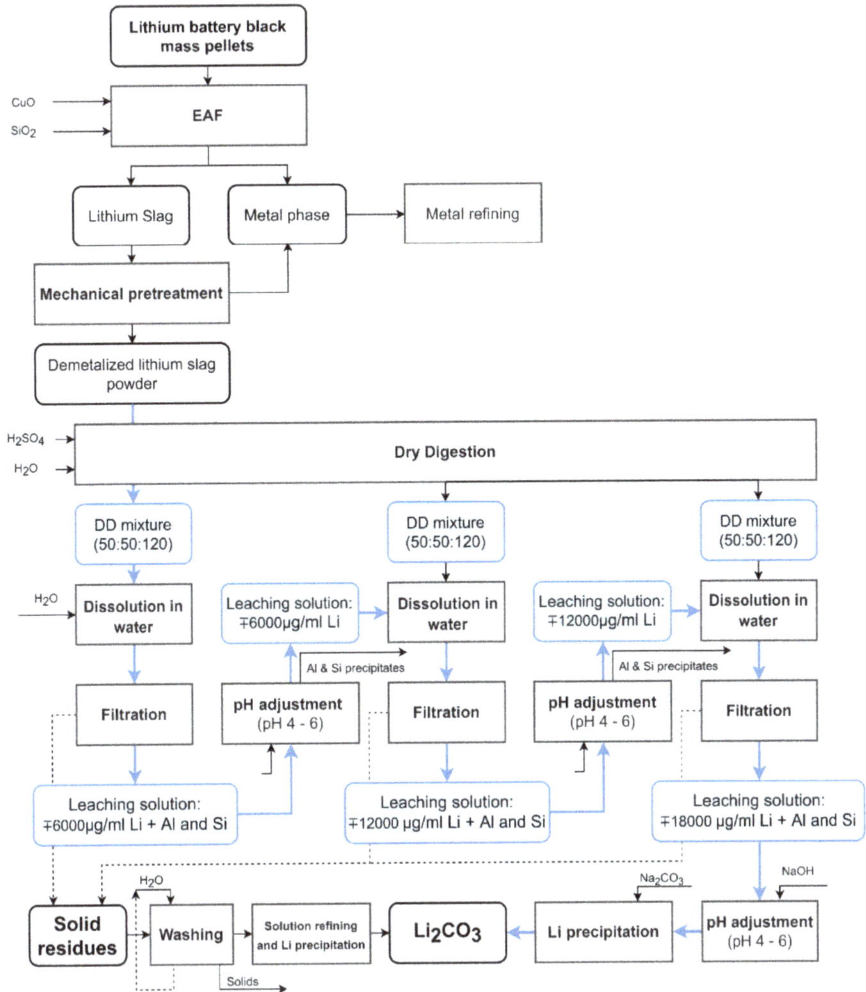

Figure 25. Proposed approach for lithium slag recycling by dry digestion method.

4. Conclusions

Lithium slag recycling methods were investigated in this paper. Chemical analysis showed that the slag produced by smelting of lithium battery black mass in EAF with the addition of CuO and SiO_2 is a valuable secondary source with a lithium content of 6.96%. The sample of Li slag suitable for leaching was prepared by crushing, grinding, and magnetic separation with the use of which the magnetic metals content was reduced to a minimum.

By direct leaching of the slag in 0.5 and 1 M H_2SO_4, it was possible to achieve a lithium recovery close to 100% after only five minutes of leaching at 80 °C and after 30 min of leaching at 20 °C. Kinetic study showed that the apparent activation energy for lithium leaching in 0.5 and 1 M H_2SO_4 is 22.48 and 23.86 kJ/mol, respectively, and the process of Li leaching is described well by Jander kinetic model, which corresponds to 3-dimensional cylindrical diffusions mechanism as the rate-controlling step. The disadvantage of direct leaching of Li slag is the high leaching efficiency of silicon, which causes the formation of gels. Gels complicate a chemical analysis of solutions, filtration, and further recovery of the products.

Therefore, the conditions for gel formation and the possibilities of gel formation preventions were investigated. The results of the observation show that gels are produced when the concentration of Si in the solution exceeds 1000 µg/mL and at the same time the pH of the solution is lower than 0.75. Thermodynamic study has shown that leaching of Li phases of the slag should be also possible with the precipitation of solid SiO_2 instead of H_4SiO_4 production followed by gelation.

The dry digestion method was investigated to prevent the formation of H_4SiO_4. By this method, it was possible to overcome the activation energy of reactions in which SiO_2 precipitates out of solution. Lithium and aluminum sulfates were also produced in this step. The optimal ratio of components in the dry digestion is 10 mL of concentrated H_2SO_4 and 24 mL H_2O per 10 g of slag. Mixtures obtained by DD were leached in a water dissolution step, where Si was leached with a much lower rate compared to the dissolution of lithium and aluminum silicates. The leaching efficiency of silicon was reduced to 1.25% by dissolving a mixture of 10:10:24 in 500 mL of water, while maintaining high leaching yields of lithium and aluminum. The solution obtained under these conditions contains 1492 µg of lithium per ml. Further increase in the lithium concentration to 5950 µg/mL (92.12%) was achieved by dissolving a mixture of 50:50:120 in 500 mL of water. The study confirmed an important role of the washing of leaching residues after dry digestion and water leaching. Remaining 7.8% of lithium were recovered by washing in 500 mL of water. Wash water containing lithium can be then reused in the next batch of water dissolution of dry digested mixture.

Author Contributions: Conceptualization, J.K., D.O., and T.H.; methodology, J.K., D.O., and C.D.; software, J.K; validation, J.K., D.O., A.M., M.S., C.V., C.D., B.F., and T.H.; formal analysis, J.K.; investigation, J.K. and D.O.; resources, J.K.; data curation, J.K. and D.O.; writing—original draft preparation, J.K.; writing—review and editing, J.K., D.O., A.M., M.S., C.V., C.D., B.F., and T.H.; visualization, J.K.; supervision, T.H.; project administration, J.K. and C.V.; funding acquisition, B.F., A.M., T.H., J.K., and C.V. All authors have read and agreed to the published version of the manuscript.

Funding: The exchange cost was funded by Deutscher Akademischer Austauschdienst (DAAD), grant number 57453240. This work was funded by the Ministry of Education of the Slovak Republic under grant MŠ SR VEGA 1/0556/20. This work was also funded by association UNIVNET, project No. 0201/0082/19.

Acknowledgments: The authors are grateful to Accurec Recycling GmbH (Germany) for providing pelletized black mass. The authors would also like to express their gratitude to the DAAD and Ministry of Education of the Slovak Republic for enabling the joint research in battery recycling.

References

1. Dańczak, A.; Klemettinen, L.; Kurhila, M.; Taskinen, P.; Lindberg, D.; Jokilaakso, A. Behavior of Battery Metals Lithium, Cobalt, Manganese and Lanthanum in Black Copper Smelting. *Batteries* **2020**, *6*, 16. [CrossRef]
2. Helbig, C.; Bradshaw, A.M.; Wietschel, L.; Thorenz, A.; Tuma, A. Supply Risks Associated with Lithium-Ion Battery Materials. *J. Clean. Prod.* **2018**, *172*, 274–286. [CrossRef]
3. Nitta, N.; Wu, F.; Lee, J.T.; Yushin, G. Li-Ion Battery Materials: Present and Future. *Mater. Today* **2015**, *18*, 252–264. [CrossRef]
4. Gu, F.; Guo, J.; Yao, X.; Summers, P.A.; Widijatmoko, S.D.; Hall, P. An Investigation of the Current Status of Recycling Spent Lithium-Ion Batteries from Consumer Electronics in China. *J. Clean. Prod.* **2017**, *161*, 765–780. [CrossRef]
5. Contestabile, M.; Panero, S.; Scrosati, B. A Laboratory-Scale Lithium-Ion Battery Recycling Process. *J. Power Sources* **2001**, *92*, 65–69. [CrossRef]
6. Pinegar, H.; Smith, Y.R. Recycling of End-of-Life Lithium-Ion Batteries, Part II: Laboratory-Scale Research Developments in Mechanical, Thermal, and Leaching Treatments. *J. Sustain. Metall.* **2020**, *6*, 142–160. [CrossRef]

7. Porvali, A.; Aaltonen, M.; Ojanen, S.; Velazquez-Martinez, O.; Eronen, E.; Liu, F.; Wilson, B.P.; Serna-Guerrero, R.; Lundström, M. Mechanical and Hydrometallurgical Processes in HCl Media for the Recycling of Valuable Metals from Li-Ion Battery Waste. *Resour. Conserv. Recycl.* **2019**, *142*, 257–266. [CrossRef]

8. European Commission. Report from the Commission to the European Parliament, the Council, the European Economic and Social Committee, the Committee of the Regions and the European Investment Bank: On the Implementation of the Strategic Action Plan on Batteries: Building a Strategic Battery Value Chain in Europe. Available online: https://eur-lex.europa.eu/legal-content/EN/TXT/HTML/?uri=CELEX:52019DC0176&from=EN (accessed on 16 June 2020).

9. Pinegar, H.; Smith, Y.R. Recycling of End-of-Life Lithium Ion Batteries, Part I: Commercial Processes. *J. Sustain. Metall.* **2019**, *5*, 402–416. [CrossRef]

10. Tarascon, J.; Armand, M. Issues and Challenges Facing Rechargeable Lithium Batteries. *Nature* **2001**, *414*, 359–367. [CrossRef]

11. Huang, B.; Pan, Z.; Su, X.; An, L. Recycling of Lithium-Ion Batteries: Recent Advances and Perspectives. *J. Power Sources* **2018**, *399*, 274–286. [CrossRef]

12. The European Parliament and the Council of the European Union. Directive 2006/66/EC of the European Parliament and of the Council on Batteries and Accumulators and Waste Batteries and Accumulators and Repealing Directive 91/157/EEC. Available online: https://eur-lex.europa.eu/legal-content/EN/TXT/PDF/?uri=CELEX:02006L0066-20131230&rid=1 (accessed on 6 May 2020).

13. European Commission. Report from the Commission to the European Parliament, the Council, the European Economic and Social Committee and the Committee of the Regions: Critical Raw Materials Resilience: Charting a Path towards greater Security and Sustainability. Available online: https://eur-lex.europa.eu/legal-content/EN/TXT/?uri=CELEX:52020DC0474 (accessed on 15 September 2020).

14. Werner, D.; Peuker, U.A.; Mütze, T. Recycling Chain for Spent Lithium-Ion Batteries. *Metals* **2020**, *10*, 316. [CrossRef]

15. Zhong, X.; Liu, W.; Han, J.; Jiao, F.; Qin, W.; Liu, T. Pretreatment for the Recovery of Spent Lithium Ion Batteries: Theoretical and Practical Aspects. *J. Clean. Prod.* **2020**, *263*, 121439. [CrossRef]

16. Liu, J.; Wang, H.; Hu, T.; Bai, X.; Wang, S.; Xie, W.; Hao, J.; He, Y. Recovery of $LiCoO_2$ and Graphite from Spent Lithium-Ion Batteries by Cryogenic Grinding and Froth Flotation. *Miner. Eng.* **2020**, *148*, 106223. [CrossRef]

17. Zhang, G.; Du, Z.; He, Y.; Wang, H.; Xie, W.; Zhang, T. A Sustainable Process for the Recovery of Anode and Cathode Materials Derived from Spent Lithium-Ion Batteries. *Sustainability* **2019**, *11*, 2363. [CrossRef]

18. Zhang, G.; He, Y.; Feng, Y.; Wang, H.; Zhang, T.; Xie, W.; Zhu, X. Enhancement in Liberation of Electrode Materials Derived from Spent Lithium-Ion Battery by Pyrolysis. *J. Clean. Prod.* **2018**, *199*, 62–68. [CrossRef]

19. Li, J.; Lai, Y.; Zhu, X.; Liao, Q.; Xia, A.; Huang, Y.; Zhu, X. Pyrolysis Kinetics and Reaction Mechanism of the Electrode Materials During the Spent $LiCoO_2$ Batteries Recovery Process. *J. Hazard. Mater.* **2020**, *398*, 122955. [CrossRef]

20. Zhong, X.; Liu, W.; Han, J.; Jiao, F.; Qin, W.; Liu, T.; Zhao, C. Pyrolysis and Physical Separation for the Recovery of Spent $LiFePO_4$ Batteries. *Waste Manag.* **2019**, *89*, 83–93. [CrossRef]

21. Ruismäki, R.; Dańczak, A.; Klemettinen, L.; Taskinen, P.; Lindberg, D.; Jokilaakso, A. Integrated Battery Scrap Recycling and Nickel Slag Cleaning with Methane Reduction. *Minerals* **2020**, *10*, 435. [CrossRef]

22. Shi, J.; Peng, C.; Chen, M.; Li, Y.; Eric, H.; Klemettinen, L.; Lundström, M.; Taskinen, P.; Jokilaakso, A. Sulfation Roasting Mechanism for Spent Lithium-Ion Battery Metal Oxides Under SO_2-O_2-Ar Atmosphere. *JOM* **2019**, *71*, 4473–4482. [CrossRef]

23. Peng, C.; Hamuyuni, J.; Wilson, B.P.; Lundström, M. Selective Reductive Leaching of Cobalt and Lithium from Industrially Crushed Waste Li-Ion Batteries in Sulfuric Acid System. *Waste Manag.* **2018**, *76*, 582–590. [CrossRef]

24. Porvali, A.; Chernyaev, A.; Shukla, S.; Lundström, M. Lithium Ion Battery Active Material Dissolution Kinetics in Fe (II)/Fe (III) Catalyzed Cu-H_2SO_4 Leaching System. *Sep. Purif. Technol.* **2020**, *236*, 116305. [CrossRef]

25. Dewulf, J.; van der Vorst, G.; Denturck, K.; van Langenhove, H.; Ghyoot, W.; Tytgat, J.; Vandeputte, K. Recycling Rechargeable Lithium Ion Batteries: Critical Analysis of Natural Resource Savings. *Resour. Conserv. Recycl.* **2010**, *54*, 229–234. [CrossRef]

26. BRGM. *Le Lithium (Li)—Éléments de Criticité*; BRGM, the French Geological Survey: Orleans, France, 2017; pp. 1–8. Available online: http://www.mineralinfo.fr/page/fiches-criticite (accessed on 19 October 2020).

27. Mathieux, F.; Ardente, F.; Bobba, S.; Nuss, P.; Blengini, G.; Alves Dias, P.; Blagoeva, D.; Torres De Matos, C.; Wittmer, D.; Pavel, C.; et al. *Critical Raw Materials and the Circular Economy—Background Report*; JRC Science-for-policy report, EUR 28832 EN; Publications Office of the European Union: Luxembourg, 2017; ISBN 978-92-79-74282-8. [CrossRef]

28. Harper, G.; Sommerville, R.; Kendrick, E.; Driscoll, L.; Slater, P.; Stolkin, R.; Walton, A.; Christensen, P.; Heidrich, O.; Lambert, S.; et al. Recycling Lithium-Ion Batteries from Electric Vehicles. *Nature* **2019**, *575*, 75–86. [CrossRef]

29. Wang, H.; Friedrich, B. Development of a Highly Efficient Hydrometallurgical Recycling Process for Automotive Li–Ion Batteries. *J. Sustain. Metall.* **2015**, *1*, 168–178. [CrossRef]

30. Sommerfeld, M.; Vonderstein, C.; Dertmann, C.; Klimko, J.; Oráč, D.; Miškufová, A.; Havlik, T.; Friedrich, B. A Combined Pyro- and Hydrometallurgical Approach to Recycle Pyrolyzed Lithium-Ion Battery Black Mass Part 1: Production of Lithium Concentrates in an Electric Arc Furnace. *Metals* **2020**, *10*, 1069. [CrossRef]

31. Voßenkaul, D.; Birich, A.; Müller, N.; Stoltz, N.; Friedrich, B. Hydrometallurgical Processing of Eudialyte Bearing Concentrates to Recover Rare Earth Elements Via Low-Temperature Dry Digestion to Prevent the Silica Gel Formation. *J. Sustain. Metall.* **2017**, *3*, 79–89. [CrossRef]

32. Ma, Y.; Stopic, S.; Gronen, L.; Friedrich, B. Recovery of Zr, Hf, Nb from eudialyte residue by sulfuric acid dry digestion and water leaching with H_2O_2 as a promoter. *Hydrometallurgy* **2018**, *181*, 206–214. [CrossRef]

33. Ma, Y.; Stopic, S.; Gronen, L.; Milivojevic, M.; Obradovic, S.; Friedrich, B. Neural Network Modeling for the Extraction of Rare Earth Elements from Eudialyte Concentrate by Dry Digestion and Leaching. *Metals* **2018**, *8*, 267. [CrossRef]

34. Davris, P.; Stopic, S.; Balomenos, E.; Panias, D.; Paspaliaris, I.; Friedrich, B. Leaching of rare earth elements from eudialyte concentrate by suppressing silica gel. *Miner. Eng.* **2017**, *108*, 115–122. [CrossRef]

35. Alkan, G.; Yagmurlu, B.; Gronen, L.; Dittrich, C.; Ma, Y.; Stopic, S.; Friedrich, B. Selective silica gel free scandium extraction from Iron-depleted red mud slags by dry digestion. *Hydrometallurgy* **2019**, *185*, 266–272. [CrossRef]

36. London Metal Exchange (LME). Available online: https://www.lme.com/Metals (accessed on 19 October 2020).

37. Shanghai Metals Market (SMM). Available online: https://price.metal.com/ (accessed on 19 October 2020).

38. Roine, A. *HSC Chemistry Thermodynamic Software*; Outotec: Pori, Finland, 2018.

39. Hem Shanker Ray, H.; Saradindukumar, R. Analysis of Kinetic Data for Practical Applications. In *Kinetics of Metallurgical Processes*; Springer Nature: Singapore, 2018; Volume 1, pp. 315–340.

40. Havlik, T. *Hydrometallurgy—Principes and applications*; Woodhead Publishing Limited: Cambridge, UK, 2008; p. 536. ISBN 978-1-84569-407-4.

41. Takáčová, Z.; Havlik, T.; Kukurugya, F.; Orac, D. Cobalt and Lithium Recovery from Active Mass of Spent Li-ion Batteries: Theoretical and Experimental Approach. *Hydrometallurgy* **2016**, *163*, 9–17. [CrossRef]

42. Iler, R.K. *The Chemistry of Silica: Solubility, Polymerization, Colloid and Surface Properties, and Biochemistry*; Wiley: New York, NY, USA, 1979.

43. Icopini, G.A.; Brantley, S.L.; Heaney, P.J. Kinetics of silica oligomerization and nanocolloid formation as a function of pH and ionic strength at 25 °C. *Geochimica Cosmochimica Acta* **2005**, *69*, 293–303. [CrossRef]

Publisher's Note: MDPI stays neutral with regard to jurisdictional claims in published maps and institutional affiliations.

Article

Investigation of Centrifugal Fractionation with Time-Dependent Process Parameters as a New Approach Contributing to the Direct Recycling of Lithium-Ion Battery Components

Tabea Sinn [1,*], Andreas Flegler [2], Andreas Wolf [2], Thomas Stübinger [3], Wolfgang Witt [3], Hermann Nirschl [1] and Marco Gleiß [1]

[1]   Karlsruhe Institute of Technology (KIT), Institute for Mechanical Process Engineering and Mechanics, Straße am Forum 8, 76131 Karlsruhe, Germany; hermann.nirschl@kit.edu (H.N.); marco.gleiss@kit.edu (M.G.)

[2]   Fraunhofer Institute for Silicate Research (ISC), Neunerplatz 2, 97082 Würzburg, Germany; andreas.flegler@isc.fraunhofer.de (A.F.); andreas.wolf2@isc.fraunhofer.de (A.W.)

[3]   Sympatec GmbH, Am Pulverhaus 1, 38678 Clausthal-Zellerfeld, Germany; TStuebinger@sympatec.com (T.S.); WWitt@sympatec.com (W.W.)

*   Correspondence: tabea.sinn@kit.edu; Tel.: +49-721-608-42410

Received: 11 November 2020; Accepted: 27 November 2020; Published: 1 December 2020

Abstract: Recycling of lithium-ion batteries will become imperative in the future, but comprehensive and sustainable processes for this are still rather lacking. Direct recycling comprising separation of the black mass components as a key step is regarded as the most seminal approach. This paper contributes a novel approach for such separation, that is fractionation in a tubular centrifuge. An aqueous dispersion of cathode materials (lithium iron phosphate, also referred to as LFP, and carbon black) serves as exemplary feed to be fractionated, desirably resulting in a sediment of pure LFP. This paper provides a detailed study of the commonly time-dependent output of the tubular centrifuge and introduces an approach aiming to achieve constant output. Therefore, three different settings are assessed, constantly low, constantly high and an increase in rotational speed over time. Constant settings result in the predictable unsatisfactory time-variant output, whereas rotational speed increase proves to be able to maintain constant centrate properties. With further process development, the concept of fractionation in tubular centrifuges may mature into a promising separation technique for black mass in a direct recycling process chain.

Keywords: fractionation; tubular centrifuge; rotational speed control; particle size analysis; lithium iron phosphate; LFP; carbon black; lithium-ion battery; direct battery recycling; recovery

---

1. Introduction

The threat of global climate change and environmental pollution are, among other reasons, fueling the electric revolution in mobility [1], which in turn increases the demand for powerful batteries. Lithium-ion batteries (LIBs) are mostly regarded as most suitable battery type for this and other applications [2]. However, the performance of every LIB suffers from ageing mechanisms which inevitably lead to its withdrawal, eventually [3]. With a foreseeable increasing number of LIBs in use and withdrawn thereafter, the amount of LIB material that can and should be recycled increases equally, since permanent disposal seems not to become a practical option for the upcoming mass of LIBs. Environmental issues again and scarce resources primarily drive the development of comprehensive, flexible and large-scale suitable recycling strategies for the valuable LIB materials [4].

At present, there are three kinds of approaches for the recycling of LIBs, that are pyrometallurgy, hydrometallurgy and direct recycling [2,3,5].

Pyrometallurgical metal recovery means the thermal decomposition of LIBs in high-temperature furnaces. This comparatively simple procedure is versatile concerning the actual feed-in composition, but the resulting gases are partly detrimental and require post-processing [6] while the obtained slag and metal alloy contain comparably few kinds of metals and in low quality [4,7].

Hydrometallurgical reclamation methods comprise the leaching of metals from active material in aqueous solutions of acids and reducing agents. Numerous steps of subsequent precipitation reactions allow one to recover a wider variety of higher quality metals compared to pyrometallurgy [8,9]. However, it is more complex and premature sorting and separation of the LIB constituents is advisable in order to significantly improve the efficiency of the hydrometallurgical process chain [2].

The two approaches named have in common to basically recover metallic LIB components, but not to preserve the original active materials and their morphology. By contrast, in direct recycling, electrode material is removed from the electrodes and reconditioned with the objective to regain active materials that can directly be reused in the manufacture of fully functional "recycling-LIBs". Among the three approaches named, it enables recovery of the highest absolute amount and diversity of LIB-constituents [3]. Pre-sorting as specifically as possible is expedient in this approach, too, but dispensing with numerous intermediate steps and expensive processing afterwards by mainly concentrating on physical separation techniques, it could also render recycling more lucrative. Manifold physical separation techniques exist that might be worth considering. They share the characteristics to exploit different physical properties of the active materials (like density, particle size, ferromagnetism or hydrophobicity). Admittedly, a complete direct recycling process chain for LIBs is complex to establish.

One central element of a direct recycling process chain after shredding and pre-sorting is the separation of the black mass components. The black mass is most commonly an aqueous dispersion of the electrode coatings comprising metal oxides, conductive carbon black (CB) and binders. There are concepts for this crucial step relying on physical separation. For example, one approach to separate CB from metal oxides in the black mass is froth flotation [10], making use of the hydrophobicity of carbon. In most cases, binders must be eliminated firstly in order to release the active material for further treatment [11]. Research is taking direction towards the development of water-soluble binders, which facilitates treatment steps in recycling chains besides possible improvements in other properties like mechanical flexibility [12].

This paper concentrates on the separation of exemplary cathode materials for a start, that are lithium iron phosphate (LiFePO$_4$, LFP) and CB besides the binders carboxymethylcellulose (CMC) and styrene butadiene rubber (SBR), dispersed in water. A novel separation strategy for these materials is presented, centrifugation of the dispersion with the intention to fractionate LFP and CB.

Centrifugation is a familiar unit operation in process engineering [13,14], but its full potential in relation to LIB recycling has not been harnessed yet. Naturally, the fundamental separation mechanism in centrifuges is the differing settling/sedimentation behavior of constituents according to the specific acting centrifugal force. Usually, like in this work, dense solid particles sediment in a liquid. The force depends on numerous properties of the particles, mainly density and size [15]. Differences in these properties could be exploited to fractionate particles of different kinds utilizing a centrifuge, as treated in the present report for LFP and CB particles. Since the resulting centrifugal forces are relatively weak for both particle species given, a strong centrifuge type has been chosen, precisely a tubular centrifuge [16–18]. Although tubular centrifuges offer the advantage of high centrifugal forces thanks to their quite simple design without a means to transport sediment out, the same also results in only semi-continuous operation and transient behavior [19,20]. With unaltered operational settings, namely feed flow rate and rotational speed, the separation becomes less efficient over time. If constant output quality is desired, this complicates process design significantly.

Earlier works like [21,22] have shown for different particulate systems that various operational settings influence the output of the tubular centrifuge and notable differences in the separation behavior can be observed. Building upon these earlier examinations, this paper's subject is the influence of different operational settings on the separation and fractionation quality for a mixed dispersion containing the cathode materials LFP and CB. Thereby the focus lies firstly on the temporal development of the separation process applying different settings. Secondly, the authors investigate a novel approach to achieve semi-continuous fractionation with constant output by adapting the rotational speed over time. A detailed comparison is made between the common approach to set constant operational parameters on the one hand, based on two exemplary cases with constantly weak and constantly strong separation conditions (low and high rotational speed), and on the other a new strategy to achieve constant output with desired properties as a third case, comprising a sequence of increasing rotational speed. This is a novelty in centrifuge operation and has not been applied for tubular centrifuges yet, neither has it been taken into consideration as a possible way to assess high-speed centrifugation to recover LIB materials. Since the focus of this paper lies on rotational speed variation, feed flow rate was kept constant for the time being.

2. Materials and Methods

2.1. Tubular Centrifuges and Fractionation by Sedimentation Behavior

As a representative of solid-bowl centrifuges, tubular centrifuges mainly consist of a narrow cylindrical rotor with bearings at its foot and top ends, a construction which allows high-speed rotation offering high centrifugal forces up to $100,000 \times g$ in some cases [22]. Naturally, the fundamental separation mechanism in tubular centrifuges is settling/sedimentation of denser constituents than the carrier medium due to the acting centrifugal force.

In this work, dense solid particles that are LIB cathode materials sediment in water. Dispersion is injected into the rotor via a nozzle at one end (in this work, at the bottom) as feed, flowing upward along the rotation axis. Particles that sediment on their way towards the outlet accumulate at the inner wall of the rotor, forming a liquid-saturated sediment. Residual dispersion, i.e., liquid and particles that have not reached the sediment during their residence time, leaves the rotor at the upper side as centrate. There is no device to transport the sediment out of the rotor, it accumulates up to a maximum extent where no particles can be separated anymore. This is the latest point to stop and open the rotor for manual sediment extraction. The centrifuge used in this work and its geometric dimensions are schematically illustrated in Figure 1a, where the cross-sectional view also outlines the characteristic cone-shaped sediment build-up in the tubular rotor explained below. Relevant geometrical dimensions are the rotor length L_R, radial range of the inlet weir R_W, radial range of the drum (to the inner wall of the rotor) R_D and the radial position of the sediment surface $R_S(l, t)$, a dimension varying with the regarded axial position l and process time t on account of the progressive sediment formation.

Sedimentation behavior, or more precisely, the specific sedimentation velocity of particles, determines which particles are separated in the centrifuge rotor during their residence time. For appropriate process design, a realistic description of the sedimentation velocities is therefore helpful (besides a description of sediment build-up, which is not further treated in this paper, refer for example to [23–28] for more details). Basically, the particles' settling velocities depend on their size and density, according to Stokes' Law [15] in a centrifugal field:

$$u_s(x) = \frac{x^2 \cdot (\rho_S - \rho_L) \cdot r \cdot \omega^2}{18 \cdot \eta_L} \tag{1}$$

$$\text{with } \omega = 2\pi \cdot n \tag{2}$$

where u_s is Stokes' settling velocity, x designates the particle diameter, ρ_S and ρ_L the solid (particle) and liquid mass density, r the radial position of the particle and ω the angular velocity, calculating with n

denoting the rotational speed, i.e., the number of revolutions per time. Real settling velocities observed may differ from this idealized treatment due to effects of particle-particle interactions that can arise as a result of manifold reasons, like particle surface properties [29], shape and concentration-dependent effective viscosity [30,31] that lead to acceleration or hindrance (hindered settling) of the particles [24]. Generally, those real settling velocities are measurable and subject to research up until now [30,32,33]. Concluding a functional correlation of experimental findings allows to simply calculate the real settling velocity $u_{real}(x,c)$ approximately by applying the following expression:

$$u_{real}(x,c) = u_s(x) \cdot h(x,c) \tag{3}$$

with $h(x,c)$ being the hindered settling function determined from experimental data, depending on particle size and particle concentration c [25,31]. Such measurements can for example be performed in a LUMiSizer laboratory centrifuge (L.U.M. GmbH, Berlin, Germany) [23,34].

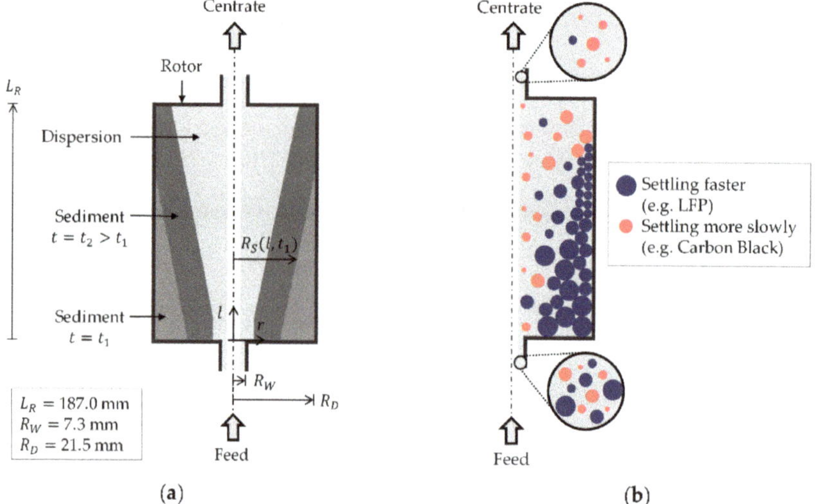

Figure 1. Schematic illustration of the tubular centrifuge and particle sedimentation, in cross-sectional view. (**a**) Tubular centrifuge with geometrical dimensions and progressive sediment build-up. Geometry data given for CEPA Z11; (**b**) Particles in the centrifuge (particle sizes exaggerated): Feed dispersion containing two particle systems, specific settling behavior and sediment build-up for almost complete fractionation, centrate containing nearly solely the particle system settling more slowly (Blue: Faster-settling component, representing LFP; Red: Component settling more slowly, representing carbon black).

A particle collective consequently shows a settling velocity distribution, according to its composition. Faster settling particles are more likely to be separated. Pursuant to their properties (cf. Section 2.2), LFP is the faster settling material compared to CB, in this work. The discrepancy in their settling behaviors indicates that LFP and CB can theoretically be completely fractionated, i.e., separated by species, applying a sedimentation-based separation apparatus like a tubular centrifuge.

The overarching objective of such a process, like in this paper, is to recover the components formerly mixed in the feed dispersion as purely as possible despite possible practical challenges. Hence, an appropriate measure to examine the success of a fractionation process is the separation efficiency. It is defined as the ratio of to the mass of material i that has been separated in the apparatus ($m_{sed,i}$) over the mass of i that has entered the centrifuge ($m_{in,i}$) (Equation (4), first part) and can also be calculated inversely applying the mass of i leaving the centrifuge in centrate ($m_{centrate,i}$) (Equation (4), second part). Assuming the volumetric flow rate of liquid to be approximately constant, calculation via the concentrations in feed, sediment or centrate ($c_{feed,i}$, $c_{sed,i}$, $c_{centrate,i}$) is also possible (Equation (5)):

$$T_i = \frac{m_{sed,i}}{m_{in,i}} = 1 - \frac{m_{centrate,i}}{m_{in,i}} \tag{4}$$

$$T_i = \frac{c_{sed,i}}{c_{feed,i}} = 1 - \frac{c_{centrate,i}}{c_{feed,i}} \tag{5}$$

In the regarded process, i can stand for the solid material overall as well as the solid components of interest, namely LFP or CB. Accordingly, the envisaged complete fractionation can be expressed in terms of the separation efficiencies of the two particulate systems to be fractionated with values $T_{LFP} = 1$ (or 100%) and $T_{CB} = 0$ (or 0%), implying that the entire amount of LFP entering the centrifuge is separated while CB is not separated at all, i.e., not found in the sediment. The means to achieve this is an appropriate setting of the process operational parameters, namely rotational speed of the centrifuge n and volumetric flow rate of the feed dispersion $\dot{V}$. In practice, this ideal case of complete fractionation is not entirely feasible since a certain mixing due to small vortexes or secondary flows in the rotor [17,21,35] cannot be excluded, which can lead to slight impurities in sediment and centrate, however. As explained in more detail below, tubular centrifuges are commonly characterized by a time-dependent separation behavior. Thus, a sensitive examination must also include the development of the separation efficiencies over process time, $T_{LFP}(t)$ and $T_{CB}(t)$. The optimal setting adapts to the time-variant apparatus behavior, leading—in case of complete fractionation again—to consistent process outputs $T_{LFP}(t) = 100\% = const.$ and $T_{CB}(t) = 0\% = const.$ over the entire process time, despite the transient separation conditions.

The reason for the time-dependency of a tubular centrifuge's separation behavior lies in the progressive sediment formation [17,20,35]. As explained above, particle collectives settle with their specific distribution of settling velocities. Generally, when injected into a tubular centrifuge, particles settling faster will be separated closer to the inlet than slower ones. With sediment gradually occupying free space and the feed flow rate set constant, the axial flow velocity of the dispersion in the rotor accelerates, shifting the balance of forces acting on the particles. As a result, the individual separation position of a particle shifts axially towards the outlet with increasing filling level, which is also why the filling of the rotor typically proceeds from the inlet towards the outlet side, leaving a truncated cone-shaped free space for the dispersion to flow through, as outlined in Figure 1. Figure 1b illustrates in more detail the axial distribution of particles with different settling velocities in sediment and dispersion for the case of a nearly complete fractionation, where only a few particles of the slower settling species are found in the sediment close to the outlet.

The shifting balance of forces for the individual particles is equivalent to time- and position-dependently weakening separation conditions and decreasing separation efficiency. Increasingly faster (i.e., usually bigger sized) particles cannot be separated anymore. [23,36] describe the separation efficiency depending of the operational parameters, that are rotational speed n and feed flow rate Q. A temporal decrease in separation efficiency can theoretically be counteracted by an appropriate raise in rotational speed [17] and/or reduction of feed flow rate.

Working further on this foundation, the effect of different operational settings on the separation efficiencies of LFP and CB particles, entering the centrifuge in a mixed dispersion, are examined in this paper. Besides detailed investigations with LFP and CB about the influence of the operational parameters n and Q, a more time-resolved study is still lacking and shall also be contributed with this work. Its purpose is to evaluate three different operational strategies. While the dependency of the feed flow rate is not yet regarded in this paper and its value remains unaltered throughout the cases, three settings for rotational speed are compared in detail. In the first case, rotational speed is set to a constant low value (20,000 rpm) with the intention to obtain no CB in the sediment at all. However, this might be accompanied by a high loss of LFP, which shall be minimized in the second case that is run at a constantly high rotational speed (40,000 rpm). This way, though, settling of CB into the sediment has to be considered. Both constant rotational speed settings do not take the temporal change in separation behavior into account. To do so, in the third case, a temporal increase of rotational speed is applied in order to cover the estimated optimal setting over the entire process time, which in a different set-up [17] indicated to be a convenient method. This rotational speed trajectory has been defined beforehand utilizing a simplified flowsheet simulation of the process, which is basically developed at the basis of [23,36]. However, the simulation is still subject to further research and will be elucidated and published separately in the future.

2.2. LIB Cathode Particle Systems and Binders Used

The feed dispersion used for all experiments is a mixed dispersion of LFP, Super C65/carbon black (CB) particles as well as binders, namely carboxymethylcellulose (CMC) and styrene-butadiene rubber (SBR), diluted with demineralized water to an overall solids content of 1.6 wt%. Table 1 below lists the composition of the solids mixture, relevant properties of all constituents and their envisaged behavior during centrifugal fractionation.

Table 1. Components in the feed dispersion and their individual properties.

Solid Constituent (Acronym)	Solids Fraction	Carbon Content	Density	Behavior during Complete Fractionation
Lithium Iron Phosphate (LFP)	85 wt%	13 wt% (Coating)	3.5 g/cm^3	Fast-settling component. Settles completely and purely
Super C65/Carbon Black (CB)	10 wt%	67 wt%	1.9 g/cm^3	Slowly settling component. Stays in centrate
Carboxymethyl-cellulose (CMC)	2.5 wt%	20 wt%	1.1 g/cm^3	Not affected by centrifugal force, but partly settling attached to particles
Styrene Butadiene Rubber (SBR)	2.5 wt%		1.0 g/cm^3	

Figure 2a depicts the particle size distributions (PSDs) of LFP and CB particles. The PSDs have been measured by means of laser diffraction (cf. Section 2.4). With regard to the PSDs of the individual particle systems, LFP is partly coarser than CB on the whole, but a large overlap from 0.2 μm up to about 10 μm is evident. Though, according to the density data in Table 1 and remarks in Section 2.1, LFP can clearly be referred to as the faster settling species. Consulting the scanning electron microscope (SEM) images in Figure 2b in addition, LFP primary particles are rather ellipsoid and roughly 500 nm in diameter, while CB primary particles are much smaller with less than 100 nm in diameter and form branched agglomerates.

The binders are assumed to behave virtually inert to the centrifugal field due to their density. Reasons also include CMC being soluble in water is not to be regarded as rigid "particles" and SBR present as only nano-sized polymer fibers. However, it cannot be excluded that they attach to LFP and CB, complicating a general assumption regarding their principal location during centrifugation and therefore the assessment of the detailed composition of samples (cf. Section 2.4).

To evaluate the separation efficiency of LFP and CB in detail, the total carbon content of samples was determined with a LECO C744 (cf. Section 2.4) system. The dispersion component containing the most carbon is CB, naturally. However, the results must be interpreted with caution since the other components comprise carbon as well. 2.2 wt% of the LFP particles is, more precisely, a carbon coating which makes up 13 wt% of the total carbon mass and the binders for their part contribute 20% to total carbon in the feed dispersion. So by and large CB makes up about two thirds of the carbon measurable in a feed sample and even complete fractionation will not deliver carbon-free samples.

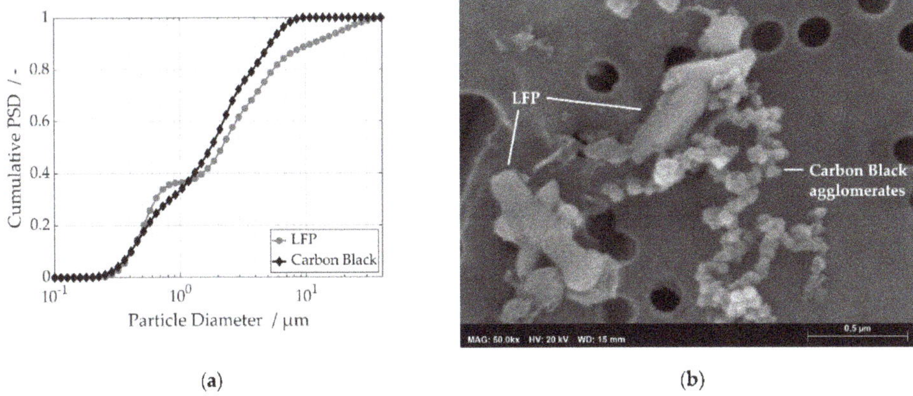

(**a**)

(**b**)

Figure 2. Particles in the feed dispersion. (**a**) Particle Size Distributions of LFP (grey) and Carbon Black (black) particles. (**b**) SEM image of LFP and Carbon Black particles on a nucleopore filter.

Figure 2b is an image of particles on a filter without liquid, but it can be seen as an indication for the formation of CB agglomerates and LFP-CB mixed-species agglomerates. Agglomerates possibly form and break at different stages of the procedure including feed preparation and centrifugation, which cannot certainly be said or quantified at present. For all repeating experiments concerning one case treated in this work, an individual batch of cathode material paste was diluted to be used as feed dispersion. PSD measurements for the three feed dispersions have been carried out as well (cf. Section 2.4) and a strong tendency to agglomerate was visible during successive measurements within feed samples of case 2 and 3.

An overview of the characteristic particle sizes x_{10}, x_{50} and x_{90}, including mean values and standard deviations, as well as measured PSDs in detail (averaged) is displayed in Figure 3.

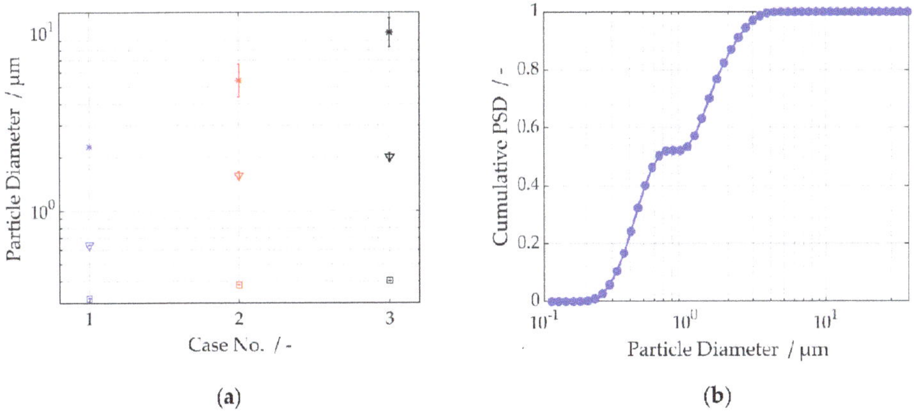

(**a**)

(**b**)

Figure 3. *Cont.*

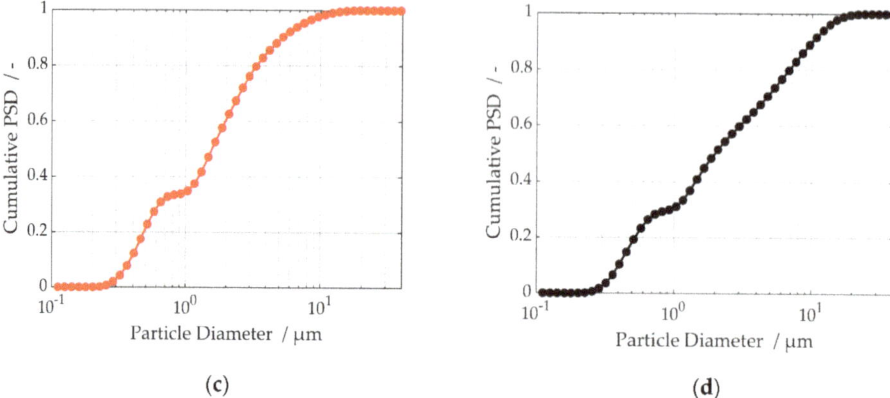

Figure 3. Feed particle size distributions used in the three cases. Blue: 20,000 rpm (Case 1); Red: 40,000 rpm (Case 2); Black: Increasing rotational speed (Case 3). (**a**) Characteristic particle sizes, squares: x_{10}; triangles: x_{50} with dashed lines as guides to the eye; stars: x_{90}. (**b**–**d**) Detailed particle size distributions for the three cases, averaged.

The high standard deviations, especially for the x_{90} values, reveal progressing agglomeration among the coarser particles. Therefore, it must be kept in mind that potential for agglomerate formation and breakage exists in this work, too, and probably differs between the three cases regarded. Mainly case 2 and 3 findings might be affected.

2.3. Experimental Set-up and Procedure

The entire experimental set-up, important parameters and sampling locations are outlined in Figure 4. Feed dispersion is stored in a continuously stirred tank and pumped into the centrifuge rotor using a membrane pump equipped with a pulsation damping pressure retention valve. For all experiments shown in this paper, the flow rate has been set to $\dot{V} = 200$ mL/min. Injection into the rotor is achieved via a nozzle attached to the rotor's bottom end. The tubular centrifuge used in this work is the model Z11 by CEPA (Carl Padberg Zentrifugenbau GmbH, Lahr im Schwarzwald, Germany), which can run on rotational speeds between 20,000 rpm and 54,000 rpm (equals $10,000 \times g$ to $70,000 \times g$ at the inner wall surface). It is equipped with a touch panel allowing to define rotational speed manually, which is utilized to enter both the constant values in the first two cases and the temporal sequence in the third case. It should be noted for the third case that the manual input of values necessitated a curtailing of the calculated rotational speed values to a feasible timetable. So, although the calculation was made for one value per second, the input was limited to one value every 30 s and the actual input was merely a step function. The rotational speed settings are displayed in Figure 5. After passing the rotor with a nominal volume of 250 mL, centrate is discharged via the run-off tray and hose connection into a collecting tank. Samples for subsequent centrate analyses were taken every three minutes at the hose connection, too. To begin an experiment, the rotor was set on the (initial) rotational speed and demineralized water was fed into the rotor. When the rotor was completely filled and water ran out, feed was switched to the dispersion tank. Experimental time started counting when dispersion entered the nozzle. Duration of every experiment in this paper was 30 min, three repeating experiments have been carried out per case. After disassembly, sediment samples were taken at the bottom (B), middle (M) and top (T) positions and resuspended in demineralized water. Feed samples (Figure 3) were taken once per case and directly out of the stirred feed tank before the experimental procedure began.

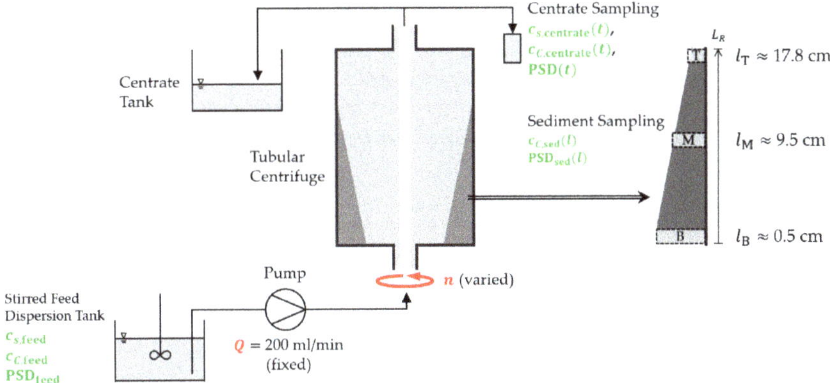

Figure 4. Outline of the experimental set-up with components (black/grey), operational parameters (**red**) and sampling locations with properties analyzed (**green**).

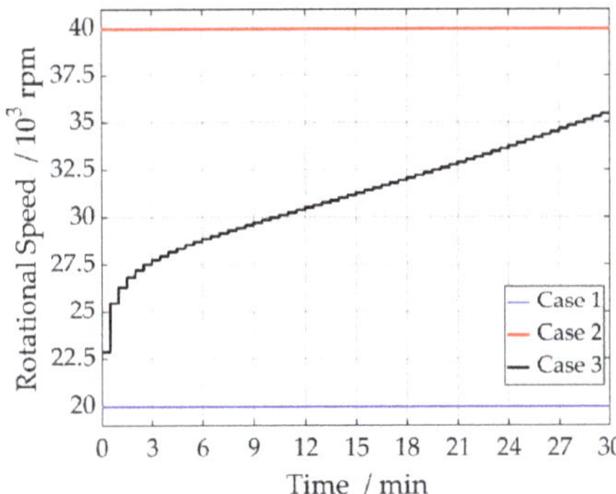

Figure 5. Rotational speed settings for the three cases examined: Blue: Case 1, constantly low; Red: Case 2, constantly high; Black: Case 3, stepwise increase.

2.4. Analytical Methods

All centrate samples have been analyzed concerning the overall solids content by means of drying and applying Equation (5) to calculate overall separation efficiency. According to capacities given, centrate and sediment samples have also been examined with regard to the particle size distribution (PSD) and carbon content. For this purpose, samples of two sampling times had to be mixed in each case, i.e., the first and second were mixed, the third and fourth, and so on. Thus, the PSD and carbon content measurements represent quasi-averaged states. For carbon content measurements, the mixed samples were dried and the solid remains grinded using a mortar.

Measurements of particle size distribution have been carried out with an advanced laser diffraction sensor based on HELOS technology (Sympatec GmbH, Clausthal-Zellerfeld, Germany). Equipped with a blue laser source and multi-element photo detectors for forward, backward and wide-angle light scattering detection, the device was utilized for at least six repeated measurements per sample. Measurements were performed in a flow-through cuvette with sample supply achieved with the wet dosing station LIQXI (also by Sympatec).

Carbon content measurements for the dry, grinded samples took place in a LECO C744 (LECO Instrumente GmbH, Mönchengladbach, Germany), including a high frequency induction furnace and non-dispersive infrared (NDIR) detection cell. Two to three repeating measurements have been executed according to the obtainable mass per sample.

3. Results

For a sound evaluation of the experimental data, all the results must be sensibly combined. To enhance clarity, the section is divided in two subsections. The first one treats the results concerning the centrate, where particular attention is paid to the time-dependent development. In the second section the results concerning the sediment measurements are presented, which are naturally linked to the centrate analysis.

3.1. Time-Dependent Centrate Analysis

Figure 6 displays the calculated separation efficiency referring to the overall solids content over the examined experimental duration for the three cases. The weak separation conditions (20,000 rpm) clearly led to the least solids yield as sediment, showing the lowest start value of 89% and dropping rapidly down to 72%. The strongest separation conditions (40,000 rpm) and the rotational speed increase yield roughly identical outputs, especially including the bars (here marking minimum and maximum of the single values calculated from measurements). However, the mean values in case 2 reveal a minor tendency to decrease from the first to the latest (93 to 90%) measurement, while case 3 lets one urmise a relatively constant separation efficiency of 91%. All in all, the high overall solid separation efficiencies in case 2 and 3 already reveal that not only LFP, but also CB has been separated from the feed dispersion to a certain extent.

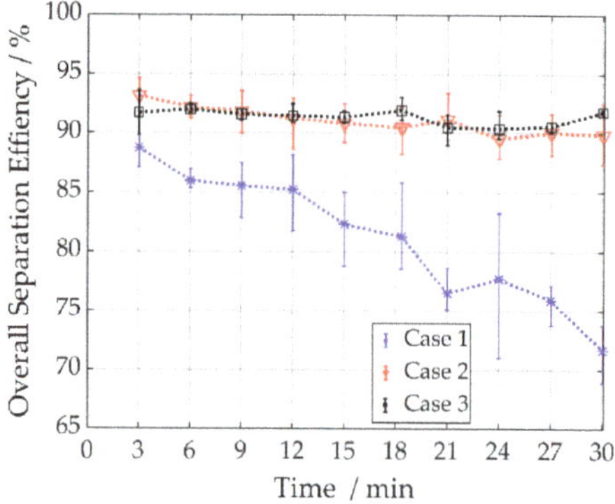

Figure 6. Overall separation efficiency for the three cases over time. Blue: 20,000 rpm; Red: 40,000 rpm; Black: Increasing rotational speed. Squares: Mean values; bars: Minimum/maximum of the three repeating experiments; dashed lines are guides to the eye.

Figure 7a sums up the PSD measurement results for the centrate samples, showing the mean results and standard deviations for x_{10}, x_{50} and x_{90} over time and per case. Case 1 evinces the most significant increase in all characteristic particle sizes over time, as well as remarkably higher x_{90} sizes in general. Case 2 centrate contains the finest particles in all respects, but a modest increase leads to nearly identical x_{10} compared to case 3 towards the end of the experimental time. The rotational speed increase in case 3 seems to yield almost constant PSD characteristics slightly larger than those in case 2. A closer look at the PSDs in detail (Figure 7b–d) evinces again the strong resemblance of centrate PSDs in case 2 and 3. In the beginning, case 2 centrate contains distinguishably finer particles, but shifts visibly approaching case 3 while the latter's PSDs remain virtually identical over the entire time. Both case 2 and 3 only comprise the fine particles from the feed PSDs (the fraction furthest to the left, cf. Figure 3) whereas case 1 centrate contains the coarser fraction of the feed to a great extent, which causes the high x_{90} values in Figure 6. The share of the coarser fraction increases over time, too. So it could be assumed from the PSDs that the centrate in case 2 and 3 contains a considerable amount of CB, namely rather the finer particles, besides probably only few and fine LFP particles from the feed dispersion. Case 1, on the contrary, might even still contain LFP particles and nearly all the CB. Carbon content measurements provide greater clarity. Results for the centrate samples are depicted in Figure 8. Except for the first value, case 1 centrate samples show the lowest carbon content decreasing inside the small range between about 49% C to 47% C. This implies that not only CB, but also a portion of LFP remains in the centrate at 20,000 rpm, even at the beginning when there is no considerable amount of sediment in the rotor. Also, the share of LFP seems to be slightly increasing over time, lowering the carbon content, but also an increasing number of CB particles must stay in centrate regarding the rapidly dropping overall separation efficiency (Figure 6), on the whole. Thus, it could be concluded that both species' separation efficiencies sink, but T_{LFP} a little faster than T_{CB} does. In contrast, case 2 shows a significant increase in carbon content over time from 47% C to nearly 55% C. This suggests that a certain share of (fine) LFP particles is present, shrinking over time. Taking into account the overall fine, but slightly coarsening centrate PSDs and the marginally decreasing overall separation efficiency, it also means that both fine CB and fine LFP particles are contained in the centrate, but with the share of CB particles growing fast, T_{CB} probably declines more rapidly than T_{LFP} does. Finally, case 3 centrate samples have generally the highest carbon contents measured. They increase to a small extent from about 52% C to 56% C, which suggests that there are still some LFP particles contained, though their fraction diminishes. Hence, the observed mostly steady centrate overall solids content and PSDs are consistently supplemented by a practically constant carbon content.

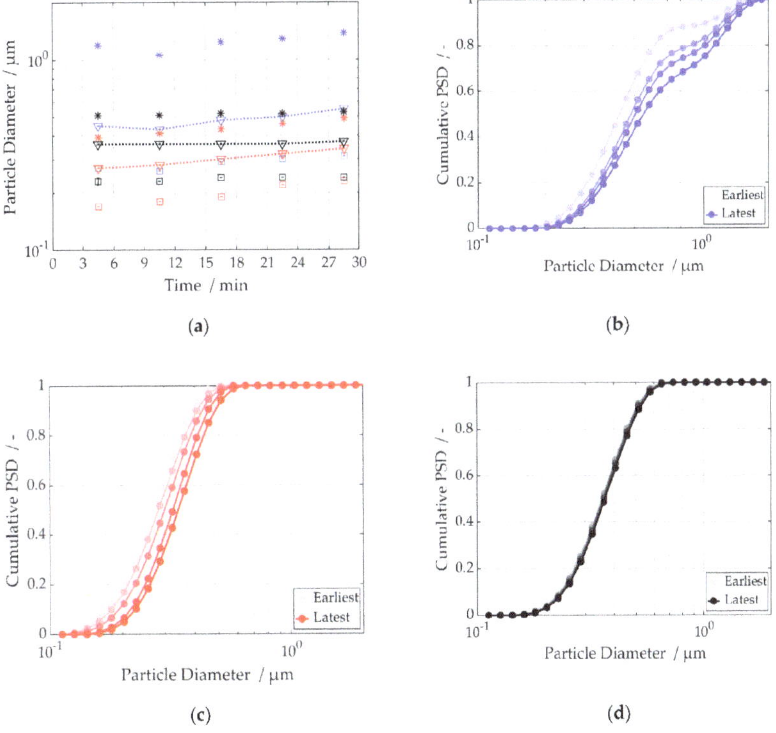

Figure 7. Centrate particle size distributions for the three cases over time. Blue: 20,000 rpm; Red: 40,000 rpm; Black: Increasing rotational speed. (**a**) Characteristic particle sizes, squares: x_{10}; triangles: x_{50} with dashed lines as guides to the eye; stars: x_{90}. (**b**–**d**) Detailed particle size distributions for the three cases over time; light color: Earliest measurement; darkening shade with progressing time until darkest color: Latest measurement.

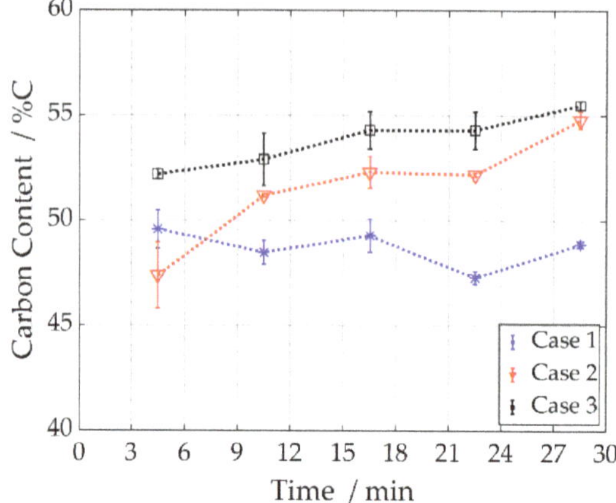

Figure 8. Centrate overall carbon content for the three cases over time. Blue: 20,000 rpm; Red: 40,000 rpm; Black: Increasing rotational speed. Squares: mean values; bars: Standard deviations; dashed lines are guides to the eye.

3.2. Sediment Analysis

The sediment analysis complicates compared to the centrate evaluation as they cover the entire sediment height, i.e., a cross-section of the integral process development, which impedes definite interpretations of the measurements.

The PSDs of the resuspended sediment samples are delicate to evaluate since a strong tendency to agglomerate is observed in most of the samples. Figure 9a depicts the mean results and standard deviations in x_{10}, x_{50} and x_{90} for the three cases at bottom, middle and top of the rotor. Case 2 and 3 show the expected trend towards finer particles in direction to the top of the rotor while case 1 exhibits rather steady characteristics over the three sampling positions. Case 3 exhibits the coarsest particles at the bottom compared to the other cases. In the middle and especially at the top, case 2 and 3 reveal quite close PSD characteristic sizes.

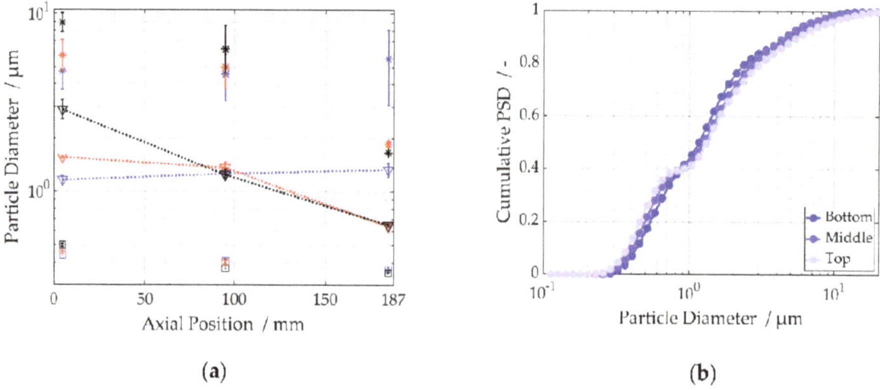

(a) **(b)**

Figure 9. *Cont.*

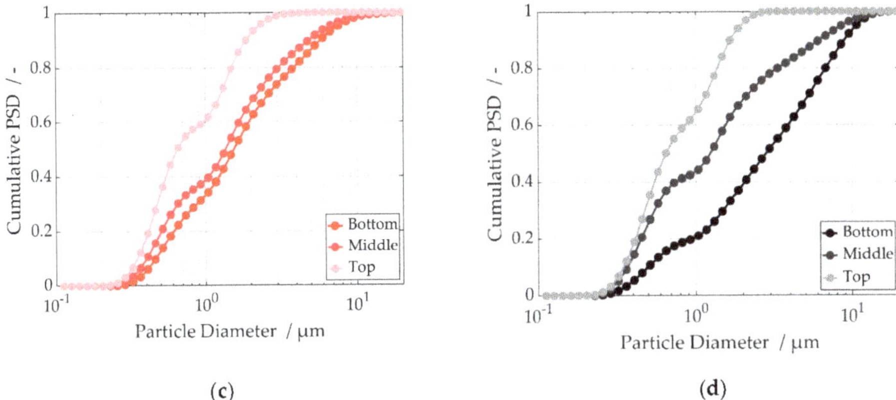

(c)   (d)

Figure 9. Sediment particle size distributions for the three cases and at bottom, middle and top axial position. Blue: 20,000 rpm; Red: 40,000 rpm; Black: Increasing rotational speed. (**a**) Characteristic particle sizes, squares: x_{10}; triangles: x_{50} with dashed lines as guides to the eye; stars: x_{90}. (**b**–**d**) Detailed particle size distributions for the three cases at axial positions **bottom** (darkest), **middle** (moderate shade), **top** (lightest).

Considering the results for carbon content in Figure 10 provides further indications about the differing sediment constitution. In good agreement with expectations, the carbon content at the bottom is less than 3% C in all cases, indicating that solely LFP particles are separated immediately after the inlet, regardless of the operational setting. All cases show an increasing carbon content along the rotor, too, according to expectations.

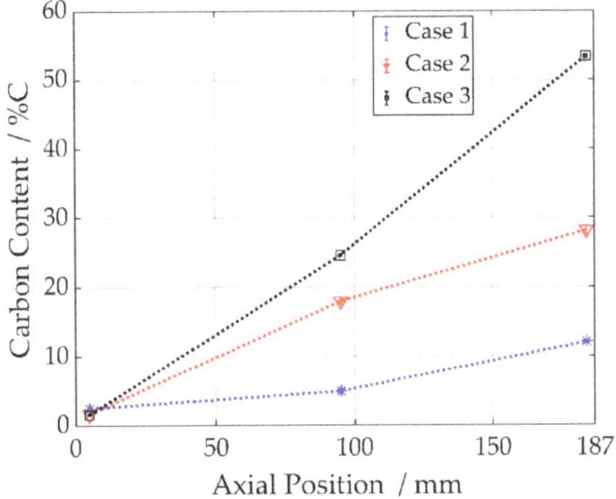

Figure 10. Sediment overall carbon content for the three cases at bottom, middle and top axial position. Blue: 20,000 rpm; Red: 40,000 rpm; Black: Increasing rotational speed. Squares: mean values; bars: Standard deviations; lines are guides to the eye.

The first case's overall low carbon contents (middle 5% C, top 12% C) suggest no or only low amounts of CB. According to the feed PSD (Figure 3) for this case, agglomerates do not play a crucial role. Separation conditions seem too weak to make numerous CB particles settle, as anticipated.

Equally according to assumptions, the carbon contents in case 2 is higher in the middle (18% C) and at the top (28% C) compared to case 1. Obviously, relatively more CB is separated under stronger conditions, besides a higher absolute yield of LFP, as the centrate results indicate.

As a quick conclusion from the centrate results, the carbon content in case 3 sediment samples should be lower than in case 2. Yet, the opposite is the case, the carbon contents are the highest with about 25% C in the middle and 53% C at the top. Due to the suggested higher CB content in centrate in case 3 (Figure 8) and not significantly different overall solids separation efficiencies (i.e., expected absolute sediment amount distributed in the rotor) between case 2 and 3 (Figure 6), these measurements are rather counterintuitive in the first instance. A possible explanation is given in Section 4.

4. Discussion

On the whole, the expectations based on the theoretical foundations are met to a great extent. Applying weak separation conditions, only the fastest settling particles are separated, that are mainly LFP and only few CB particles, if at all. This operational procedure results in a sediment containing practically only LFP (besides the carbon coating and binders). However, although a sediment very rich in LFP is desirable in terms of maximum LFP recovery for further recycling processing of the active material, the absolute yield of LFP is comparably low, regarding the dropping overall separation efficiency in Figure 6. This flaw may for example be tackled recirculating the centrate or setting up a centrifuge cascade, but these approaches would enhance the outlay in directions that are not intended at the actual state of this work. Under constantly strong separation conditions, a greater share of CB particles is separated alongside a larger absolute amount of LFP, compared to weaker separation conditions. So, the advantage of gaining an enhanced absolute active material yield is at the expense of a measurable contamination with CB. All things considered, both process modes with constant rotational speed resulted approximately in the anticipated temporal behavior: Overall separation efficiency decreases with ongoing time, centrate PSD contains coarser particles and centrate carbon contents offer logical supplements to these measurements, confirming the first assumptions.

With the intention to combine the best of both strategies, a rotational speed increase is examined as a new approach to counteract the temporal behavior. Centrate measurements suggest that almost constant separation conditions can be maintained with this procedure. The overall separation efficiency, centrate carbon content and especially centrate PSDs reveal approximately no variation over time. However, the overall separation efficiency is on a level too high to imply that only LFP is separated. Consequently, the rotational speed curve is expedient for keeping the centrate properties constant, but precisely the values set are presumably too high, and the accurate rotational speed curve lies possibly in a sense shifted versus the one applied. Reasons may lie in vortexes (back-)mixing the dispersion inside the rotor to some extent, which are not regarded in the centrifuge model for the time being, as well as other simplifications and imperfections of the model that served as basis to determine the rotational speed curve.

An increasing carbon content in sediment over the rotor length is also observed in all three cases and again meets expectations. On first examination, though, the comparably high carbon contents in middle and top sediment samples in the case with adapted rotational speed appear counterintuitive since these measurements contradict the intention to achieve complete fractionation of LFP and CB, which seems to be roughly fulfilled considering the centrate results. Of course, small vortexes and disturbances may complicate factual pure fractionation, but the strategy appears to work regarding the nearly constant and high carbon content in centrate. Against this background, the measured carbon contents are distinctly above expectations and not evident to explain. Nevertheless, it should firstly be stressed that all carbon content measurements only provide relative values and definite absolute conclusions cannot be drawn from the information provided by all given options for sample analysis. Secondly, endeavoring for an explanation, a closer look on the feed PSDs might be worthwhile.

As a result of the individual preparation of the feed dispersion for each of the three cases, they do not have identical PSDs, as formerly shown in Figure 3. As stated there, the feed dispersions in case 2 and 3 show a striking coarse fraction and strong tendency to agglomerate. It cannot be stated with certainty which amount of agglomerates entered the process and whether the agglomerates are stable or break during their residence time inside the rotor.

Anyhow, agglomerates seem to have entered the centrifuge in case 2 and 3. They might be more likely to break with stronger centrifugal forces acting on them, i.e., at high rotational speeds, like in case 2 or the end of case 3. If they are broken, CB and LFP particles settle individually at their specific velocities. Conversely, if rotational speed is low, like in the first part of the curve in case 3, agglomerates possibly do not break or take longer time to do so. Breaking after some time means that CB particles contained in an agglomerate are carried along a part of the way to the sediment until they are released to settle at their individual velocity. Both variants enhance the likelihood that CB particles reach sediment. Diluted for PSD measurements, resuspended sediment samples have a great tendency to agglomerate (again).

The essential difference between constantly high and increasing rotational speed is that in the former case, the likelihood for agglomerates to break is higher from the beginning and therefore, more individual particles settle separately during the entire process time. By contrast, applying an increasing rotational speed the forces are probably only strong enough to cause breach in the later course and for a certain duration unbroken agglomerates settle. According to theory, CB particles or comparably small individual particles are less likely to arrive at sediment than LFP particles or comparably great agglomerates, so probably a higher CB reaches sediment when agglomerates containing CB are not broken. This entire explanation is to be regarded as a hypothesis that requires further examination.

Apart from the controversial results for sediment carbon content, the reported results can be seen as a first proof that an appropriate increase of rotational speed can counteract the otherwise declining separation efficiency and keep centrate properties constant. This motivates to continue working on the approach to separate LFP and CB through a centrifugation step and recover LFP in the sediment as purely as possible. The latter point inevitably requires a verified explanation of the carbon contents measured and expedient adaptions in process design. It is intended to install an ultrasound flow-through cell before the centrifuge inlet to guarantee agglomerate-free feed. The precise rotational speed curve applied seems to lead approximately to the desired outcome. Still, small variances and the too high overall separation efficiency show that a more sensitive rotational speed setting would be preferable. The great benefit of the centrifuge model is therefore supposed to disclose when it is combined with on-line measurement techniques to realize a model-based control concept that the authors aim to present in future publications. According to this concept, rotational speed shall be adapted based on model predictions and precisely tuned corresponding to on-line measurements (PSD and centrate composition, like [37]). In this way, the authors aim to be able to recover an approximately pure LFP sediment that can eventually be re-used to manufacture new LIB cathodes. There is a good prospect that once the entire procedure is validated to work for LFP-based cathode material, centrifugation may also be applied to fractionate anode material (graphite), or generally black mass whose constituents are processible in water, which would contribute to a new, more environmentally sound recovery process chain.

5. Conclusions

Centrifugation was investigated as a technique for a physical separation step of black mass into its components as part of a direct recycling process chain for lithium-ion batteries. A tubular centrifuge is used to examine fractionation of an exemplary cathode material dispersion containing LFP and carbon black particles. Commonly operated at fixed settings, tubular centrifuges do not deliver constant output properties, but separation conditions that diminish over time, which would not serve the purpose to recover pure fractions of LFP and carbon black. According to theoretical foundations, LFP should be recovered as sediment, while carbon black stays in the centrate. Three operational strategies have therefore been examined concerning the temporal development of the centrate and sediment properties, two cases of constantly low (20,000 rpm) and constantly high (40,000 rpm) rotational speed, as well as a rotational speed increase based on a process model. The latter was desired to achieve constant centrate properties and a sediment rich in LFP. Expectations have been met concerning centrate. Overall solids content, particle size distributions and carbon content showed the anticipated temporal dependency when constant rotational speed was applied, whereas they remained mostly steady with increasing rotational speed. In that respect, rotational speed increase confirms to serve as suitable strategy to control the centrifuge output. However, sediment measurements reveal discrepancies from the expectations that may have arisen due to agglomerates in the feed dispersion. Once this issue is overcome, centrifugal fractionation may become a promising separation technique for lithium-ion battery active materials. In order to establish a usable set-up, the aim is to develop a model-based control concept for the operational settings of the centrifuge, which relies on on-line measurement tools for particle size distributions and composition of the dispersion as well.

Author Contributions: Conceptualization and funding acquisition, cooperation of all authors; formal analysis, software, validation, visualization, writing—original draft, T.S. (Tabea Sinn); data curation, investigation, methodology and project administration, T.S. (Tabea Sinn), A.F., T.S. (Thomas Stübinger) and A.W.; supervision, W.W., H.N. and M.G.; writing—review & editing, A.F., T.S. (Thomas Stübinger), A.W., H.N. and M.G. Resources provided by the contributing institutions thanks to funding. All authors have read and agreed to the published version of the manuscript.

Funding: This research was funded by the German Federal Ministry of Education and Research (BMBF), grant number 033RK065A-D.

Acknowledgments: The authors would like to acknowledge and thank Carl Padberg Zentrifugenbau GmbH (CEPA, Lahr im Schwarzwald, Germany) for providing a free exemplary of tubular centrifuge model Z11 and service, which made the experiments in this paper possible. Furthermore, they acknowledge support by the KIT-Publication Fund of the Karlsruhe Institute of Technology.

Conflicts of Interest: The authors declare no conflict of interest. The funders had no role in the design of the study; in the collection, analyses, or interpretation of data; in the writing of the manuscript, or in the decision to publish the results.

References

1. Bauer, C.; Hofer, J.; Althaus, H.J.; Del Duce, A.; Simons, A. The environmental performance of current and future passenger vehicles: Life cycle assessment based on a novel scenario analysis framework. *Appl. Energy* **2015**, *157*, 871–883. [CrossRef]
2. Harper, G.; Sommerville, R.; Kendrick, E.; Driscoll, L.; Slater, P.; Stolkin, R.; Walton, A.; Christensen, P.; Heidrich, O.; Lambert, S.; et al. Recycling lithium-ion batteries from electric vehicles. *Nature* **2019**, *575*, 75–86. [CrossRef] [PubMed]
3. Gaines, L. The future of automotive lithium-ion battery recycling: Charting a sustainable course. *Sustain. Mater. Technol.* **2014**, *1–2*, 2–7. [CrossRef]
4. Lv, W.; Wang, Z.; Cao, H.; Sun, Y.; Zhang, Y.; Sun, Z. A Critical Review and Analysis on the Recycling of Spent Lithium-Ion Batteries. *ACS Sustain. Chem. Eng.* **2018**, *6*, 1504–1521. [CrossRef]
5. Ciez, R.E.; Whitacre, J.F. Examining different recycling processes for lithium-ion batteries. *Nat. Sustain.* **2019**, *2*, 148–156. [CrossRef]
6. Hanisch, C.; Loellhoeffel, T.; Diekmann, J.; Markley, K.J.; Haselrieder, W.; Kwade, A. Recycling of lithium-ion batteries: A novel method to separate coating and foil of electrodes. *J. Clean. Prod.* **2015**, *108*, 301–311. [CrossRef]
7. Sun, L.; Qiu, K. Vacuum pyrolysis and hydrometallurgical process for the recovery of valuable metals from spent lithium-ion batteries. *J. Hazard. Mater.* **2011**, *194*, 378–384. [CrossRef]
8. Tanong, K.; Coudert, L.; Mercier, G.; Blais, J.F. Recovery of metals from a mixture of various spent batteries by a hydrometallurgical process. *J. Environ. Manag.* **2016**, *181*, 95–107. [CrossRef]
9. Yang, Y.; Xu, S.; He, Y. Lithium recycling and cathode material regeneration from acid leach liquor of spent lithium-ion battery via facile co-extraction and co-precipitation processes. *Waste Manag.* **2017**, *64*, 219–227. [CrossRef]
10. Zhan, R.; Oldenburg, Z.; Pan, L. Recovery of active cathode materials from lithium-ion batteries using froth flotation. *Sustain. Mater. Technol.* **2018**, *17*, 9. [CrossRef]
11. Li, X.; Zhang, J.; Song, D.; Song, J.; Zhang, L. Direct regeneration of recycled cathode material mixture from scrapped LiFePO4 batteries. *J. Power Sources* **2017**, *345*, 78–84. [CrossRef]

12. Guerfi, A.; Kaneko, M.; Petitclerc, M.; Mori, M.; Zaghib, K. LiFePO4 water-soluble binder electrode for Li-ion batteries. *J. Power Sources* **2007**, *163*, 1047–1052. [CrossRef]

13. Anlauf, H. Recent Developments in Research and Machinery of Solid–Liquid Separation Processes. *Dry Technol.* **2006**, *24*, 1235–1241. [CrossRef]

14. Leung, W.W. *Centrifugal Separations in Biotechnology*; Elsevier Science: Amsterdam, The Netherlands, 2007.

15. Stokes, G.G. On the effect of the internal friction of fluids on the motion of pendulums. *Transact. Camb. Philos. Soc.* **1850**, *9*, 8–106.

16. Flegler, A.; Schneider, M.; Prieschl, J.; Stevens, R.; Vinnay, T.; Mandel, K. Continuous flow synthesis and cleaning of nano layered double hydroxides and the potential of the route to adjust round or platelet nanoparticle morphology. *RSC Adv.* **2016**, *6*, 57236–57244. [CrossRef]

17. Konrath, M.; Brenner, A.K.; Dillner, E.; Nirschl, H. Centrifugal classification of ultrafine particles: Influence of suspension properties and operating parameters on classification sharpness. *Spt. Purif. Technol.* **2015**, *156*, 61–70. [CrossRef]

18. Spelter, L.E.; Meyer, K.; Nirschl, H. Screening of Colloids by Semicontinuous Centrifugation. *Chem. Eng. Technol.* **2012**, *35*, 1486–1494. [CrossRef]

19. Konrath, M.; Hackbarth, M.; Nirschl, H. Process monitoring and control for constant separation conditions in centrifugal classification of fine particles. *Adv. Powder Technol.* **2014**, *25*, 991–998. [CrossRef]

20. Spelter, L.E.; Steiwand, A.; Nirschl, H. Processing of dispersions containing fine particles or biological products in tubular bowl centrifuges. *Chem. Eng. Sci.* **2010**, *65*, 4173–4181. [CrossRef]

21. Spelter, L.E.; Schirner, J.; Nirschl, H. A novel approach for determining the flow patterns in centrifuges by means of Laser-Doppler-Anemometry. *Chem. Eng. Sci.* **2011**, *66*, 4020–4028. [CrossRef]

22. Konrath, M.; Gorenflo, J.; Hübner, N.; Nirschl, H. Application of magnetic bearing technology in high-speed centrifugation. *Chem. Eng. Sci.* **2016**, *147*, 65–73. [CrossRef]

23. Gleiß, M. *Dynamische Simulation der Mechanischen Flüssigkeitsabtrennung in Vollmantelzentrifugen*; KIT Scientific Publishing: Karlsruhe, Germany, 2018. [CrossRef]

24. Gleiss, M.; Nirschl, H. Modeling Separation Processes in Decanter Centrifuges by Considering the Sediment Build-Up. *Chem. Eng. Sci.* **2015**, *38*, 1873–1882. [CrossRef]

25. Michaels, A.S.; Bolger, J.C. Settling rates and sediment volumes of flocculated kaolin suspensions. *Ind. Eng. Chem. Funda.* **1962**, *1*, 24–33. [CrossRef]

26. Sambuichi, M.; Nakakura, H.; Osasa, K. Zone settling of concentrated slurries in a centrifugal field. *J. Chem. Eng. Jpn.* **1991**, *24*, 489–494. [CrossRef]

27. Stickland, A.D. Compressional rheology: A tool for understanding compressibility effects in sludge dewatering. *Water Res.* **2015**, *82*, 37–46. [CrossRef]

28. Tiller, F.M.; Khatib, Z. The theory of sediment volumes of compressible, particulate structures. *J. Colloid Interface Sci.* **1984**, *100*, 55–67. [CrossRef]

29. Tsai, F.Y.; Jhang, J.H.; Hsieh, H.W.; Li, C.C. Dispersion, agglomeration, and gelation of LiFePO4 in water-based slurry. *J. Power Sources* **2016**, *310*, 47–53. [CrossRef]

30. Balbierer, R.; Gordon, R.; Schuhmann, S.; Willenbacher, N.; Nirschl, H.; Guthausen, G. Sedimentation of lithium–iron–phosphate and carbon black particles in opaque suspensions used for lithium-ion-battery electrodes. *J. Mater. Sci.* **2019**, *54*, 5682–5694. [CrossRef]

31. Richardson, J.; Zaki, W. The sedimentation of a suspension of uniform spheres under conditions of viscous flow. *Chem. Eng. Sci.* **1954**, *3*, 65–73. [CrossRef]

32. Bülow, F.; Nirschl, H.; Dörfler, W. On the settling behaviour of polydisperse particle clouds in viscous fluids. *Eur. J. Mech. B Fluids* **2015**, *50*, 19–26. [CrossRef]

33. Garrido, P.; Concha, F.; Bürger, R. Settling velocities of particulate systems: 14. Unified model of sedimentation, centrifugation and filtration of flocculated suspensions. *Int. J. Miner. Process.* **2003**, *72*, 57–74. [CrossRef]

34. Lerche, D. Dispersion Stability and Particle Characterization by Sedimentation Kinetics in a Centrifugal Field. *J. Dispers. Sci. Technol.* **2002**, *23*, 699–709. [CrossRef]

35. Spelter, L.E.; Nirschl, H. Classification of Fine Particles in High-Speed Centrifuges. *Chem. Eng. Technol.* **2010**, *33*, 1276–1282. [CrossRef]

36. Gleiß, M.; Sinn, T.; Nirschl, M. A dynamic model to predict separation behaviour and sediment build-up of colloidal silica particles within tubular centrifuges. *Adv. Powder Technol.* **2020**, under review.

37.	Winkler, M.; Sonner, H.; Gleiss, M.; Nirschl, H. Fractionation of ultrafine particles: Evaluation of separation efficiency by UV/vis spectroscopy. *Chem. Eng. Sci.* **2020**, *213*, 115374. [CrossRef]

Publisher's Note: MDPI stays neutral with regard to jurisdictional claims in published maps and institutional affiliations.

Article

Li-Distribution in Compounds of the Li$_2$O-MgO-Al$_2$O$_3$-SiO$_2$-CaO System—A First Survey

Thomas Schirmer [1,*], Hao Qiu [2], Haojie Li [3], Daniel Goldmann [2] and Michael Fischlschweiger [3]

[1]   Department of Mineralogy, Geochemistry, Salt Deposits, Institute of Disposal Research, Clausthal University of Technology, Adolph-Roemer-Str. 2A, 38678 Clausthal-Zellerfeld, Germany

[2]   Department of Mineral and Waste Processing, Institute of Mineral and Waste Processing, Waste Disposal and Geomechanics, Clausthal University of Technology, Walther-Nernst-Str. 9, 38678 Clausthal-Zellerfeld, Germany; hao.qiu@tu-clausthal.de (H.Q.); daniel.goldmann@tu-clausthal.de (D.G.)

[3]   Technical Thermodynamics and Energy Efficient Material Treatment, Institute for Energy Process Engineering and Fuel Technology, Clausthal University of Technology, Agricolastr 4, 38678 Clausthal-Zellerfeld, Germany; haojie.li@tu-clausthal.de (H.L.); michael.fischlschweiger@tu-clausthal.de (M.F.)

*   Correspondence: thomas.schirmer@tu-clausthal.de; Tel.: +49-5323-722917

Received: 20 October 2020; Accepted: 30 November 2020; Published: 4 December 2020

Abstract: The recovery of critical elements in recycling processes of complex high-tech products is often limited when applying only mechanical separation methods. A possible route is the pyrometallurgical processing that allows transferring of important critical elements into an alloy melt. Chemical rather ignoble elements will report in slag or dust. Valuable ignoble elements such as lithium should be recovered out of that material stream. A novel approach to accomplish this is enrichment in engineered artificial minerals (EnAM). An application with a high potential for resource efficient solutions is the pyrometallurgical processing of Li ion batteries. Starting from comparatively simple slag compositions such as the Li-Al-Si-Ca-O system, the next level of complexity is reached when adding Mg, derived from slag builders or other sources. Every additional component will change the distribution of Li between the compounds generated in the slag. Investigations with powder X-Ray diffraction (PXRD) and electron probe microanalysis (EPMA) of solidified melt of the five-compound system Li$_2$O-MgO-Al$_2$O$_3$-SiO$_2$-CaO reveal that Li can occur in various compounds from beginning to the end of the crystallization. Among these compounds are Li$_{1-x}$(Al$_{1-x}$Si$_x$)O$_2$, Li$_{1-x}$Mg$_y$(Al)(Al$_{3/2y+x}$Si$_{2-x-3/2y}$)O$_6$, solid solutions of Mg$_{1-(3/2y)}$Al$_{2+y}$O$_4$/LiAl$_5$O$_8$ and Ca-alumosilicate (melilite). There are indications of segregation processes of Al-rich and Si(Ca)-rich melts. The experimental results were compared with solidification curves via thermodynamic calculations of the systems MgO-Al$_2$O$_3$ and Li$_2$O-SiO$_2$-Al$_2$O$_3$.

Keywords: lithium; thermodynamic modeling; engineered artificial minerals (EnAM); melt experiments; PXRD; EPMA

---

1. Introduction

With respect to the development in electromobility as well as to the changes in circular energy systems, Li-ion batteries are of central importance. To safeguard raw material sources especially for critical elements such as Co, Ni and Li as key components of this technology, efficient recycling processes are essential. One of the most important routes to recycle these battery types is the pyrometallurgical processing, which can deal with a broad range of input material. While Co, Ni and other valuable heavy metals such as Cu report into the alloy melt, Li is transferred at least in major amounts into the slag phase of this process.

A resource and energy efficient recovery of Li from the slag could be accomplished, if Li were concentrated in specific Li-rich artificial mineral phases, which could then be separated after crystallization and cooling of the slag by means of mineral processing technologies, generating concentrates for following hydrometallurgical processing.

Previous research has shown that Li can be recovered in the form of the $LiAlO_2$ crystals through flotation from a remaining silicate slag matrix [1]. The hydrometallurgical processing of Li enriched silicate slag has also shown that Li recovery can reach 80–95% [2].

As long as the complete system, based on a Li_2O-Al_2O_3-SiO_2-CaO mixture, does not contain any other element, the results are promising. As soon as other elements are added, new phases start to crystallize.

Besides Li and Al, reporting from the Li-ion battery input into the slag, Si, Ca and often Mg (at least partly from dolomite as slag building component) are introduced as slag builders to ensure an optimized split between metal alloy melt, slag and dust phase in the pyrometallurgical process.

Until now, the thermodynamics of the overall process have not been investigated sufficiently and therefore for extended systems such as Li_2O-MgO-Al_2O_3-SiO_2-CaO this work serves as a starting point. Consequently, this should allow understanding some basic principles and giving further insights into these slag-systems. Additionally, a solid ground should be provided for further research on these slag systems, because in the future more complex slag systems, e.g., Mn-containing mixtures, should be investigated since they will represent future inputs to this recycling route especially for the NCM-type batteries.

The Umicore Battery Recycling Process is a vital pyrometallurgical process developed for the recovery of NiMH and spent lithium-ion batteries [3]. From the composition of a slag with the compounds Li_2O-MgO-Al_2O_3-SiO_2-CaO and high aluminum content, it is observed that Li is present in the slag in the form of the $LiAlO_2$ [2], which would facilitate subsequent recovery by flotation. At the same time, the spinel phase appears in all three Umicore slags, and, in one of the Umicore slags, Li is even partially dispersed in the spinel phase [3].

Even though spinel phases appear in different slags if bivalent ions such as those of Mg are present, there is little published research on the impact of spinel on the formation of separate $LiAlO_2$ crystals because of the scavenging of Al from the melt and the formation of $Mg_{1-(3/2y)}Al_{2+y}O_4$/$LiAl_5O_8$ solid solution.

In this study, three synthetic slags with different contents of MgO based on the Li_2O-MgO-Al_2O_3-SiO_2-CaO oxide system were prepared using pure chemical reagents. The degree of supercooling was then reduced by controlling and cooling the melt slowly to obtain thoroughly crystallized synthetic slags for research. The synthetic slags were then analyzed by X'Ray powder diffraction (PXRD) and electron probe microanalysis (EPMA) for mineralogical studies and finally compared to the solidification curves obtained by thermodynamic calculation. This served as a starting point for studying the influence of spinel formation and understanding important phase reactions in the five-component oxide system Li–Mg–Al–Si–Ca.

To increase the knowledge on the behavior of slag systems and the options to predict and stimulate the creation of artificial mineral phases, an interdisciplinary approach was taken, comprising thermodynamical modeling, pyrometallurgical processing, mineralogical analysis and prediction and testing of mineral processing technologies. In this paper, the focus is put on mineralogical analysis in connection with thermodynamic modeling.

2. Background

To better understand the results presented in this article, the existing information about the compounds of important binary and ternary systems containing Li_2O, MgO, Al_2O_3, SiO_2 and CaO is summarized. This information serves as the starting point to analyze and improve the existing data and develop respective thermodynamic modeling strategies.

2.1. Important Binary Phase Systems Containing Li

In the systems Li_2O-CaO and Li_2O-MgO, except for limited solid solution, no explicit phase reactions are reported (e.g., Konar et al. [4]).

In the system Li_2O-Al_2O_3, several stable lithium aluminate compounds are described: Li_5AlO_4, $LiAlO_2$ and $LiAl_5O_8$ [5,6]. Additionally, a high temperature compound $LiAl_{11}O_{17}$ at $0.8 < Al_2O_3 < 0.92$ and $>2200\ °C$ is mentioned [5]. The compounds $Li_2Al_4O_7$ synthesized by Kale et al. [7] and Li_3AlO_3 were found to be instable by Kale et al. [7] and are not part of the data published by Konar et al. [5]. In this phase diagram, there is also a thermal barrier at the mole fraction of $Al_2O_3 = 0.5$ ($LiAlO_2$), so that at $0.18 < Al_2O_3 < 0.5$ the resulting mixture is Li_5AlO_4/$LiAlO_2$ and at $0.5 < Al_2O_3 < 0.82$ the resulting mixture is $LiAlO_2$/$LiAl_5O_8$. The two compounds important for this work, $LiAlO_2$ and $LiAl_5O_8$, both have polymorphs. According to Konar et al. [5], $LiAlO_2$ comprises a tetragonal γ-phase (high temperature) and a cubic α-phase (low temperature) modification and $LiAl_5O_8$ generally crystallizes in a spinel (high temperature) and a low temperature primitive cubic form [8]. According to Li et al. [9], $LiAlO_2$ comprises four polymorphs: a tetragonal γ-phase, a rombohedral α-phase, an orthorhombic β-phase and two phases of high temperature.

In the system MgO-Al_2O_3, the only binary compound is cubic $MgAl_2O_4$ (spinel) with the idealized composition at a mole fraction of $Al_2O_3 = 0.5$. At this ratio, there is also a thermal barrier. In the area of mole fraction $0 < MgO < 0.05$ in the temperature range 1900–2800 $°C$, solid MgO can form a solid solution with Al_2O_3 [5]. The region of mole fraction $0.5 < Al_2O_3 < 0.96$, particular important for this study, comprises a complete solid solution, so that an Al-rich melt can be in equilibrium with a spinel relatively enriched in Mg [10].

The system Li_2O-SiO_2 comprises the binary compounds Li_8SiO_6, Li_4SiO_4, $Li_6Si_2O_7$, Li_2SiO_3 and $Li_2Si_2O_5$ [11]. Additionally, the prediction shows two thermal barriers at the composition Li_4SiO_4 and Li_2SiO_3.

2.2. Important Ternary Phase Systems Containing Li

In the system Li_2O-MgO-Al_2O_3, three important primary crystallization fields can be predicted [5]: spinel ($MgAl_2O_4$), MgO and γ-$LiAlO_2$. Interesting isopleths are spinel-$LiAl_5O_8$, spinel-$LiAlO_2$, spinel-Li_2O and MgO-$LiAlO_2$. From this intersects, it can be concluded that a limited amount (i.e., maximum mole fraction = 0.31) of $LiAlO_2$ can be dissolved in MgO. Additionally, the compounds $LiAl_5O_8$ and $MgAl_2O_4$ can be combined to an ideal spinel solid solution [5].

The system Li_2O-Al_2O_3-SiO_2 contains the Li-bearing binary systems Li_2O-SiO_2 and Li_2O-Al_2O_3, as described in Section 2.1 [12]. With respect to the present work, the primary crystallization fields of $LiAlO_2$, $LiAl_5O_8$, eucryptite ($LiAlSiO_4$) and spinel are of particular interest. The compound $LiAlO_2$, described in Section 2.1, can additionally incorporate Si according to an substitution of $Li^+ + Al^{3+} = Si^{4+} + v$ (vacancy) so that the general formula is α ($LiAl^{4+}$, vSi^{4+}])O_2 and γ (Li, v)^{Li}[Al^{3+}, Si^{4+}]$^M O_2$ [12]. The compound eucryptite can be derived from SiO_2 via a substitution of $Li^+ + Al^{3+} = Si^{4+} + v$ [12] and crystallizes as quartz in the trigonal system, whereas a low temperature α-polymorph is disordered and a β-polymorph is ordered. Additionally, this compound can incorporate Mg and be broken up into the compounds $LiAlO_2$, $Mg_{0.5}AlO_2$ and SiO_2 [13]. The spinel crystallizes in a cubic system and can have a very variable chemistry with respect to the Al/Mg ratio and the solid solution with $LiAl_5O_8$ (see Section 2.1).

3. Materials and Methods

3.1. Materials

Chemicals

The chemicals used for producing synthetic slags are lithium carbonate (Merck, purum), calcium oxide (Sigma-Aldrich, reagent grade, St. Louis, MO, United States), silicon dioxide (Sigma-Aldrich, purum p.a.,

St. Louis, MO, United States), aluminum oxide (Merck) and magnesium oxide (98% wt.%, Roth, Karlsruhe, Germany). All chemicals ordered via Merck KGaA, Darmstadt, Germany.

3.2. Methods

3.2.1. Experiments

The chemical compositions of input materials for the synthesis of slags are listed in Table 1. The chemicals were manually mixed in a mortar and grinded in a disc mill for 5 min. Each sample was placed in a Pt-Rh crucible and heated in a chamber furnace (Nabertherm HT16/17, Nabertherm GmbH, Lilienthal, Germany) in an air atmosphere. The heating regime is shown in Figure 1. A heating rate of 2.89 °C/min was first employed to reach 720 °C, which is the melting temperature of Li_2CO_3, and then a heating rate of 1.54 °C/min was used to aid in the decomposition of Li_2CO_3 and to reach the target temperature. Samples were kept at 1600 °C for 2 h. Thereafter, the samples were cooled to 500 °C at a cooling rate of 0.38 °C/min and quenched in water.

Table 1. Calculated Theoretical Chemical Bulk Composition of the Samples According to the Weighed Quantities.

Sample	Li_2CO_3	$CaCO_3$	SiO_2	Al_2O_3	MgO
Content	%	%	%	%	%
1	22.87	22.40	16.36	32.97	5.40
2	22.51	21.55	16.08	32.17	7.70
3	22.32	21.24	15.63	31.54	9.27

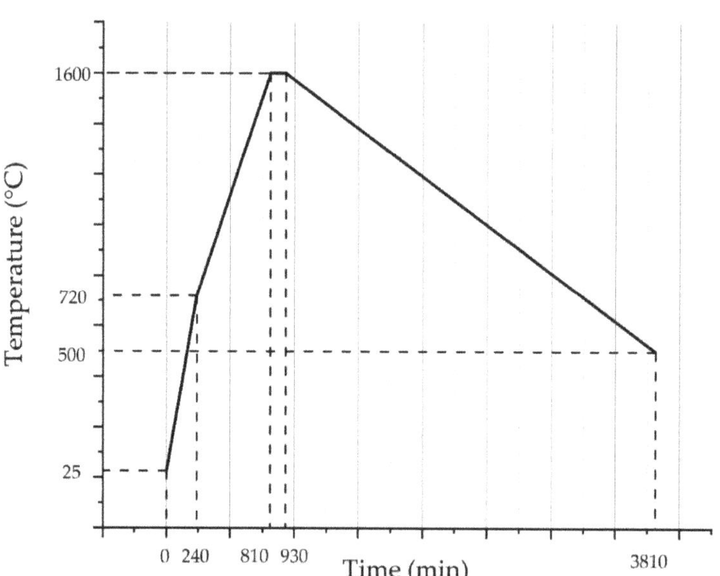

Figure 1. Schematic diagram of temperature regime.

3.2.2. Chemical Bulk Analysis

The element content was determined with ICP optical emission spectrometry (ICP-OES 5100, Agilent, Agilent Technologies Germany GmbH & Co. KG, Waldbronn, Germany). Samples were melted with lithium tetra borate in a platinum crucible at 1050 °C, and then the samples were leached with dilute hydrochloric acid to measure the content of Al, Ca, Mg and Si. To measure other elements, the samples were mixed with nitric acid and digested at 250 °C and under a pressure of 80 bar in an autoclave (TurboWAVE, MLS, Leutkirch im Allgäu, Germany).

3.2.3. Mineralogical Investigation

An overview of the mineralogical composition was provided by powder X-Ray diffraction (PXRD), using a PANalytical X-Pert Pro diffractometer, equipped with a Co-X-Ray tube (Malvern Panalytical GmbH, Kassel, Germany). For identification of the compounds, the pdf-2 ICCD XRD database, the American Mineralogist Crystal Structure Database [14] and the RRUFF-Structure database [15] were assessed.

The analysis of single crystals and grains was carried out with electron probe microanalysis (EPMA). EPMA is a standard method to characterize the chemical composition in terms of single spot analysis or element distribution patterns, accompanied by electron backscattered Z (ordinal number) contrast (BSE(Z)) or secondary electron (SE) micrographs. To carry out EPMA measurements, the sample was prepared as polished block in epoxy resin, coated with carbon and characterized using a Cameca SX^{FIVE} FE Field Emission) electron probe, equipped with five wavelength dispersive (WDX) spectrometers (CAMECA SAS, Gennevilliers Cedex, France). The following elements/(lines) were used to quantify the measurement points: Na (Kα), Mg (Kα), Al (Kα), Si (Kα), K (Kα), Ca (Kα), Ti (Kα), Mn (Kα) and Fe (Kα). To calibrate the wavelength dispersive X-ray fluorescence spectrometers (WDRFA), an appropriate suite of standards and analyzing crystals was used. The reference materials were provided by P&H Developments Ltd. (Glossop, Derbyshire, UK) and Astimex Standards Ltd. (Toronto, ON, Canada). The beam size was set to 0, leading to a beam diameter of substantially below 1 μm (100–600 nm with field emitters of Schottky-type, e.g., Jercinovic et al. [16]). To evaluate the measured intensities, the X-PHI-Model was applied [17].

Lithium, one of the key elements in this study, cannot be directly analyzed since EPMA uses X-ray fluorescence to detect the elements in the sample and the extremely low fluorescence yield and long wavelength of Li Kα makes the direct determination of this element nearly impossible. With the reasonable assumption that other refractory light elements such as Be and B are not present in the investigated material and volatile elements and compounds such as F, H_2O, CO_2 or NO_3^- are effectively eliminated during the melt experiment, Li can be calculated using virtual compounds, as depicted in as described in Section 4.3.1. If necessary, the balanced Li concentration was included into the matrix correction calculation. To access the analytical accuracy with respect to Li-containing compounds, the international reference material spodumene (Astimex) and the in-house standard LiAlO$_2$ were used (Table 2). Additionally, Li containing crystalline phases identified by X-ray diffraction (PXRD) could be referenced to the EPMA result.

Table 2. Repeated Measurements on Two Li-Compounds. Spod, Spodumene; %StdDev, Percentage Standard Deviation of the Measured Points (pepeats: $n = 5$); R, Recovery; LiAl, LiAlO$_2$.

wt.%	Average Spod.	%StDev., Spod.	Ref. Spod.	R (%)	Average LiAl	%StDev., LiAl	Ref. LiAl	R (%)
Al	15.04	0.35	14.4	104	41.24	0.22	40.9	101
Mg	0.00	n. a.	0.0	n. a.	0.01	n. a.	0.0	n. a.
Ti	0.00	n. a.	0.0	n. a.	0.00	n. a.	0.0	n. a.
Mn	0.05	n. a.	0.0	n. a.	0.00	n. a.	0.0	n. a.
Fe	0.02	n. a.	0.0	n. a.	0.03	n. a.	0.0	n. a.
Ca	0.01	n. a.	0.0	n. a.	0.01	n. a.	0.0	n. a.
K	0.00	n. a.	0.0	n. a.	0.00	n. a.	0.0	n. a.
Si	28.71	0.56	30.0	96	0.01	n. a.	0.0	n. a.
Na	0.10	2.83	0.09	112	0.00	n. a.	0.0	n. a.

3.2.4. Thermodynamic Modeling

For a better understanding of the experimental mechanisms investigated in the Li_2O-MgO-Al_2O_3-SiO_2-CaO system, the thermodynamic modeling of the phase behavior and the solidification in subsystems is of high relevance and hence applied in this work. Especially the knowledge of the phase behavior of the MgO-Al_2O_3 subsystem and the phases solidified at respective temperatures of certain component concentrations of the Li_2O-Al_2O_3-SiO_2 subsystem is important and contributes to the clarification and understanding of primary crystallization mechanisms figured out by the mineralogical characterization. On principal, based on already existing experimental data and thermodynamic studies, which are stated below, an optimized database for the subsystem was completed and applied to calculate the respective phase and solidification behavior. Generally, all calculations, i.e., for the binary MgO-Al_2O_3 and the ternary Li_2O-Al_2O_3-SiO_2 subsystems, were performed with the modified quasi-chemical model (MQM) [18–20] and the compound-energy formalism (CEF) [21], implemented in Factsage [22].

Specific insights into the database adaption regarding the two subsystems are presented subsequently. The thermodynamic database for the oxides such as MgO and Al_2O_3 comes from the FT oxide database [22] without any modification. Regarding the ternary subsystem Li_2O-Al_2O_3-SiO_2, the thermodynamic properties of SiO_2, Al_2O_3 and the mullite solid solution were used from the FT oxide [22] database without any modification. However, for the Gibbs energy of the Li_2O, the optimized value from [11] was integrated into the database. For compounds such as Li_2SiO_3, Li_4SiO_4, $Li_6Si_2O_7$, $Li_2Si_2O_5$-LT (low-temperature form) and $Li_2Si_2O_5$-HT (high-temperature form), the thermodynamic data were taken from [11]. The standard formation enthalpy of Li_8SiO_6 was optimized in this work with a value of 3, 521, 499.2 J/mol. Furthermore, for the binary compounds in the Li_2O-Al_2O_3, the standard formation enthalpy of the Li_5AlO_4 was optimized to 2,389,980 J/mol. The standard entropy of $LiAl_{11}O_{17}$ was optimized to a value of 350.55 $Jmol^{-1}K^{-1}$. The ternary compounds including the α- and β-eucryptite solid solutions, β-spodumene solid solution and α-$LiAlO_2$ solid solution were obtained from [12] without any modification. However, the Gibbs energy of the end member $G^0_{VaAlO_2}$ in the γ-$LiAlO_2$ solid solution was calculated with the assumption that the reciprocal energy of endmember is zero, while the other three endmembers were obtained directly from [12].

Based on these data, the CALPHAD calculations were performed for the subsystems, which are used for further explanations and discussions in connection with the new experimental findings in the next section.

4. Results

This section presents the measurement results of the melt experiments from PXRD and EPMA. First, three PXRD measurements from experiments with different Mg-concentration are compared (Section 4.2). In Section 4.3, an overview of the material with BSE(Z) micrographs and detailed spatially resolved quantitative point measurements and element distribution profiles recorded with EPMA are presented. Additionally, in Section 4.4, experimental findings are compared with thermodynamic model predictions for the relevant subsystems.

4.1. Bulk Chemistry of the Melt Experiments

The measurement results presented in Table 3 show that 14–20% of Li is lost during the melting and cooling of the material. The same applies for Na, which always appears as contaminant in open systems due to the overall availability (air, dust, skin, clothing, etc.).

Table 3. Comparison of the Bulk Chemical Composition Measured with ICP-OES of the Four Melt Experiments, Given in Mole Percent. The bold emphazises the Li-loss which is important to see (Li is volatile).

	Raw Mix (Mole Fraction)			Product (Mole Fraction)			Recovery %		
	V1	**V2**	**V3**	**V1**	**V2**	**V3**	**V1**	**V2**	**V3**
Al_2O_3	32.90	32.10	31.48	34.21	33.64	33.63	3.99	4.79	6.86
CaO	22.35	21.50	21.19	22.79	22.57	22.13	1.97	4.95	4.44
Li_2O	22.81	22.46	22.28	20.00	18.69	17.57	**−12.35**	**−16.77**	**−21.11**
MgO	5.32	7.59	9.14	5.32	8.27	9.90	−0.11	9.03	8.32
SiO_2	16.32	16.05	15.60	17.43	16.59	16.53	6.81	3.40	5.98
Na_2O	0.3	0.3	0.3	0.3	0.23	0.2	−14.0	−21.4	−28.6

4.2. PXRD Comparison of the Three Melt Experiments

The results of the PXRD measurements are presented in Figure 2, showing an overview of the diffractograms of all experiments (above) and three enlarged sections, showing important spinel and lithium aluminate diffraction peaks.

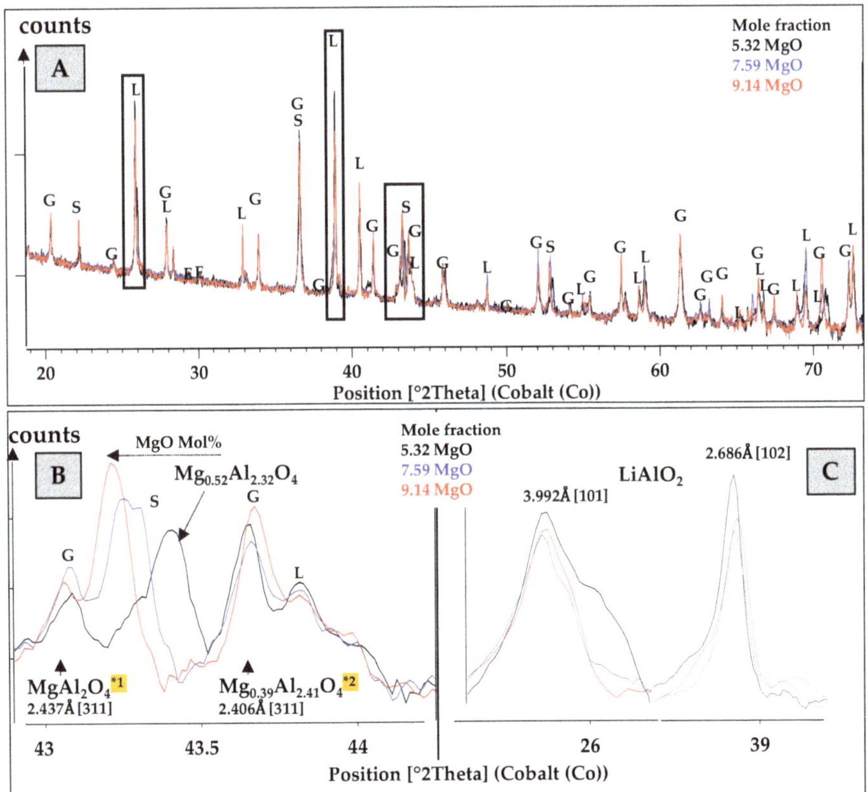

Figure 2. (**A**) PXRD of the solidified melt. G, gehlenite; S, spinel; L, $LiAlO_2$; E, eucryptite. (**B**) Enlarged section of the main spinel peak, * 1,: position of the main peak of $MgAl_2O_4$ from the ICCD-PDF2 No. 00-021-1152; * 2, position of the main peak of an Al-rich spinel from the ICCD-PDF2 No. 00-048-0528, peaks; $Mg_{0.52}Al_{2.32}O_4$, average composition of a spinel grain of the melt experiment with 5.32 Mol% Mg, determined with EPMA (see Section 4.2). (**C**) Enlarged sections of the first two main $LiAlO_2$ peaks.

The overview of all XRD measurements show the compounds gehlenite, spinel, $LiAlO_2$ and eucryptite (Figure 2A), whereas eucryptite is at the detection limit (<2–5 wt.%). The enlarged section of the 2-theta region of the main spinel peaks gives an indication of the changing composition of the spinel with the change of the Mg content (Figure 2B). Because of the high Al-concentration, the main (100%) spinel peaks of all experiments lie between those of the standard spinel $MgAl_2O_4$ and an aluminum-dominated $Mg_{1-(3/2y)}Al_{2+y}O_4$. Additionally, there is an indication of increasing spinel content with rising Mg concentrations. The Li-Al-oxide peaks are best explained with the diffraction pattern of $LiAlO_2$ (ICCD PDF2 No. 00-038-1464). The comparison of the two main peaks of the three experiments gives a hint that the amount of crystalline $LiAlO_2$ could be negatively correlated with the amount of Mg in the melt because the highest main peak intensities were measured in the sample with the lowest Mg concentration (Figure 2C).

4.3. EPMA Results

The main compounds of the melt experiments, determined with EPMA, were:

- Spinel: $Mg_{1-(3/2y)}Al_{2+y}O_4$
- Lithium aluminate (LiAl): $Li_{1-x}(Al_{1-x}Si_x)O_2$
- Eucryptite-like lithium alumosilicate (ELAS): $Li_{1-x}Mg_y(Al)(Al_{3/2y+x}Si_{2-x-3/2y})O_6$
- Gehlenite-like calcium-alumosilicate (GCAS): $Ca_2AL_2SiO_7$ with minute amounts of Mg

The compound (GCAS) is an end member of the melilite-like calcium-alumosilicate (MCAS), which is used for this phase with higher amounts of ions in addition to Ca:

- Melilite-like calcium-alumosilicate (MCAS): $(Na,Ca,Li)_2(Al,Mg,Li)(Al,Si)_2O_7$, which according to the calculations (Section 4.3.3) can also be a potential host for Li

An overview recorded with BSE(Z) shows a matrix of bright Ca-alumosilicate (GCAS/MCAS) interspersed with dendritic or massive dark LiAl. Within this mixture, large idiomorphic or hypidiomorphic crystals of spinel can be identified (Figures 3 and 4).

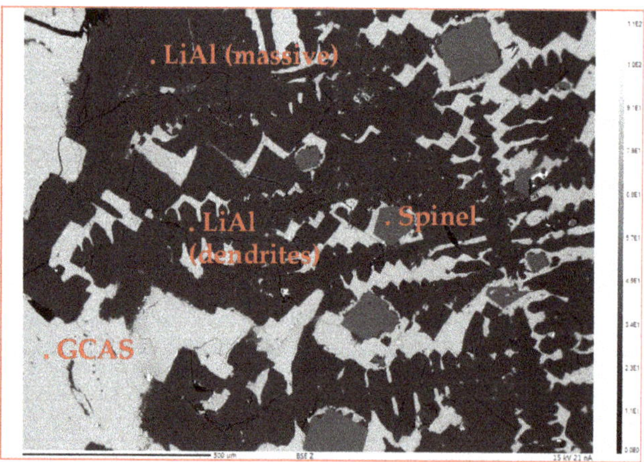

Figure 3. Electron micrograph (BSE(Z) of the solidified melt. Medium grey grains, spinel; dark gray sections and dendrites, LiAl surrounded by Ca-alumosilicate (GCAS, light grey sections); black, pores or preparation damage.

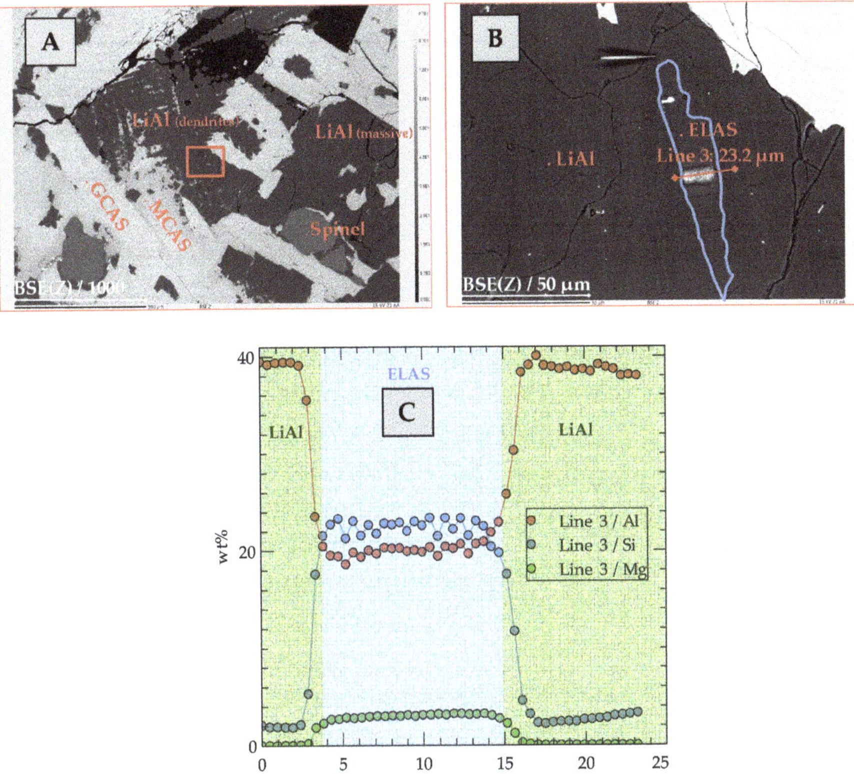

Figure 4. (**A**): Overview of the solidified melt (backscattered electron micrographs BSE(Z)). Medium grey grains, spinel; dark gray sections and dendrites, LiAl surrounded by Ca-alumosilicate (GCAS, MCAS, light grey sections); black, pores or preparation damage; red square, detail presented in (**B**). (**B**) Enlarged section from the red square in (**A**): the blue rim marks the grain of ELAS where the scan of Line 3 (red line) was measured. (**C**) Quantitative line scan of Line 3 (red line in (**B**)).

4.3.1. Lithium Aluminate (LiAl) and Lithium-Alumosilicate (ELAS)

The LiAl can be classified into two morphologic forms: massive and dendrite-like (Figures 3 and 4A). A closer look into the massive LiAl reveals thin lath-shaped grains of ELAS or a corresponding melt (Figure 4B). A line scan over such a lath-shaped grain reveals a quite homogeneous composition with more or less sharp borders to the surrounding LiAl (Figure 4C). The ELAS can be described as a mixture of the virtual compounds $LiAlO_2$, $Mg_{0.5}AlO_2$ and SiO_2, as listed in Table 4.

Mult. depicts a factor to multiply the three components to generate an optimized ELAS or $Li_{1-x}Mg_y(Al)(Al_{3/2y+x}Si_{2-x-3/2y})O_6$ similar to the average measured concentrations (except for Li) on Line 3 (Figure 4). The Li value results from the multiplications and this was used to calculate the formula of the analyzed ELAS in the sample. In a similar manner, the Si-containing LiAl with the general formula $Li_{1-x}(Al_{1-x}Si_x)O_2$ can be calculated as a mixture of SiO_2 and $LiAlO_2$. The calculated formulas of the ELAS and the LiAl are:

$$ELAS: (Li_{0.96}Mg_{0.24})(Al)(Al_{0.45}Si_{1.55})O_6$$

$$LiAl: (Li_{0.94})(Al_{0.94}Si_{0.06})O_2$$

The Si concentration in the dendritic LiAl is distinctively lower as in the massive crystals (compare Tables 4 and 5). The calculated formula of the LiAl in this case is:

$$\text{LiAl: } (\text{Li}_{0.97})(\text{Al}_{0.97}\text{Si}_{0.03})\text{O}_2$$

Table 4. Calculation of virtual compound ratios and average composition of the ELAS and the LiAl on Line 3, shown in Figure 4. Opt., calculated ideal composition; Meas., measured average; Mult., factor for multiplication of the virtual compounds; (Calc.), calculated values (Li, O); %StdDev, percentage standard deviation of the measured points (LiAl (Meas.), repeats, $n = 23$).

wt.%	Virtual Compounds			ELAS (Opt.)	ELAS, Meas.	ELAS %StDev.	LiAl (Opt.)	LiAl (Meas.)	LiAl %StDev.
	LiAlO$_2$	Mg$_{0.5}$AlO$_2$	SiO$_2$						
Al	40.9	37.9	0.0	20.1	20.1	3.2	38.9	38.9	1.4
Mg	0.0	17.1	0.0	3.1	3.0	7.5	0.0	0.0	n. a.
Ti	0.0	0.0	0.0	0.0	0.0	n. a.	0.0	0.0	n. a.
Mn	0.0	0.0	0.0	0.0	0.1	n. a.	0.0	0.0	n. a.
Fe	0.0	0.0	0.0	0.0	0.1	n. a.	0.0	0.1	n. a.
Ca	0.0	0.0	0.0	0.0	0.0	8.9	0.0	0.0	n. a.
K	0.0	0.0	0.0	0.0	0.0	n. a.	0.0	0.0	n. a.
Si	0.0	0.0	46.7	23.2	22.4	3.6	2.5	2.5	18.3
Na	0.0	0.0	0.0	0.0	0.0	n. a.	0.0	0.0	n. a.
O (Calc.)	48.5	45.0	53.3	50.2	49.5	n. a.	49.0	49.1	n. a.
Li (Calc.)	10.5	0.0	0.0	3.4	3.4	n. a.	10.01	10.01	n. a.
Mult.	0.33	0.18	0.49		← Multiplication factors for ELAS (Opt.)				
Mult.	0.95	0	0.049		← Multiplication factors for LiAl (Opt.)				
Sum	100	100	100	100	98.7		100.4	100.6	

Table 5. Calculation of Virtual Compound Ratios and Average Composition of the LiAl in the Dendrites (Dend.) Shown in the BSE(Z) Micrograph of Figure 3. Opt., Calculated Ideal Composition; Meas., Measured Average; Mult., Factor for Multiplication of the Virtual Compounds; (Calc.), Calculated Values (Li, O); %StdDev, Percentage Standard Deviation of the Measured Points (LiAl (Dend.) (Meas.), Repeats, $n = 4$).

wt.%	Virtual Compounds		LiAl (Opt.)	LiAl (Dend.) (Meas.)	LiAl (Dend.) % StDev.
	LiAlO$_2$	SiO$_2$			
Al	40.9	0.0	40.3	40.3	0.4
Mg	0.0	0.0	0.0	0.0	n. a.
Ti	0.0	0.0	0.0	0.0	n. a.
Mn	0.0	0.0	0.0	0.0	n. a.
Fe	0.0	0.0	0.0	0.0	n. a.
Ca	0.0	0.0	0.0	0.0	n. a.
K	0.0	0.0	0.0	0.0	n. a.
Si	0.0	46.7	1.1	1.1	12.3
Na	0.0	0.0	0.0	0.0	n. a.
O (Calc.)	48.5	53.3	49.0	49.1	n. a.
Li (Calc.)	10.5	0.0	10.36	10.36	n. a.
Mult.	0.98	0.024	← Multiplication factors for LiAl (Opt.)		
Sum	100	100	100.8	100.9	

4.3.2. Spinel

Spinel as the first crystallizing compound obeys the crystallization equilibrium inasmuch as the composition of the spinel with the highest Al content is connected with the corresponding Al-rich melt. The Mg concentrations in the measured profile (Figure 5B) are increasing from the center to the rim of the crystal. A look at the ratio of Mg/Al in a line scan through a spinel crystal starting at the center of the grain in the direction to the rim shows no increase within a first region. After this first region, the ratio increases. Closer to the rim, the ratio decreases sharply and directly at the rim (a few µm) the ratio development of the two elements is reversed again (Figure 5B).

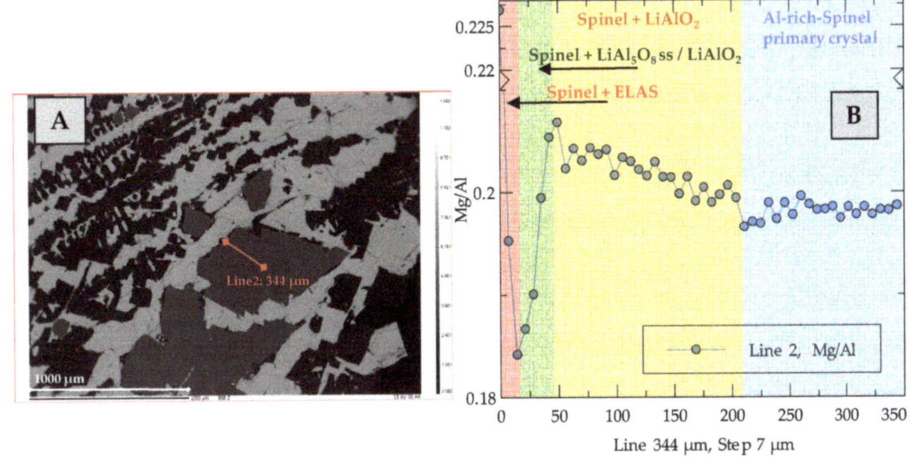

Figure 5. (**A**): Electron micrograph of a spinel crystal (medium grey), partly with a thin coat of LiAl (dark grey sections) surrounded by GCAS (light grey sections). Black, pores or preparation damage. (**B**) Development of the Mg/Al ratio from the center to the rim of a spinel grain along the red line in Figure 5A.

4.3.3. Ca-Alumosilicate (GCAS/MCAS)

The matrix component of the melt experiments (e.g., Figure 3 or Figure 4A light grey sections) can be generally expressed as $X_2YZ_2O_7$, where X can be Na^+ and Ca^{2+}; Y can be Al^{3+}, Mg^{2+} and Fe^{2+}; and Z can be Al^{3+} and Si^{4+}. The coordination of X is 8, and Y and Z are tetrahedral [23]. This Ca-alumosilicate compound generally is known as melilite. The investigated material comprises two types of Ca-alumosilicate:

- GCAS: High Al, low Si, very low Mg and virtually no Na
- MCAS: Low Al, high Si, ~3 wt.% Mg and 0.7–2.3 wt.% Na/Li is plausible

These two types are difficult to distinguish in the BSE(Z)-micrograph (Figure 4A) because of the almost same light grey shade (very similar mean atomic number). Because Na is not part of the initial materials (impurity), the concentration of the MCAS compound can be considered rather low and represents the eutectic residual melt. Nevertheless, this compound is interesting to assess a potential Li incorporation into the matrix of Ca-alumosilicate. In theory, Li^+ can be present in 4 or 8 coordination, whereas the ionic radius is very similar to Mg (4-coordination) or Na (8-coordination) (e.g., the ionic radii are published by Shannon (1976) [24]). The MCAS possesses a lower total sum of the measured concentrations and excess Si when calculating the chemical formula of the MCAS using the general melilite based on seven oxygen atoms.

Table 6a depicts how a calculation of virtual Ca-alumosilicate (CAS) can be conducted using five virtual compounds, namely $Li_2Si_3O_7$, $Na_2Si_3O_7$, $Ca_2Al_2SiO_7$ $Ca_2MgSi_2O_7$ and $Ca_3Si_2O_7$, assuming (limited) solid solution between those compounds. The Li value resulting from the multiplications was used to calculate a chemical formula of the analyzed MCAS. According to this calculation, the $Li_2Si_3O_7$ makes up about 1 wt.% of the total composition (see Table 6b).

Table 6. (a) Multiplication Factors (Mult.) for Calculation of an Optimized GCAS and MCAS. (b) Average Composition of the GCAS and MCAS Ca-Alumosilicate Solid Solution in Single Point Analysis, Compared with the Optimized Compounds Calculated with the Factors of Table 6a. Opt., Calculated Ideal Composition; Meas., Measured Average; (Calc.), Calculated Values (Li, O); %StdDev, Percentage Standard Deviation of the Measured Points (Repeats, $n = 7$ (GCAS), $n = 6$ (MCAS)).

wt.%	Virtual Compounds				
	$Li_2Si_3O_7$	$Na_2Si_3O_7$	$Ca_2AL_2SiO_7$	$Ca_2MgSi_2O_7$	$Ca_3Si_2O_7$
Al	0.0	0.0	19.7	0.0	0.0
Mg	0.0	0.0	0.0	8.9	0.0
Ti	0.0	0.0	0.0	0.0	0.0
Mn	0.0	0.0	0.0	0.0	0.0
Fe	0.0	0.0	0.0	0.0	0.0
Ca	0.0	0.0	29.2	29.4	41.7
K	0.0	0.0	0.0	0.0	0.0
Si	40.1	34.8	10.2	20.6	19.5
Na	0.0	19.0	0.0	0.0	0.0
O (Calc.)	53.3	46.2	40.8	41.1	38.8
Li (Calc.)	6.6	0.0	0.0	0.0	0.0
Multiplicator					
GCAS	0.00	0.00	0.98	0.03	0.00
MCAS	0.092	0.084	0.40	0.33	0.093
Sum	100	100	100	100	100

(a)

wt.%	GCAS (Opt.)	GCAS (meas.)	GCAS %StDev.	MCAS (Opt.)	MCAS Meas.	MCAS %StDev.
Al	19.2	19.2	2.8	7.87	7.87	10.7
Mg	0.3	0.3	46.0	2.94	2.94	5.2
Ti	0.0	0.0	n. a.	0.00	0.01	n. a.
Mn	0.0	0.0	n. a.	0.00	0.02	n. a.
Fe	0.0	0.1	n. a.	0.00	0.07	n. a.
Ca	29.4	29.4	0.4	25.25	25.25	4.0
K	0.0	0.0	n. a.	0.00	0.02	n. a.
Si	10.6	10.2	4.4	19.29	19.29	3.5
Na	0.0	0.0	n. a.	1.59	1.59	23.5
O (Calc.)	41.1	40.6	n. a.	42.25	42.29	n. a.
Li (Calc.)	0.0	0.0	n. a.	0.61	0.61	n. a.
Sum	100.6	99.7		99.8	100.0	

(b)

The calculated formulas of the GCAS and the MCAS are:

$$GCAS: \quad Ca_{2.02}(Al_{1.96}Mg_{0.03})(Al_{1.96}Si)O_7$$

$$MCAS: \quad (Na_{0.18}Ca_{1.67}Li_{0.15})(Al_{0.52}Mg_{0.32}Li_{0.08})(Al_{0.17}Si_{1.82})O_7$$

4.4. Comparison of Experimental Findings with Thermodynamically Modeled Subsystems

Based on the respective phases of interest, the relevant subsystems are $MgO\text{-}Al_2O_3$ and $Li_2O\text{-}Al_2O_3\text{-}SiO_2$. The modeled phase diagrams are presented in Figures 6 and 7, respectively. In Figure 6, a comparison between the modeled phase equilibria and the experimental data is made. In Figure 6, the composition of the initial melt and the composition of different spinel grains from two line scans and several single spot measurements, analyzed experimentally at room temperature (RT), are presented in an overlay with the thermodynamic phase equilibrium data for the subsystem $MgO\text{-}Al_2O_3$. The composition of the initial melt is the starting point of the spinel crystallization. It can be seen that all measured spinel grains show a significant higher Mg concentration compared to the initial Mg concentration in the melt.

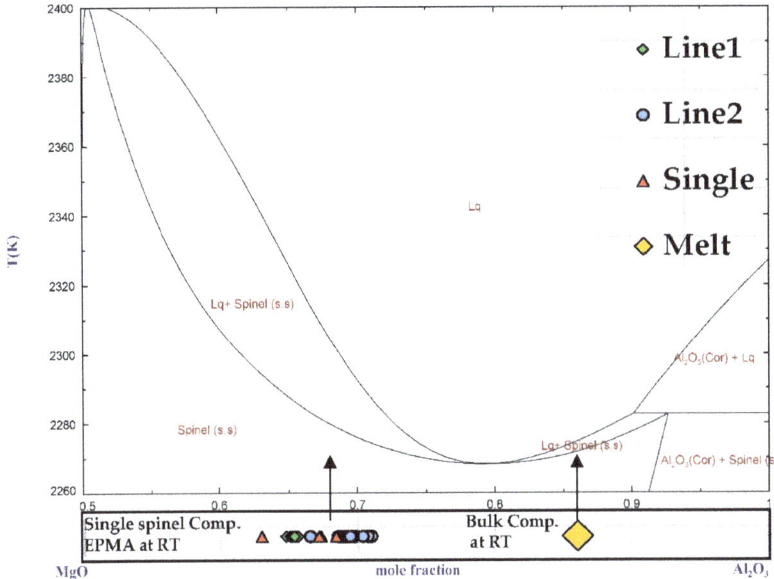

Figure 6. The calculated MgO-Al$_2$O$_3$ phase diagram at 1 atm total based on [22]. Lq, Liquid; spinel (s.s), spinel solid solution; Al$_2$O$_3$(Cor), corundum. In this diagram, the composition of the initial melt and the composition of different spinel grains from two line scans and several single spot measurements, analyzed at room temperature (RT), are presented. The composition of the initial melt is the starting point of the spinel crystallization.

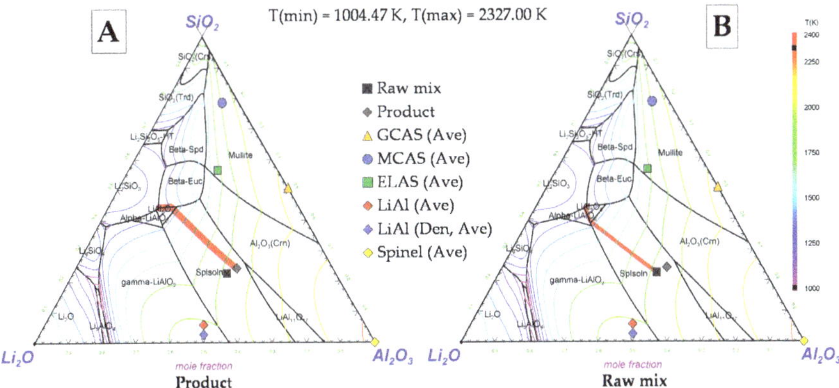

Figure 7. Calculated Li$_2$O-Al$_2$O$_3$-SiO$_2$ liquidus projection at 1 atm total pressure based on [22] is shown. Red line, equilibrium solidification paths starting at the initial point of the "product" (**A**) and the "raw mix" (**B**). The initial points represent the respective component concentrations in the liquid phase. Isothermal lines are drawn in Kelvin at every 100 K. In this diagram, the average compositions of the single compounds, analyzed with EPMA at room temperature (see Tables 4, 5 and 6b), and the bulk chemistry of the "raw mix" and the "product" are presented.

Figure 7 shows the thermodynamic calculated Li$_2$O-Al$_2$O$_3$-SiO$_2$ subsystem. The equilibrium solidification paths for the "raw mix" (Figure 7B) and the "product" (Figure 7A) composition are calculated and presented in the respective ternary phase diagram. Additionally, the average compositions of the single compounds, analyzed with EPMA at room temperature (see Tables 4, 5 and 6b), and the bulk chemistry of the "raw mix" and "product" are visualized in an overlay with the modeling results in Figure 7.

The "raw mix" and the "product" composition is in the spinel solid solution area, which is concluded by the thermodynamic modeling results based on the subsystem. Hence, the thermodynamic modeled solidification predicts spinel as the primary crystallizing phase (see Figure 8). After decreasing the temperature continuously under assumed equilibrium conditions, the solidifications of different phases are shown in Figure 8, for the "product" (Figure 8A) and "raw mix" (Figure 8B) initial concentrations, respectively. The thermodynamic prediction of the subsystem solidification shows that spinel as primary crystal is formed in solid solution with high temperature $LiAl_5O_8$ for both initial compositions. With progressing solidification, low temperature $LiAl_5O_8$ is also crystallizing out of solution. This finding holds true for both initial compositions. With increasing solidification progress, $LiAlO_2$ and Li_2SiO_3 are formed with a very low amount of eucryptite, for the "raw mix" configuration, while, for the "product" composition (Figure 8A), eucryptite is formed in a higher amount without any $LiAlO_2$.

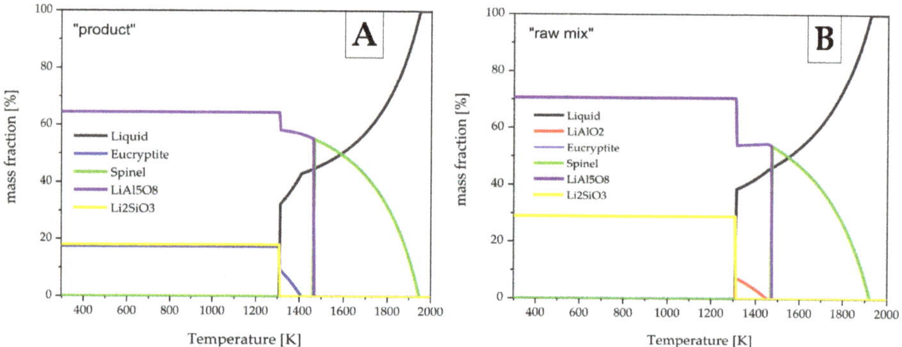

Figure 8. Calculated equilibrium solidification curves of the "product" (**A**) and the "raw mix" (**B**) for Li_2O-Al_2O_3-SiO_2 system. Calculated point interval is 5 K.

5. Discussion

The experimental investigation of solidified melt in connection with thermodynamic modeling of chemical reactions and solidification is an important tool to investigate how a slag system behaves and how it can be engineered. These investigations and the obtained results can serve as starting point for understanding efficient design of experiments to generate the desired phases. With a combination of thermodynamic calculation and mineralogical investigation, the probability that the artificial slag contains the desired phases can be maximized. Therefore, one purpose of the experiments carried out in this project was a first survey of the mineral compounds and the morphology of a solidified melt with the basic components Li_2O, MgO, Al_2O_3, SiO_2 and CaO with a melt composition in the primary crystallization field of spinel in the subsystem Li_2O, Al_2O_3 and SiO_2. Another purpose was to investigate the influence of the Mg content on the ratio of the mineral compounds. The results of these experiments are also intended to serve as a basis for further thermodynamic modeling. Additionally, the applicability of the combination of PXRD and EPMA to this research topic was assessed. This includes the calculation of the lithium containing mineral compounds on basis of the EPMA result without access to measured lithium concentrations. In the following, the different identified phases are discussed:

5.1. Spinel-Like Oxides

The experiments show idiomorphic phenocrysts of spinel as the first crystallizing compound with decreasing temperature. The spinel crystals are surrounded by massive hypidiomorphic crystallites of LiAl and melilite-like alumosilicate (GCAS/MCAS). Additionally, LiAl forms dendritic elongated structures of hypidiomorphic crystallites. The changes in the chemistry of a single spinel crystal (Figure 5B) can help to explain a part of the crystallization curve of the melt. This is also used to validate thermodynamic model predictions of the three-component subsystem. Starting in the primary crystallization field of spinel, an Al-rich $Mg_{1-(3/2y)}Al_{2+y}O_4$ starts to form. These crystals are in equilibrium with a corresponding melt (Figure 5B (blue area) and Figure 6). The EPMA reveals that, compared with the Mg/Al ratio of the melted material, all measured spinel grains are enriched in Mg. This observation shows the complex spinel behavior, which cannot be explained with the simple binary phase diagram MgO-Al_2O_3. Nevertheless, the measured Mg/Al ratio increases (Figure 5B, yellow area). This can be explained with the composition of the melt reaching the phase boundary between spinel and LiAl. Through scavenging of Al from the melt during formation of LiAl, the Mg concentration in the melt increases and therefore the spinel–melt equilibrium changes. At a later stage of the crystallization process, the Al concentration in the crystal increases again, an indication that now a solid solution between $Mg_{1-(3/2y)}Al_{2+y}O_4$ and Mg-free $LiAl_5O_8$ forms (Figure 5B, green area). At the end of the crystallization, the Mg concentration rises again (Figure 5B, red area). This is an indication that the crystallization leaves the crystallization path between $LiAl_5O_8$ and spinel in direction of the crystallization path between eucryptite (or ELAS) and spinel. Therefore, the crystallization of the spinel would no longer include the aluminum-rich $LiAl_5O_8$ and the relative Mg concentration of the crystallizing spinel would increase, although a part of the Mg is incorporated into the ELAS. The crystallization path concluded by experimental observations of the developing spinel composition is on principal in good correlation with the thermodynamically predicted solidification phases in the early stages (Figures 7 and 8). However, for lower temperatures, the solidification predictions deviate from the experimental findings, which is due to non-equilibrium cooling conditions and hence phase generation. The modeled results show that small deviations in the initial concentration in the spinel solid solution field can result in strong deviations regarding appearing solid phases and solidification path behavior.

The PXRD patterns of the investigated melts with increasing Mg concentration show a displacement of the spinel main peak. The angular position of the main peak of these spinel variations is between the simple $MgO \times Al_2O_3$ compound and a pattern of an Al-rich spinel with the formula $Mg_{0.39}Al_{2.41}O_4$ and weakly correlates with the Mg concentration as:

$$Mg_y = 20.198 \times d_{311} - 48.247 \tag{1}$$

5.2. LiAl and ELAS

The formation of dendrites is an indication for rapid crystallization of LiAl in a small temperature interval from a supercooled melt and/or (macro)segregation (for macro segregation, see, e.g., Ahmadein et al. [25]). Due to the rather long cooling cycle (two days, Figure 1), undercooling may be improbable but cannot completely be excluded. Nevertheless, it is plausible that a segregation of an Al-Li-rich melt occurs from which the first generation of LiAl crystals forms. The Si concentrations of the dendritic LiAl is lower than in the massive LiAl, indicating a different origin, thus a different melt as well (compare LiAl in Tables 4 and 5). A few parts of the massive LiAl contain small lath-shaped grains of ELAS. This compound can be derived from eucryptite and can contain up to 18 wt.% $Mg_{0.5}AlO_2$. These grains are an indication of segregation of Si- and Mg-rich phases (melt) from the Al-rich LiAl-melt, as described above. Interestingly, the representing point of this calculated compound is not located in the primary crystallization field of pure $LiAlSiO_4$. This is due to the lower calculated Li content because of the Li-free $Mg_{0.5}AlO_2$ compound.

5.3. Ca-Alumosilicate

The matrix of the material consists of slightly hypidiomorphic grains of Ca-alumosilicate. The morphology of the crystals indicates that the formation starts together or slightly after the beginning of the crystallization of LiAl, which itself often shows hypidiomorphic growth. The chemical composition of these Ca-alumosilicates starts with nearly ideal gehlenite (GCAS) with minute amounts of impurities such as Mg. The other type of Ca-alumosilicate (MCAS) incorporates higher amounts of impurities such as Na and Mg (Table 6b). Because of the presence of an alkaline element such as Na, the latter compound seems to represent the end of the crystallization, i.e., the residual eutectic melt. Interestingly, this compound delivers a total of distinctively less than 100 wt.% (element concentrations calculated as simple oxide compounds) and possesses an excess of Si after calculation of the melilite formula. This is an indication that another silicious component is present in the crystal structure. Because the sample contains no free SiO_2 (like quartz) and the analysis shows no additional element for calculation of a silicious component, incorporation of Li into the crystal or glassy structure as shown in Table 6b is plausible. After incorporation of a virtual compound $Li_2Si_3O_7$, a formula can be calculated indicating a consistent crystal-like chemistry or a stoichiometric glass.

The mineralogical characterization of a melt as presented above provides a basis for refining the thermodynamic model, showing the real assemblage of components and the real chemical composition of the compounds/phases. An example would be ELAS. The eucryptite compound, used for the thermodynamic calculation, is ideal $LiAlSiO_4$. EPMA reveals that the real eucryptite-like alumosilicate (ELAS) can be expressed with $(Li_{0.96}Mg_{0.24})(Al)(Al_{0.45}Si_{1.55})O_6$. This compound contains Mg, which has to be taken into account when using this compound to predict a crystallization curve. The same is valid for the lithium aluminate compound (LiAl, $Li_{1-x}(Al_{1-x}Si_x)O_2$) that contains Si. Another important property is the inherent potential kinetic inhibition of the phase reactions in the system of interest. The morphology of the slag including structure and habitus corresponds to the crystallizing reactions during the cooling process. Additionally, the total chemistry and the spatial resolved development of element ratios can be used to explain the solidification process.

6. Conclusions and Outlook

In this work, an experimental investigation of a Li_2O-MgO-Al_2O_3-SiO_2-CaO system was carried out in combination with thermodynamic modeling of relevant subsystems. Based on bulk chemistry analysis, PXRD and EPMA, the crystallization paths of various phases were reconstructed and explained. It was shown that spinel is always the primary crystallizate. Furthermore, depending on minute variations in the chemistry of the melt, the result of the thermodynamically predicted further phase development can be substantially different. In this case, comparing the solidification of the raw material and the product the unpredictable loss of Li during the melt experiment seems to offer the possibility of a complete suppression of the $LiAlO_2$ formation in favor of $Mg_{1-(3/2y)}Al_{2+y}O_4$/$LiAl_5O_8$ solid solution, although this was not observed in the experiments. The eucryptite and Li-silicate compound are the ends of the solidification in both scenarios. Nevertheless, a knowledge and/or control of all reaction parameters such as partial pressures of all elements (particularly, Li here) and compounds, grain size distribution, morphology and chemistry of the raw material is crucial to develop an efficient and reproducible slag modification process. The solidification route of the system could be qualitatively predicted by thermodynamic modeling of the Li_2O-Al_2O_3-SiO_2 subsystem with the result that minute variations of the initial chemistry can lead to different solidification paths.

Additionally, a relative Mg enrichment of spinel grains could be observed experimentally. Furthermore, the development of the composition in single spinel grains during spinel grains give an indication of the existing solid solution $Mg_{1-(3/2y)}Al_{2+y}O_4$/$LiAl_5O_8$, which was only predicted and not verified in the past.

The results presented in this article show that Li cannot be incorporated into a single early crystallizing compound in an easy way. The investigations show that Li is present in LiAl, ELAS and with higher uncertainty in spinel (as solid solution $Mg_{1-(3/2y)}Al_{2+y}O_4$/$LiAl_5O_8$) and MCAS. To modify this complex multi-component system (oxides of Li, Mg, Al, Si and Ca) to gain desired mineral compounds requires, besides experimental work (melt experiments, component printing and combinatorial thin film deposition), new thermodynamic modeling strategies even for higher component systems, especially with a good quantitative predictability for the phase fractions.

Furthermore, future research work will concentrate on the development of phase separation processes, predominantly by flotation for the main identified Li-bearing phases described in this paper (basic research on the way) and on the extension of component mixtures in the slag building process, advancing step by step into slag systems expected in melting processes of actual and future battery systems.

Author Contributions: T.S. conceived the paper. T.S., H.L. and M.F. conducted the literature review. All experiments were designed and performed by H.Q. and D.G. The chemical bulk analysis was executed by the analysis laboratory of the institute. The phase analysis (PXRD and EPMA) and the mineralogical investigation were conducted by T.S. Thermodynamic modeling was conducted by H.L. and M.F. Interpretation and discussion were conducted by all authors. All authors have read and agreed to the published version of the manuscript.

Funding: This research was funded by the Clausthal University of Technology in the course of a joint research project "Engineering and processing of Artificial Minerals for an advanced circular economy approach for finely dispersed critical elements" (EnAM).

Acknowledgments: We acknowledge support by Open Access Publishing Fund of Clausthal University of Technology.

Conflicts of Interest: The authors declare no conflict of interest. The funders had no role in the design of the study, in the collection, analyses, or interpretation of data, in the writing of the manuscript, or in the decision to publish the results.

References

1. Haas, A.; Elwert, T.; Goldmann, D.; Schirmer, T. Challenges and research needs in flotation of synthetic metal phases, EMPRC 06/25-06/26/2018. In *Proceedings of the European Mineral Processing & Recycling Congress*; GDMB Verlag GmbH, Ed.; GDMB: Clausthal-Zellerfeld, Germany, 2018; ISBN 978-3-940276-84-1.
2. Elwert, T.; Goldmann, D.; Schirmer, T.; Strauß, K. *Recycling von Li-Ionen-Traktionsbatterien*; Recycling und Rohstoffe: Neuruppin, Germany, 2012; pp. 679–690.
3. Elwert, T.; Strauss, K.; Schirmer, T.; Goldmann, D. Phase composition of high lithium slags from the recycling of lithium ion batteries. *World Metall. ERZMETALL* **2012**, *65*, 163–171.
4. Konar, B.; Dong-Geun, K.; Jung, I.-H. Coupled Phase Diagram Experiments and Thermodynamic Optimization of the Binary Li_2O-MgO and Li_2O-CaO Systems and Ternary Li_2O-MgO-CaO System. *Ceram. Int.* **2017**, *43*, 13055–13062. [CrossRef]
5. Konar, B.; Van Ende, M.-A.; Jung, I.-H. Critical Evaluation and Thermodynamic Optimization of the Li_2O-Al_2O_3 and Li_2O-MgO-Al_2O_3 Systems. *Metall. Mater. Trans. B* **2018**, *49*, 2917–2944. [CrossRef]
6. Aoyama, M.; Yusuke, A.; Koji, I.; Sawao, H.; Shinobu, H.; Iwamoto, Y. Synthesis and Characterization of Lithium Aluminate Red Phosphors. *J. Lumin.* **2013**, *135*, 211–215. [CrossRef]
7. Kale, M.A.; Joshi, C.P.; Moharil, S.V. Combustion Synthesis of Some Compounds in the Li_2O-Al_2O_3 System. *Int. J. Self-Propagating High-Temp. Synth.* **2012**, *21*, 19–24. [CrossRef]
8. Datta, R.K.; Rustum, R. Phase Transitions in $LiAl_5O_8$. *J. Am. Ceram. Soc.* **1963**, *46*, 388–390. [CrossRef]
9. Li, X.; Kobayashi, T.; Zhang, F.; Kimoto, K.; Sekine, T. A New High-Pressure Phase of $LiAlO_2$. *J. Solid State Chem.* **2004**, *177*, 1939–1943. [CrossRef]
10. Lewin, E.M.; Robbins, C.R.; McMurdie, H.F.; Reser, M.K. *Phase Diagrams for Ceramists. Volume 1: Figures 1-2066. 5*; American Ceramic Society: Columbus, OH, USA, 1985; ISBN 10: 0916094049.
11. Konar, B.; Van Ende, M.-A.; Jung, I.-H. Critical Evaluation and Thermodynamic Optimization of the Li-O, and Li_2O-SiO_2 Systems. *J. Eur. Ceram. Soc.* **2017**, *37*, 2189–2207. [CrossRef]
12. Konar, B.; Dong-Geun, K.; Jung, I.-H. Critical Thermodynamic Optimization of the Li_2O-Al_2O_3-SiO_2 System and Its Application for the Thermodynamic Analysis of the Glass-Ceramics. *J. Eur. Ceram. Soc.* **2018**, *38*, 3881–3904. [CrossRef]
13. Iwatsuki, M.; Norio, T.; Fukasawa, T. Analysis of a Magnesium-Containing: β-Eucryptite Solid Solution by the Lattice Constant Method and Its Application. *Bull. Chem. Soc. Jpn.* **1975**, *48*, 1217–1221. [CrossRef]
14. Downs, R.T.; Hall-Wallace, M. The American Mineralogist Crystal Structure Database. *Am. Mineral.* **2003**, *88*, 247–250.

15. Lafuente, B.; Downs, R.T.; Yang, H.; Stone, N. The power of databases: The RRUFF project. In *Highlights in Mineralogical Crystallography*; Armbruster, T., Danisi, R.M., Eds.; W. De Gruyter: Berlin, Germany, 2016; pp. 1–30.

16. Jercinovic, M.J.; Williams, M.L.; Allaz, J.; Donovan, J.J. Trace analysis in EPMA. *IOP Conf. Ser. Mater. Sci. Eng.* **2012**, *32*, 012012. [CrossRef]

17. Merlet, C. Quantitative Electron Probe Microanalysis: New Accurate Φ (ρz) Description. In *Electron Microbeam Analysis*; Boekestein, A., Pavićević, M.K., Eds.; Springer: Vienna, Austria, 1992; Volume 12, pp. 107–115. [CrossRef]

18. Pelton, A.D.; Degterov, S.A.; Eriksson, G.; Robelin, C.; Dessureault, Y. The modified quasichemical model I—Binary solutions. *Metall. Mater. Trans. B* **2000**, *31*, 651–659. [CrossRef]

19. Pelton, A.D.; Chartrand, P. The modified quasi-chemical model: Part II. Multicomponent solutions. *Metall. Mater. Trans. A* **2001**, *32*, 1355–1360. [CrossRef]

20. Pelton, A.D. A general "geometric" thermodynamic model for multicomponent solutions. *Calphad* **2001**, *25*, 319–328. [CrossRef]

21. Barry, T.I.; Dinsdale, T.A.T.; Gisby, J.A.; Hallstedt, B.; Hillert, M.; Jansson, B.; Taylor, J.R. The compound energy model for ionic solutions with applications to solid oxides. *J. Phase Equilibria* **1992**, *13*, 459–475. [CrossRef]

22. Bale, C.W.; Bélisle, E.; Chartrand, P.; Decterov, S.A.; Eriksson, G.; Gheribi, A.E.; Hack, K.; Jung, I.H.; Kang, Y.B.; Melançon, J.; et al. FactSage thermochemical software and databases, 2010–2016. *Calphad* **2016**, *54*, 35–53. [CrossRef]

23. Okrusch, M.; Matthes, S. *Mineralogie: Eine Einführung in die Spezielle Mineralogie, Petrologie und Lagerstättenkunde*, 9th ed.; Springer Spektrum: Berlin, Germany, 2014.

24. Shannon, R.D. Revised effective ionic radii and systematic studies of interatomic distances in halides and chalcogenides. *Acta Crystallogr. Sect. A* **1976**, *32*, 751–767. [CrossRef]

25. Ahmadein, M.; Wu, M.; Ludwig, A. Analysis of Macrosegregation Formation and Columnar-to-Equiaxed Transition during Solidification of Al-4wt.% Cu Ingot Using a 5-Phase Model. *J. Cryst. Growth* **2015**, *417*, 65–74. [CrossRef] [PubMed]

Publisher's Note: MDPI stays neutral with regard to jurisdictional claims in published maps and institutional affiliations.

Article

Optimization of Manganese Recovery from a Solution Based on Lithium-Ion Batteries by Solvent Extraction with D2EHPA

Nathália Vieceli [1],*, Niclas Reinhardt [2], Christian Ekberg [1] and Martina Petranikova [1]

[1] Department of Chemistry and Chemical Engineering, Industrial Materials Recycling and Nuclear Chemistry, Chalmers University of Technology, SE-41296 Gothenburg, Sweden; che@chalmers.se (C.E.); martina.petranikova@chalmers.se (M.P.)
[2] MEAB Metallextraktion AB, Datavägen 51, S-43632 Askim, Sweden; niclas@meab-mx.se
* Correspondence: nathalia.vieceli@chalmers.se

Abstract: Manganese is a critical metal for the steelmaking industry, and it is expected that its world demand will be increasingly affected by the growing market of lithium-ion batteries. In addition to the increasing importance of manganese, its recycling is mainly determined by trends in the recycling of iron and steel. The recovery of manganese by solvent extraction has been widely investigated; however, the interaction of different variables affecting the process is generally not assessed. In this study, the solvent extraction of manganese from a solution based on lithium-ion batteries was modeled and optimized using factorial designs of experiments and the response surface methodology. Under optimized conditions (O:A of 1.25:1, pH 3.25, and 0.5 M bis(2-ethylhexyl) phosphoric acid (D2EHPA)), extractions above 70% Mn were reached in a single extraction stage with a coextraction of less than 5% Co, which was mostly removed in two scrubbing stages. A stripping product containing around 23 g/L Mn and around 0.3 g/L Co can be obtained under optimized conditions (O:A of 8:1, 1 M H_2SO_4 and around 13 min of contact time) in one stripping stage.

Keywords: lithium-ion battery; battery recycling; manganese recovery; solvent extraction; D2EHPA; factorial design of experiments

Citation: Vieceli, N.; Reinhardt, N.; Ekberg, C.; Petranikova, M. Optimization of Manganese Recovery from a Solution Based on Lithium-Ion Batteries by Solvent Extraction with D2EHPA. *Metals* **2021**, *11*, 54. https://doi.org/10.3390/met11010054

Received: 1 December 2020
Accepted: 24 December 2020
Published: 29 December 2020

Publisher's Note: MDPI stays neutral with regard to jurisdictional claims in published maps and institutional affiliations.

1. Introduction

Manganese is one of the most abundant metals in the Earth's crust; however, manganese is highly dispersed (low-grade), and minerals are widely distributed. The identified manganese resources are concentrated in a few countries—the main manganese mining areas are in China, South Africa, Australia, and Gabon [1–3].

The main end use of manganese is in the steel industry, which accounts for 90% of the world´s manganese demand. Manganese is also widely used in ironmaking and alloys with aluminum, magnesium, and copper [3–6]. Non-metallurgical applications account for only 5–10% of the manganese consumption, which is used in electrical systems, in the chemical industry, in the ceramic and glass production, and in the agricultural sector [7]. In electrical systems, manganese dioxide is used for cathodic depolarizer in dry cells, alkaline batteries, and lithium-ion batteries (LIBs) [4].

Natural manganese dioxide is used in dry cells, while high-grade synthetic manganese dioxide is produced chemically or by electrolysis to be used in alkaline batteries and LIBs [4]. Lithium manganese spinels (such as $LiMn_2O_4$) and layered lithium–nickel–manganese–cobalt (NMC) oxide systems have an important role in the development of advanced rechargeable lithium-ion batteries, with cost and environmental advantages [8]. Thus, nowadays, most automakers and some electronics makers use some version of NMC system in their LIBs [9].

In this context, the United States of America Department of Defense has recently classified manganese as one of the most critical mineral commodities for the United States because it is essential for important industrial sectors, has no substitutes, and has a potential for supply disruptions, since the country is strongly dependent on imports [10]. Additionally, the United States included electrolytic manganese metal in the National Defense Stockpile in 2019 as a critical material for defense purposes [2].

Although it is expected that steel will continue leading the manganese demand, the consumption of manganese in batteries applications is projected to grow fast in the next decade, boosted by the rapid growth in the lithium-ion battery market, which is expected to increase from $35 billion USD in 2020 and reach $71 billion USD in 2025 [11,12]. Thus, electrolytic manganese dioxide (EMD) for the battery industry is expected to be the fastest-growing segment of the manganese market [13], increasing the manganese production along with the global demand for batteries [14].

EMD is generally produced from high-grade manganese ores [15], and in general, converting manganese ores to EMD involves a high-temperature pyrometallurgical process, which has some drawbacks such as environmental impacts, high-energy consumption, and high costs. Furthermore, because the roasting process decreases the oxide content in the ore, EMD producers face competition from chemical and steel industry buyers of high-grade manganese ores [16]. In this context, the recovery of manganese from spent LIBs can help decrease supply risks and impacts linked to the primary production of manganese. However, although there is an increasing importance of manganese, its recycling is mainly determined by trends in the recycling of iron and steel, and in general, materials are not recycled specifically due to their manganese content [2,17,18]. Moreover, when it comes to LIBs recycling, the presence of manganese in the leaching solutions has been linked to a decrease in the selective separation of cobalt and nickel, and for this reason, manganese should be previously recovered [19,20].

The recovery of manganese from primary and secondary resources by solvent extraction has been investigated by several authors [14,20–26]. Table 1 (on the next page) summarizes the optimal extraction conditions described in some studies focused on the extraction of manganese from different feed solutions, including from leach solutions from spent LIBs. It is possible to highlight that bis(2-ethylhexyl) phosphoric acid (D2HEPA) is the most widely used extractant to recover Mn from liquors from LIBs as well as from other solutions.

Although several studies on the recovery of manganese by solvent extraction have been published, the effect of different variables affecting the process is generally approached using one-factor-at-a-time, which does not allow identifying interaction effects among them. In this context, the main goal of this study was to optimize the solvent extraction of manganese using the factorial design of experiments and response surface methodologies to assess and model the effects of the variables affecting the process. The optimization of the recovery of manganese was studied using a synthetic solution based on an acid leach from spent LIBs. The results can support further investigations focused on the recovery of manganese from spent LIBs, which can be considered an important secondary resource of a critical material for many important industrial sectors.

Table 1. Summary of conditions for the manganese solvent extraction from published studies (corresponding to the best conditions reported).

Extractant	Saponification	Modifier	O:A	Optimum pH	Temperature (°C)	Contact Time (min)	Feed	Initial Composition (g/L)				%E (Mn)	Reference
								Mn	Co	Ni	Li		
0.4 M D2EHPA	-	-	2:1	3.2	25	15	Leach solution produced from spent LIBs (acid leaching with H_2SO_4 and H_2O_2)	3.66	19.33	5.19	3.58	-	[20]
15% D2EHPA	60% (with 0.5 M ammonia)	5% TBP	1:1	2.25	25	5	Leaching liquor of spent LIBs	5.91	24.79	6.24	6.68	99.9	[26]
4 M (D2EHPA/Mn molar)	65% (with NaOH 5 M)	10% TBP	1:1	3.8	room	10	Electrodic LIB powder pre-leached with H_2SO_4	4.6	21.8	2.7	3.2	~90	[25]
0.05 M NaD2EHPA (best results)	-	5% TBP	9:8	2.7	30	5	Stock solution with Mn and Co (0.01 M)	0.01 M	0.01 M	-	-	99.94%	[27]
15% Cobalt loaded D2EHPA	70–75% (with NaOH 10 M)	5% TBP	1:1	3.2	25	5	Sulfuric acid leaching liquor of mixed types of cathode materials (real sample) waste cathode materials	6.31	6.45	6.89	1.6	99%	[24]
20% PC88A/25% Versatic 10	-	-	1:1	4.5	room	5	Leaching solution from spent LIBs	11.7	11.4	12.2	5.3	99.5%	[23]
25% Cobalt loaded D2EHPA	-	1-decanol	1:1	3.5	25	5	Cobalt electrolyte solution	0.8	55.7	-	-	100% (70% in one stage)	[28]
10% D2EHPA	-	5%TBP	1:1	3.5	40	10	Synthetic laterite solution containing Ni, Co, Mn, Mg, Zn, and Cu	2	0.3	3	-	99%	[29]

Table 1. *Cont.*

Extractant	Saponification	Modifier	O:A	Optimum pH	Temperature (°C)	Contact Time (min)	Feed	Initial Composition (g/L)				%E (Mn)	Reference
								Mn	Co	Ni	Li		
30% D2EHPA	20% (with NaOH 10 M)	5% TBP	1:1	2.6–2.7	room	15	Leaching solution from spent LIBs, treated with H_2SO_4 and H_2O_2	2	0.3	3	-	Removal of Mn and Cu	[22]
40% D2EHPA	-	-	1:1	3.5	room	10	Leaching acid solution from cathode material	9.18	11.32	11.51	1.76	~100	[21]
0.4 M D2EHPA	-	-	2:1	3.2	25	15	Leach solution produced from spent LIBs (acid leaching with H_2SO_4 and H_2O_2)	3.66	19.33	5.19	3.58	-	[20]
20% D2EHPA (0.6 M)	70–75% (with NaOH 10 M)	-	1:2	4–5	25	5	Co, Ni, and Li were removed by precipitation	5.27	5.84	4.93	1.25	97%	[30]
D2EHPA	-	-	-	2.5–3.5	n.i.	n.i.	Leach liquor from LIBs	5–30	5 to 45	5 to 30	1 to 10	100%	[31]
25% D2EHPA (Cyanex 272 was also tested)	-	-	1:1.5	2.7	5 and 25	n.i.	Synthetic sulfuric acid solutions (Ca, Mn, Na, and Mg)	0.58–5.3	-	-	-	65%	[32]
D2EHPA	-	-	1:1–1:5	2.2–2.3	40	continuous	Kakanda tailings (Cu and Co recovery in RDC)	1.3	3	-	-	70–90%	[33]
20% D2EHPA	-	-	1:1	2.2–2.3	n.i.	continuous	Cobalt bearing feed from a cobalt refinery in South Africa. Fe and Cu were first precipitated	0.1	5.5	-	-	100%	[34]

n.i.: not informed.

2. Materials and Methods

Bis(2-ethylhexyl) phosphoric acid (D2EHPA, 97%, Sigma Aldrich, Germany) was used as solvent extraction reagent as it was supplied, without any additional purification. Isopar L (Exxon Mobil, USA) was used as diluent. A synthetic solution was prepared based on the chemical composition of an original solution obtained through the acid leaching of spent lithium-ion batteries with sulfuric acid, which was investigated in detail in previous work (unpublished results). The synthetic solution was prepared using sulfates ($NiSO_4.6H_2O$, $CoSO_4.7H_2O$, $MnSO_4.H_2O$, Li_2SO_4, Sigma Aldrich, Germany) and Milli-Q water. Impurities typically present in acid leach solutions from LIBs such as Cu and Al were not included into the synthetic solution because they are generally removed using conventional purification processes, for example, cementation and purification, before the solvent extraction.

Preliminary extraction tests, scrubbing, and stripping tests were performed in glass vials (3.5 mL) using a shaking machine (IKA-Vibrax, Germany) operating with 1000 vibrations per min to promote the contact between phases. The experiments were performed at room temperature. Specific conditions used in the preliminary tests are reported in the Results section. The extraction and stripping of manganese and cobalt were optimized using factorial designs of experiments and response surfaces. These methodologies are explained in detail by Montgomery [35]. For the factorial design of experiments of the extraction phase, tests were carried out using plastic containers (50 mL), in which the stirrer from a mixer-settler device was coupled. The stirring speed was set at 1000 rpm, and the tests were also performed at room temperature.

The pH of the aqueous phase was measured using a pH meter (Metrohm 827 pH lab, Switzerland), and the electrode was regularly calibrated before and during the experimental procedures. The pH was adjusted whenever it was needed with 5 M or 10 M NaOH to minimize the dilution effect of the feed solution. Samples from the aqueous phase were taken 10 min after finishing the contact time at the established pH to obtain a complete separation of phases. Chemical analysis was performed by Inductively Coupled Plasma— Optical Emission Spectroscopy (ICP-OES, iCAP™ 6000 Series, USA) using samples from the aqueous phase, which were diluted in 0.5 M nitric acid. The extraction efficiency of metals was determined by Equation (1):

$$\%E = 100 * \frac{D_X}{D_X + \left(V_{aq}/V_{org}\right)} \tag{1}$$

where V_{aq} and V_{org} represent the volume of the aqueous phase and the volume of the organic phase, respectively, and D_X is the distribution ratio, which describes the ratio between the concentration of a certain metal (X) in the aqueous phase and in the organic phase and it can be determined by Equation (2). In some cases, the log D is used to assist the interpretation of results.

$$D_x = C_{X\ organic}/C_{X\ aqueous} \tag{2}$$

The separation factors (β) between two elements (X and Y) can be calculated using Equation (3), and it is determined by the division of the distribution ratio of each element, being normally greater than one. This equation was used to determine the separation factor of manganese in preference to other metals.

$$\beta = D_X/D_Y \tag{3}$$

Experimental Design

A full 2^k factorial design of experiments was used to fit a second-order linear regression model to the experimental results. To estimate the experimental uncertainty, four additional experiments were performed under the same conditions at the central level of the factors (n_C, central point). The effects of three factors (k = 3), each one with two levels (2^3 factorial design), on the process response (y, manganese extraction or cobalt extraction) were studied. The factors and levels were selected based on results from preliminary tests and on the literature review.

Experimental design of the extraction stage: The factors investigated in the design of experiments to model the extraction stage were equilibrium pH (x_1), organic to aqueous ratio, O:A (x_2) and molar concentration of D2EHPA (x_3). Each factor was varied in two levels.

Experimental design of the stripping stage: To model the stripping stage, the effect of the following three factors was evaluated: molar concentration of sulfuric acid (x_1), organic to aqueous ratio, O:A (x_2) and stripping time (x_3). Each factor was varied in two levels.

Axial points were included ($2k$ axial points) in both designs to estimate the quadratic terms of the models, setting up a central composite design. Tests were performed in random order. The distance of the axial points from the central point was $\alpha = 1$ (face-centered central composite design). The standard, high, and low levels of the factors are presented in Table 2.

Table 2. Factors considered in the factorial design of experiments of the extraction and stripping stages and respective levels.

Stage	Factors	Unit	Levels		
			Low (−1)	Standard (0)	High (+1)
Extraction	Equilibrium pH (x_1) *	dimensionless	2.5	3.25	4.0
	Organic to aqueous phase, O:A (x_2)	dimensionless	0.5	1.25	2
	Concentration of D2EHPA (x_3)	M	0.4	0.5	0.6
Stripping	Concentration of H_2SO_4 (x_1)	M	0.05	1.025	2
	Organic to aqueous phase, O:A (x_2)	dimensionless	1	4.5	8
	Stripping time (x_3)	min	2	13.5	25

* Equilibrium pH after a contact time of 10 min, with a maximum variation of ±0.05 from the value defined in the design.

The process response, y, was used to fit the coefficients of a linear second-order regression model, using the linear least squares method. Only statistically significant variables were considered in the models (p-value smaller than the significance level of 0.05). Analysis of variance (ANOVA) was used to assess the significance of the fitted model. The variance of the response accounted for the models was evaluated by the coefficient of determination (R^2), and the existence of pure quadratic curvature was determined by hypothesis testing. Response surfaces and contour plots were used to assist the optimization of the processes.

3. Results and Discussion

3.1. Preliminary Tests of Extraction

Preliminary tests were performed to determine the best conditions to be further investigated in the factorial design of experiments. The extraction of Mn, Ni, Co, and Li at different contact times can be observed in Figure 1. The mechanism of extraction of manganese using D2HEPA is very fast. The extraction of Mn was about 60% after only 5 min of contact time, and after 10 min, the extraction achieves the maximum values (approximately 70%). The coextraction of Co, Ni, and Li is slightly higher after 5 min of contact time, but it is still lower than 20%. At 10 min of contact time, the increase in the extraction of Mn resulted in a decrease of the coextraction of the other metals. The coextraction of Co, Ni, and Li after 10 min of contact time was around 11, 5, and 3%, respectively. This is in accordance with the results reported in the literature. Chen et al. [24] studied the extraction of manganese from the leaching liquor of spent LIBs using cobalt-loaded D2EHPA, and they reported that the equilibrium was achieved after only 3 min. Hossain et al. [28] also observed that the kinetics of the manganese extraction using Co-D2EHPA was fast, and the equilibrium was achieved in 5 min. Thus, low contact times are required for the extraction of manganese.

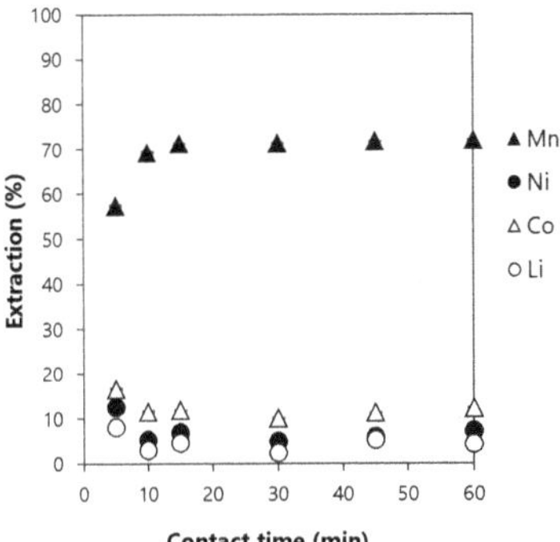

Figure 1. Extraction of metals at different leaching times. Conditions: O:A of 1:1; 0.5 M bis(2-ethylhexyl) phosphoric acid (D2EHPA) and pH of 3.5.

3.2. Effect of the Concentration of Modifier (% Volume of TBP)

Preliminary tests using TBP (tributyl phosphate, Sigma Aldrich, Germany) as a modifier were performed to evaluate its potential to increase the extraction of manganese as well as its separation from the other metals. The extraction of Mn, Ni, Co, and Li without using TBP and when volumetric concentrations of 2.5%, 5%, and 10% TBP were used can be seen in Figure 2, where the error bars represent the standard deviation of triplicates. The extraction of Mn had a slight increase when the concentration of TBP was increased until 5%. However, the coextraction of all other metals also increased when TBP was used as a modifier. For all evaluated metals, the extraction decreased when 10% of TBP was used. Considering that no formation of a third phase was observed, it was decided not to use TBP in the next tests.

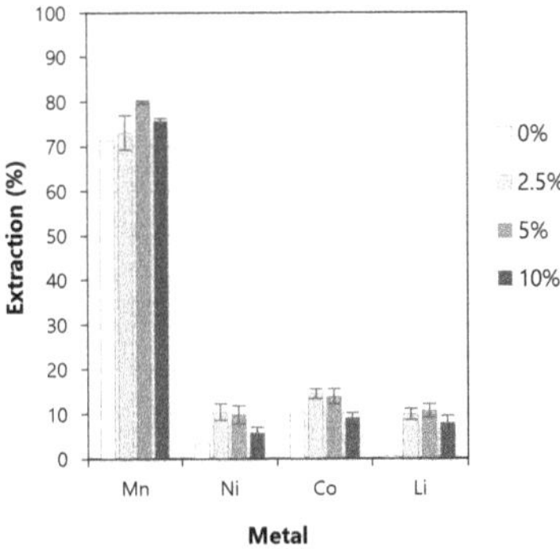

Figure 2. Extraction of metals using different volumetric concentrations of TBP as a phase modifier. Conditions: contact time of 10 min, 0.5 M D2EHPA, equilibrium pH of 3.5, organic to aqueous ratio (O:A) of 1:1. Error bars represent the standard deviation of triplicates.

3.3. Effect of the pH on the Extraction of Metals

The extraction of Mn, Co, Li, and Ni for three different molar concentrations of D2HEPA (0.4, 0.5, and 0.6 M) at different pH values can be seen in Figure 3. Some tests were performed using 0.2 M D2EHPA, but in this case, the extraction of manganese never exceeded 30%, and since this concentration is lower than the ones usually reported in the literature, further tests using 0.2 M D2EHPA were not performed. The initial pH of the synthetic solution based on the composition of the LIBs leach liquor was 3.8. After contacting the synthetic solution with the extractant, the pH of the aqueous phase decreased to about 2. This behavior was expected, considering the mechanism of extraction of metals using D2EHPA (Equation (4)) described by Zhang and Cheng [14], which results in a decrease in the pH.

$$M^{2+} + 2\overline{(HA)_2} \leftrightharpoons \overline{MA_4H_2} + 2H^+ \tag{4}$$

where M represents the metal, $\overline{(HA)_2}$ represents D2EHPA in the organic phase, and $\overline{MA_4H_2}$ represents the metal–organic complex [14].

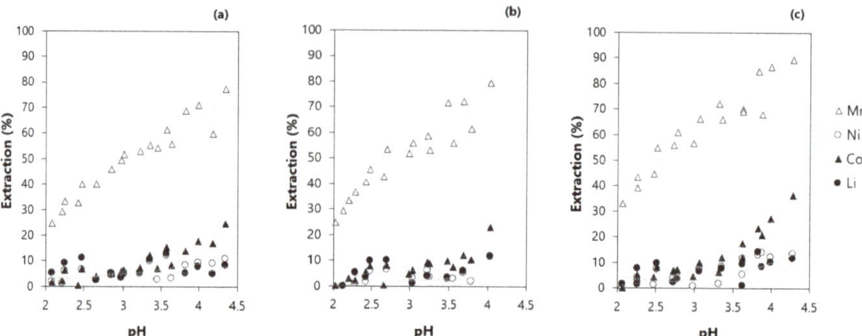

Figure 3. Extraction of metals using different molar concentrations of D2EHPA: (**a**) 0.4 M D2EHPA, (**b**) 0.5 M D2EHPA, (**c**) 0.6 M D2EHPA. Conditions: O:A of 1:1, contact time of 10 min.

The extraction of manganese increased with the pH for the three different concentrations of D2EHPA, but when the pH was increased to about 4, the coextraction of other metals was also more pronounced, mainly of cobalt. The increase in the molar concentration of D2HEPA also promoted an increase in the extraction of manganese, which was more pronounced when 0.6 M D2EHPA was used.

3.4. Effect of the Organic to Aqueous Ratio (O:A)

Preliminary tests were performed to evaluate the effect of the O:A ratio on the extraction of metals (Figure 4). The extraction of manganese increased with the O:A ratio (Figure 4a); however, the coextraction of cobalt also increased with the O:A ratio. For this reason, O:A ratios from 0.5 to 2 were further investigated in the factorial design of experiments. The isotherm representing the distribution of manganese in the aqueous and organic phase can be seen in Figure 4b. The extraction of manganese can be theoretically achieved after two extraction stages using an O:A ratio of 1.25.

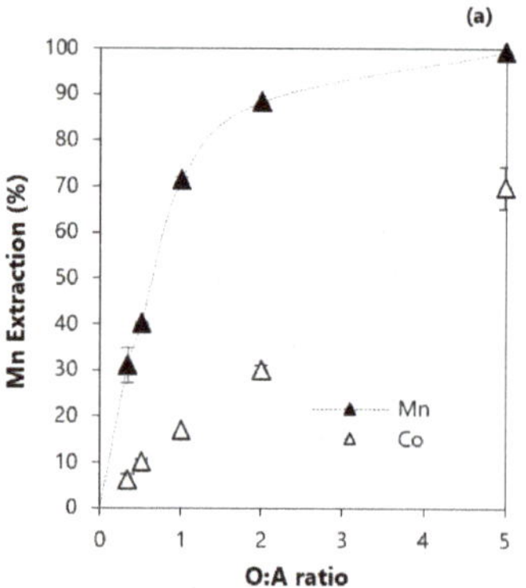
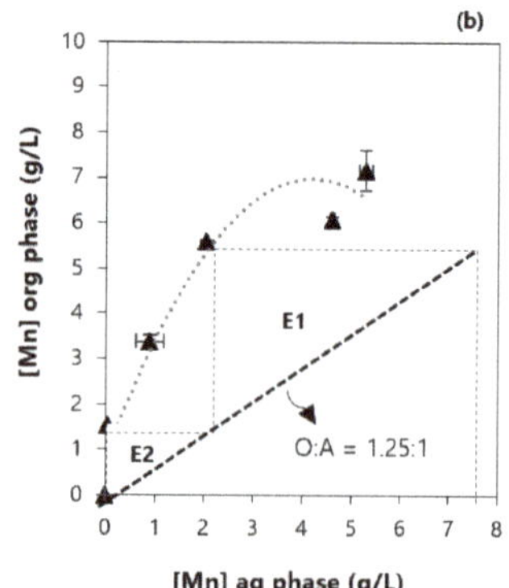

Figure 4. (**a**) Extraction of manganese and cobalt using different O:A ratios and (**b**) McCabe–Thiele diagram of the Mn extraction. Conditions: equilibrium pH of 3.5, 0.5 M D2EHPA, contact time of 10 min. Error bars represent the standard deviation of triplicates.

3.5. Extraction Stage: Factorial Design of Experiments and Regression Model

The conditions of the factorial design of experiments and respective responses (manganese and cobalt extraction) for each experiment are presented in Table 3. Tests from 1 to 8 correspond to the base 2^3 design. Tests from 9 to 12 are the replicates in the central point of the design and were used to determine the experimental error. Tests from 13 to 18 are the axial points added to the design. All the tests were performed at room temperature using a contact time of 10 min. The concentrations of metals in the raffinate and in the organic phase are reported in the Supplementary Materials (Table S1), as well as the extraction of Ni and Li, which in general remain at low values. The Supplementary Material (Table S2) also reports the distribution ratios (D) and separation factors (β).

Table 3. Conditions of the experimental design and results for the extraction of manganese and cobalt.

Run Order	Std Order	Coded Variables			Real Variables			Response (Extraction)	
		x_1	x_2	x_3	pH	O:A	D2EHPA	Mn (%)	Co (%)
6	1	−1	−1	−1	2.5	0.5	0.4	20	2
11	2	1	−1	−1	4	0.5	0.4	51	12
10	3	−1	1	−1	2.5	2	0.4	61	4
4	4	1	1	−1	4	2	0.4	92	23
14	5	−1	−1	1	2.5	0.5	0.6	30	4
5	6	1	−1	1	4	0.5	0.6	57	8
12	7	−1	1	1	2.5	2	0.6	79	1
13	8	1	1	1	4	2	0.6	97	44
18	9	0	0	0	3.25	1.25	0.5	72	4
8	10	0	0	0	3.25	1.25	0.5	73	5
7	11	0	0	0	3.25	1.25	0.5	73	5
9	12	0	0	0	3.25	1.25	0.5	70	4
15	13	−1	0	0	2.5	1.25	0.5	48	1
16	14	1	0	0	4	1.25	0.5	88	25
2	15	0	−1	0	3.25	0.5	0.5	38	9
1	16	0	1	0	3.25	2	0.5	91	16
17	17	0	0	−1	3.25	1.25	0.4	63	7
3	18	0	0	1	3.25	1.25	0.6	81	3

The adjusted regression model (y) for the extraction of manganese and the extraction of cobalt are represented by Equations (5) and (6), respectively. The models are only valid for the range of values tested in this study, and they only include factors with a statistically significant effect on the responses ($\alpha = 0.05$).

$$Mn\ (\%) = 72.0 + 14.7\ x_1 + 22.3x_2 + 5.7x_3 - 7.4x_2^2 \tag{5}$$

$$Co\ (\%) = 6.6 + 10.0\ x_1 + 5.2x_2 + 6.0x_1x_2 + 3.7x_1x_2x_3 + 4.7x_1^2 \tag{6}$$

The results of the analysis of variance of the fitted models for the extraction of manganese and cobalt are presented in Table 4, which was adapted from the ANOVA table from the Regression Analysis tool of Excel (Analysis ToolPak add-in). The replicates in the central level of the design allow estimating the experimental pure error and decomposing the Residual Sum of Squares (RSS) into the Sum of Squares due to Pure Error (SSPE) and the Sum of Squares due to Lack of Fit (SSLOF). The presence of curvature was verified for both models using the pure curvature testing (p-value = 0.048 and 0.046 for manganese and cobalt, respectively). The significance of the fitted models is indicated by the results of the F-test. The model adequacy was assessed by the Lack of Fit (LOF) test, but the results were lower than the significance level ($\alpha = 0.05$) for both models, given the low experimental error in the central point of the design and a small variance of the experimental error when compared to the residual error.

Table 4. Results of the analysis of variance of the fitted models for the extraction of manganese and cobalt.

Response	Source	Degree of Freedom	Sum of Squares	Mean Square	F-Value	p-Value
Manganese extraction	Regression	10	7964.8	796.5	43.6	2.4×10^{-5}
	Residual	7	127.9	18.3	-	-
	Lack of fit	4	120.4	30.1	12.2	3.4×10^{-2}
	Pure error	3	7.4	2.5	-	-
	Totals	17	8092.7	-	-	-
Cobalt extraction	Regression	10	1988.3	198.8	17.6	4.9×10^{-4}
	Residual	7	79.2	11.3	-	-
	Lack of fit	4	78.0	19.5	49.2	4.6×10^{-3}
	Pure error	3	1.2	0.4	-	-
	Totals	17	2067.5	-	-	-

Pareto charts of the standardized effects of the variables on the responses are presented in Figure 5a for the manganese extraction and in Figure 5b for the cobalt extraction. The standardized effects were calculated by dividing each coefficient by its standard error. The standardized effects correspond to the *t*-statistic values. A variable is considered statistically significant if its *p*-value is smaller than the defined significance level (0.05 for a confidence level of 95%). The significance level is identified in the graphs by dashed lines (2.36 at abscissa) and it corresponds to the 0.975 quartile in the Student's distribution, with seven degrees of freedom (total number of estimated coefficients subtracted from the total number of experiments). Thus, the effect of variables and their interactions is more significant as they are to the right of the red dashed line.

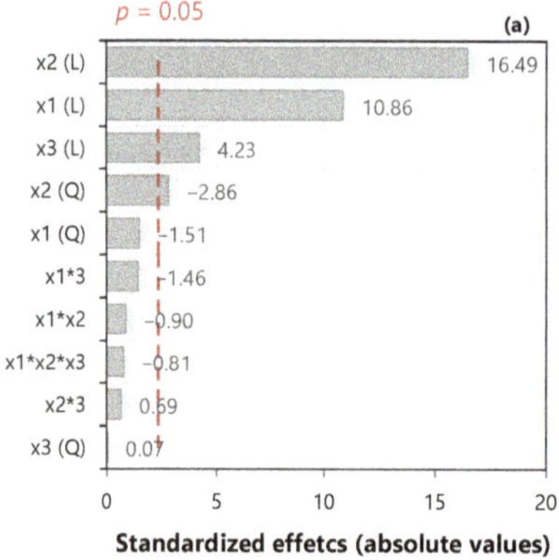

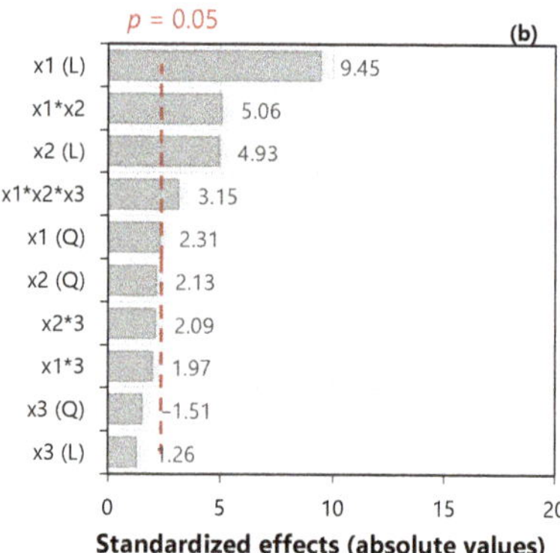

Figure 5. Pareto charts of the absolute values of the standardized effects of the factors for the regression model for (**a**) manganese extraction and (**b**) cobalt extraction with a significance level α = 0.05. Legend: x_1: pH, x_2: O:A ratio, x_3: molar concentration of D2EHPA, (Q): quadratic terms, (L): linear terms.

The variables with higher effects on the manganese extraction were x_2 (O:A ratio), x_1 (pH), and x_3 (molar concentration of D2EHPA). The quadratic effect of the factor x_2 is also significant in the extraction of manganese. Then, it can be concluded that the extraction of manganese increases with the increase of the pH, extractant concentration, and the O:A ratio. The quadratic terms x_1^2 and x_3^2, as well as all the interactions, did not present a significant effect on the manganese extraction in the range of values tested in this work (at a confidence level of 95%).

Regarding the extraction of cobalt (Figure 5b), the main effects were accounted for the variables x_1 (pH), x_2 (O:A ratio) and the interactions of x_1x_2 and $x_1x_2x_3$, with a positive effect on the response with the increase of their levels. The quadratic terms x_1^2, x_2^2, and x_3^2, the factor x_3 (molar concentration of D2EHPA), as well as the interactions x_1x_3 and

x_2x_3 did not present a significant effect on the extraction of cobalt in the range of values considered for a confidence level of 95%.

The coefficient of determination (R^2) was used to assess the goodness of fit of the models. The model for the manganese extraction presented an $R^2 = 0.98$ and for the cobalt extraction an $R^2 = 0.96$. This coefficient indicates that 98% and 96% of the response variability is explained by the fitted models, respectively. The relation between the experimentally observed responses for the extraction of manganese (Figure 6a) and cobalt (Figure 6b) is represented in the scatter plots below. This relation demonstrates that the adjusted models can provide a good fit to the experimental results under the range of values considered in the study.

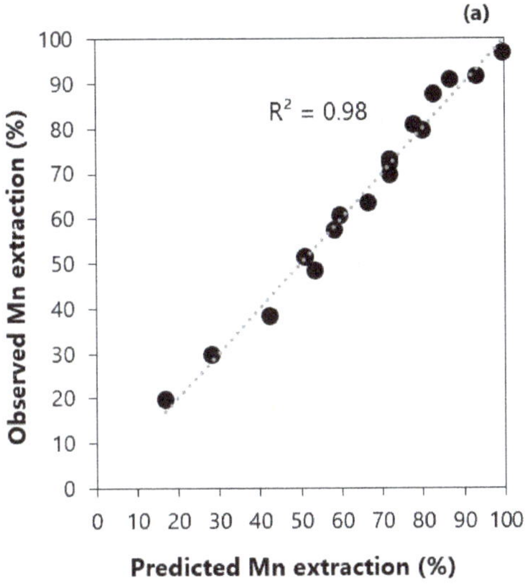
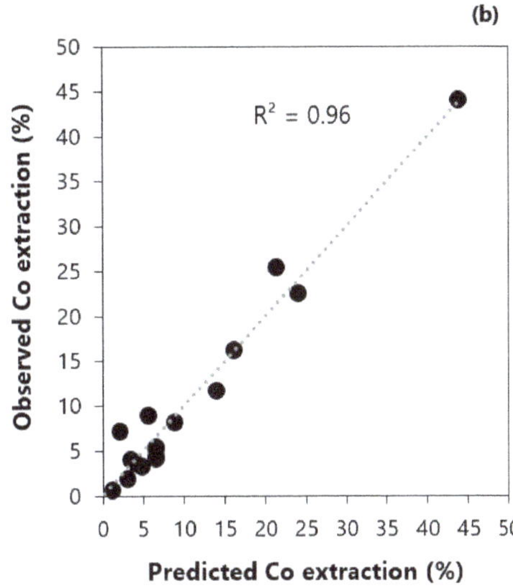

Figure 6. Responses predicted by the model versus experimentally observed: (a) manganese extraction and (b) cobalt extraction.

3.6. Response Surfaces: Extraction of Manganese and Cobalt

To help to understand the effect of the different factors on the extraction of manganese and cobalt, response surfaces were used. They were depicted using contour plots to show a clear representation of the surfaces. Contour plots are represented by a set of lines of constant response, being constructed in planes defined by pairs of variables. Therefore, each line represents a particular response of the fitted model.

The contour plots representing the manganese extraction when the factor x_1 (pH) was fixed at its low level (-1, pH = 2.5), standard level (0, pH = 3.2), and high level (+1, pH = 4) can be seen in Figure 7a–c, respectively. The responses for the extraction of cobalt under these same conditions are represented in Figure 7d–f. To construct the contour plots, the level of the factors x_2 (O:A ratio) and x_3 (molar concentration of D2EHPA) was changed from the low to the high level. The responses (y = % extraction) are represented by legends on the left of each graph. Results are only valid in the range of values considered in this study.

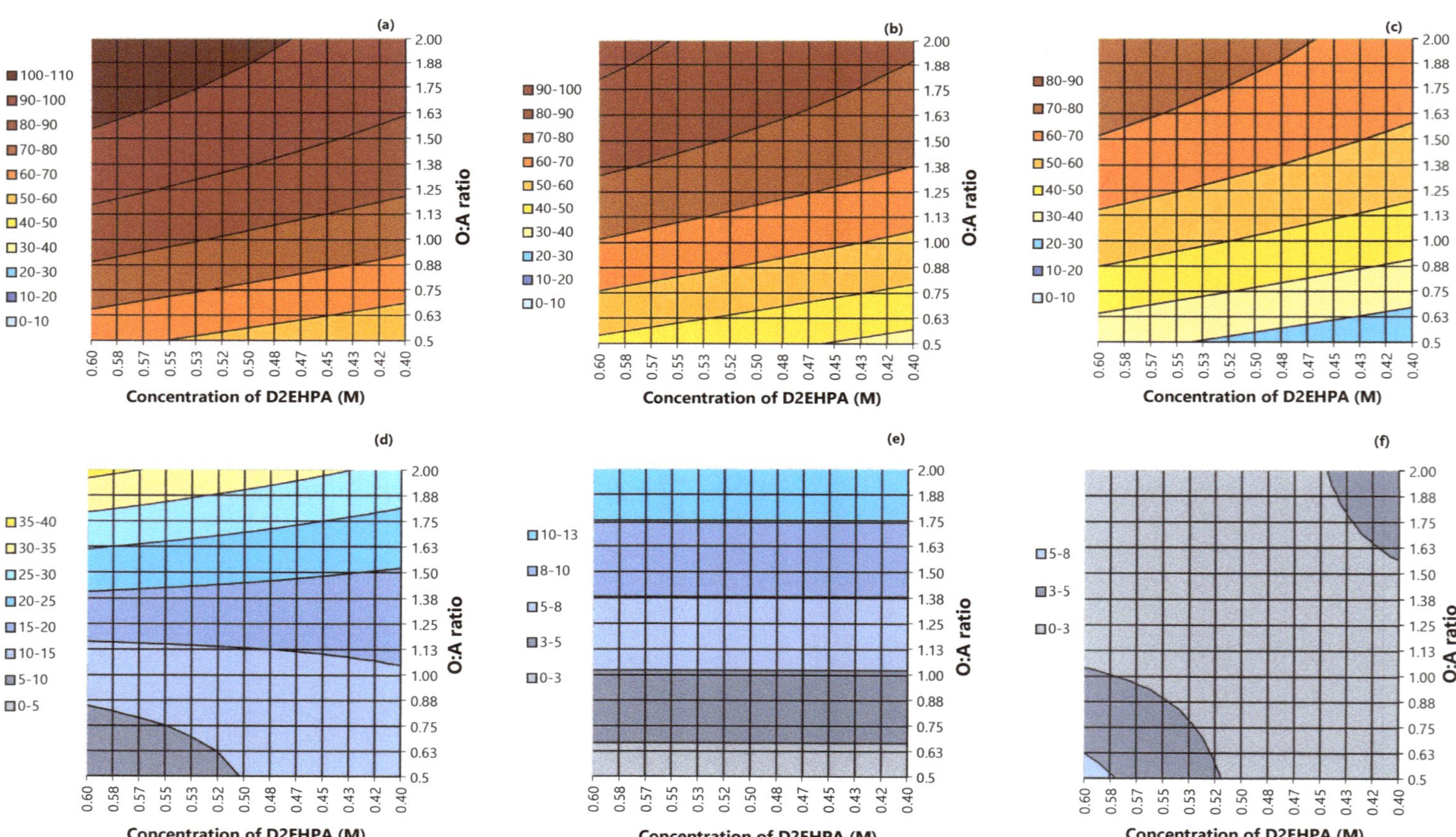

Figure 7. Contour plots representing the (**a**–**c**) extraction of manganese and the (**d**–**f**) coextraction of cobalt when the pH was set at 2.5 (**a,d**), at pH of 3.25 (**b,e**) and at pH 4 (**c,f**).

The extraction of manganese when the pH was set at 2.5 is represented in Figure 7a. High manganese extractions can be achieved for any level of concentration of D2EHPA provided that the O:A ratio is also at a high level, which is explained by the highest effect of the O:A ratio on the response. At the lowest pH, the lowest extraction of manganese was verified at the lowest level of the O:A ratio (0.5:1) and at the lowest concentration of extractant (0.4 M). On the other hand, when the pH was 2.5, the highest extraction of manganese was observed at the highest level of the O:A ratio (2:1) and at the highest level of concentration of D2EHPA (0.6 M). However, when the pH was 2.5, the extraction of manganese never exceeded 70–80%, which can be explained by the mechanism of the reaction of D2HPA, by which an increase in the concentration of H^+ ions will move the equilibrium to the left side, hiding the formation of products. When the pH was set at 2.5, it is possible to observe in Figure 7d that the extraction of cobalt was kept at a very low level and never exceeded 5%, which was reached only when high concentrations of D2EHPA or high O:A ratios were employed.

The behavior of the extraction of manganese when the pH was 2 was similar to the one when the pH was 3.25, as can be observed in Figure 7b. However, the increase in the pH resulted in an increase in the highest extraction of manganese, which was raised to 80–90%. The lowest extraction of manganese at pH 3.25 was also obtained when the concentration of D2EHPA and the O:A ratio were at their lowest levels (0.4 M and 0.5:1, respectively). The highest extraction of manganese at pH 3.25 was achieved when the other two factors were at the highest level (0.6 M and 2:1). Extractions of manganese above 70% can be obtained for the whole range of values tested for the concentration of D2EHPA, provided that the O:A ratio is at least 1.4:1. When the pH was set at the standard level (3.25), the extraction of cobalt is mainly dependent on the O:A ratio (Figure 7e). Thus, it is possible to keep the coextraction of cobalt below 8% provided that the O:A ratio does not exceed around 1.4:1.

Contour plots representing the extraction of manganese when the pH was set at 4 can be seen in Figure 7c. The extraction of manganese reached higher values when the other two factors were combined at a higher pH, which is explained by the significant effect of the pH on the response, as it was discussed in the regression analysis. At the highest pH, the extraction of manganese was always above 50%. The lowest extraction was obtained when the concentration of D2HEPA and the O:A ratio were at the lowest level (0.4 M and 0.5:1, respectively). When both factors were increased to the highest level, the extraction of manganese achieved the maximum results. It is important to highlight that for certain conditions, the fitted model slightly overestimated the responses (above 100%). The coextraction of cobalt also increased to higher values when the pH was set at the highest level (Figure 7f), which is also compatible with the significant effect of the pH on the cobalt response, which was observed in the regression analysis. The highest coextraction of cobalt was observed when the concentration of D2EHPA and the O:A ratio were at their highest levels (0.6 M and 2:1, respectively) and achieved around 35%. At pH 4, the coextraction of cobalt remained at lower levels when both the O:A ratio and concentration of D2EHPA were set at lower levels.

Considering the results using the fitted models, to keep the coextraction of cobalt low even though obtaining high extractions of manganese, the pH, O:A ratio, and concentration of D2EHPA should be kept at intermediate levels. For this reason, the next stages (scrubbing and stripping) were studied using a loaded organic obtained at the central level of the tested factors (pH of 3.25, O:A 1.25:1, and 0.5 M D2EHPA). The concentration of the loaded organic obtained at these conditions to be used in the next stages was compatible with the results of the factorial design of experiments.

3.7. Scrubbing of the Loaded Organic

According to Ritcey and Ashbrook [36], scrubbing usually refers to the removal of unwanted coextracted species in the loaded organic. The purpose of scrubbing the organic phase is to replace coextracted or mechanically entrained Co, Ni, or Li together with Mn [20]. Although it can be considered an important stage to purify the loaded organic and selectively remove some undesired metals, the scrubbing stage was not studied in detail in this work, and the scrubbing conditions proposed by Peng et al. [20] were used. Thus, the loaded organic obtained using the standard conditions of the factorial design of experiments was scrubbed twice with a pure solution containing 4 g/L Mn prepared using $MnSO_4 \cdot H_2O$, without pH adjustment (pH: 4.4) for 10 min at an O:A ratio of 10:1. The final composition of the scrubbing solutions (1 and 2) after contact with the loaded organic and the resultant organic phase is presented in Table 5.

Table 5. Composition of the scrubbing solutions and the resultant organic phase after two scrubbing stages with 4 g/L Mn (O:A of 10:1, contact time of 10 min).

Solution	Concentration (g/L)			
	Mn	Co	Ni	Li
Feed solution	7.4	18.7	7.2	1.1
Aqueous phase (after extraction)	2.1	18.0	7.0	1.0
Scrubbing solution 1 (aqueous phase)	0.8	3.0	0.3	0.1
Scrubbing solution 2 (aqueous phase)	2.1	1.9	<0.1	<0.1
Organic phase	4.7	0.1	0.1	<0.1

3.8. Stripping Stage: Factorial Design of Experiments and Regression Model

The experimental conditions of the factorial design for the stripping of the loaded organic and respective responses are presented in Table 6. The final concentrations of manganese and cobalt (g/L) in the stripping product were considered as the process responses. All experiments were performed at room temperature after two scrubbing stages (detailed in Section 3.7).

Table 6. Conditions of the experimental design and results for the stripping of cobalt and manganese.

Random Order	Std Order	Coded Variables			Real Variables			Response	
		x_1	x_2	x_3	$[H_2SO_4]$	O:A	Time	Mn (g/L)	Co (g/L)
9	1	−1	−1	−1	0.05	1	2	4	0.06
14	2	1	−1	−1	2	1	2	4	0.05
4	3	−1	1	−1	0.05	8	2	11	0.31
2	4	1	1	−1	2	8	2	19	0.26
15	5	−1	−1	1	0.05	1	25	5	0.08
11	6	1	−1	1	2	1	25	5	0.07
3	7	−1	1	1	0.05	8	25	10	0.41
8	8	1	1	1	2	8	25	28	0.42
16	9	0	0	0	1.025	4.5	13.5	17	0.26
5	10	0	0	0	1.025	4.5	13.5	16	0.24
7	11	0	0	0	1.025	4.5	13.5	16	0.24
1	12	0	0	0	1.025	4.5	13.5	17	0.27
18	13	−1	0	0	0.05	4.5	13.5	9	0.15
12	14	1	0	0	2	4.5	13.5	17	0.26
10	15	0	−1	0	1.025	1	13.5	5	0.07
6	16	0	1	0	1.025	8	13.5	23	0.36
13	17	0	0	−1	1.025	4.5	2	14	0.21
17	18	0	0	1	1.025	4.5	25	22	0.34

The regression models for the stripping of manganese and cobalt are represented by Equations (7) and (8), respectively, and only factors with a statistically significant effect on the responses were inserted in the models ($\alpha = 0.05$). The models are only valid for the range of values tested in this study.

$$Mn\ (g/L) = 16.9 + 3.4x_1 + 6.8x_2 + 2.0x_3 + 3.3x_1x_2 - 4.0x_1^2 - 3.1x_2^2 \tag{7}$$

$$Co\,(g/L) = 0.25 + 0.14x_2 + 0.04x_3 + 0.03x_2x_3 \tag{8}$$

The results of the analysis of variance of the models are presented in Table 7. The presence of curvature was verified only for the model representing the manganese stripping with the pure curvature testing (p-value = 0.04). The results of the F-test can be related to the significance of the fitted models. The model adequacy was assessed by the LOF test, but the result for the manganese stripping was lower than the significance level (α = 0.05), which can be related to the low experimental error in the central point of the design.

Table 7. Results of the analysis of variance of the fitted models for the stripping of manganese and cobalt.

Response	Source	Degree of Freedom	Sum of Squares	Mean Square	F-Value	p-Value
Concentration of manganese	Regression	10	880.2	88.0	20.4	3.0×10^{-4}
	Residual	7	30.2	4.3	-	-
	Lack of fit	4	28.7	7.2	13.9	2.8×10^{-2}
	Pure error	3	1.5	0.5	-	-
	Totals	17	910.4	-	-	-
Concentration of cobalt	Regression	10	0.2	2.4×10^{-2}	21.4	2.6×10^{-4}
	Residual	7	7.93×10^{-3}	1.1×10^{-3}	-	-
	Lack of fit	4	7.33×10^{-3}	1.8×10^{-3}	9.0	5.1×10^{-2}
	Pure error	3	6.08×10^{-4}	2.0×10^{-4}	-	-
	Totals	17	0.2	-	-	-

Pareto charts of the standardized effects of the variables on the responses are presented in Figure 8. A significant effect on the stripping of manganese (Figure 8b) was accounted for the three main variables: x_1 (concentration of H_2SO_4), x_2 (O:A ratio), and x_3 (stripping time). The interaction effect of x_1 and x_2 was also significant, as well as the effect of the quadratic term x_1^2. Thus, the stripping of manganese will increase with the increase of the levels of these three variables. The quadratic terms x_2^2 and x_3^2, as well as all the other interactions, did not have a significant effect on the manganese stripping, considering the range of values tested at a confidence level of 95%. Only the variables x_2 (O:A ratio) and x_3 (stripping time) had a positive and significant effect on the stripping of cobalt (Figure 8b). Thus, the concentration of acid did not show a significant effect on the stripping of cobalt in the tested range nor did it have all the interactions and quadratic terms (at a confidence level of 95%).

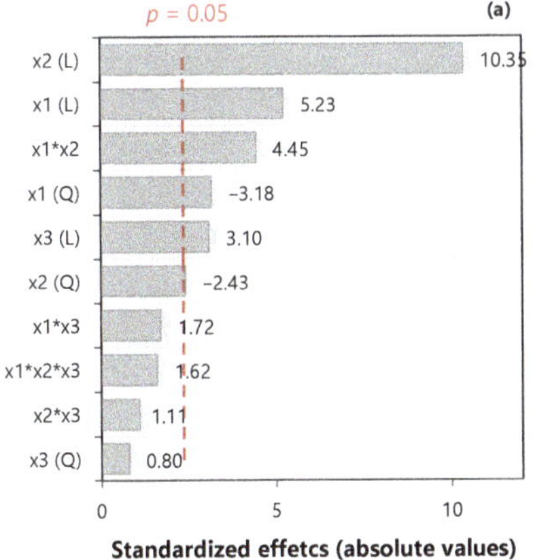

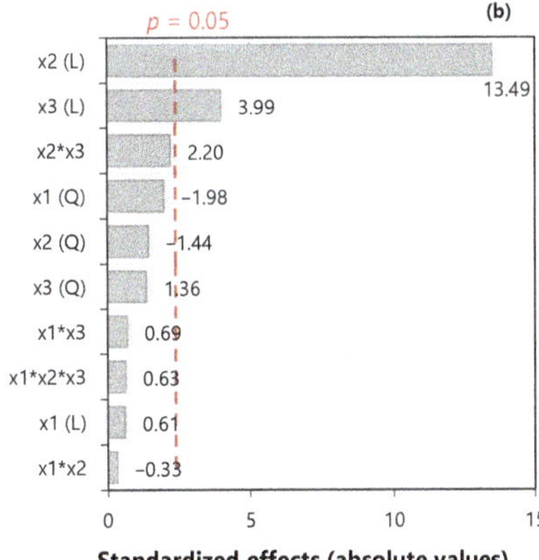

Figure 8. Pareto charts of the absolute values of the standardized effects of the factors for the regression model for the (**a**) manganese stripping and (**b**) for the cobalt stripping. Significance level α = 0.05. Legend: x_1: molar concentration of H_2SO_4, x_2: O:A ratio, x_3: stripping time, (Q): quadratic terms, (L): linear terms.

Both models presented an $R^2 = 0.97$, which is indicative that a large proportion of the variance of the response can be explained by the independent variables, considering the range of values tested in the experiments. The relation between the experimentally observed responses and those obtained using the fitted model for the stripping of manganese and cobalt are represented in Figure 9a,b, respectively, which illustrates how the models provide a good fit to the experimental results.

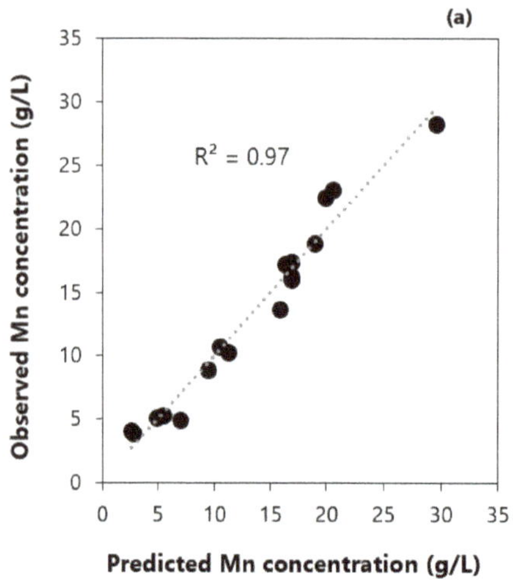

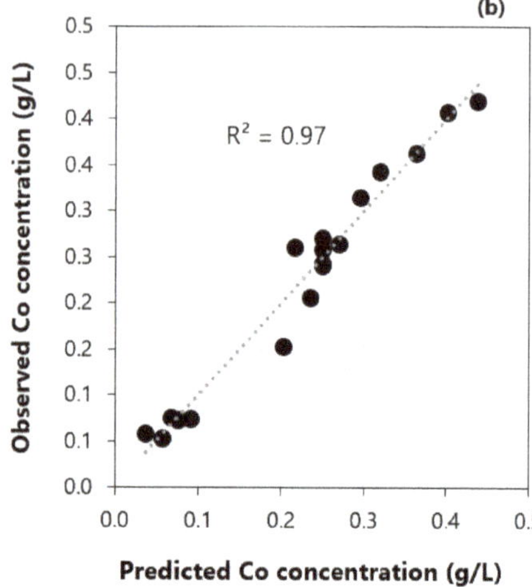

Figure 9. Responses predicted by the model versus experimentally observed: (**a**) manganese stripping and (**b**) cobalt stripping.

3.9. Response Surfaces: Stripping of Manganese and Cobalt

The contour plots in Figure 10a–c represent the response surfaces of the manganese stripping when the factor x_2 (O:A ratio) was set at its low level (−1, O:A = 1:1), standard level (0, O:A = 4.5:1), and high level (+1, O:A = 8:1), respectively. The stripping of cobalt for different combinations of O:A ratio and time is represented by the contour plots in Figure 10d, given that the concentration of sulfuric acid did not have a significant effect on it. The values of the response (y) are represented by legends on the left side of each graph. Results are only valid in the range of values considered in this study. The concentrations of metals remaining in the organic phase and in the stripping product for each test are reported in the Supplementary Materials (Table S3).

When the O:A used in the stripping was 1:1 (Figure 10a), a low concentration of manganese was obtained and never exceeded 10 g/L, which was expected given the larger volume of aqueous phase. At the lowest concentration of H_2SO_4 (0.05 M), the lowest concentration of manganese in the stripping product was verified at the lowest stripping time (2 min), being lower than 3 g/L Mn. With the increase in the concentration of H_2SO_4 and in the leaching time, a slight increase in the concentration of manganese was observed (maximum of 10 g/L).

The stripping behavior of manganese when the O:A ratio was set at 4.5:1 can be observed in Figure 10b. At this O:A ratio, the lowest concentration of manganese was around 8–10 g/L, and it was reached when the concentration of H_2SO_4 was the lowest (0.05 M) at the shortest stripping time (2 min). Increasing the concentration of acid from 0.9 to 2 M and the stripping time from 15 to 25 min promoted an increase in the concentration of manganese, which reached around 20 g/L.

The concentration of manganese was the highest when the O:A ratio was set at 8:1 (Figure 10c) and it was higher than 10 g/L for all tested conditions. The concentration of manganese reached higher values when the other two variables (time and concentration of acid) were combined at the highest O:A ratio, which is related to the highly significant effect of the O:A ratio on the response, as previously discussed in the regression analysis. When the concentration of acid was at the lowest level (0.05 M) and the stripping time was also at the lowest level (2 min), the concentration of manganese was around 10 g/L. When both factors were increased to their highest levels, the concentration of manganese achieved the maximum results (23–25 g/L). In Figure 10a–c, it is also possible to observe how the concentration of acid (x_1) has a more pronounced effect on the concentration of manganese in the stripped product, which was also represented by a quadratic term in the model, causing a curvature in the response surface. Thus, a slight increase in the concentration of acid can cause a higher effect on the concentration of manganese.

The stripping of cobalt (Figure 10d) was mainly affected by the O:A ratio and by the leaching time, while the concentrations of H_2SO_4 tested in this study did not have a significant effect on the concentration of cobalt in the stripped liquor. The concentration of cobalt increased along with the O:A ratio and the stripping time, but it never exceeded 0.5 g/L. Thus, it can be concluded that very high concentrations of manganese in the stripping product (>23 g/L) can be obtained using high O:A ratios and concentrations of sulfuric acid of around 1 M. However, the stripping time should not exceed around 13 min, in order to keep the concentration of cobalt at a low level (<0.3 g/L). Additionally, the fitted models can support the optimization of the stripping process.

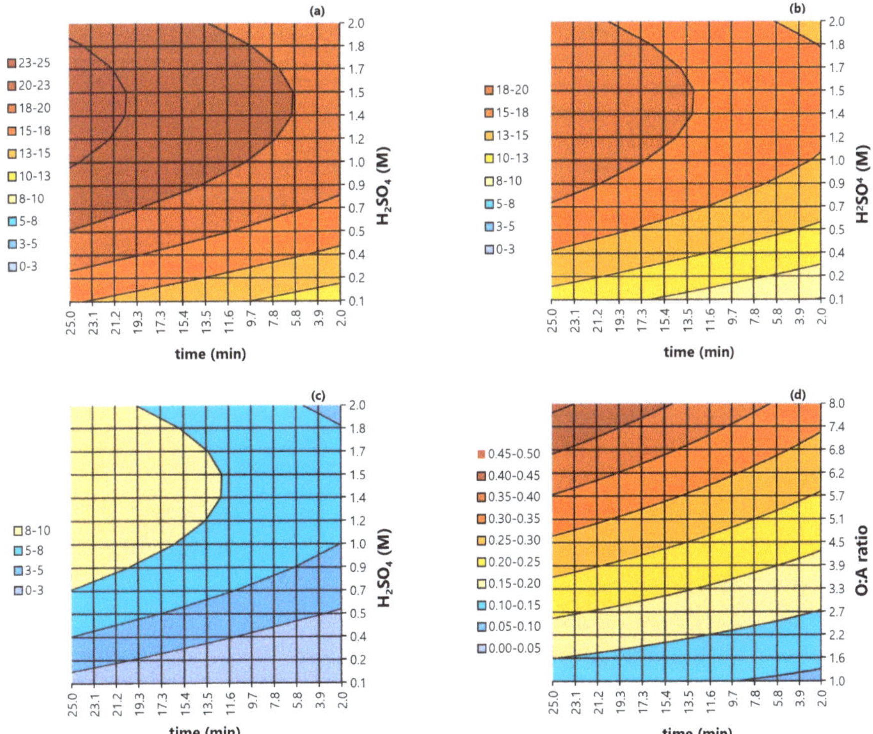

Figure 10. Contour plots representing the (**a**–**c**) stripping of manganese (**a**) when the O:A was set at 8:1, (**b**) when the O:A ratio was set at 4.5:1, and (**c**) when the O:A ratio was set at 1:1. (**d**) represents the stripping of cobalt at different combinations of stripping time and O:A ratios.

The fitted models can help to optimize the solvent extraction of manganese and can also assist with the construction of distribution isotherms and McCabe–Thiele diagrams, which are very helpful to predict the distribution of metals in both phases of the system (aqueous and organic) and to theoretically determine the number of required stages. The distribution isotherms for the stripping of manganese and cobalt, whose results were determined using the fitted models, are presented in the Supplementary Materials (Figure S1).

4. Conclusions

The recovery of manganese from a solution based on lithium-ion batteries was investigated using the factorial design of experiments and the response surface methodologies in order to assess the effect of different factors on the solvent extraction of manganese. These methodologies were also used to optimize the extraction and stripping stages, aiming to minimize the coextraction of cobalt. Preliminary tests were performed to determine the experimental conditions to be further investigated in the factorial design of experiments. The use of a modifier (TBP) was tested, but the formation of a third phase was not observed, and for this reason, additional tests with a modifier were not performed. The extraction of manganese using D2EHPA was fast, and maximum results were achieved after 10 min of contact time.

The factors evaluated in the extraction stage were the equilibrium pH, the molar concentration of D2EHPA, and the organic to aqueous ratio. Under optimized conditions (O:A of 1.25:1, pH 3.25, and 0.5 M D2EHPA), extractions above 70% Mn were reached in a single extraction stage with a coextraction of around only 5% Co, which was mostly removed in two scrubbing stages. Other combinations of factors can also result in high extractions of manganese and low coextractions of cobalt. In general, the coextraction of lithium and nickel remained low. The variables considered for the optimization of the stripping stage were the concentration of sulfuric acid, the organic to aqueous ratio, and the stripping time. A stripping product containing around 23 g/L Mn and around 0.3 g/L Co can be obtained under optimized conditions (O:A of 8:1, 1 M H_2SO_4, and around 13 min of contact time) in a single stripping stage. Increasing the number of extraction stages can promote an increase in the concentration of manganese loaded in the organic phase and should be further investigated in up-scale tests using mixer-settlers. Moreover, the fitted models for the extraction and stripping stages can help optimize these processes and can also assist with the construction of McCabe–Thiele diagrams to predict the number of stages required to maximize the recovery of manganese.

The results obtained can support further investigations on the recovery of manganese from spent lithium-ion battery solutions, which are an important secondary resource of manganese, using solvent extraction with D2EHPA. Moreover, the use of methodologies to model and optimize the process can assist the process management, considering that multiple combinations of factors can result in high extractions of manganese and low coextractions of other metals. Knowing these alternatives can help to better design the process to reduce the consumption of energy and reagents, minimizing costs and environmental impacts.

Supplementary Materials: The following are available online at https://www.mdpi.com/2075-4701/11/1/54/s1, Table S1. Conditions of the experimental design and concentrations of metals in the raffinate and in the organic phase after one extraction stage. Contact time of 10 min. Legend: [aq]: concentration of metal in aqueous phase, [org] concentration of meta in organic phase, Table S2. Conditions of the experimental design, distribution ratios (D) and separation factors (β) after one extraction stage. Contact time of 10 min, Table S3. Conditions of the experimental design and concentrations of metals remaining in the organic phase and in the stripping product. Legend: [aq]: concentration of metal in aqueous; phase, [org] concentration of metal in organic phase, Figure S1. Distribution isotherms of (**a**) manganese stripping and (**b**) cobalt stripping obtained using the fitted models. Conditions used as input in the fitted models: stripping time: 13.5 min (coded variable: 0), O:A ratio: 8:1 (coded variable: +1), concentration of H_2SO_4: 1 M (coded variable: 0).

Author Contributions: Data curation, formal analysis, investigation, visualization, writing—original draft preparation, N.V.; Conceptualization, methodology, validation, writing—review and editing, N.V., N.R., C.E., M.P.; resources, N.R., C.E., M.P.; project administration and supervision, M.P.; funding acquisition, C.E., M.P.; All authors have read and agreed to the published version of the manuscript.

Funding: This research was supported by VINNOVA (reference number of project: 2019-02069).

Institutional Review Board Statement: Not applicable.

Informed Consent Statement: Not applicable.

Data Availability Statement: The data presented in this study are available in supplementary material here.

Conflicts of Interest: The authors declare no conflict of interest.

References

1. The Royal Society of Chemistry, Manganese. Available online: https://www.rsc.org/periodic-table/element/25/manganese (accessed on 21 November 2020).
2. Corathers, L.A.; U.S. Geological Survey. Manganese. In *Mineral Commodity Summaries*; 2020; p. 105. Available online: https://pubs.usgs.gov/periodicals/mcs2020/mcs2020.pdf (accessed on 10 October 2020).
3. South Australia, Government of South Australia, Department of Energy and Mining, Manganese. Available online: https://energymining.sa.gov.au/minerals/mineral_commodities/manganese (accessed on 21 November 2020).
4. Cardarelli, F. *Materials Handbook: A Concise Desktop Reference*, 2nd ed.; Springer: London, UK, 2008; ISBN 978-1-84628-668-1.
5. United States Government Printing Office. *Strategic and Critical Materials. Hearings before a Subcommittee of the Committee on Military Affairs, United States Senate, Seventy-Seventh Congress, First Session, Relative to Strategic and Critical Materials and Minerals*; Printed for the use of the Committee on Military Affairs, Washington: Fairfax, VA, USA, 1941.
6. *Manganese Reserves and Resources of the World and Their Industrial Implications*; National Academies Press: Washington, DC, USA, 1981; ISBN 978-0-309-32998-9.
7. Maynard, J.B. Manganese Rocks and Ores. In *Isotope Geochemistry: The Origin and Formation of Manganese Rocks*; Elsevier: Amsterdam, The Netherlands, 2017; ISBN 9780128031650.
8. Thackeray, M.M.; Croy, J.R.; Lee, E.; Gutierrez, A.; He, M.; Park, J.S.; Yonemoto, B.T.; Long, B.R.; Blauwkamp, J.D.; Johnson, C.S.; et al. The quest for manganese-rich electrodes for lithium batteries: Strategic design and electrochemical behavior. *Sustain. Energy Fuels* **2018**, *2*, 1375–1397. [CrossRef]
9. Olivetti, E.A.; Ceder, G.; Gaustad, G.G.; Fu, X. Lithium-Ion Battery Supply Chain Considerations: Analysis of Potential Bottlenecks in Critical Metals. *Joule* **2017**, *1*, 229–243. [CrossRef]
10. Cannon, W.F.; Kimball, B.E.; Corathers, L.A. *Manganese*; Schulz, K.J., DeYoung John, H., Jr., Seal, R.R., II, Bradley, D.C., Eds.; The U.S. Geological Survey: Reston, VA, USA, 2017; series number 1802; ISBN 978-1-4113-3991-0. Available online: https://pubs.er.usgs.gov/publication/pp1802L (accessed on 21 November 2020).
11. Statista, Projected Size of the Global Lithium-Ion Battery Market from 2020 to 2025. Available online: https://www.statista.com/statistics/1011187/projected-global-lithium-ion-battery-market-size/ (accessed on 21 November 2020).
12. *Roskill Manganese Outlook to 2030*, 16th ed. Available online: https://roskill.com/market-report/manganese/ (accessed on 25 November 2020).
13. Chow, N. Manganese ore for lithium batteries. *Metal Powder Rep.* **2012**, *67*, 34–36. [CrossRef]
14. Zhang, W.; Cheng, C.Y. Manganese metallurgy review. Part I: Leaching of ores/secondary materials and recovery of electrolytic/chemical manganese dioxide. *Hydrometallurgy* **2007**, *89*, 137–159. [CrossRef]
15. Biswal, A.; Chandra Tripathy, B.; Sanjay, K.; Subbaiah, T.; Minakshi, M. Electrolytic manganese dioxide (EMD): A perspective on worldwide production, reserves and its role in electrochemistry. *RSC Adv.* **2015**, *5*, 58255–58283. [CrossRef]
16. Moore Stephans. Manganese. Is It the Forgotten Battery Mineral? Snapshot Worldwide Manganese Ore Prices. Available online: https://www.moorestephens.com.au/MediaLibsAndFiles/media/australia.moorestephens.com/Images/Profile%20Photos%20(Contact%20boxes)%20110w%20x%20110h%20px/Western%20Australia/Manganese-Moore-Stephens-Report.pdf (accessed on 25 November 2020).
17. Jones, T.S. *Manganese Recycling in the United States in 1998*; US Department of the Interior, US Geological Survey: Reston, VA, USA, 2001. [CrossRef]
18. Bullis, L.H.; Mielke, J.E. *Strategic and Critical Materials*; Routledge: New York, NY, USA, 2019; ISBN 9780367288822.
19. Granata, G.; Pagnanelli, F.; Moscardini, E.; Takacova, Z.; Havlik, T.; Toro, L. Simultaneous recycling of nickel metal hydride, lithium ion and primary lithium batteries: Accomplishment of European Guidelines by optimizing mechanical pre-treatment and solvent extraction operations. *J. Power Sources* **2012**, *212*, 205–211. [CrossRef]
20. Peng, C.; Chang, C.; Wang, Z.; Wilson, B.P.; Liu, F.; Lundström, M. Recovery of High-Purity MnO2 from the Acid Leaching Solution of Spent Li-Ion Batteries. *JOM* **2020**, *72*, 790–799. [CrossRef]

21. Yang, Y.; Xu, S.; He, Y. Lithium recycling and cathode material regeneration from acid leach liquor of spent lithium-ion battery via facile co-extraction and co-precipitation processes. *Waste Manag.* **2017**, *64*, 219–227. [CrossRef] [PubMed]

22. Wang, F.; Sun, R.; Xu, J.; Chen, Z.; Kang, M. Recovery of cobalt from spent lithium ion batteries using sulphuric acid leaching followed by solid–liquid separation and solvent extraction. *RSC Adv.* **2016**, *6*, 85303–85311. [CrossRef]

23. Joo, S.-H.; Shin, S.M.; Shin, D.; Oh, C.; Wang, J.-P. Extractive separation studies of manganese from spent lithium battery leachate using mixture of PC88A and Versatic 10 acid in kerosene. *Hydrometallurgy* **2015**, *156*, 136–141. [CrossRef]

24. Chen, X.; Chen, Y.; Zhou, T.; Liu, D.; Hu, H.; Fan, S. Hydrometallurgical recovery of metal values from sulfuric acid leaching liquor of spent lithium-ion batteries. *Waste Manag.* **2015**, *38*, 349–356. [CrossRef] [PubMed]

25. Pagnanelli, F.; Moscardini, E.; Altimari, P.; Abo Atia, T.; Toro, L. Cobalt products from real waste fractions of end of life lithium ion batteries. *Waste Manag.* **2016**, *51*, 214–221. [CrossRef] [PubMed]

26. Li, J.; Yang, X.; Yin, Z. Recovery of manganese from sulfuric acid leaching liquor of spent lithium-ion batteries and synthesis of lithium ion-sieve. *J. Environ. Chem. Eng.* **2018**, *6*, 6407–6413. [CrossRef]

27. Devi, N.B.; Nathsarma, K.C.; Chakravortty, V. Separation of divalent manganese and cobalt ions from sulphate solutions using sodium salts of D2EHPA, PC 88A and Cyanex 272. *Hydrometallurgy* **2000**, *54*, 117–131. [CrossRef]

28. Hossain, M.R.; Nash, S.; Rose, G.; Alam, S. Cobalt loaded D2EHPA for selective separation of manganese from cobalt electrolyte solution. *Hydrometallurgy* **2011**, *107*, 137–140. [CrossRef]

29. Cheng, C.Y. Purification of synthetic laterite leach solution by solvent extraction using D2EHPA. *Hydrometallurgy* **2000**, *56*, 369–386 [CrossRef]

30. Chen, X.; Zhou, T.; Kong, J.; Fang, H.; Chen, Y. Separation and recovery of metal values from leach liquor of waste lithium nickel cobalt manganese oxide based cathodes. *Sep. Purif. Technol.* **2015**, *141*, 76–83. [CrossRef]

31. Ahan, S.C.; Jeon, B.K.; Kim, B.E.; Lee, M.J.; Sonu, C.H. Method for Recovering Valuable Metals from Lithium Secondary Battery Wastes; 2010. Available online: https://patents.google.com/patent/WO2012050317A3/en (accessed on 25 November 2020).

32. Jouni, P.; Erkki, P. *Recovery of Manganese from Mixed Metal Solutions by Solvent Extraction with Organophosphorus Acid Extractants* Saint-Petersburg Mining University: St Petersburg, Russia, 2008.

33. Feather, A.; Sole, K.C.; Dreisinger, D.B. Pilot-plant evaluation of manganese removal and cobalt purification by solvent extraction. In *Proceedings of the ISEC'99*; Bercelona, Spain, 1999, Society of Chemical Industry (SCI): London, UK, 2001; Volume 2, ISBN 0-901001-83-8. Available online: http://www.solventextract.org/documents/1999/ISEC-1999-Proceedings-Vol2.pdf (accessed on 25 November 2020).

34. Cole, P.M. The introduction of solvent-extraction steps during upgrading of a cobalt refinery. *Hydrometallurgy* **2002**, *64*, 69–77 [CrossRef]

35. Montgomery, D.C. *Design and Analysis of Experiments*, 8th ed; John Wiley & Sons, Inc.: Hoboken, NJ, USA, 2012; ISBN 978-1-118-14692-7.

36. Ritcey, G.M.; Ashbrook, A.W. *Solvent Extraction: Principles and Applications to Process Metallurgy*; Elsevier: Amsterdam, The Netherlands, 1984; ISBN 0444417702.

metals

Article

A Novel Pyrometallurgical Recycling Process for Lithium-Ion Batteries and Its Application to the Recycling of LCO and LFP

Alexandra Holzer *, Stefan Windisch-Kern, Christoph Ponak and Harald Raupenstrauch

Chair of Thermal Processing Technology, Montanuniversitaet Leoben, Franz-Josef-Strasse 18, 8700 Leoben, Austria; stefan.windisch-kern@unileoben.ac.at (S.W.-K.); christoph.ponak@unileoben.ac.at (C.P.); harald.raupenstrauch@unileoben.ac.at (H.R.)
* Correspondence: alexandra.holzer@unileoben.ac.at; Tel.: +43-3842-402-5803

Abstract: The bottleneck of recycling chains for spent lithium-ion batteries (LIBs) is the recovery of valuable metals from the black matter that remains after dismantling and deactivation in pre-treatment processes, which has to be treated in a subsequent step with pyrometallurgical and/or hydrometallurgical methods. In the course of this paper, investigations in a heating microscope were conducted to determine the high-temperature behavior of the cathode materials lithium cobalt oxide (LCO—chem., $LiCoO_2$) and lithium iron phosphate (LFP—chem., $LiFePO_4$) from LIB with carbon addition. For the purpose of continuous process development of a novel pyrometallurgical recycling process and adaptation of this to the requirements of the LIB material, two different reactor designs were examined. When treating LCO in an Al_2O_3 crucible, lithium could be removed at a rate of 76% via the gas stream, which is directly and purely available for further processing. In contrast, a removal rate of lithium of up to 97% was achieved in an MgO crucible. In addition, the basic capability of the concept for the treatment of LFP was investigated whereby a phosphorus removal rate of 64% with a simultaneous lithium removal rate of 68% was observed.

Keywords: lithium-ion batteries (LIBs); recycling; pyrometallurgy; critical raw materials; lithium removal; phosphorous removal; recovery of valuable metals

Citation: Holzer, A.; Windisch-Kern, S.; Ponak, C.; Raupenstrauch, H. A Novel Pyrometallurgical Recycling Process for Lithium-Ion Batteries and Its Application to the Recycling of LCO and LFP. *Metals* **2021**, *11*, 149. https://doi.org/10.3390/met11010149

Received: 11 December 2020
Accepted: 11 January 2021
Published: 14 January 2021

Publisher's Note: MDPI stays neutral with regard to jurisdictional claims in published maps and institutional affiliations.

1. Introduction

The development of lithium-ion batteries (LIBs) has experienced an enormous upswing in recent years, which is, in addition to portable devices, mainly due to the steadily increasing demand in the electric vehicle (EV) sector. According to forecasts, this trend will continue in the coming years [1,2]. Further prognoses predict that sales of LIBs are expected to increase from 160 GWh in 2018 to over 1.2 TWh in 2030 [1]. Their use in electrical appliances, EVs and stationary storage is due to their advantages over other storage media, such as high energy density, long service life and high operating voltage [3,4]. Since consumed LIBs contain a large number of valuable metals, recycling has a considerable environmental impact in view of the conservation of valuable resources [5]. In addition to this idea of resource protection, waste reduction and the energy-efficient and economical treatment of hazardous substances are also driving recycling efforts [6]. The timeliness and necessity of recycling LIBs is further underlined by the 2020 list of critical raw materials published by the European Commission. Among others, cobalt, lithium and phosphorus can be found [7].

A major challenge with regard to recycling is posed by the strongly fluctuating waste stream. This is the product of the requirements of the countless applications for energy storage and the resulting multitude of electrode materials of LIBs [8]. In the respective literature there is a variety of different recycling processes, which can basically be divided into preparation for recycling, pre-treatment and main processing, including pyro- and hydro-metallurgy. In the first mentioned area, the processes of discharging and dismantling can be found [5]. The aim of the pre-treatment is to improve the recovery rate, to adapt the waste stream to the downstream process step and to reduce the energy consumption of the following pyro- or hydro-metallurgical process [6,9]. In Europe, there are several companies that already perform the preparation and pre-treatment of spent LIBs on a larger scale, like Accurec Recycling GmbH, Duesenfeld GmbH or Redux GmbH [10–12]. The latter starts the recycling process with collection and temporary storage, followed by manual sorting. As of this point in time there is still a considerable safety risk due to the residual charge of the LIBs. They are completely discharged, and the energy gained is fed back into the operating network. Subsequently, components such as electronics, cables, plastics, aluminum, and iron are dismantled and sorted. During the subsequent deactivation, the coating of the conductor foils is dissolved and the separator as well as the electrolyte are removed. During the mechanical treatment, the remaining components such as iron, aluminum, copper and the fine material (also called active material or black matter) of cathode and anode material are separated. The separation of the individual fractions is carried out with a magnetic separator, air separator and sieving [13]. The resulting black matter can be further treated in a pyro- or hydro-metallurgical process.

In pyrometallurgical treatment of LIBs, the physiochemical transformation temperatures above 1400 °C are used to recover the valuable metals [14]. As a partial step in an overall process, pyrometallurgy is a suitable instrument for purifying the feed stream of substances undesirable for hydrometallurgy. Fluorine, chlorine, graphite, phosphorus, etc., pose a particular challenge to hydrometallurgy. Pyrometallurgical processes are generally robust against impurities and organic contaminants, because volatile components can be evaporated [5]. Graphite from the anode can be used as a reducing agent and burned in various processes in the presence of oxygen, thus helping to maintain the process temperature. Since the reaction kinetics in pyrometallurgical processes increase extremely due to the high temperatures, productivity is higher compared to hydrometallurgy [15]. Although the large number of research activities in recent years has focused on hydrometallurgy [9], there is significant scientific output in the field of pyrometallurgy, some of which is already being applied on an industrial scale. Several recent reports claim that large-scale pyrometallurgical processes have greater potentials in terms of sustainability than their hydrometallurgical counterparts [16–21]. Industrial scale processes are those that have more than 1000 t/a recycling capacity. In Europe, the companies Umicore, Accurec and Nickelhütte Aue should be mentioned here, and outside the EU, for example, SungEel, Kyoei Seiko and Dowa. The overall processes usually lead via a mechanical and/or thermal step to pyro- and hydro-metallurgy [5]. The pyrometallurgical step is typically based on shaft furnaces or electric arc furnaces for melting this feedstock [22]. A direct comparison of the recycling efficiency of the individual processes is often very difficult, since the reference basis of the values given is usually not given or only partially given. However, it can be stated that recycling routes which include a pyrometallurgical step have the highest overall recycling efficiency, in some cases exceeding 50% [5]. Since pyrometallurgical processes are operated at high temperatures, their energy requirements are correspondingly high. In addition, large quantities of waste gas are produced which have to be treated. A disadvantage of current pyrometallurgical processes is the slagging of lithium, the recovery of which in turn requires an enormous hydrometallurgical effort [9,23]. The economic efficiency of lithium recovery depends on the concentration in the slag. As a rule, in the co-processing of LIBs in metallurgical plants, the lithium is diluted to such an extent that recovery is not economically feasible [24]. In recent years, a number of advances have been made in the field of slag post-treatment. These research ventures on a non-industrial scale focus, for example, on the concentration of Li in the slag by selective addition of slag-forming agents during the pyrometallurgical process and subsequent hydrometallurgical treatment [25,26]. Recent progress has also been made in the area of early-stage lithium extraction. In this process, sulphate roasting treatment was used to convert the cathode material from NMC batteries into a water-soluble lithium sulphate (Li_2SO_4) and a water-insoluble oxide (NiCoMn-oxide) [5]. However, depending on the price of lithium, processes specially developed for LIB recycling may in future be quite economical in terms of lithium recovery [24]. Various advantages and disadvantages also result from the different interconnection types of the overall process. For example, the primary energy consumption via pyrometallurgical routes is higher, but the resulting additional costs are more than compensated by lower operating costs in the hydrometallurgical step [5]. The recycling of P from LIBs is described in the literature in very few publications. Most of them are related to the hydrometallurgical process route, other processes deal with the regeneration of the cathode material [27].

Hydrometallurgical processes are highly selective and can therefore achieve high purities [15]. Leaching is the key process in hydrometallurgy to convert the metals to ions in a solution. This can be divided into bio leaching with metabolic excrements of microorganisms or fungi and chemical leaching with organic or inorganic acids [28–30]. Subsequently, the valuable metals are separated and recovered from the leaching solution. Since the structure of the leaching solution is complicated, it is usually necessary to use several different methods from the portfolio of solvent extraction, chemical precipitation and electrochemical deposition [28]. Hydrometallurgical methods result in extremely good recycling rates of up to 100% [28,31]. They also require a high level of equipment and a large number of process steps, which usually results in a correspondingly high volume of polluted wastewater. In order to operate the process economically, it is very important to separate and concentrate as many metals and impurities as possible in advance. For each additional metal, at least 1–2 additional process steps would be required, which is only economical if the metal value or quantity is correspondingly high [15].

Especially with regard to the raw materials contained in LIBs, which are included in the list of critical raw materials of the European Commission [7], and from an ecological point of view, a sustainable handling of spent LIBs is essential. According to Elwert et al. [32], recycling processes specialized in LIBs will gain more and more importance in the future. This is due to the increasing rate of return of spent LIBs to the waste stream, more regulations by the authorities and also decreasing amounts of valuable nickel and cobalt for direct use in nickel and cobalt producing plants. Furthermore, the growing market for LFP and the increasing interest in lithium recovery also plays a major role. Of particular importance in terms of regulation is the recently published European Commission proposal to revise EU Directive 2006/66/EC, which sets recovery rates of up to 70% for Li and 95% for other valuable metals such as Co, Ni and Cu by 2030 [33], which forces recyclers to increase recovery rates and their process efficiency.

It can be summarized that there is a multitude of different recycling processes and methods, which are characterized by their positive properties in certain areas but also have individual disadvantages. In the field of pyrometallurgy, lithium slagging and in particular the absence of possibilities to recover from the slag with reasonable effort can be identified as a bottleneck.

The novel pyrometallurgical recycling process presented in this paper is characterized by the recovery of an alloy with a simultaneous utilization of lithium and phosphorus via the gas flow. The following points provide a more detailed insight into the theoretical considerations and practical implementations for the most efficient recovery of valuable metals from LIBs using this process. Initially, appropriate analyses were carried out to better understand the behavior of cathode materials in high-temperature applications under reducing conditions. To determine the lithium removal rate without the presence of phosphorus, the cathode material lithium cobalt oxide (LCO) was examined in an experiment. In addition to the successive optimization of the reactor concept and adaptation to the waste stream from spent LIBs, another experiment with LCO in a modified setup was performed and compared to the previous one. To verify the basic suitability of the pyrometallurgical apparatus for the simultaneous removal of phosphorus and lithium via the gas flow, experiments were carried out with the cathode material lithium iron phosphate (LFP).

2. Process Concept and Methods

2.1. Used Materials

In total, three different experiments were carried out with two types of feedstock. As Windisch-Kern et al. [34] have already described, experiments with black mass from a pre-processing step have already shown that lithium could be removed to a considerable extent. As this has raised additional questions, a detailed investigation of the pure cathode materials, i.e., LCO, lithium nickel manganese cobalt oxide (NMC—chem., $LiNi_{0.33}Mn_{0.33}Co_{0.33}O_2$), lithium nickel cobalt aluminum oxide (NCA—chem., $LiNi_{0.8}Co_{0.15}Al_{0.05}O_2$) and LFP, was indispensable. For the purpose of clarifying the questions dealt with in this paper, the materials LCO and LFP were used, which were produced by the Chinese company Gelon Energy Corporation. The appearance of this feedstock can be described as a fine, black powder. Since this carbo-thermal process requires a reducing agent and the graphite bed in the reactor is only used for energy input, graphite powder from coke pellets of a steel mill is added. The graphite cubes with a side length of 2.5 cm come from electrodes of a steel mill and have an average purity of 99% with an electrical resistance of 4–8 $\mu\,\Omega$m and a density of 1.55–1.75 g cm^{-3}. [35] The amount of graphite powder required for the reduction was determined by stoichiometry of the respective cathode material. For this purpose the weighed mass of the cathode material was multiplied by the molar ratio of $LiCoO_2$ or $LiFePO_4$. After determining the moles O by multiplying the mass O by the relative atomic mass of O, the necessary mass of C was calculated by using the relative mass of C and assuming that a conversion to CO takes place in the reactor. The corresponding percentage C requirement is finally obtained by a rule of three of the masses of O and C. Table 1 shows the composition of the input materials determined from their stoichiometric composition.

Table 1. Composition of the mixture of cathode material and graphite powder in wt.%.

Compound	Li	Co	Fe	P	C
LCO-C	5.67	48.17	-	-	20.00
LFP-C	3.34	-	26.90	14.92	24.00

The products obtained from the experiments were examined by ICP-OES and ICP-MS by means of aqua regia digestion according to ÖNORM EN 13657:2002-12.

2.2. Material Specific Investigations

Since the behavior of the individual cathode materials at high temperature applications is hardly or not at all described in the literature, detailed investigations were undertaken at the Chair of Thermal Processing Technology. These included analyses in a Hesse Instruments EM 201 with an HR18-1750/30 furnace heating microscope. The results should be used for planning the process control in the following experiments in the inductively heated reactor, which is presented in Section 2.3. Furthermore, a better understanding of the behavior of the cathode materials should be gained. To be able to simulate the planned process as detailed as possible, graphite powder was added to the cathode material. The addition of graphite powder was carried out to an extent of 10 wt.% under the assumption that C is converted to CO_2 and transported away via the argon-purged atmosphere. The mixture of the corresponding cathode material and graphite powder was examined with at least one reproduction experiment. For better comparability a uniform heating rate was always set, which corresponds to the maximum possible with the heating microscope used. This is primarily to ensure the shortest possible residence time in the furnace chamber since interactions of LCO with the furnace material consisting of Al_2O_3 have been determined and damage to this should be prevented as far as possible. Up to a temperature of 1350 °C a heating rate of 80 °C/min was selected, from 1350 to 1450 °C 50 °C/min and up to 1700 °C a heating rate of 10 °C/min with a holding time of 15 min at 1700 °C was dialed. To avoid oxidation with the ambient air, the reactor was flushed with argon at a flow rate of 2 L/min. A maximum furnace temperature of 1700 °C was chosen, which allows for an approximate sample temperature of 1630 °C.

Figure 1 illustrates the standardized sample preparation. The material is centrally positioned in a cylinder with a diameter of 3.5 mm and a height of 2.5 mm on an Al_2O_3 platelet with an approximate weight of 0.1 g.

Figure 1. Structure of the sample on the white sample plate made of Al_2O_3 before the trial.

2.3. Reactor Concept

The novel reactor concept, which was constructed at the Chair of Thermal Processing Technology of the Montanuniversitaet Leoben, is based on the inductive heating of graphite pieces in a packed bed reactor. The cornerstone of knowledge generation in this field was laid at the chair already in 2012, by the EU subsidized project RecoPhos for the pyrometallurgical treatment of sewage sludge ash for simultaneous recovery of phosphorus and the contained valuable metals. Its results are described by Schönberg et al. [36] and Samiei et al. [37]. Based on this and corresponding follow-up projects, also in the field of basic oxygen furnace slag (BOFS) treatment, a batch operated post-lab-scale plant and a pilot-scale plant as a continuous process have been developed and built. This knowledge advantage was used to adapt the mechanism for pyrometallurgical recovery of valuable metals from processed LIB material in two ways. On the one hand, the theoretical idea of the continuous reactor, which should be conceptually similar to the set-up from research work in the field of sewage sludge ash and BOFS utilization, is applied. On the other hand, the post-lab-scale setup developed in the subject area mentioned above can initially be used for first experiments without further adaptations. In the long term, the realization of larger scales and corresponding throughput of recycled material as a continuous unit is planned. Intensive research activities on a small scale are indispensable for the most efficient implementation of gradual scale-ups to industrial maturity. In view of the process development as well as the knowledge gained about the input material, the previously mentioned apparatus in batch operation, the so-called InduMelt plant, is used for this purpose. These two process concepts and their respective challenges and developments are explained in detail below.

2.3.1. Continuous Reactor Concept

In order to treat the expected future waste stream from used LIBs, a technology with correspondingly high throughput rates is required. The currently pursued approach at the Chair of Thermal Processing Technology is based on a continuous reactor concept which currently exists as a pilot plant with a material throughput rate of 10 kg/h. Even if, according to initial findings from investigations of the LIB black matter, the design must differ from that used for sewage sludge ash and BOFS, the basic principle remains the same. Ponak [38] describes the so-called InduRed reactor as an cylindrical arrangement of refractory materials filled with pieces of electrode graphite which allow a horizontal and radial homogeneous temperature distribution when heated by the induction coils, as seen in Figure 2.

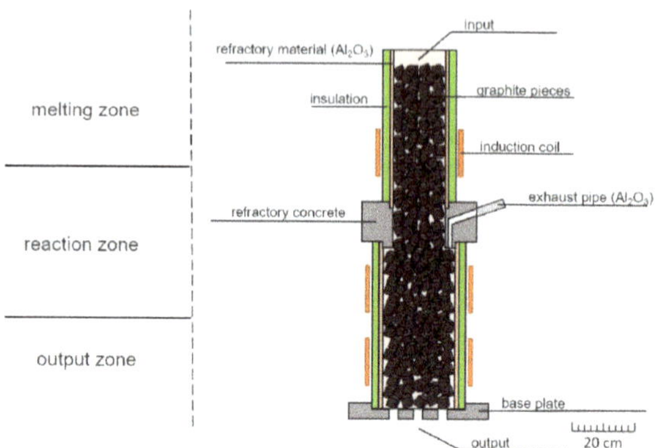

Figure 2. Schematic illustration of the InduRed reactor for the continuous treatment of sewage sludge ashes and BOFS [39].

The process starts with the material feed from a feed vessel above the reactor via a screw conveyor and a low volume of argon. Inert gas purging is highly relevant at this position, especially at higher temperatures, as the graphite bed is protected against oxidation by possible false air and, mainly, to direct small ash or black matter particles directly onto the graphite surface. In the first zone, the melting zone, the fine-grained material inserted is heated to melting temperature without reaching the reduction temperature of the critical component phosphorus. The resulting molten film then moves through the reactor to the reduction zone. In this zone the corresponding energy is induced so that the reduction temperature is reached close to the implemented gas flue. At this point, it has to be mentioned that the graphite pieces are not supposed to participate in the reduction reactions and serve only as a susceptor material. Added carbon powder functions as a reductant. Through this reaction process, phosphorus is converted into the gaseous phase and can be removed directly from the reactor via the gas flue by means of a negative pressure-generating induced draft fan. Downstream there is a post-combustion chamber in which external air or oxygen are used to convert elemental phosphorus to P_2O_5. The subsequent hydrolysis finally enables the production of phosphoric acid. The remaining material in the reactor moves on to the discharge zone, where the third and last coil provides enough energy that the phosphorus-free material does not reach the solidification temperature and finally leaves the reactor via the reactor floor. The resulting material can be divided into a metal and a slag fraction, which, however, are not yet separated from each other in the current expansion stage and are collected in a vessel below the reactor output.

The advantages of this apparatus are manifold. In comparison with an electric arc furnace (EAF), no molten bath of metal is formed so that the P_2 (g)–Fe (l) contact possibility and in further consequence the formation of iron phosphide can be decreased immensely. This fact is promoted by a thin molten film, which massively shortens the distance of mass transport. In this case it is particularly important for the diffusion of P and its removal as gas. The graphite pieces offer a large surface area for reactions and by coupling into the induction field, the heat for the endothermic reduction reactions is permanently provided directly at the respective particle surface. Even if the energy demand is increased, the main form of the reduction reaction is direct reduction, resulting in a lower carbon demand. [38] In the course of the reaction processes in the reactor, a very low oxygen partial pressure and a correspondingly high CO to CO_2 ratio is established, which in turn promotes the reduction reactions [40].

In order to use this process also for the waste stream from spent LIBs, the reactor design and the corresponding post-treatment of the output streams must be adapted. The input material for the planned continuous process comes from a pre-treatment plant, which is a fine fraction as low in Cu and Al as possible consisting of a mixture of cathode and anode materials. After being fed into the pyrometallurgical reactor, the material should react according to the principle described above. The most important difference is that the idea of the treatment of this material is to remove not only phosphorus but also lithium from the reactor via the gas flow. An initial concept for the post-treatment of the liquid fraction, which leaves the reactor chamber via its bottom, provides for an oxygen inlet. Thus, in accordance with the different oxygen affinities, for example, the input stream of NMC, LCO, NCA and LFP should result in the purest possible CoNiFe alloy. Oxygen-affine elements such as Mn and Al, as well as the residues of P and Li that are not removed via the gas phase, are to be slagged. The resulting products can therefore either be sold on the market as raw materials as required, or further broken down into their constituent parts in further post-treatment steps, for example via the hydrometallurgical route. A further additional important step to be investigated is the post-treatment of the resulting gas fraction. In particular, it will be necessary to implement a corresponding process for the separation of Li and P and consequently to treat them further according to the resulting qualities. As the points just described show, a combination with other processes should be aimed for. With regard to the overall reactor design, an adaptation of the current development will be essential. This includes issues such as the optimal refractory material for the reactor wall or a possible need to expand the gas extraction system.

2.3.2. Batch Reactor Concept

Based on the technology described in Section 2.3.1, the process design shall be adapted to the requirements of the black matter out of LIBs. For this purpose, in-depth tests were carried out for a better understanding of the material to be processed, which are partly described in Windisch-Kern et al. [34]. Since the scale of a continuous pilot plant for experiments of this kind would firstly be too complex and secondly would not correspond to the research status at the Chair of Thermal Processing Technology in the LIB field, the experiments were carried out on a post-laboratory scale. For this purpose, the reactor concept of the unit operating in batch mode was adopted from the developments in the field of sewage sludge ash and BOFS, as shown in Figure 3a and hereinafter referred to as Design 1. The system behind it is similar to the continuous concept, with the difference that the material to be investigated is already in the reactor at the beginning of the experiment and there is no material output via the ground. Most of the material melted during the experiment is accumulated and collected at the bottom of the reactor or adheres to the cube surface as spherical formations. In addition, the gas outlet is also not subjected to negative pressure, so that the resulting gases leave the reactor without constraint. For the construction of the reactor, an Al_2O_3 ring with a diameter of 20 cm, Al_2O_3 mortar and refractory concrete were used. The graphite bed provides a cube surface of 1725 cm^2 for the transfer of the induced heat. An insulation around the reactor has the function of the protection of the induction coil, to reduce the heat losses and to enable as good a separation as possible from ambient air.

To enable a qualitative measurement of the exhaust gas flow, a gas scrubber was additionally installed at the outlet of the gas flue (Figure 3b). This was realized with a bubbling frit in which the exhaust gas is enriched in a 2.5 molar H_2SO_4 solution. The temperature was measured on the outside of the reactor by two category S-thermocouples and inside the reactor by two category K-thermocouples. Due to the expected breakage of the second mentioned thermocouples, they are only used to find a correlation between the outside temperature and the inside temperature.

Preliminary tests have shown that LCO, with its high cobalt content, is highly reactive to the crucible material of Al_2O_3. In addition, sampling proved to be particularly difficult because it was not possible to separate the black mass clearly from the mortar. This makes it almost impossible to close the mass balance in the future. Another disadvantageous fact of this reactor concept is that, due to its position, the highest induction of the current takes place in the upper part of the reactor. Because of the inevitable turbulence in the reactor during the experiment due to the gases, the material accumulates at the bottom of the reactor, so the energy supply position is suboptimal. To take into account the mentioned disadvantages, a new design was developed, which is shown in Figure 4 and hereinafter referred to as Design 2.

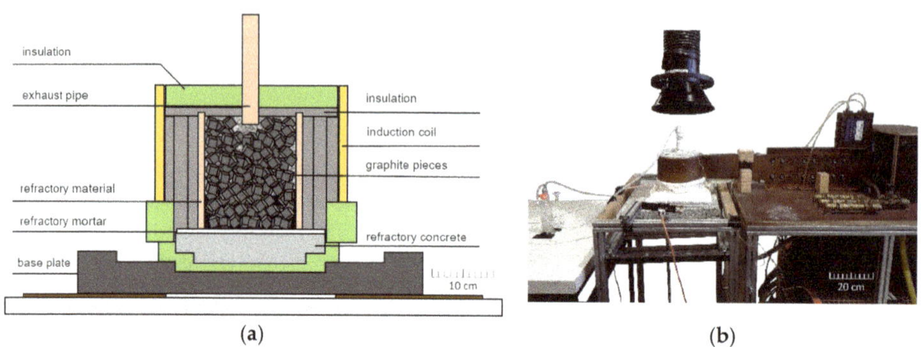

Figure 3. (**a**) Schematic representation of the original InduMelt plant (Design 1) [34]; (**b**) overall setup in test operation.

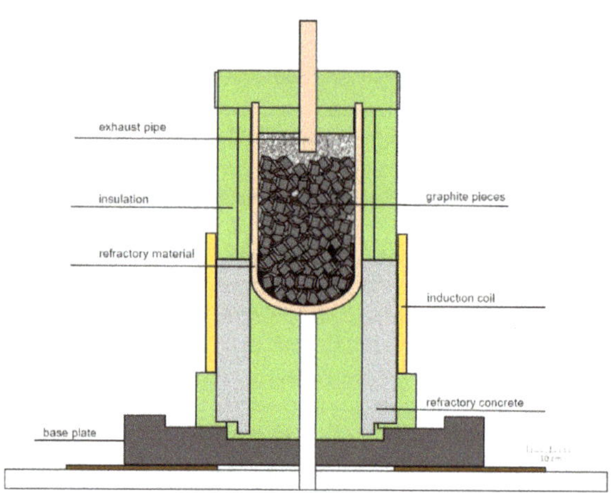

Figure 4. Schematic representation of the new reactor design (Design 2).

This is a cylindrical crucible with a half-arc bottom made of MgO. It was placed centrally on a refractory concrete structure in a way that only the lower part of the MgO crucible is within the induction coil. Appropriate insulation made of refractory matting should reduce the heat loss and thus the energy requirement and protect as far as possible against the ingress of false air from the environment. To be able to make a qualitative statement about the escaping gas during the test, an exhaust pipe made of Al_2O_3 was again implemented. The temperature was measured by a category S-thermocouple from below and in the reactor by two category K-thermocouples.

For a direct comparison of the different reactor concepts of the InduMelt plant, the same feedstock, the mixture LCO-C mentioned in Section 2.1 with a quantity of 550 g, was examined in both crucible concepts. To ensure that reproducible initial conditions prevailed in both designs, the charging of the cubes and the sample was also performed uniformly in all experiments. Thus, at the beginning, 15 cubes were positioned in the reactor and one third of the sample was charged onto them. After positioning a K-thermocouple, another 10 cubes were performed followed by addition of another third of the sample. This was repeated a second time to finally fill the reactor with 11 cubes after positioning the second

K-thermocouple. This filling quantity also represents the maximum possible capacity of Design 1. The content of Design 2 is approximately 25% larger which, however, was not utilized due to the aforementioned comparability with Design 1. After the experiment, all components of the reactor are weighed. The adhesions to the graphite cubes are removed by light mechanical processing. These adhesions are consequently separated into fractions larger and smaller than 1 mm by means of a sieve tower, together with the remaining finer fraction that may be produced. With the aid of a magnet, these are further separated into magnetic and non-magnetic, with the former finally being assigned to metal and the latter referred to as slag. Larger pieces of metal are collected together after checking with a magnet. The same is done with larger non-magnetic pieces, which are again referred to as slag. The individual fractions are finally weighed and analyzed. The main difference in sampling between Design 1 and Design 2 is the collection of the diffused areas of the grout or reactor adhesions, which will be discussed in more detail in Section 3.2.1.

The aim is to determine the interaction between the cathode material respectively the reaction products of which and the corresponding crucible material and to compare them with each other. On the other hand, the individual transfer coefficients should provide information on whether the choice of the crucible material affects the recovery rates of the individual species.

In a third trial, the basic suitability of the overall reactor concept for the treatment of LFP with the aim of removing Li and P from the material was investigated. For this purpose, Design 1 from Figure 3a was selected again in which a quantity of 394.5 g of the mixture LFP-C from Section 2.1 was charged into the reactor.

3. Results and Discussion

3.1. High-Temperature Properties of the Cathode Materials Used

In a first step to determine the behavior of cathode materials from LIBs at temperatures above 1600 °C and under reducing conditions, experiments were performed in a heating microscope. Figure 5a provides a picture of the result of the experiment with the mixture with LCO. A strong dark blue coloration was observed on the platelet. This may be due to a reaction between the Al_2O_3 platelet and cobalt to form cobalt aluminate with its typical blue appearance [41]. The product of the melting process under reducing conditions is a metal structure, which can be classified as strongly magnetic after examination with a magnet. This magnetism could also be detected in experiments with LFP, which is shown in (b). In contrast to the experiment described above, there is no blue coloration, but a brown to reddish appearance.

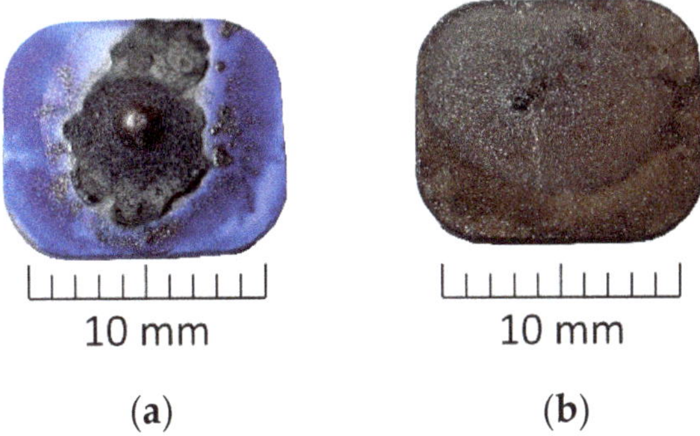

Figure 5. Condition of the sample during examination under the heating microscope: (**a**) after analysis for LCO-C; (**b**) after analysis for LFP-C.

Figure 6 displays the recording of the replication experiment LCO-C via optical measurement in the heating microscope. In Figure 6 the cross-sectional area of the experiments with LCO-C (black dotted line) and its repetition LCO-C-Re (green dotted line) over temperature during the experiment is shown. This cross-sectional area is the size of the sample cylinder detected by the heating microscope respectively its change with temperature increase.

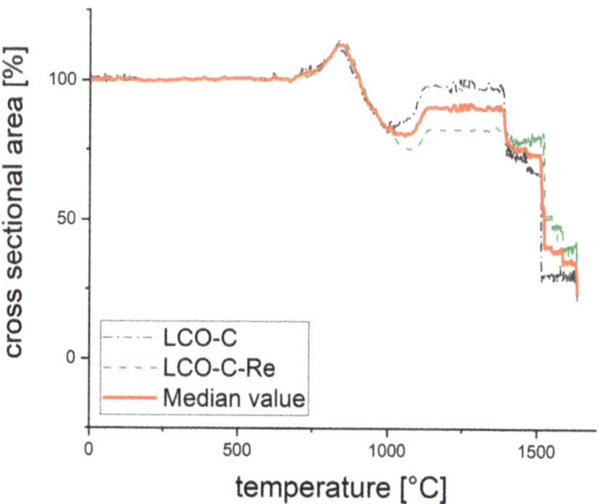

Figure 6. Results of the heating microscope of LCO-C: trend of the cross-sectional area of the experiments LCO-C and LCO-C-Re during heating and the median value of the both graphics.

It should be mentioned that by comparing the recorded images with the corresponding values of the cross-sectional area, faulty measurements caused by incorrect detection of the baseline by the heating microscope were removed from the data series. Since the basic behavior at the individual temperatures is nearly identical and differs only by different cross-sectional areas, a mean value was determined, which represents the red line. When looking at Figure 6, the first noteworthy surface changes can be observed from 675 °C onwards. From 675 °C to 845 °C a growth of the cross-sectional area was detected, which subsequently decreased again to 1054 °C with single deflections to about 80% of the original area. Up to a temperature of 1127 °C an increase in magnification was detected, which remained relatively constant with single deflections up to 1380 °C. From this temperature on, the cross-sectional area decreased continuously with a smaller slope in the range of 1393 °C to 1507 °C and a significant decrease up to 1525 °C. Up to the end there was a further decrease of the cross-sectional area, which, however, when looking at the single images from the heating microscope, can be traced back to the continuous distribution of the molten material on the platelet.

Figure 7 illustrates the results of the experiments in the heating microscope with the mixture LFP-C. The previously mentioned measurement error was particularly striking in the first experiment of LFP-C (black dotted line in Figure 7) in the range from 1163 °C to about 1400 °C. Nevertheless, a corresponding trend could be determined by correctly measuring individual values in some cases, which in turn could be confirmed by a repeated measurement of LFP-C-Re (green dotted line in Figure 7). Again, the mean value is shown in the red curve.

The results show that up to a temperature of approximately 920 °C the cross-sectional area first rises slightly and then falls back to just below 100% of the initial value. Afterwards, a pulsating enlargement of the surface takes place which decreases at about 1200 °C. After a further pulsating behavior between 1240 and 1310 °C the area is continuous again to remain relatively constant from about 1410 °C on.

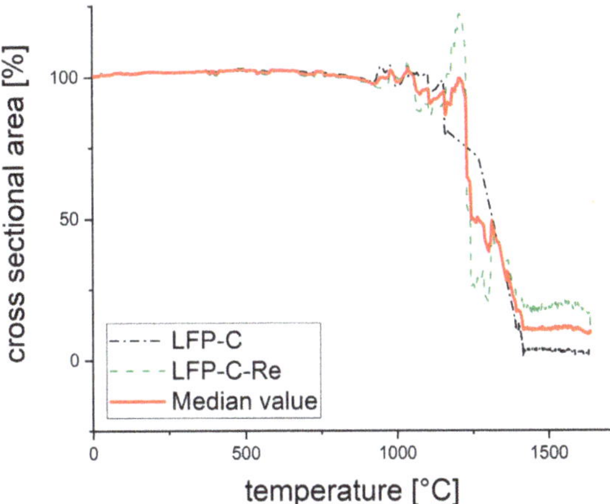

Figure 7. Results of the heating microscope of LFP-C: trend of the cross-sectional area of the experiments LFP-C and LFP-C-Re during heating and the median value of the both graphics.

3.2. InduMelt Experiments: Process Development and Suitability of the Different Reactor Concepts

The main part of the experimental investigation of LIB cathode materials was examined in the InduMelt plant. For this purpose, three experiments were carried out, which will be considered separately below according to their purpose. According to the results of the analyses, as described in Section 3.1, the maximum necessary process temperatures for the tests in the InduMelt plant were set at 1525 °C for LCO-C and 1400 °C for LFP-C. This is due to the fact that the sample material should be completely liquid at this point in time according to the heating microscope.

3.2.1. Results from Experiments with LCO-C in Both Reactor Designs

The first experiment, which is described in detail below, was carried out in Design 1. The entire experiment lasted nearly 8 h. Figure 8a shows the power input of the induction unit and the corresponding temperatures over the test time. Two type S-thermocouples (S-TC 1 and S-TC 2 in Figure 8a) were used on the reactor surface and two type K-thermocouples inside the reactor. The latter were initially installed at different heights in the reactor, one in the area of the first cube layer (K-TC bottom) and one in the upper area (K-TC top). In contrast to the S-thermocouples, the measurement results of the K-thermocouples are subject to considerable fluctuations due to melting of their insulation, influences of the material in the reactor, etc. However, an approximate temperature spread of the different thermocouple types of 500 °C could be determined up to the end.

During the process, noticeable anomalies were documented. Starting at 0.95 h and a K-TC bottom temperature of 450 °C a strong formation of condensate in the exhaust pipe to the gas scrubber was observed. This was attributed to the drying of the mortar. The temperature spread of the S-thermocouples at 1.75 h (124 °C S-TC 1) can be explained by a slight realignment of S-TC 1. Especially interesting was the continuously increasing white smoke from the exhaust pipe, which started at 5.10 h and 1180 °C internal temperature and stopped at 1341 °C. This resulted in a continuous white deposit in the exhaust pipe of the scrubber, as shown in Figure 8b. After the white smoke formation stopped, the acid in the scrubber gradually changed from transparent to a slightly yellow liquid. The assumption that the white deposits are Li or a corresponding compound could be confirmed after analysis of the liquid in the scrubber, which are summarized in Table 2.

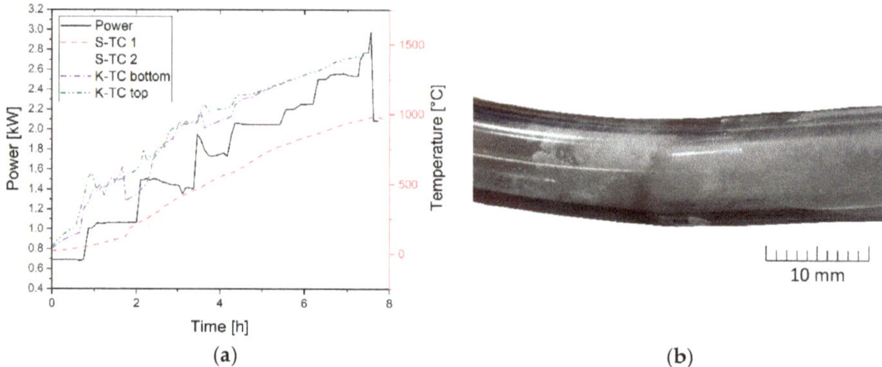

Figure 8. Experimental performance of LCO-C in Design 1: (**a**) comparison of power and temperature over time; (**b**) deposits in the exhaust pipe of the gas scrubber.

Table 2. Results from the gas scrubber respectively from its frit liquid after suction of the exhaust gas in Design 1 LCO-C in mg/L.

Fraction	Li	Co
Frit liquid	1650	0.64

Sampling after the experiment revealed a total of 4 fractions, the results of which are shown in Table 3.

Table 3. Analysis results of the fractions of the experiment LCO-C in Design 1.

Fraction	Weight (g)	Li (wt.%)	Co (wt.%)
Slag	3.20	5.62	0.12
Mortar	145.60	4.52	0.12
Metal	251.70	0.01	100.00
Powder	37.30	1.49	53.4

The fraction defined therein as slag could be identified as dark to light grey non-magnetic pieces smaller than 10 mm with minimal metallic inclusions, as illustrated in Figure 9a. The mortar shown represents the part into which the test material has diffused. This is optically visible by a dark discoloration of the originally white mortar. In Figure 9b the reactor is demonstrated from below after the concrete floor has been separated. During sampling, care was taken to find the clearest possible separation between the white mortar and the diffused areas, but this proved to be very difficult. The largest product of the experiment in terms of mass was the metal fraction, which could be obtained in pieces larger than 10 mm. The metal piece shown in Figure 9c serves as an example. The analysis showed an impressive purity of 100% Co and an impurity of only 0.01% Li. It should be noted that there may be some variation in sampling and digestion errors, resulting in the overall result not reaching exactly 100%. The fourth fraction was a powder with particles smaller than 1 mm, as can be seen in Figure 9d, which was mostly magnetic. This property is also confirmed by analyses with a cobalt content of 53.4%.

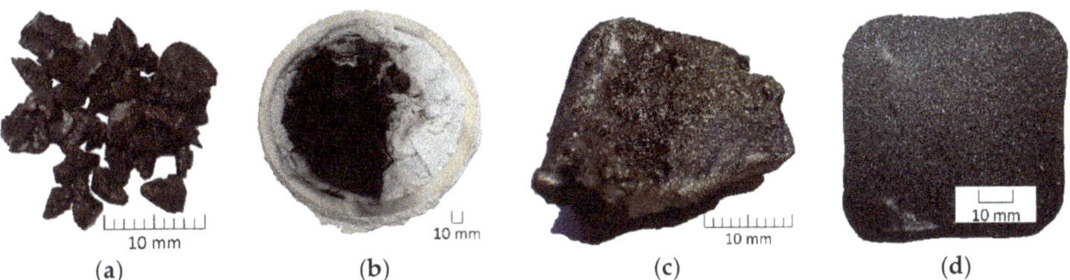

Figure 9. Products of the experiment LCO-C in Design 1: (**a**) slag; (**b**) ceramic ring and mortar seen from the bottom; (**c**) metal; (**d**) powder.

Taking into account the respective weighed masses and the analysis results, the transfer coefficients of the individual elements of the fractions were calculated. In detail, the analyses from ICP-MS and ICP-OES of the individual fractions were converted to mass percent and multiplied by the weighed mass at sampling. By adding the respective element masses, a total mass per element could be determined. This represents the amount that was still detectable in the fractions in the reactor after the test. Afterwards, a comparison of the masses before and after the experiment was carried out. The difference was assumed to be a transfer into the gas flow leaving the reactor during the experiment or a transfer into the individual solid fractions. The results of this calculation can be seen in Figure 10.

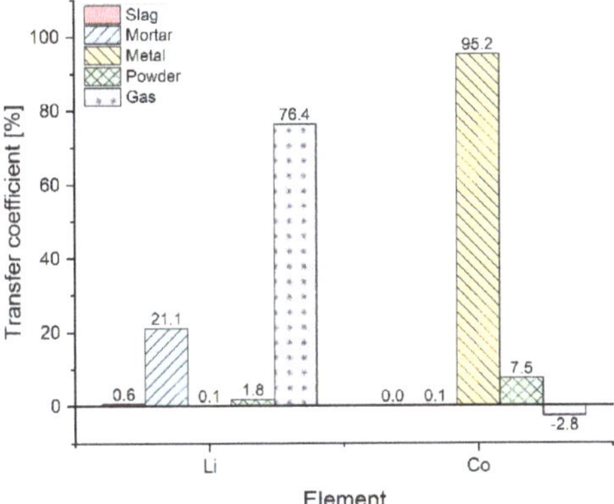

Figure 10. Transfer coefficients of the elements into the individual fractions in % of the experiment in Design 1.

At this point it should be mentioned that the transfer coefficients determined must be seen as initial guide values and internal comparison values and must be confirmed accordingly by repeated experiments in the optimum reactor setup. Nevertheless, the trend was also observed in experiments with NMC and NCA, as described in Windisch-Kern et al. [34]. The result of the transfer coefficients in Figure 10 shows a Li removal rate of over 76% into the gas stream from the cathode material used. Based on the thermokinetic consideration of LCO by Kwon et al. [16] and assuming that most of the Li has left the reactor during the phase of white smoke (approximately 1160–1340 °C), it can be assumed that it is Li_2O. The transfer of over 21% Li into the mortar can be considered as an undesirable result. The percentage of Li in the slag is not negligible in the analyses (Table 2) with 5.6%, but due to the small quantity of slag it is insignificant for the total consideration with 0.6%. The small amount of Li in the metal (0.1%) is a great result with regard to the purest possible metal fraction. A further potential for improvement can be seen when considering the Li content of 1.8% in the powder, whereby this value can possibly be lowered with a longer holding time of the final temperature. The result of Co can be interpreted as extremely promising. Only 7.5% is found in the powder and 95.2% in the metal, which can be directly transferred for further use in the corresponding metal industry. The resulting difference to 100% can be explained by the extremely difficult sampling, especially the identification of the individual fractions and the subsequent weighing. At this point, the proportion in the slag and mortar can also be neglected with less than 0.1%. The comparison of these results with other processes is difficult at this point because the composition of the input material differs significantly from a real waste stream of LIB. Nevertheless, by using pure cathode material, without impurities such as Al or Cu, a value of the theoretically maximum possible removal rate of Li can be determined. Vest [15] describes a Li_2O transfer rate from the waste stream of LIB of 40.5% in their process based on an electric arc furnace. Even though a direct comparison with this value is not possible, the gap between Vest's result and the theoretically possible value in this method shows an enormous potential.

Much more remarkable in this context is the result of Design 2. The temperature record of the LCO-C experiment in Design 2, which can be viewed in Figure 11 in combination with the power input over time with the same naming as in the previous experiment described above, shows that the temperature is highest in the lower part of the reactor. Thus, the goal of the reactor design of a more targeted temperature provision in the lower area could be realized.

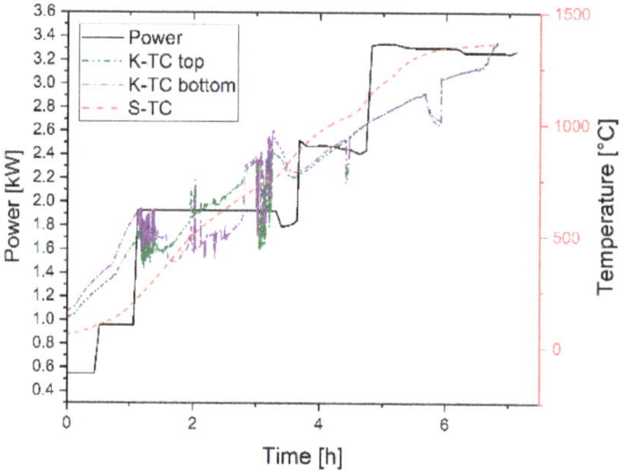

Figure 11. Experimental performance of LCO-C in Design 2, comparison of power and temperature over time.

The extreme fluctuations of the K-thermocouples between the test duration of approximately 1 to 3 h could be explained in retrospect in such a way that after the insulation around the thermocouple wires had melted, they reconnected at a higher point in the reactor. The initial theory could be confirmed when the reactor was opened after the experiment, because the thermocouples could be found in the upper part of the reactor free of cubes and no longer in the cube bed. Again, a white smoke formation with the same deposits in the exhaust pipe of the gas scrubber could be detected. This phenomenon occurred at an S-thermocouple temperature range from 1165 to 1340 °C. Again, as the smoke intensity decreased, a successive discoloration of the acid in the scrubber from transparent to a light yellow was to be determined. The results of the acid analysis from the gas scrubber can be taken from Table 4. The high value of Li confirms the impression, as already assumed in the experiment in Design 1, that Li can be removed from the reactor via the gas flow.

Table 4. Results from the frit liquid of the gas scrubber after suction of the exhaust gas in Design 2 LCO-C in mg/L.

Fraction	Li	Co
Frit liquid	1230	1.2

The results of the investigation of the test material LCO-C in Design 2 can be seen in Table 5. Essentially, the analysis differs from the experiment in Design 1 only in that there was no mortar due to the construction. When the MgO crucible was weighed after the experiment, it was found to be 81.2 g heavier than the initial weight. This could be attributed to adhesions on the crucible, which were mechanically extracted as completely as possible. The result was a fine powder. Despite considerable mechanical effort, only 3.8 g could be removed from the crucible without damage, which will be referred to as crucible adhesion in the following.

Table 5. Results from the experiment LCO-C in Design 2.

Material	Weight (g)	Li (wt.%)	Co (wt.%)
Slag	1.3	5.59	3.3
Crucible adhesion	3.8	6.29	27.2
Metal	242.2	0.12	93.9
Powder	35.2	0.69	66.6

The fractions did not differ in their appearance from those in Figure 9a,c,d. Only the metal pieces were larger, as shown in Figure 12. Analysis of the metal fraction revealed a purity of Co of 93.9% with a negligible amount of 0.12% Li.

Figure 12. Metal fraction from the LCO-C experiment in Design 2.

The transfer coefficients, which are illustrated in Figure 13, were determined from the results in Table 5, taking into account the corresponding weight changes of the individual fractions during the experiment.

This result represents a unique selling point in the pyrometallurgical processing of cathode material from LIB. Compared to Design 1, even though only 85.9% of the Co was transferred to an almost pure Co-metal phase, more than 97% of the Li were removed from the material and the reactor via the gas flow. Thus, a nearly 21% higher Li removal rate could be achieved in Design 2.

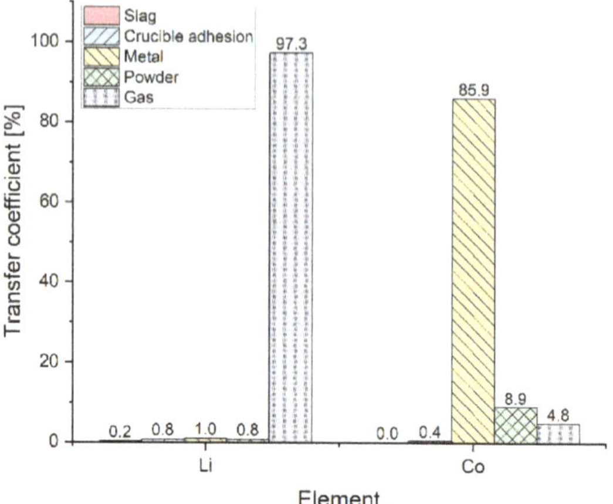

Figure 13. Transfer coefficients of the elements into the individual fractions in % of the experiment in Design 2.

Since the attribution of the Co not found to the gas phase is rather questionable to this extent and cannot be traced back exclusively to errors in sampling and analysis, a closer look at the results is necessary. In addition to the result display in Figure 13, another variant is possible. It is assumed that the difference of the weighed crucible adhesion (81.2 g) to the extracted amount (3.8 g) consists of the same composition as the extracted fraction. This results in a lithium removal rate of 81.7% with a Co value that was not found (i.e., attributed to the gas phase) of −3.12%. This variant cannot clarify the difference to 100%, but by combining the two methods, one obtains a range in which the transfer coefficients move.

Besides this result, Design 2 also turns out to be a better choice when considering the interactions between the sample material and the reactor. As shown in Figure 14a, a massive attack of the reactor wall was observed in Design 1 (Al_2O_3) with ring diameter reductions of up to 0.2 mm. On the other hand, Figure 14b shows the reactor in Design 2 (MgO) after the experiment. From the difference between the weighing before and after the test, it is known that the reactor was over 81 g heavier afterwards. The theory of adhesion to the reactor can also be seen in the illustration, whose boundary is marked with the red arrow.

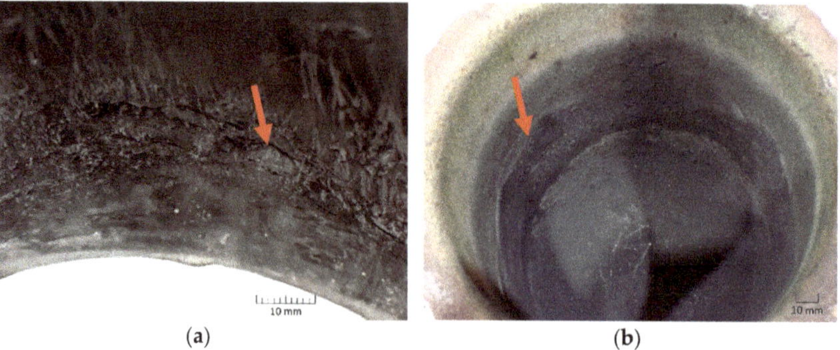

(a)   (b)

Figure 14. Visual appearance of the crucibles after the experiments: (**a**) traces of attack on the Al_2O_3 crucible wall in Design 1; (**b**) appearance of the MgO crucible after experiment 2 with obvious adhesions.

This factor is particularly important for long-term experiments in a continuous set-up. If the sample material and the Al_2O_3-ceramics are in contact for a longer period of time, this attack would lead to a destruction of the reactor. In addition to the advantages already mentioned, Design 2 also features a much simpler construction and therefore easier sampling. A drawback of Design 2 is the higher energy input required, which can be seen in the comparison of the power curve of Figures 8 and 11. However, this can be solved by an improved positioning of the induction coil. Furthermore, the consequences of adhesions in continuous operation must be investigated.

Even though the target temperatures were not reached in the experiments with LCO-C, it can still be assumed that a sufficiently high temperature was reached. Firstly, the temperature in the reactor was measured in the space between the refractory mat and the graphite cubes, as mentioned above, which implies that the temperature in the cube bed must have been even higher due to the heat input in it. In addition, if the temperature was too low, the appearance of the products would be different. An example of this is the metal from Co, as shown in Figures 9c and 12, whose melting point is known to be 1495 °C.

3.2.2. Results from Experiments with LFP-C

The test with LFP-C was carried out in Design 1. Since the temperature measurement via the K-thermocouples was already faulty from the beginning of the experiment and a repair was no longer possible at this point in time, only the curves of the recordings from the S-thermocouples are visible in Figure 15. However, the delta value to the K-thermocouples should be similar to that in the LCO-C experiment in Design 1, whereby approximately 500 °C can be added to the value of the S-thermocouples to determine the internal temperature at higher temperatures.

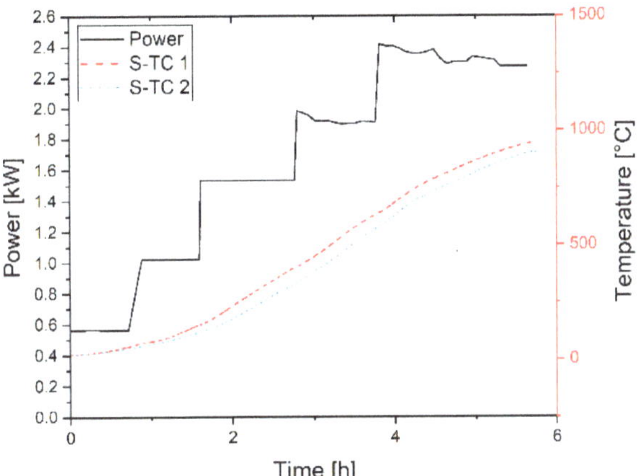

Figure 15. Experimental performance of LFP-C in Design 1, comparison of power and temperature over time.

During the heating process, smoke development was particularly noticeable at an outside temperature of approximately 750 °C (approximate internal temperature of 1210 °C), which completely ignited after a short time. This flame, which is an indication of the reaction of phosphorus with oxygen [42], could finally be detected constantly up to an outside temperature of approximately 860 °C (approximate internal temperature of 1310 °C) as shown in Figure 16a. During this time the acid in the gas scrubber changed its color to a brownish liquid. The results of the exhaust gas analysis can be taken from Table 6.

Figure 16. Products of the experiment LFP-C in Design 1: (**a**) flame formation in exhaust gas flow; (**b**) slag; (**c**) ceramic ring and mortar seen from the bottom; (**d**) metal.

Table 6. Results from the frit liquid of the gas scrubber after suction of the exhaust gas in Design 1 for LFP-C in mg/L.

Fraction	Li	Fe	P
Frit liquid	2.0	1.5	200

In the exhaust gas analysis only a small value for P and a very small amount of Li could be found, which is possibly due to the formation of a flame out of the exhaust pipe. In order to be able to make a statement about the efficiency of the reactor concept for the recycling of LFP, the results of the ICP-OES or ICP-MS must be examined more closely.

A total of 5 fractions could be detected during sampling. The appearance of the fraction classified as non-magnetic slag (Slag 1) differed significantly from that in Figure 9a. Individual spheres with a diameter of up to 5 mm were detected, as shown in Figure 16b. In comparison with the appearance of the magnetic metal fraction in Figure 16d, the difficulty of clearly classifying the individual fractions is obvious. The analysis showed that the Fe content in the slag was even higher than in the material identified as metal. In addition, however, a significant amount of Li (3.19%) and P (15.9%) was also analyzed. Slag 2 in Table 7 is a non-magnetic powder with particles smaller than 1 mm the appearance of which is the same as in Figure 9d. The same applies to the magnetic material identified as powder. In this experiment, care was again taken during sampling to separate as much diffused areas of the sample material from the mortar. These brownish areas are also visible in Figure 16c, which shows the ceramic ring with the mortar after removal of the refractory concrete. Analysis of the metal displayed in Figure 16d shows only 51 wt.% Fe and over 8 wt.% P. This low Fe content suggests that no complete reduction has occurred. A further indication for the correctness of this assumption is the fact that during mechanical processing of the fraction with a hammer, the spheres disintegrated into a powder already with a small amount of force.

Table 7. Results of the individual fractions after the experiment LFP-C in Design 1.

Fraction	Weight (g)	Li (wt.%)	Fe (wt.%)	P (wt.%)
Slag 1	36.1	3.19	53.60	15.90
Slag 2	16.3	4.95	0.61	7.89
Mortar	66.0	0.90	3.29	1.66
Metal	96.5	0.89	51.00	8.24
Powder	70.5	1.91	35.90	8.43

Referring to the respective masses of the individual fractions, the transfer coefficients can be taken from Figure 17.

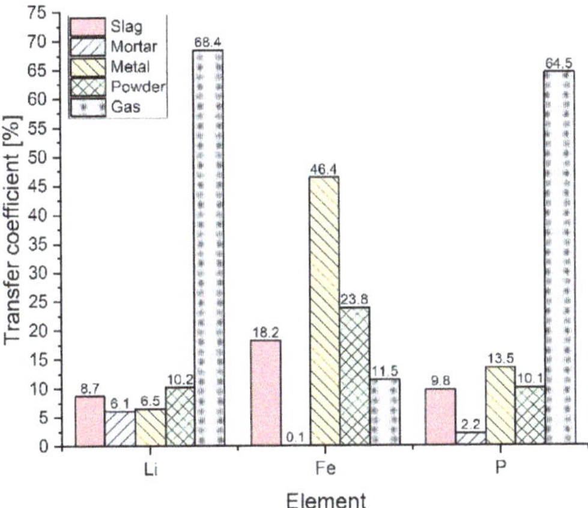

Figure 17. Transfer coefficients of the elements into the individual fractions in % of the experiment in Design 1.

Of particular interest are the removal rates of Li (68.4%) and P (64.5%) over the gas flow. The remaining amount of Li here is distributed relatively evenly among the other fractions with a slight concentration on the remaining powder. As already mentioned, it is reasonable to assume that no complete reduction of Fe has occurred. This is also reflected in a considerable value of phosphorus in the metal fraction.

A direct comparison of the experiments in Design 1 shows that the Li removal rate for LFP-C of 68.4% is 8% lower than in the experiment with LCO-C. However, the parallel removal of phosphorus of 64.5% represents a respectable result. It also can be seen that in both the LCO-C test with a transfer coefficient of 21.1% (Figure 10) for Li and the LFP-C test with 6.1% (Figure 17), a significant amount of Li was transferred to the mortar. If the results of LCO-C in Design 2 are also included, it can be assumed that gasification rates for LFP-C are better in this construction method. Looking at the transfer coefficients in Figure 17, a significant portion of the Fe (11.5%) is attributed to the gas flow. De facto, this is the amount that was not recovered during sampling compared to the input amount. The correctness or falsification of this classification and the above mentioned assumptions must be investigated in further experiments. Particular attention must be paid to safety in the pyrometallurgical removal of phosphorus from LIB, consisting of the cathode material LFP, in the exhaust gas post-processing. In addition to oxidation in the air, as shown in Figure 16a, the high toxicity [43] is also of particular importance. These factors must be given special consideration in future developments of the reactor concept presented.

4. Conclusions

Within the scope of this paper, the suitability of a new pyrometallurgical recycling process associated with materials from LIBs for the recovery of valuable metals was investigated. With the background of a continuous adaptation of the reactor concept to the waste stream from spent LIBs, two different reactor designs were used, in each of which the cathode material LCO with carbon addition was examined for better comparability. In a third trial, the basic capability of the technology for the treatment of LFP was also examined. In addition, knowledge about the behavior of the examined cathode materials used in high-temperature applications was investigated in an upstream step in a heating microscope.

The transfer coefficients determined in the experiments of the novel pyrometallurgical recycling process serve exclusively as a comparison of the efficiency of the presented reactor concepts or as a first benchmark of the basic suitability of the process for the treatment of LFP.

From the experiments in the heating microscope the maximum necessary temperatures for the transformation into a molten phase could be determined. This state of aggregation is necessary in the long run to meet the requirements of the theoretically determined principle of a continuous process. For the experiments with LCO a temperature of 1525 °C and for LFP 1400 °C could be determined.

In the trial LCO-C in Design 1, 95.2% Co of the original input fraction was converted into the metal, with a purity of Co of 100%. Due to the high Li content in the mortar only 76.4% of the Li could be transferred into the exhaust gas flow. This contrasts with the result of the experiment with LCO-C in Design 2. Its analysis shows a metal purity of 93.9% Co and a remarkable lithium removal rate in a range from 81.7% to 97.3%. In addition to this impressive lithium removal rate, Design 2 has also proven to be the better choice for future use due to its interaction with the feed material. For example, massive interactions and attacks on the Al_2O_3 crucible have been detected in Design 1, whereas there were no physical damages with the MgO crucible in Design 2. Although the danger of destruction of the reactor wall during long-term experiments in a future continuous process has been averted, the effect of the detected adhesions on the MgO crucible still needs to be investigated in further tests. However, the initially formulated goal of a more targeted heat supply in the lower part of the crucible was achieved by Design 2.

The experiment with LFP-C was performed in Design 1 and achieved a lithium removal rate of 68.4% with parallel phosphorus removal of 64.5%. Since the results of the LCO-C experiments showed that a higher lithium removal could be achieved when using Design 2, a repetition of the LFP-C experiment in this setup can be expected to result in a higher removal rate. In addition, since sampling in this experiment has proven to be particularly difficult due to the appearance of the fractions, and since detailed examination of the results has revealed questions that need to be clarified, such as the undetectable amount of Fe, further investigations are indispensable.

Nevertheless, it can be summarized that Design 2 with its MgO crucible has proven to be a better choice with regard to its suitability for pyrometallurgical treatment of material from LIBs. This is due to its inertness to the sample material as well as the higher Li removal rate determined. In addition, it was found that the use of the technology is also suitable for the cathode material LFP and that considerable P and Li removal rates have already been achieved. However, in order to be able to treat the fluctuating waste stream from spent LIBs with an appropriate efficiency, in-depth investigations are needed beforehand to gain knowledge of the behavior of all common cathode materials. In order to increase efficiency the fraction referred to as slag must also be subjected to more detailed investigations in the future, for example with an XRD analysis. From this, knowledge of the phases present is to be generated and the formation of these is to be suppressed with targeted measures. In the current development phase, this has not yet been the focus of research. Furthermore it is necessary to identify the influence of additional fractions such as Cu and Al from conductor foils on the process.

Compared to commercial techniques used today for recycling spent LIBs, the simultaneous recovery of lithium and phosphorus via the process presented in this paper is its most significant advantage. This first potential assessment for pyrometallurgical recovery of Li would also theoretically meet the requirements of the proposed amendment to the EU Directive 2006/66/EC of a Li recovery up to 70% by 2030. Aspects such as the economics, energy efficiency and environmental impact of this intermediate step in the overall recycling chain, as well as possible recovery rates of a waste stream of spent LIBs in the new reactor design, need to be determined in further studies.

Author Contributions: Conceptualization, A.H. and S.W.-K.; methodology, A.H.; investigation, A.H., S.W.-K. and C.P.; resources, A.H., S.W.-K. and C.P.; writing—original draft preparation, A.H.; writing—review and editing, A.H., S.W.-K., C.P. and H.R.; visualization, A.H.; supervision, C.P. and H.R.; project administration, C.P.; funding acquisition, H.R. All authors have read and agreed to the published version of the manuscript.

Funding: This research was funded by the Zukunftsfonds Steiermark with funds from the province of Styria, Austria, grant number GZ: ABT08-189002/2020 PN:1305.

Institutional Review Board Statement: Not applicable.

Informed Consent Statement: Not applicable.

Data Availability Statement: The data presented in this study are available on request from the corresponding author.

Conflicts of Interest: The authors declare no conflict of interest.

References

1. Pillot, C. The Rechargeable Battery Market and Main Trends 2018–2030. In Proceedings of the ICBR 2019, Lyon, France, 18–20 September 2019.
2. Palacín, M.R.; de Guibert, A. Why do batteries fail? *Science* **2016**, *351*, 1253292. [CrossRef]
3. Zhang, X.; Xie, Y.; Lin, X.; Li, H.; Cao, H. An overview on the processes and technologies for recycling cathodic active materials from spent lithium-ion batteries. *J. Mater. Cycles Waste Manag.* **2013**, *15*, 420–430. [CrossRef]
4. Ordoñez, J.; Gago, E.J.; Girard, A. Processes and technologies for the recycling and recovery of spent lithium-ion batteries. *Renew. Sustain. Energy Rev.* **2016**, *60*, 195–205. [CrossRef]
5. Sojka, R.; Pan, Q.; Billman, L. Comparative Study of Lithium-Ion Battery Recycling Processes. In Proceedings of the ICBR 2020, Salzburg, Austria, 16–18 September 2020.
6. Li, L.; Zhang, X.; Li, M.; Chen, R.; Wu, F.; Amine, K.; Lu, J. The Recycling of Spent Lithium-Ion Batteries: A Review of Current Processes and Technologies. *Electrochem. Energy Rev.* **2018**, *1*, 461–482. [CrossRef]
7. Blengini, G.A.; Latunussa, C.E.; Eynard, U.; Torres de Matos, C.; Wittmer, D.; Georgitzikis, K.; Pavel, C.; Carrara, S.; Mancini, L.; Unguru, M.; et al. *Study on the EU's list of Critical Raw Materials (2020) Final Report*; Publications Office of the European Union: Luxembourg, 2020. [CrossRef]
8. Arthur D Little. Available online: https://www.adlittle.com/en/insights/viewpoints/future-batteries (accessed on 26 December 2020).
9. Huang, B.; Pan, Z.; Su, X.; An, L. Recycling of lithium-ion batteries: Recent advances and perspectives. *J. Power Sources* **2018**, *399*, 274–286. [CrossRef]
10. Duesenfeld GmbH, Ecofriendly Recycling of Lithium-Ion Batteries. Available online: https://www.duesenfeld.com/recycling_en.html (accessed on 26 December 2020).
11. Accurec Recycling GmbH, Lithium Batterie Recycling. Available online: https://accurec.de/lithium?lang=de (accessed on 26 December 2020).
12. Redux GmbH, Redux Smart Battery Recycling. Available online: https://www.redux-recycling.com/de (accessed on 26 December 2020).
13. Arnberger, A.; Coskun, E.; Rutrecht, B. Recycling von Lithium-Ionen-Batterien. In *Recycling und Rohstoffe*; Thiel, S., Thomé-Kozmiensky, E., Goldmann, D., Eds.; TK Verlag: Nietwerder, Germany, 2018.
14. Liu, C.; Lin, J.; Cao, H.; Zhang, Y.; Sun, Z. Recycling of spent lithium-ion batteries in view of lithium recovery: A critical review. *J. Clean. Prod.* **2019**, *228*, 801–813. [CrossRef]
15. Vest, M. Weiterentwicklung des pyrometallurgischen IME Recyclingverfahrens für Li-Ionen Batterien von Elektrofahrzeugen. Ph.D. Thesis, RWTH Aachen University, Aachen, Germany, 28 January 2016.
16. Kwon, O.; Sohn, I. Fundamental thermokinetic study of a sustainable lithium-ion battery pyrometallurgical recycling process. *Resour. Conserv. Recycl.* **2020**, *158*, 104809. [CrossRef]
17. Abdou, T.R.; Espinosa, D.C.R.; Tenório, J.A.S. Recovering of Carbon Fiber Present in an Industrial Polymeric Composite Waste through Pyrolysis Method while Studying the Influence of Resin Impregnation Process: Prepreg. In *REWAS 2016*; Kirchain, R.E., Blanpain, B., Meskers, C., Olivetti, E., Apelian, D., Howarter, J., Eds.; Springer International Publishing: Cham, Switzerland, 2016; pp. 313–318. [CrossRef]
18. Beheshti, R.; Tabeshian, A.; Aune, R.E. Lithium-Ion Battery Recycling Through Secondary Aluminum Production. In *Energy Technology 2017*; Zhang, L., Drelich, J.W., Neelameggham, N.R., Guillen, D.P., Haque, N., Zhu, J., Eds.; The Minerals, Metals & Materials Series; Springer International Publishing: Cham, Switzerland, 2017; Volume 135, pp. 267–274. [CrossRef]
19. Gao, R.; Xu, Z. Pyrolysis and utilization of nonmetal materials in waste printed circuit boards: Debromination pyrolysis, temperature-controlled condensation, and synthesis of oil-based resin. *J. Hazard. Mater.* **2019**, *364*, 1–10. [CrossRef]
20. Li, J.; Wang, G.; Xu, Z. Environmentally-friendly oxygen-free roasting/wet magnetic separation technology for in situ recycling cobalt, lithium carbonate and graphite from spent $LiCoO_2$/graphite lithium batteries. *J. Hazard. Mater.* **2016**, *302*, 97–104. [CrossRef]
21. Xiao, J.; Li, J.; Xu, Z. Novel Approach for in Situ Recovery of Lithium Carbonate from Spent Lithium Ion Batteries Using Vacuum Metallurgy. *Environ. Sci. Technol.* **2017**, *51*, 11960–11966. [CrossRef]
22. Werner, D.; Peuker, U.A.; Mütze, T. Recycling Chain for Spent Lithium-Ion Batteries. *Metals* **2020**, *10*, 316. [CrossRef]

23. Yin, H.; Xing, P. Pyrometallurgical Routes for the Recycling of Spent Lithium-Ion Batteries. In *Recycling of Spent Lithium-Ion Batteries*; An, L., Ed.; Springer International Publishing: Cham, Switzerland, 2019; Volume 4, pp. 57–83. [CrossRef]
24. Elwert, T.; Frank, J. Auf dem Weg zu einem geschlossenen Stoffkreislauf für Lithium-Ionen-Batterien. In *Recycling und Sekundärrohstoffe*; Thomé-Kozmiensky, E., Holm, O., Friedrich, B., Goldmann, D., Eds.; Thomé-Kozmiensky Verlag GmbH: Neuruppin, Germany, 2020.
25. Bai, Y.; Muralidharan, N.; Sun, Y.K.; Passerini, S.; Stanley Whittingham, M.; Belharouak, I. Energy and environmental aspects in recycling lithium-ion batteries: Concept of Battery Identity Global Passport. *Mater. Today* **2020**, *41*, 304–315. [CrossRef]
26. Guoxing, R.; Songwen, X.; Meiqiu, X.; Bing, P.; Youqi, F.; Fenggang, W.; Xing, X. Recovery of Valuable Metals from Spent Lithium-Ion Batteries by Smelting Reduction Process Based on MnO$_2$-SiO$_2$-Al$_2$O$_3$ Slag System. In *Advances in Molten Slags, Fluxes, and Salts, Proceedings of the 10th International Conference on Molten Slags, Fluxes and Salts 2016, Seattle, WA, USA, 22–25 May 2016*; Reddy, R., Chaubal, P., Pistorius, P.C., Pal, U., Eds.; Springer International Publishers: Cham, Switzerland, 2016; pp. 210–218. [CrossRef]
27. Forte, F.; Pietrantonio, M.; Pucciarmati, S.; Puzone, M.; Fontana, D. Lithium iron phosphate batteries recycling: An assessment of current status. *Crit. Rev. Environ. Sci. Technol.* **2020**. [CrossRef]
28. Yao, Y.; Zhu, M.; Zhao, Z.; Tong, B.; Fan, Y.; Hua, Z. Hydrometallurgical Processes for Recycling Spent Lithium-Ion Batteries: A Critical Review. *ACS Sustain. Chem. Eng.* **2018**, *6*, 13611–13627. [CrossRef]
29. Meshram, P.; Pandey, B.D.; Mankhand, T.R.; Deveci, H. Comparision of Different Reductants in Leaching of Spent Lithium Ion Batteries. *JOM* **2016**, *68*, 2613–2623. [CrossRef]
30. Ghassa, S.; Farzanegan, A.; Gharabaghi, M.; Abdollahi, H. Novel bioleaching of waste lithium ion batteries by mixed moderate thermophilic microorganisms, using iron scrap as energy source and reducing agent. *Hydrometallurgy* **2020**, *197*, 105465. [CrossRef]
31. Zheng, X.; Zhu, Z.; Lin, X.; Zhang, Y.; He, Y.; Cao, H.; Sun, Z. A Mini-Review on Metal Recycling from Spent Lithium Ion Batteries. *Engineering* **2018**, *4*, 361–370. [CrossRef]
32. Elwert, T.; Römer, F.; Schneider, K.; Hua, Q.; Buchert, M. Recycling of Batteries from Electric Vehicles. In *Behaviour of Lithium-Ion Batteries in Electric Vehicles. Green Energy and Technology*; Pistoia, G., Liaw, B., Eds.; Springer International Publishing AG: Cham, Switzerland, 2018; pp. 289–321. [CrossRef]
33. Republik Österreich Parlament. Available online: https://www.parlament.gv.at/PAKT/EU/XXVII/EU/04/37/EU_43776/imfname_11029480.pdf (accessed on 21 December 2020).
34. Windisch-Kern, S.; Holzer, A.; Ponak, C.; Raupenstrauch, H. Pyrometallurgical lithium-ion-battery recycling: Approach to limiting lithium slagging with the InduRed reactor concept. *Processes* **2021**, *9*, 84. [CrossRef]
35. HENSCHKE GmbH, Graphite Electrode Specification. Available online: https://henschkegmbh.de/index.php?option=com_content&view=article&id=55&Itemid=63&lang=en (accessed on 29 August 2019).
36. Schönberg, A.; Samiei, K.; Kern, H.; Raupenstrauch, H. Der RecoPhos-Prozess—Rückgewinnung von Phosphor aus Klärschlammasche. *Osterr. Wasser Abfallwirtsch.* **2014**, *66*, 403–407. [CrossRef]
37. Samiei, K.; Schönberg, A. *Basic Design of the InduCarb Reactor: Power Input for a Homogeneous Temperature Distribution*; Chair of Thermal Processing Technology (Montanuniversitaet Leoben): Leoben, Austria, 2014; unpublished.
38. Ponak, C. Carbo-Thermal Reduction of Basic Oxygen Furnace Slags with Simultaneous Removal of Phosphorus via the Gas Phase. Ph.D. Thesis, Montanuniversitaet Leoben, Leoben, Austria, 2 September 2019.
39. Ponak, C.; Windisch, S.; Mally, V.; Raupenstrauch, H. Recovery of Manganese, Chromium, Iron and Phosphorus from Basic Oxygen Furnace Slags. In *Optimum Utilization of Resources and Recycling for a Sustainable Solution*; GDMB Verlag GmbH: Clausthal-Zellerfeld, Germany, 2019; Volume 3, pp. 1311–1319.
40. Ponak, C.; Montanuniversitaet Leoben, Leoben, Austria. Personal communication, 2020.
41. Taguchi, M.; Nakane, T.; Hashi, K.; Ohki, S.; Shimizu, T.; Sakka, Y. Reaction temperature variations on the crystallographic state of spinel cobalt aluminate. *Dalton Trans.* **2013**, *42*, 7167–7176. [CrossRef]
42. Klemenc, A. *Anorganische Chemie auf Physikalisch-Chemischer Grundlage*; Springer Vienna: Vienna, Austria, 1951; pp. 201–208.
43. Sedlmeyer, J. Über Phosphorvergiftungen. *Dtsch. Z. Gesamte Gerichtl. Med.* **1932**, *19*, 365–383. [CrossRef]

 *metals*

Article

Early-Stage Recovery of Lithium from Tailored Thermal Conditioned Black Mass Part I: Mobilizing Lithium via Supercritical CO$_2$-Carbonation

Lilian Schwich *, Tom Schubert and Bernd Friedrich

IME, Institute for Process Metallurgy and Metal Recycling, RWTH Aachen University, 52056 Aachen, Germany; tom.schubert@rwth-aachen.de (T.S.); bfriedrich@ime-aachen.de (B.F.)
* Correspondence: lschwich@ime-aachen.de; Tel.: +49-2418-095-194

Abstract: In the frame of global demand for electrical storage based on lithium-ion batteries (LIBs), their recycling with a focus on the circular economy is a critical topic. In terms of political incentives, the European legislative is currently under revision. Most industrial recycling processes target valuable battery components, such as nickel and cobalt, but do not focus on lithium recovery. Especially in the context of reduced cobalt shares in the battery cathodes, it is important to investigate environmentally friendly and economic and robust recycling processes to ensure lithium mobilization. In this study, the method early-stage lithium recovery ("ESLR") is studied in detail. Its concept comprises the shifting of lithium recovery to the beginning of the chemo-metallurgical part of the recycling process chain in comparison to the state-of-the-art. In detail, full NCM (Lithium Nickel Manganese Cobalt Oxide)-based electric vehicle cells are thermally treated to recover heat-treated black mass. Then, the heat-treated black mass is subjected to an H$_2$O-leaching step to examine the share of water-soluble lithium phases. This is compared to a carbonation treatment with supercritical CO$_2$, where a higher extent of lithium from the heat-treated black mass can be transferred to an aqueous solution than just by H$_2$O-leaching. Key influencing factors on the lithium yield are the filter cake purification, the lithium separation method, the solid/liquid ratio, the pyrolysis temperature and atmosphere, and the setup of autoclave carbonation, which can be performed in an H$_2$O-environment or in a dry autoclave environment. The carbonation treatments in this study are reached by an autoclave reactor working with CO$_2$ in a supercritical state. This enables selective leaching of lithium in H$_2$O followed by a subsequent thermally induced precipitation as lithium carbonate. In this approach, treatment with supercritical CO$_2$ in an autoclave reactor leads to lithium yields of up to 79%.

Keywords: battery recycling; lithium-ion batteries; metallurgical recycling; metal recovery; recycling efficiency; carbonation; lithium phase transformation; autoclave; supercritical CO$_2$

Citation: Schwich, L.; Schubert, T.; Friedrich, B. Early-Stage Recovery of Lithium from Tailored Thermal Conditioned Black Mass Part I: Mobilizing Lithium via Supercritical CO$_2$-Carbonation. *Metals* **2021**, *11*, 177. https://doi.org/10.3390/met1020177

Received: 19 November 2020
Accepted: 18 January 2021
Published: 20 January 2021

1. Introduction

The need for lithium recovery from LIBs is a crucial topic in terms of increased electromobility since lithium is and will remain a relevant element also in next-generation batteries. Lithium is currently industrially, not recycled. Hydrometallurgical research focuses on recovering lithium at the end of the processes; thus, impurities from process additives are possible, and moreover, reagents like Na$_2$CO$_3$ are needed for generating a marketable lithium product. The present study aims to present a method to mobilize lithium without using expensive or environmentally harmful additives: The early-stage lithium recovery ("ESLR") method. This "ESLR" particularly requires a suitable thermal pretreatment, and other elements can then be integrated into existing metal refining processes. The full "ESLR" process investigated here are shown in Figure 1.

For this research's purpose, the publication is structured into three parts: first, state-of-the-art processes for LIBs recycling are contrasted to innovative research for lithium phase transformation. Second, our own research results based on experimental studies are presented and subsequently evaluated in terms of lithium yield and purity. Concludingly, the obtained results are discussed by showing their scientific findings and process technology relevance in comparison to the state-of-the-art.

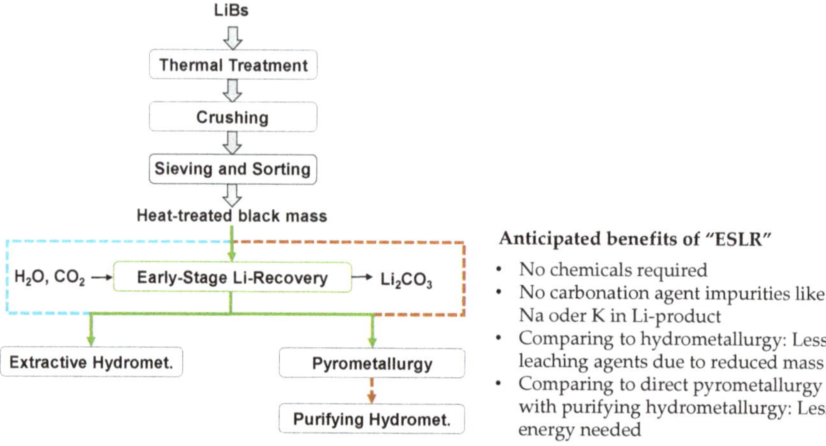

Figure 1. General flowchart of the early stage Li-recovery discussed in this study and the process benefits at a glance.

1.1. State-of-the-Art in Recycling Li-ion Batteries

LIBs recycling comprises different modules and sequences, leading to alternative process paths. Statements regarding future-dominant process pathways are afflicted with uncertainties due to location and know-how aspects and also because of the heterogeneous and changing scrap stream compositions [1]. However, the available processes, until the point of having generated marketable products, can be divided into preconditioning and metallurgical extraction [2]. The pretreatment steps, in turn, can be asserted to different sectors: deactivation/discharging [3], mechanical processing as dismantling of EV modules and packs to cell level, comminution and sorting by size or physical properties [4] and finally a thermal treatment [5]. Within the metallurgical techniques, there are mainly hydro- and pyrometallurgical processes available [6]. They both comprise benefits and drawbacks; for example, in hydrometallurgy also ignoble elements, like Fe, Al and C can be recovered, but on the other hand, the processing goes along with comparatively slow kinetics [6]. Depending on individual core objectives, the cells can be charged into a smelter without any pretreatment [3], but regarding a circular economy approach, it is beneficial to consider pretreatment steps [7] in order to maximize resource efficiency. Besides conventional industrial treatments, different studies are in place to give an overview also on innovative emerging recycling paths [8,9] and also approaches to evaluate the environmental impacts of different paths [10–12]. First, the available processes for recycling Li-ion batteries are described, and second, innovative processes for CO_2-promoted lithium phase transformations are shown. Therefore, first, indirect carbonation principles and studies are outlined, second, literature on direct carbonation is presented and third, the role of CO_2 in a supercritical state is pointed out. Goal of this detailed elaboration is a monitoring of gaps in literature regarding efficient lithium recovery from LIBs.

1.1.1. Thermal Preconditioning

Thermal pretreatments can be carried out, for example, as pyrolysis. Here, the cells are deactivated in the absence of oxygen at temperatures of typically 600 °C [13]. Pyrolysis (as well as classical incineration) are thermal pretreatments allowing for a safe cell deactivating and facilitating further downstream recycling without risking a so-called thermal runaway [14,15]. Through chemical cracking and such removal as organic compounds in gaseous form, which originate primarily from binder, electrolyte, and separator, takes place [13,16]. A major advantage of thermal treatments comprises a safe cell deactivating, thus contributing to risk mitigation in the context of fire incidents. This thermal runaway can occur, for example, during scrap transport, storing, but also by mechanical processing due to this mechanical, electrical or thermal abuse [17]. Several studies report the second advantage of thermal pretreatments, namely an improved detaching of black mass from the cell's current collector foils [7,16,18–21]. Additionally, suitable mechanical preconditioning concepts are required for efficient downstream processing, especially for hydrometallurgical treatments [3]. A mechanical process consists of comminution and sorting for splitting black mass and other cell components, such as casing and current collector foils. Hence, by subsequent mechanical postprocessing, aluminum and copper foils, along with the metallic casing, either aluminum or steel, can be separated as marketable products from the black mass. Black mass then contains all electrochemical active electrode materials [22]. Due to different battery systems on the market, black mass always has different chemical compositions [22].

The separation into individual fractions by means of sieving or physical separation techniques contributes to higher yields of the valuable components and, finally, increases process recycling efficiency [14]. In this way, copper, aluminum and steel can be integrated into their specific recycling processes. Regarding the extracted black mass, two processing alternatives are in place: hydrometallurgical and pyrometallurgical treatments. In the following, these two methods and their challenges for lithium recycling are compared.

1.1.2. Lithium Behavior in Pyro- and Hydrometallurgical Recycling Steps and Need for Early-Stage Li-Separation

Smelting of possibly pelletized black mass with the addition of SiO_2 as slag additive in an electric arc furnace has shown that lithium accumulates both in slag and flue dust [23]. Due to its ignoble character, extraction via a metal phase is not possible. As can be seen in Figure 2, a negligible proportion of approximately 0.35% of lithium is accounted for in the alloy produced. Depending on the selected slag system and the amount of slag, increased accumulation in the slag or flying dust can be realized (see Figure 2). Since the slag has a solubility limit for lithium oxide, according to Vest [24], the evaporation of lithium takes place when the corresponding concentration is exceeded. Due to re-oxidation processes, lithium oxide is accumulated in the flue dust (see Figure 2 below, according to Vest [24]). When operating at the lab-scale, smaller quantities of slag are generated, leading to a larger proportion of lithium transferred to the flue dust. The two Sankey diagrams in Figure 2 show a broad distribution of lithium between the three phases slag, flue dust and partly alloy, which is valid for both smelting setups. In order to extract lithium from the produced slag, energy-intensive crushing, classifying and hydrometallurgical purifying are required, but the costs for these treatment steps are currently not covered by lithium's raw material price [3].

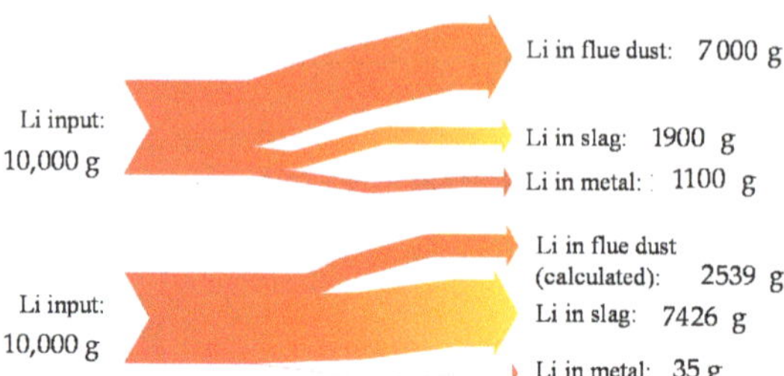

Figure 2. Lithium distribution after the smelting of black mass in an electric arc furnace. Above: process design aiming for a lithium enrichment in flue dust, based on [23,24]. Below: process design aiming for a lithium enrichment in slag [23]. Reproduced with permission from Stallmeister, C., Schwich, L. and Friedrich, B., Early-Stage Li-Removal—Vermeidung von Lithiumverlusten im Zuge der Thermischen und Chemischen Recyclingrouten von Batterien; published by Thomé-Kozmiensky Verlag GmbH, 2020.

When treating the extracted black mass by hydrometallurgical processing, lithium is not always enriched in one single product fraction, neither, but can be distributed during the multi-step precipitation series in all filter cakes [25]. Firstly, a black mass is typically leached in mineral acid. For this purpose, it is beneficial to conduct the thermal pretreatment as described easing the dissolution process [26]. Here, Shin et al. report the binder's (polyvinylidene fluoride, PVDF) property of not dissolving in acidic solution and disturbing the filtration process after leaching [27]. Yang et al. have shown a strategy to separately incinerate and then hydrometallurgically treat spent anode material in order to purify the C-fraction from lithium impurities, recovering lithium by means of Na_2CO_3 [28]. When treating the black mass from both cathode and anode, the target products such as copper, iron and aluminum, cobalt, nickel and manganese are cemented or precipitated one after the other. Lithium salt recovery, e.g., as carbonate (Li_2CO_3), is the last process step in the hydrometallurgical process chain, as suggested by Wang et al. [25]. Wang et al. have proven a lithium leaching efficiency of 98.5% [25]. For obtaining Li_2CO_3 in the last process step, a carbonation additive like sodium carbonate is used. During the precipitation stages, as can be seen from the data in Figure 3, approx. 27% of lithium remains in other filter cakes. Thus, the purity of the obtained copper, Fe/Al and Ni-Co products is reduced, and the yield of recovered lithium suffers. Moreover, a share of lithium remains in wastewater leading to a complex circuit with continuous neutralization salt removal, again including lithium losses. This complex processing is up to now industrially only viable by recovering the valuable metals nickel, copper and cobalt [3].

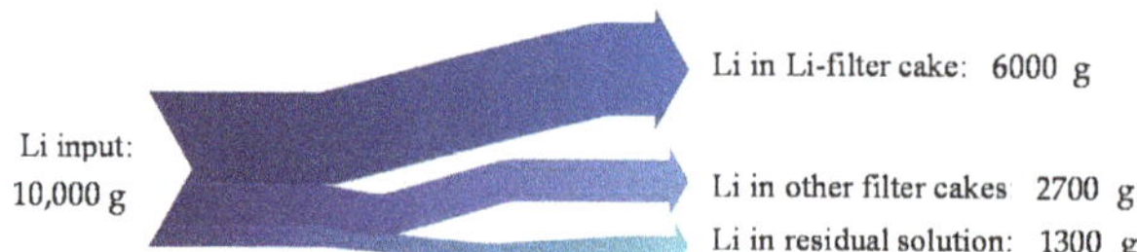

Figure 3. Lithium distribution after leaching trials of extracted black mass (current trials at IME), based on [23] (data used in [23] was based on experiments by Wang et al. [25]). Reproduced with permission from Stallmeister, C., Schwich, L. and Friedrich, B., Early-Stage Li-Removal—Vermeidung von Lithiumverlusten im Zuge der Thermischen und Chemischen Recyclingrouten von Batterien; published by Thomé-Kozmiensky Verlag GmbH, 2020.

Hence, the process's bottleneck lies in lithium extraction as a solid product instead of lithium leachability (leaching efficiency). A wide range of hydrometallurgical studies shows high leaching efficiencies using different solvents [29–34], for example, by using

H_2SO_4, a lithium leaching efficiency of 96.7% [35], and by HCl 99.2% [36]. In [37], leaching in citric acid and precipitating lithium by sodium phosphate leads to a leaching efficiency of 99% Li and a recovery as Li_3PO_4 of 89%. In [38–41] also a lithium precipitation of solid Li_2CO_3 by Na_2CO_3 is reported, reaching yields of 80% [38], 90% [39] and 99% [40]. In [38], a corresponding purity of 96.97% is reached; in [39], no purity is given; in [40], the lithium filter cake consists of 10.13 wt.% Li. With a molar ratio of $Li/Li_2CO_3 = 18.79\%$, this would mean a Li_2CO_3 content of 53.91%, assuming 10.31 wt.% exists as pure Li_2CO_3. When using cathode black mass only, a recovery of Li as Li_2CO_3 of 98.22% with a purity of 99.9% is reported [41]. This means that reaching both a high Li_2CO_3 purity with a high yield is not straightforward but achievable. Nonetheless, lithium recovery always requires additives, which can be avoided by direct H_2O-leaching. Moreover, the volume of required leaching agents can be lowered by H_2O-leaching before entering conventional hydrometallurgy due to a mass reduction. In this study, H_2O-leaching and using supercritical CO_2 are assessed for environmentally friendly and additive-free lithium recovery.

1.1.3. Liquid–Gas Carbonation (Indirect Carbonation)

The hypothesis of this work is a mobilizing of lithium by an "Early-Stage Li-Recovery". Different methods may be applied for this phase transformation to Li-carbonate, which are addressed in the following paragraphs. The use of CO_2 for carbonation purposes has been examined for non-lithium materials in numerous studies, e.g., [42–45], and even treating of battery materials with CO_2 is possible, as shown above by Hu et al. and Zhang et al. [46,47]. Generally, Kunzler et al. investigated the parameters influencing indirect carbonation, which is understood as a reaction between dissolved elements and CO_2. In contrast to that, direct carbonation is defined as a gas–solid reaction for generating carbonates [48]. Kunzler et al. found a correlation between extraction efficiency and grain size of the target metals, solid to liquid ratio, concentration and hence pH value of the leaching agent, and temperature [48].

When aiming for indirect CO_2-driven reactions, Hu et al. have conducted a combination of a reductive thermal treatment and H_2O-leaching in combination with CO_2-gas [46]. Here, only cathode material from NMC (Lithium Nickel Manganese Cobalt Oxide)-cells was mixed with lignite as a reducing agent. According to the maximum reported leaching efficiency of 85% for Li, the optimal thermal-treatment temperature is at 650 °C, and the optimal s/l ratio is 10 mL/g (1:10 (g/mL)). Zhang et al. have developed this study further using the same parameters but also investigating optimal CO_2 flow rate and leaching time, leading to a Li-recovery of 85% [47]. In summary, when treating the cathode mass only and adding reducing agents like lignite or carbon black to the thermal treatment, lithium yields of 85% [46,47]. However, if not applying CO_2 during leaching, the leaching efficiency of lithium is 40% [46,47]. This indicates that the mechanism of indirect carbonation is decisive, but in [46,47], no CO_2 atmosphere was used during the thermal treatment; instead, a solid C-carrier was added.

A similar approach is pursued by Jandová et al., where a lithium-containing solution, not stemming from batteries, is treated with CO_2 [49]. Then, the solution is heated until the lithium concentration reaches 12–13 g/L, and afterward treated with CO_2-gas at a temperature of 40 °C for 2 h to generate $LiHCO_3$. Lithium hydrogen carbonate provides a higher solubility in comparison to the first, formed Li_2CO_3. Finally, the lithium solution is boiled to produce Li_2CO_3 [46,47,49]. Moreover, an indirect carbonation approach for non-battery materials gives insights into general mechanisms when purging CO_2 into aqueous solutions [45]. Within these aqueous treatments, CO_2 dissolves as [45,50]:

$$CO_2 + H_2O \leftrightarrow H_2CO_3 \leftrightarrow H^+ + HCO_3^- \leftrightarrow CO_3^{2-} + 2\,H^+ \tag{1}$$

The more H-cations are released, the stronger is the resulting acidification [50]. These reactions are to be understood as a function of temperature, pressure and pH [45]. With increasing pH, the dominantly existing phases alternate in the following sequence: H_2CO_3, HCO_3^- and CO_3^{2-}, hence the higher the pH-value, the more H^+-ions are released, contributing to a lowered pH. Especially, CO_3^{2-} is dominant in a pH-area of 10 onwards, whereas HCO_3^- is dominant in an area from 6 to 9, as can be seen in Figure 4:

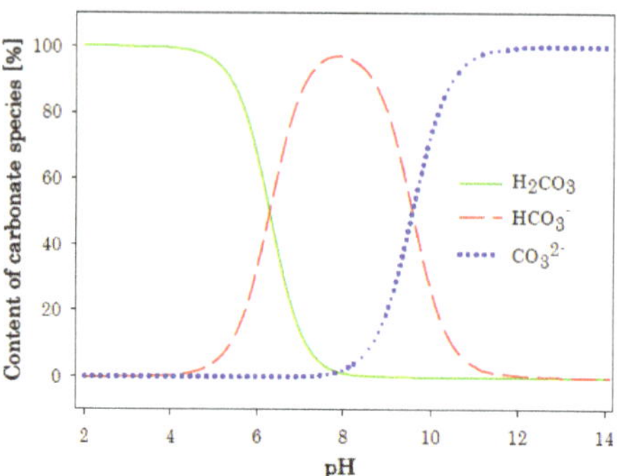

Figure 4. Available CO_2-based phases in aqueous solution over the pH [45]. Reproduced with permission from Haug, T. A., Dissolution and carbonation of mechanically activated olivine—Investigating CO_2 sequestration possibilities; published by Haug, T.A., 2010.

A reaction between HCO_3^-/CO_3^{2-} and lithium requires the presence of Li^+ in the solution. In Table 1, possible lithium phases and their solubility is presented. Connected to that, the chemical reaction formula is presented describing the dissolution of lithium phases in an aqueous solution, without and with CO_2-gas purging.

Table 1. Solubility of selected lithium phases at 20 and 100 °C.

Phase	Solubility at 20 °C	Solubility at 100 °C
LiOH	110 g/L [1]	161 g/L [1]
LiF	1.2 g/L [1]	1.34 g/L [1]
LiHCO₃	55 g/L [2]	57.4 g/L [1]
Li₂CO₃	13.3 g/L [1]	7.2 g/L [1]

[1] [51], [2] [52] at 18 °C.

Yi et al. have also reported the conversion from Li_2CO_3 in aqueous solution into $LiHCO_3$ by CO_2-based carbonation, followed by a chemical purification of the solution and subsequent crystallization of Li_2CO_3 from a $LiHCO_3$ solution by boiling [53].

If lithium is present as Li_2CO_3, it decomposes according to Equation (2) [54]:

$$Li_2CO_3 \leftrightarrow Li^+_{(aq)} + CO_3^{2-}_{(aq)} \tag{2}$$

The carbonate ion ($CO_3^{2-}_{(aq)}$) is the conjugate base of a weak acid (carbonic acid) [55]. Hence, H^+-ions are attracted at neutral or acidic areas and consumed from H_2O, which generally is present in the ionic form [55–57]. This chemical behavior equals the property of Li_2CO_3 to be alkaline (see Equation (7)) [58]. In combination with CO_2-gas, Yi et al. also report the chemical steps between Li_2CO_3 dissolution consuming H^+-ions from the H_2CO_3-decomposition (see Equation (3)) [53] and precipitation of solid Li_2CO_3 according to Equation (4) to Equation (7) [53]:

$$CO_2 + H_2O + Li_2CO_3 \leftrightarrow 2\,LiHCO_3 \tag{3}$$

$$LiHCO_3 \leftrightarrow HCO_3^- + Li^+ \tag{4}$$

$$HCO_3^- \leftrightarrow CO_3^{2-} + H^+ \tag{5}$$

$$HCO_3^{2-} + H^+ \leftrightarrow H_2O + CO_2 \tag{6}$$

$$Li_2CO_3 \leftrightarrow Li^+ + CO_3^{2-} \tag{7}$$

Hereby, the possible recombinations between aqueous CO_2-phases and lithium ions in aqueous phases are shown. These combinations can be transferred to other lithium phases, liberating lithium cations in aqueous solution, too:

If lithium is present as LiF is a black mass, it dissolves in aqueous media according to Equation (8) [59]:

$$LiF + H_2O \leftrightarrow HF + LiOH \tag{8}$$

According to the definition of strong acids and bases [60], HF ($pK_S = 3.17$ [61]) is a strong acid, whereas LiOH ($pK_b = -0.36$ [62]) is very strong base. As a resulting pH-value for dissolving 0.26 g/L at 25 °C pH = 7–8.5 is reported [63].

If lithium is formed as LiOH in a black mass, it dissociates in an aqueous solution according to Equation (9) [64–66]:

$$LiOH + H_2O \leftrightarrow Li^+ + OH^- + H_2O \leftrightarrow LiOH \cdot H_2O \tag{9}$$

the following reaction can take place if CO_2 is applied to the system [67]:

$$2\,LiOH \cdot H_2O + CO_2 \leftrightarrow Li_2CO_3 + 3\,H_2O \tag{10}$$

If lithium is present as Li_2O in a black mass, it dissociates to LiOH in aqueous solutions according to Equation (11) [68]. $LiOH_{(aq)}$ is generally stable as lithium hydroxide octahydrate ($LiOH \cdot 8H_2O$) [61].

$$Li_2O + H_2O \leftrightarrow 2\,LiOH \tag{11}$$

This passage has shown that no study on indirect carbonation by CO_2 using whole LIBs black mass, meaning anode and cathode material, is in place. In contrast to that, in the study, real industrial heat-treated black mass from anode and cathode was used without adding a reducing agent. Moreover, this gives the first-time overview of all possible lithium reactions when considering battery materials, which is crucial to extract hypotheses on ongoing mechanisms.

1.1.4. Solid–Gas Carbonation (Direct Thermal Carbonation)

Direct carbonation describes solid–gas reactions for generating a carbonate phase [48]. In this study, it will be investigated by using SCO_2. There are different studies in literature optimizing a reductive thermal treatment of black mass for mobilizing lithium via subsequent H_2O-leaching [46,47,69–74]. It should be recalled that in [46,47], which were already discussed in chapter 1.1.3, a combination of direct carbonation and indirect carbonation was performed: On one hand, a reductive thermal treatment with adding a carbon-reducing agent like lignite or carbon black contributes to the formation of Li_2CO_3, hence direct carbonation. On the other hand, CO_2 was added during leaching or Na_2CO_3 was used after a first filtration, both representing indirect carbonation. Therefore, a classification into studies with direct and indirect carbonation is not always straightforward.

However, in all reported studies [46,47,69–74], first, the battery cells are shredded, and, after extracting, a black mass is thermally treated. Battery systems used are LCO-cathode based [69,72,74], LMO-cathode based [71,74], or NMC-cathode based [73,74]. In [73], the only cathode material is used. Most studies focus on a thermal treatment in an inert atmosphere, like a vacuum, Ar or N_2 [69,71,72,74,75] or in the air [70] instead of CO_2. CO_2 was only used in [73]. The performed reductive roasting reportedly also contributes to the carbonation of lithium by precipitating it from the black mass matrix [19,70,72,74,76].

Since the focus of this study is the use of NMC-cathode based black mass, the matching studies are reported in detail: CO_2-gas purging at 600, 700 and 800 °C (direct carbonation) of NMC-cathode material was performed by Wang et al. for 120 min., but lithium yields are not given. Instead, it is stated that 1.735 g lithium of 1.95 g lithium was transferred to a solution after water leaching [73]. However, no optimal temperature for carbonation is given; instead, a spectrum of 650–800 °C is reported, and no discussion on lithium purity is in place. Xiao et al. treated black mass from NMC-cathode based cells. According to the best-case scenario for pyrolysis and H_2O-leaching, 66% of the lithium is recovered. The best-case scenario here implies vacuum pyrolysis at 700 °C for 30 min. in combination with H_2O-leaching for 30 min. and an s/l ratio of 1:40 (g/mL) (25 g/L) [74]. Nevertheless, no details on the procedure of NCM-black mass are given, especially regarding lithium recovery.

Since [73] is the only study in place using CO_2 for direct carbonation, no detailed yields are quoted from the literature.

In terms of chemical reactions involving lithium phases, only a few data are given, mostly based on thermodynamic simulations [76–78]. Nonetheless, gas–solid reactions are known, e.g., from CO_2 absorbing for air purification. Here, a reaction between LiOH and CO_2 for generating lithium carbonate is targeted [67].

$$2\,LiOH + CO_2 \rightarrow Li_2CO_3 + H_2O \tag{12}$$

A similar reaction is possible if Li_2O is present [79]:

$$Li_2O + CO_2 \rightarrow Li_2CO_3 \tag{13}$$

The reaction is rather a surface reaction, after which CO_2 diffuses into the inner part of the Li_2O particles. This diffusive process is temperature-supported. For example, at 600 °C, the diffusion takes place 10 times faster than at 500 °C [79]. In addition, once lithium carbonate is formed, it remains stable in the CO_2 atmosphere, even though being in a liquid state, until 1611 °C, before decomposing into Li_2O [80]. This passage has shown that there is a lack of process details in terms of direct carbonation LIBs black mass by CO_2-gas purging. Moreover, the available studies focus on shredding before thermal treatment. In this study, the thermal treatment is performed before a mechanical treatment.

Finally, at this point, no study has reported investigating the influence of solid/gas mechanisms (direct thermal carbonation) in contrast to liquid/gas mechanisms (indirect carbonation) in terms of carbonation either by using supercritical or ambient pressure CO_2 and by keeping the process parameters equal. The present research aims to give answers to this question.

1.1.5. The Role of Supercritical CO_2 ($SCCO_2$)

When using CO_2 for phase transformations, the combination of the liquid and gaseous phase properties is advantageous. For CO_2, the supercritical state is reached at a temperature of at least 31 °C and a pressure of 73.8 bar [81]. Here, the physical properties of CO_2 can be described by a density according to the liquid state and as viscosity equally to the gaseous state, enabling a high efficiency [82]. Chen et al. report on using supercritical CO_2 ($SCCO_2$) for carbonating spodumene-based lithium; thus, this study refers to primary lithium production [83]. In their study, also sodium carbonate (Na_2CO_3) is used as a carbonation agent, and CO_2 is added, aiming for a higher carbonate dissolution. It is reported to precipitate as Li_2CO_3 when reducing the liquid volume, according to Equation (14) [83]:

$$2\,LiHCO_{3(aq)} \leftrightarrow Li_2CO_{3(s)} + \{CO_2\} + H_2O_{(aq)} \tag{14}$$

Bertau et al. have dealt with a similar research topic: They suggested a treatment of Zinnwaldite, a lithium ore located in Germany, with $SCCO_2$. It is a promising solution for primary lithium recovery as carbonate [84]. Specific benefits comprise the avoidance of additional chemicals, such as Na_2CO_3, and significant lithium losses, the economic viability due to the low CO_2 price, and the high selectivity by transforming only alkali metal compounds [84]. Moreover, Liu et al. investigated the possibilities of recovering LIB-electrolyte components by means of $SCCO_2$ [85]. Grützke et al. also aimed for LIB-electrolyte recycling using $SCCO_2$, but in contrast to Liu et al., who used synthetic electrolyte components, whole end-of-life NMC-cells were discharged, deep-frozen and manually opened to extract the electrolyte by supercritical CO_2 [86,87]. SCCO$_2$, at, e.g., 40 °C and 80 bar in flow-through mode, is used for extracting the electrolyte along with the CO_2-stream in a cryogenic trap. A patented technique by Sloop describes a treatment of full batteries with SCCO$_2$, where lithium carbonate from the electrolyte is recovered in the frame of electrolyte removal. The electrolyte removal is reached by dissolving it in the stream of CO_2 [81,88].

Rothermel et al. rather focused on graphite recycling options and the achievable graphite purity by making use of supercritical CO_2-supported electrolyte recovery (SCCO$_2$) [89]. In 2019, Bertau et al. also reported options to recover lithium from battery black mass [90]. The so-called COOL process, consisting of discharging, mechanical extraction of black mass and a SCCO$_2$ treatment, obtained lithium carbonate with a purity of >99.5% and yield of 60%, but the yield is referring to primary ore treatments, so no information on lithium from black mass is in place [90]. In this context, another benefit is highlighted: consuming CO_2 instead of producing CO_2 in the context of rising industrial, environmentally harmful CO_2-emissions [90]. However, no details about the best-case treatment parameters and the resulting lithium yields in terms of black mass carbonation are given. An earlier patent by Bertau et al. describes lithium recovery from so-called lithium-containing battery residues by SCCO$_2$ for obtaining lithium carbonate [91]. For this material, lithium yields of >90% are reported by using electrodialysis and subsequent addition of carbonates like Na_2CO_3 or K_2CO_3, but facts on the corresponding treatment details and parameters are not given. An exemplary process with s/l = 1:40 (g/mL) (50 g/2 L), 4 h at 230 °C and 100 bar is described, reporting on a leaching efficiency of 95%. The lithium extraction via electrodialysis takes place in a Li_2SO_4-solution recovering 98% of the lithium in solution [91]. A lithium yield based on the final, solid product is not given.

This passage has shown that supercritical CO_2 plays a role in lithium recovery from different materials, but for LIB-black mass, there is a lack of knowledge regarding decisive process details. This article focuses on lithium carbonation from black mass by means of supercritical CO_2.

2. Materials and Methods

2.1. Recycling Concept with Integrated Early-Stage Li-Recovery

Under the view of the current process-related drawbacks in conventional recycling processes, and especially in terms of the need for lithium recovery, this study suggests the strategy of an early-stage Li-recovery ("ESLR") process. The method "early-stage" is studied here, describing lithium carbonation before entering acidic leaching or smelting, hence at an earlier position in the recycling chain. This treatment prevents lithium distribution, the further use of additives needed for hydrometallurgical treatments, and costly refining from a slag. The "ESLR" process comprises the following steps, as presented in Figure 5b: After the cells have been deactivated by means of a thermal treatment followed by mechanical processing (shredding and sieving to <1 mm), lithium is enriched in the heat-treated black mass, along with other electrode elements, such as Co, Ni, Mn, and C and partly Al- and Cu foil fragments. This heat-treated black mass is then leached in H_2O- to transform and extract the lithium phases and, thus, separated them from the other electrode elements.

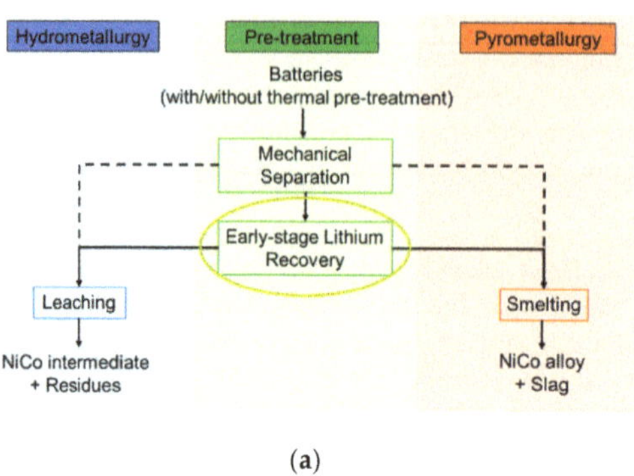

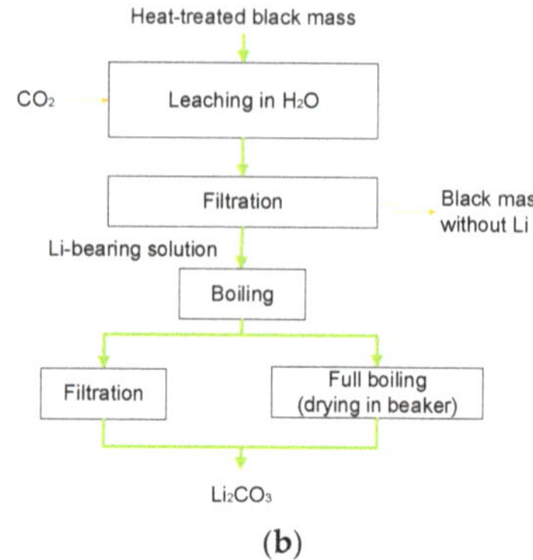

(a)

(b)

Figure 5. (**a**) Recycling strategy for lithium-ion batteries (LIBs) at IME with a focus on alternative processing using "ESLR", (**b**) detailed process steps of the "ESLR" process by autoclave/supercritical CO_2-carbonation.

As mentioned in Section 1, this phase transformation can be realized by treatment with supercritical CO_2 as indirect or direct carbonation. Lithium in the heat-treated black mass is converted into water-soluble lithium hydrogen carbonate ($LiHCO_3$) and lithium carbonate (Li_2CO_3). Indirect carbonation and H_2O-leaching occur simultaneously because the heat-treated black mass is fed in the reactor along with deionized water. In direct carbonation, the heat-treated black mass is subjected to neutral leaching in deionized H_2O after the phase transformation of lithium compounds. In both cases, lithium dissolves into an aqueous solution, and the Li-reduced, heat-treated black mass is separated by a first filtration. The first filtration's products comprise:

1. A filter cake with mainly carbon, nickel, cobalt, manganese, aluminum and copper fragments and a share of lithium ($\rightarrow$ C-filter cake);
2. A lithium-bearing filtrate (solution)

The Li-containing filtrate must finally be boiled since lithium's solubility decreases from 13.3 g/L at 20 °C to 7.2 g/L by heating to 100 °C [25]. Moreover, thus, the carbonate precipitation is supported. Either a second filtration step separates the solid carbonate (filter cake) from the residual solution, or the solution is further boiled and is left in the air to dry the carbonate.

2.2. Material Characterization

The black mass used in the present study to validate the "ESLR" process has been generated by thermal treatment of whole NCM-traction cells. Therefore, a real industrial heat-treated black mass was obtained by thermal treatment at different atmospheres: Ar 95% Ar + 5% O_2 and CO_2. For Ar-pyrolysis, the temperatures targeted are 509 °C and 603 °C. For Ar + O_2-pyrolysis, the temperature targeted is 501 °C. For CO_2-pyrolysis, the targeted temperature is 466 °C. These atmospheres have been generated by dynamic pyrolysis in a sealed reactor. It should be noted that due to the experimental setup, not all thermal treatment temperatures have been identical. The thermally treated cells were then subjected to a shredding and sieving process for extracting the heat-treated black mass. The composition of the heat-treated black mass is shown below (see Table 2):

Table 2. Chemical composition of the heat-treated black mass used for the autoclave trials pyrolyzed in Ar-atmosphere.

Al	Co	Cu	F	Fe	Li	Mn	P	C	Ni
				wt.%					
2.10	11.7	0.88	4.10	0.00	3.69	8.91	0.44	33.9	11.5

Dynamic particle analysis with QICPIC/L02 (Sympatec GmbH, Clausthal-Zellerfeld, Germany) of the heat-treated black mass reveals that according to the distribution sum (Q_3), the $d_{99.3}$-value is 101.74 µm. This means that 99.3% of the heat-treated black mass has a grain size smaller than 101.74 µm (see Figure 6). Here, the material's distribution density (q_3*), see Equation (15), can also be extracted, reaching a global maximum at ~95 µm. According to DIN ISO 9276-1, the distribution density is defined as the first derivation of the distribution sum:

$$q_3* = qr\,(x) = \frac{dQ_r(x)}{dx} \tag{15}$$

Furthermore, the distribution sum of the particles represents the number of all particles and not their volume share in the powder. Hence, there are many small particles in the heat-treated black mass, but in contrast to that, the volume of big particles (>101.74 µm) could take up more than 99.3%. Moreover, the largest particle detected in the heat-treated black mass comprises a diameter (EQPC) of 653.74 µm with a FERET_MAX value of even 748.13 µm. EQPC is defined as

$$xEQPC = \sqrt[2]{\frac{A}{\pi}} \tag{16}$$

Hence, it describes the diameter of a circle whose projection surface (shadow) is identical to the particle. FERET_MAX, on the other hand, detects the maximum diameter of a particle by analyzing it from 20 different perspectives between 0 and 180°. This value deviates from the EQPC, especially in terms of irregularly shaped particles; hence, FERET_MAX takes bigger values, especially when being distinct from a circular form. The maximum detectable particle size with this method comprises 1252 µm.

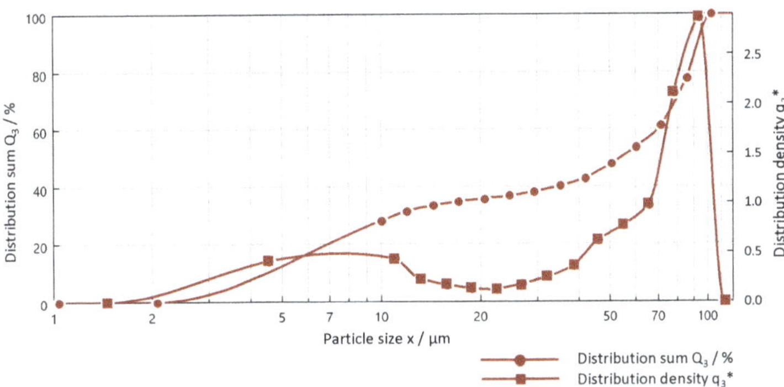

Figure 6. Dynamic particle analysis of CO_2-pyrolyzed black mass. A total of 32,650 particles were counted and hence considered for the evaluation. The distribution sum and distribution density in this diagram is based on the heat-treated black mass comprises a diameter (EQPC)-value of the particles, which is indicated as particle size *x*.

More details on the quantitative grain size detected are shown in Figure 7 and can be extracted from the chemical composition of the particles. EDS analyses were carried out for elemental mapping, as well as for point analysis on the black powder, using a ZeissGemini-FE-SEM, equipped with an Oxford UltimMax170 detector (Carl Zeiss AG, Oberkochen, Germany). In Figure 7a,b, the results of the elemental mapping (EDX-layered image) are depicted. Here, the appearance of metallic Al-flakes is shown in Figure 7a; second, a heat-treated black mass particle and a graphite particle are shown in Figure 7b. Moreover, third, Figure 7c is a 1 mm-scale distance shot showing the heterogeneity of the heat-treated black mass in terms of metallic aluminum fragments, heat-treated black mass particles and graphite powder. It should be noted that lithium cannot be displayed by this method. In addition, Figure 7c reveals that the liberation of heat-treated black mass from the aluminum current collectors could be realized since the particles do not show direct physical contact.

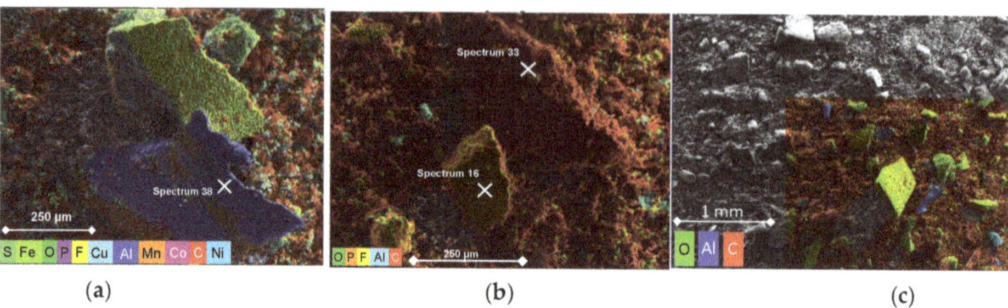

(a) (b) (c)

Figure 7. EDS analysis of CO_2-pyrolyzed EV-battery black mass. (**a**) results of the elemental mapping (EDX-layered image) of an aluminum particle, (**b**) results of the elemental mapping (EDX-layered image) of both an NMC-and-graphite particle, (**c**) Distance shot of the heat-treated black mass from elemental mapping (EDX-layered image).

The following Table 3 shows the chemical composition of the taken spectra. Here, the chemical composition of the Al-flake can be seen by Spectrum 38. However, there are some oxide-graphite-mix particles visible, which results in the chemical composition of just 58.28 wt.%. Moreover, spectrum 16 shows that the dominant particle composition is a Ni-Mn-Co-Oxide, hence, heat-treated black mass particles. Spectrum 33 reflects a graphite particle (C = 46.46 wt.%), which also shows a high amount of oxygen. This may be explained by lithium oxides, hydroxides or carbonates, which cannot be shown here. This theory can be supported by the lithium intercalation in the graphite matrix due to charging.

Table 3. Chemical composition of the heat-treated black mass spectra taken by EDS-analysis. Only elements >1 wt.% are shown quantitatively.

Element	C	O	F	P	Mn	Co	Ni	Al
Unit					wt.%			
Spectrum 38	31.9	7.35	<1 wt.%	<1 wt.%	<1 wt.%	<1 wt.%	<1 wt.%	58.28
Spectrum 16	32.83	12.1	21.9	1.81	8.59	11.38	10.62	<1 wt.%
Spectrum 33	46.46	33.51	1.47	<1 wt.%	<1 wt.%	<1 wt.%	<1 wt.%	17.95

These findings are essential to understand the properties of NCM-heat-treated black mass: An important result is the revelation of the material's heterogeneity. Hence, each sample taken shows different chemical compositions. Considering the colorful mixture of relatively big Al-particles, cathode material particles and smaller graphite/soot particles, this statement is supported. Moreover, it can be seen that the Al-foils are liberated from the cathode material. Hence, a thermal pretreatment removes binders and therefore loosens adhesions. This enhances the subsequent leaching efficiency, as indicated before. Moreover, it is shown that fluorine enriches the cathode material particles.

2.3. Neutral Leaching Reference Tests in Deionized H_2O

Neutral leaching comprises reference trials to evaluate the carbonation success. It was performed in a 1 L glassware beaker filled with deionized water. Here, 20 g of the heat-treated black mass is inserted and magnetically stirred at 350 rpm for the defined leaching time. After H_2O-leaching, a first filtration obtains a C-filter cake and a Li-bearing solution, then boiled to recover lithium as Li_2CO_3 ($\rightarrow$ Li-filter cake). Neutral leaching is performed at room temperature to increase lithium dissolution. Lithium contents in both filter cakes and solutions are measured by ICP-OES (inductively coupled plasma optical emission spectrometry). Li is the crucial indicator for carbonation success, and hence, lithium yields are calculated as follows:

$$n_{Li} = \frac{Li_{Li_2CO_3 - fc}\,(g)}{Li_{total}\,(g)} \tag{17}$$

where Li_{total} is the lithium mass in the input, and $Li_{-Li2CO3-fc}$ is the lithium mass in the lithium carbonate filter cake. Since the input material shows deviations in terms of its chemical composition, the Li_{total} value does not equal the share of the original input analysis. For this reason, the lithium values in the input are calculated as follows:

$$Li_{total}(g) = Li_{C-fc}\,(g) + Li_{solution}\,(g) \tag{18}$$

Here, C-fc represents the carbon filter cake, also indicated as "heat-treated black mass without lithium," and "$Li_{solution}$" corresponds to "Li-bearing solution". This calculation is to be contrasted to the leaching efficiency (LE). When calculating n_{Li} based on $Li_{solution}$ [g], the yields in this study were ~10% higher. For example, in one trial with a yield of 79% based on Equation (17), the LE was 88%. The reason for this inaccuracy cannot be determined at this point, but since the deviation was systematic, the authors decided to use the lower values for conservative result interpretation.

2.4. Carbonation by Supercritical CO_2

The unit operation used is a batch 1 l Büchi autoclave reactor operated with deionized water (indirect carbonation) or without any liquid (direct carbonation). The maximum operating temperature is 250 °C, and the maximum applicable pressure is 200 bar. After sealing, a stirrer constantly mixes the powder (50 g heat-treated black mass per trial (T0–T9) and 20 g heat-treated black mass per trial (T10–T22) or the suspension in the reactor. The gas flows into the reactor occurs via a valve into H_2O if the trial is conducted with an aqueous medium. As soon as the supercritical conditions are reached, a processing time of 120 min is started. The defined holding time of 120 min can be attributed to preliminary studies using autoclave-induced carbonation [92]. The general parameters for the autoclave trials can be seen in Figure 8a. Moreover, Figure 8b shows pictures of the unit operation.

Pressurestart	50 bar
Heating rate	10 °C/min.
Tmax	230 °C
Pmax **(spectrum of trials)**	95.4–138.3 bar
Holding time	120 min.

(a)

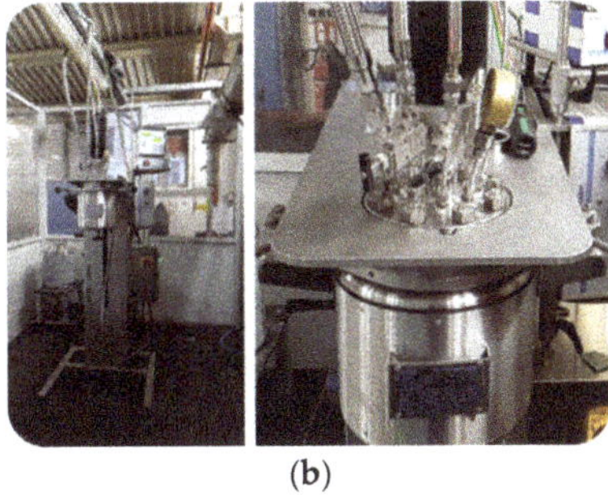

(b)

Figure 8. (a). Fixed process parameters and reached pressures for an autoclave trial (combination of heat-treated black mass with deionized H_2O and CO_2 gas) in case of trials T1–T9. After the starting pressure of 50 bar, the reactor is heated until reaching 230 °C, resulting in different P_{max}. **(b)**: Used autoclave reactor at IME, RWTH Aachen University.

After leaching, either during autoclave treatment, if H_2O is used in the autoclave or after the autoclave, if no H_2O is used in the autoclave, a first filtration is performed. The experimental procedure is identical to the procedure described in Section 2.3: Boiling is performed to reduce the volume of liquid and to precipitate Li from the solution. Lithium is obtained in the form of solid Li_2CO_3 in a subsequent (second) filtration step or by full boiling until obtaining a solid Li_2CO_3 product within the beaker. The (second) filtration step is performed as follows: The Li-bearing solution is boiled until reaching 100 mL and then is filtrated, obtaining Li_2CO_3. The Li_2CO_3-filter cake is then washed with pure ethanol since lithium carbonate does not show solubility in ethanol. It is dried for at least 24 h and then weighed. The full boiling (indicated as "drying in a beaker") is performed as follows: The Li-bearing solution is boiled until no liquid is left, why the weight of the empty beaker is to be measured before and after performing full boiling. In addition, weighing is done after a drying time of at least 24 h, too. The difference in weight equals the solid carbonate obtained.

An overview of the experimental series described in Sections 2.3 and 2.4 is given in Table 4. Parameter set 1.A, 1.O and 1.C (reference trials) represents H_2O-leaching without CO_2 addition. Hereby, insights into the mass of water-soluble Li-phases already present in the heat-treated black mass are provided. Enhanced leaching efficiencies are obtained by combining neutral leaching with CO_2-carbonation. (see experimental series 2.A, 2.O and 2.C). Moreover, carbonation trials were conducted in the autoclave as well by using argon (Ar) as process gas aiming for the same excess pressures as needed for the supercritical state (73.8 bar) (see Table 4, Parameter set 3.A). Ar as inert gas can deliver knowledge on the main mechanism for ongoing phase transformations: Either the presence of CO_2 or the extreme pressure. Parameter set 4.A detects the influence of an autoclave setup without H_2O and with CO_2 to find out whether gas–solid or gas–liquid reactions dominate in carbonation.

Table 4. Parameter matrix for combining pyrolysis conditions with autoclave conditions in this study. Reference trials represent H_2O-leaching without autoclave or CO_2-incorporation.

		Pyrolysis Conditions		
		Ar-Atmosphere	95 % Ar + 5 % O_2-Atmosphere	CO_2-Atmosphere
Autoclave conditions	H_2O-leaching (reference trials)	1.A	1.O	1.C
	$SCCO_2 + H_2O$	2.A	2.O	2.C
	$Ar + H_2O$	3.A	n/a	n/a
	CO_2 + dry autoclave	4.A	n/a	n/a

The labels of the trials are to be understood as follows: *"number.letter"*, where the *number* stands for an experimental series: 1 = neutral leaching in H_2O without an autoclave, 2 = autoclave operated with $SCCO_2 + H_2O$, 3 = autoclave operated with $Ar + H_2O$, 4 = autoclave operated with $SCCO_2$ and without H_2O. The *letter* stands for the pyrolysis atmosphere: A = Ar-pyrolysis, O = 95 vol % Ar + 5 vol % O_2-pyrolysis, C = CO_2-pyrolysis. In this study, only experimental series 1 and 2 take different pyrolysis atmospheres into account. Hence, the fields of experimental series 3.O/3.C and 4.O/4.C are not experimentally conducted yet (n/a).

3. Results

In this Section, the results of the trial series 1.A—4.A, 1.O/2.O and 1.C/2.C are discussed by lithium yields. Moreover, Sankey diagrams of the lithium distribution and bar charts on the impurities of the lithium filter cake are shown for trials 1.A.1 and T2 (2.A). The evaluation calculations are performed, as described in Section 2.3.

3.1. Neutral Leaching in Deionized H_2O

For a profound understanding of water-soluble lithium phases already present in the heat-treated black mass, the following results can be obtained. In total, 40 experiments were conducted in experimental series 1.A, 1.O, and 1.C. The amount per parameter set comprises one trial, except for trials 1.A.2-1.A.4, 1.A.5-1.A.7, 1.A.28 and 1.A.40. Since the results are in good accordance, the other trials have not been repeated. All trials have in common that when charging heat-treated black mass in H_2O, the pH value of the solution has become alkaline (pH = 11–12).

Figure 9 accordingly reveals how lithium yields from leaching heat-treated black mass in H_2O (neutral leaching) depend on six main parameters:

1. **Washing of C-filter cake with deionized water**: if this parameter is performed, it is important to keep the washing volume constant. In this study, 200 mL of deionized water are used;
2. **Filtration of Li-filter cake or full boiling**: full boiling describes the removal of H_2O in the laboratory beaker. Filtration stands for filtering the precipitating Li_2CO_3 at a minimum liquid volume. Hence, there are losses in the residual filtrate. Filtration is conventionally used after acidic leaching to avoid a co-precipitation of acid components and chemical additives;
3. **Leaching time**: 5, 30, 90 and 120 min;
4. **Particle size of heat-treated black mass**: <1 mm vs. <90 µm. The particles <90 µm are obtained by additional grinding of the heat-treated black mass;
5. **Solid/liquid ratio (g/mL)**: 1:10, 1:15, 1:22.5 and 1:30;
6. **Pyrolysis temperature**: 501 vs. 603 °C in Ar-pyrolysis

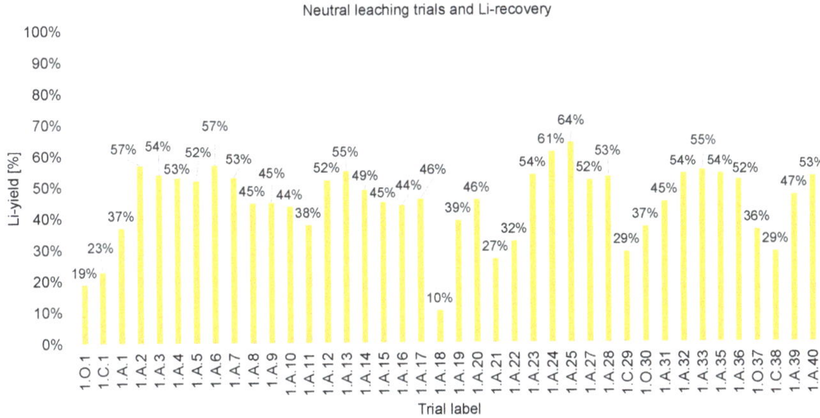

Figure 9. Lithium yields when applying no carbonation (neutral leaching) dependent on the parameters used. Trials 1.A.26 and 1.A.34 are left out of this overview since they comprise kinetic trials, for which a yield calculation is not possible due to heat-treated black mass losses during sampling.

Here, a variation of the selected parameters has an impact on the lithium yield. To evaluate the key influencing factors, representative trials are extracted and shown in detail. For efficiency reasons, only the results of experimental series 1.A are depicted for the evaluation according to the parameters selected. In terms of these influencing factors, the following conclusions are possible:

1. **Washing of C-filter cake with deionized water:**

The detailed observation of trials whose parameter combination was equal apart from the washing of the C-filter cake shows that washing is highly beneficial. During the filtration of the C-filter cake, there are physical depositions of the Li-bearing solution left in the C-filter cake. Hence, washing with deionized water liberates the C-filter cake from remaining lithium ions (see Figure 10a).

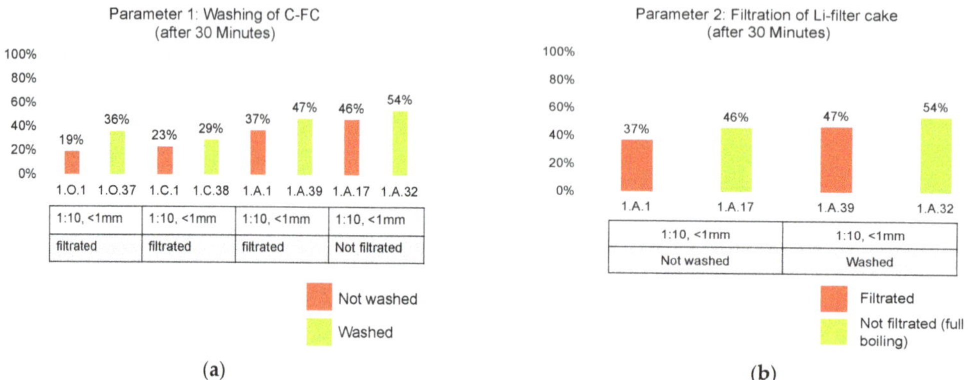

Figure 10. Detailed observation on achievable lithium yields by neutral leaching (H_2O) of heat-treated black mass. (**a**) Parameter 1: washing of C-filter cake. (**b**) Parameter 2: filtration of Li-filter cake or full boiling.

2. **Filtration of Li-filter cake or full boiling:**

Filtration is, according to Figure 10b, not an adequate tool when applying neutral leaching in H_2O. However, this analysis only focuses on lithium distribution. Hence, no information on filter cake impurities, e.g., F is given here.

3. **Leaching time:**

The optimal leaching time cannot be extracted from the performed trials, as can be seen in the range between 5 and 90 min (see Figure 11a) and between 30 and 120 min (see Figure 11b). Here, no significant improvement of dissolved lithium is achieved when comparing trials with constant parameters except for the leaching time.

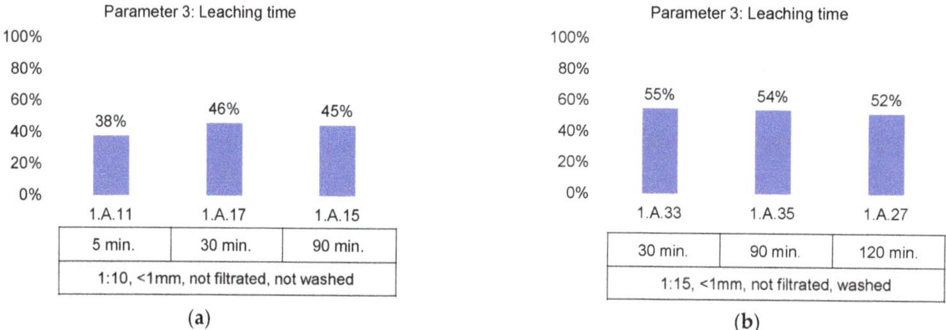

Figure 11. Detailed observation on achievable lithium yields by neutral leaching (H_2O) of black mass. The considered parameter is parameter 3: leaching time. (**a**) Dissolution and lithium recovery by comparing trials with a leaching time between 5 and 90 min. (**b**) Dissolution and lithium recovery by comparing trials with a leaching time between 30 and 120 min.

Reduced lithium shares over time can be explained by slight deviations in the chemical composition. Hence, deviations in lithium yields are also possible, also due to different lithium phases in the heat-treated black mass. Therefore, a kinetic trial can be found in Figure 12. It can be seen that lithium compounds in the heat-treated black mass of the Ar-pyrolyzed battery cells dissolve as ions instantly. Although the lithium yield can be found below Figure 12, it cannot be directly transferred to the other neutral leaching trials since the amount of heat-treated black mass, and therefore the lithium-bearing input is reduced each time a sample is taken. Samples were taken from the leaching liquor by using a particle filter, why redirecting the lost particles to the liquid was not possible. In an upscale setup, this mass reduction would show a lower impact. The calculation of lithium yields is based on a reduced leaching liquor volume by sample extraction. The last sample is taken, at 125 min, shows increased lithium mass, which can be explained by analytical deviations.

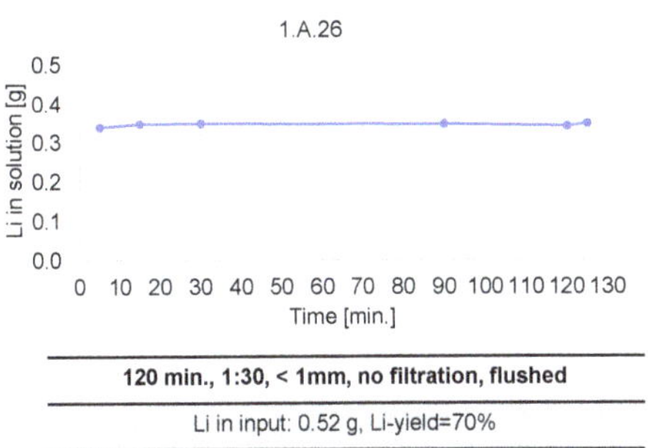

Figure 12. Kinetic trial for lithium dissolution in deionized H_2O at an s/l ratio of 1:30.

4. Particle size of heat-treated black mass:

As already reported, 99.3% of black mass particles have a grain size below 101.74 μm. In order to reduce the grain size of the few particles above this threshold, the heat-treated black mass was ground in a planetary mill. The aim of this approach was to detect the correlation between smaller grain size and an eased liberation of lithium compounds in neutral leaching. In Figure 13a, no difference with or without grinding is detected. In Figure 13b, this trend shows slightly irregular behavior when comparing trial 1.A.19 to trial 1.A.20. However, in no parameter combination and hence, trial pair compared, grinding to <90 μm has shown an improved lithium yield. This can be explained by the grain size distribution shown before: The majority of the particles shows grain sizes below 100 μm.

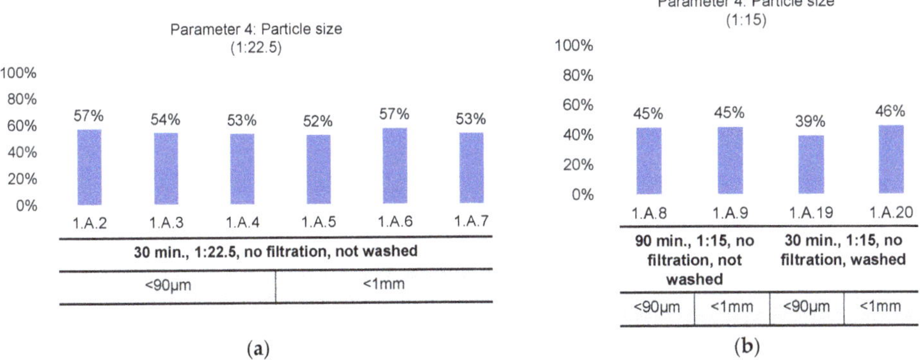

Figure 13. Detailed observation on achievable lithium yields by neutral leaching (H_2O) of heat-treated black mass. The considered parameter is parameter 4: particle size. (**a**) Dissolution and lithium recovery by comparing trials with a solid/liquid ratio of 1:22.5. (**b**) Dissolution and lithium recovery by comparing trials with solid/liquid ratio of 1:15.

Comparing trials 1.A.2-1.A.4 with trials 1.A.5-1.A.7 and trial 1.A.8 with trial 1.A.20 reveals that the grain size of the heat-treated black mass does not influence lithium yields. This is particularly interesting for residual lithiation in the anode. It proves that the degree of liberation of lithium is not enhanced by grain size reduction to <90 μm.

5. **Solid/liquid ratio (g/mL):**

At a constant grain size and leaching time, higher lithium yields are obtained at a solid/liquid ratio of 1:30 in comparison to 1:10 and 1:15 (see Figure 14a). In addition, an improved lithium recovery is possible when comparing a solid/liquid ratio of 1:30 and 1:22.5 (see Figure 14b). Although the leaching time has not shown an impact, the highest yield is reached with a solid/liquid ratio of 1:30 for 120 min. (see trial 1.A.25). When considering the solubility product of lithium carbonate in the water at 20 °C (13.3 g/L) and a lithium share of 3.7 wt%, a liquid volume of 294 mL is required for full dissolution, assuming an inserted heat-treated black mass weight of 20 g. Thus, if all lithium is present as lithium carbonate, a solid/liquid ratio of 1:15 g/mL is needed. Since the input material (heat-treated black mass) shows deviations in lithium shares and phases, the findings of a 1:30 solid/liquid ratio are supported. This means that an excess of H_2O is needed for high lithium dissolution. In Figure 14, examples of the solid/liquid ratio's impact on lithium yields are given. More trial comparisons would be 1.A.11 with a yield of 38% at a ratio of 1:10, and 1.A.10 with a yield of 44% at a ratio of 1:15. There are also trial combinations where an increase of the solid/liquid ratio leads to equal lithium yields (e.g., 1.A.15 with 1.A.9, leading to 45% lithium yield), but generally, yields of >60% can be reached only when having 20 g/600 mL (1:30).

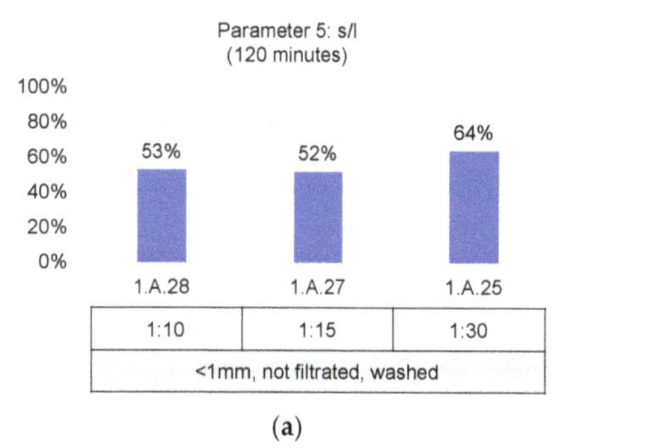

(**a**)

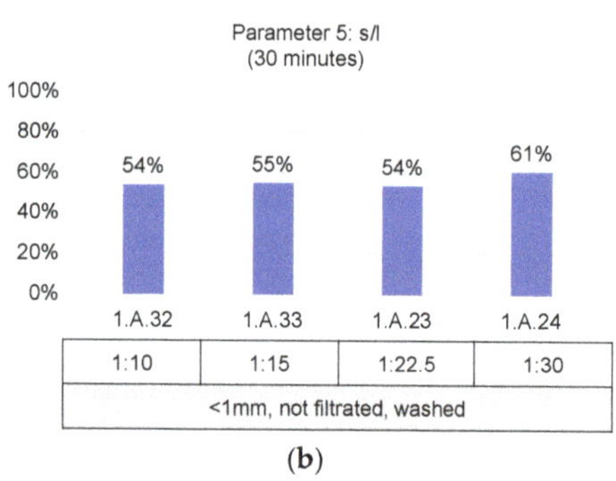

(**b**)

Figure 14. Detailed observation on achievable lithium yields by neutral leaching (H_2O) of heat-treated black mass. The considered parameter is parameter 5: particle size. (**a**) Dissolution and lithium recovery by comparing trials with a solid/liquid ratio of 1:22.5. (**b**) Dissolution and lithium recovery by comparing trials with solid/liquid ratio of 1:15.

6. **Pyrolysis temperature:**

The pyrolysis temperature plays an important role in lithium recovery, as can be seen in Figure 15. Here, the difference between a 501 and a 603 °C pyrolyzed material is pointed out. Reaching higher temperatures leads to different phase transformations within the battery cells. The impact on lithium leaching efficiency and lithium yield as solid lithium carbonate is proven by different scenarios:

Here, both grain size and leaching time do not show a significant impact on the yield. The solid/liquid ratio, along with the washing of the C-filter cake and the solid–liquid separation method (filtration vs. full boiling), seems to play an important role in this context. Up to 64% of lithium can be recovered as lithium carbonate. In addition, the parameter *pyrolysis temperature* has an impact on the lithium yield. Lithium yields by leaching heat-treated black mass without preliminary pyrolysis were not satisfying; hence these first trials are not shown in this manuscript. It must be recalled that the pyrolysis trials at Ar-atmosphere were operated at higher temperatures than the CO_2 and $Ar + O_2$ pyrolysis trials.

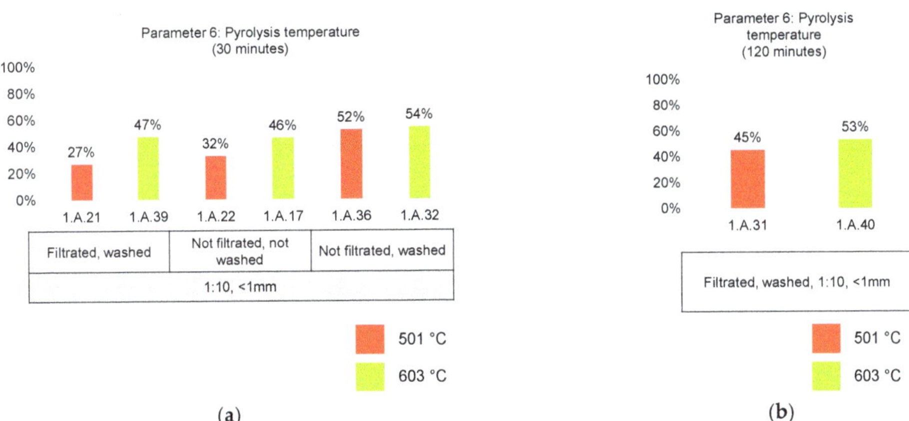

Figure 15. Detailed observation on achievable lithium yields by neutral leaching (H_2O) of heat-treated black mass. The considered parameter is parameter 6: particle size. (**a**) Dissolution and lithium recovery by comparing trials with a solid/liquid ratio of 1:22.5. (**b**) Dissolution and lithium recovery by comparing trials with solid/liquid ratio of 1:15.

For evaluating autoclave trials in terms of lithium mobilization, the lithium yields from neutral leaching are to be contrasted to the lithium yields from autoclave trials using the same parameters (see Figure 16). Since the autoclave trials were operated at a holding time of 120 min, the following diagram points out the achievable maximum lithium yields dependent on the pyrolysis temperature/atmosphere and solid/liquid ratio examined in the autoclave trials. Hereby, a direct comparison between neutral leaching (experimental series 1.A, 1.O and 1.C) and autoclave carbonation (2.A) can be performed.

Atmosphere	Ar	Ar	95 % Ar + 5 % O_2	CO_2
Targeted temperatures on cell surface during thermal pre-treatment	509 °C	603 °C	501 °C	466 °C

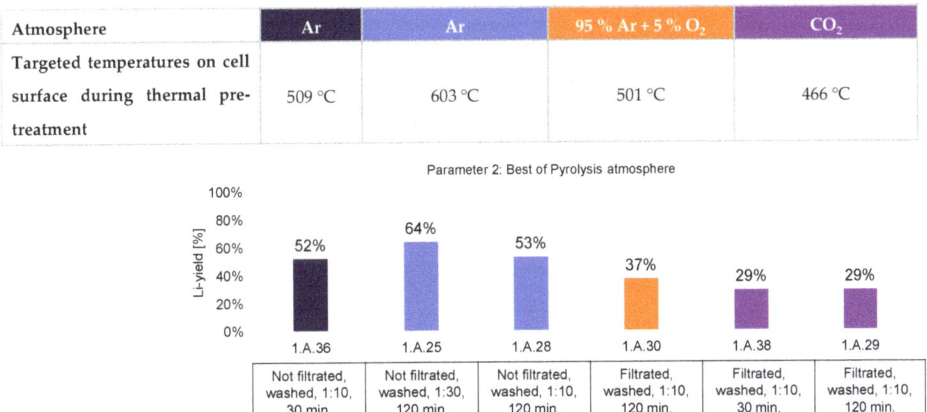

Figure 16. Best of lithium yields dependent on the pyrolysis atmospheres and temperatures. 1.A.36 has not been leached for 120 min, yet.

Since the focus of this study was the Ar-pyrolyzed material since showing the best neutral leaching results, only for this material the solid/liquid ratios were examined in the autoclave trials (1:10, 1:15, 1:30) (series 2.A, 2.O and 2.C). The CO_2- and Ar + O_2 -the pyrolyzed black mass was only treated in the autoclave carbonation set up with a solid/liquid ratio of 1:10 (2.O and 2.C). Hence, Figure 17 sums up the maximum yields of neutral leaching dependent on the pyrolysis parameters examined so far:

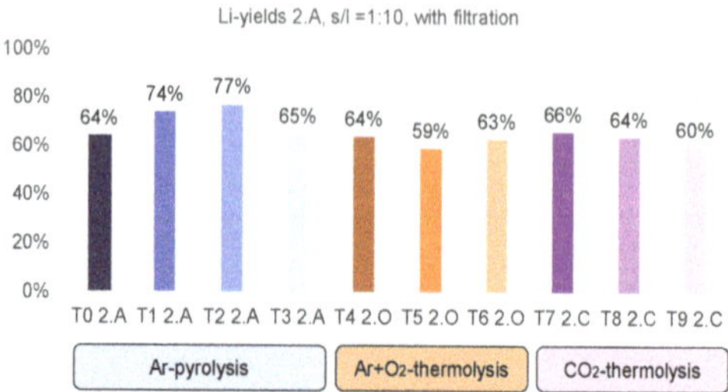

Figure 17. Lithium yields obtained by autoclave carbonation with a solid/liquid ratio of 1:10 for trials series 2.A (T0–T3, blue), 2.O (T4–T6, orange) and 2.C (T7-9, violet). Trial T0 2.A stands for an Ar-pyrolysis at 509 °C, whereas T1–T3 2.A stands for an Ar-pyrolysis at 603 °C.

3.2. Carbonation by Supercritical CO_2

Finally, the obtained lithium yields when using autoclave treatments with an s/l ratio of 1:10 for lithium carbonation can be derived from Figure 17.

Hence, a direct comparison between the atmospheres of the thermal treatments shows the following results: The 509 °C Ar-pyrolysis, that the autoclave can make a 12% difference in lithium yield. In comparison to the 603 °C Ar-pyrolysis, this difference can reach up to 24% with the correct parameter combination (120 min.). For the Ar + O_2-pyrolysis, which is here indicated as thermolysis since comprising O_2 in the atmosphere, the increased lithium yield comprises up to 27%. For the CO_2-pyrolysis, the obtained difference in lithium yield comprises up to 37%. This indicates higher lithium yields for reductive pyrolysis atmosphere (CO_2 vs. Ar-atmosphere at ~500 °C) and a stronger impact of autoclave carbonation when dealing with a not fully decomposed heat-treated black mass. This correlation needs further investigations in the future.

The elemental lithium distribution and the lithium carbonate impurities are shown exemplarily for the trial series 1.A with a solid/liquid ratio of 1:10. In Figure 18, the largest part of lithium remains in the heat-treated black mass after leaching. Moreover, the main impurity of the recovered lithium carbonate is fluorine, followed by phosphorous. This can be explained by the presence of LiF in the heat-treated black mass. It should be noted that the value "Li in filtrate" does only occur within neutral leaching and autoclave trials T0–T9 and T21 and T22, which have conducted a filtration.

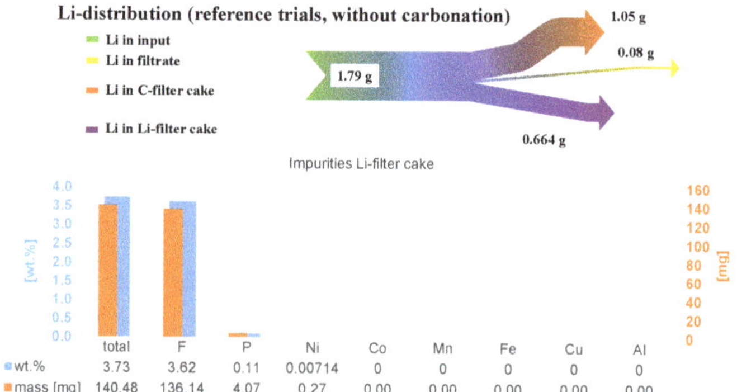

	total	F	P	Ni	Co	Mn	Fe	Cu	Al
wt.%	3.73	3.62	0.11	0.00714	0	0	0	0	0
mass [mg]	140.48	136.14	4.07	0.27	0.00	0.00	0.00	0.00	0.00

Figure 18. Lithium distribution without autoclave carbonation as exemplary data from the parameter set 1.A.1 by ICP-OES. (**Above**): Ar-pyrolysis in combination with neutral leaching at a solid/liquid ratio of 1:10. (**Below**): Matching impurities within the lithium filter cake by ICP-OES and lithium carbonate impurities.

Figure 19 shows the improvement in Li distribution when applying autoclave carbonation. Trials series 2.A was selected since the neutral leaching trials of Ar-pyrolyzed active mass at 600 °C has shown the best yields. Trial series 2.A represents Ar-pyrolysis, with a CO_2 + H_2O autoclave-reaction, and also a solid/liquid ratio of 1:10 in the autoclave. In this case, the share in the residual heat-treated black mass filter cake is significantly lower, which is a proof-of-concept of the carbonation mechanism within the autoclave.

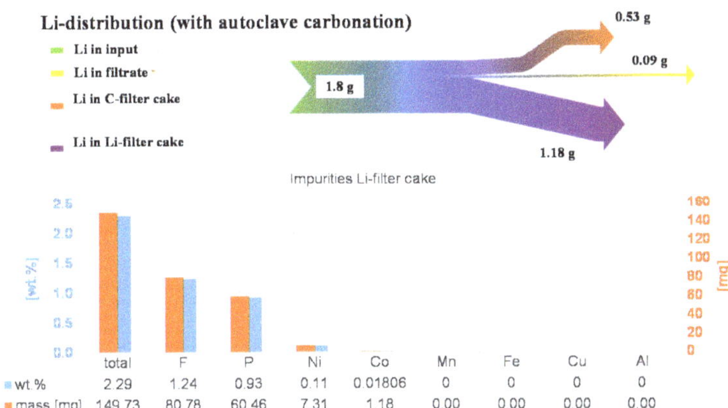

Figure 19. Lithium distribution with autoclave carbonation as exemplary data from the parameter set 2.A (T2 2.A) by ICP-OES. (**Above**): Ar-pyrolysis in combination with neutral leaching with carbonation by supercritical CO_2 + aqueous medium at a solid/liquid ratio of 1:10). (**Below**): matching impurities within the lithium filter cake.

Hence, impurities in the range of 2–4 wt.% can be derived. An XRD-evaluation gives more information on the arising phases within the heat-treated black mass, the C-filter cake and the lithium carbonate filter cake (see Figure 20). This is represented here exemplarily by the 603 °C-Ar-pyrolyzed samples, thus for trial series 2.A, also with a solid/liquid ratio of 1:10. One main finding is the removal of Li_2CO_3, present in the heat-treated black mass, from the C-filter cake. This is an indicator for the removal of water-soluble compounds. In contrast to Figure 20, XRD-evaluations of CO_2-pyrolyzed black mass at 466 °C and Ar + O_2-pyrolyzed black mass at 501 °C also detect LiNiMnO- and the NiO, which stands for an incomplete decomposition of transition metal oxides.

Small amounts of fluorine can be found in the Li-filter cake in the form of LiF. It can be seen that especially fluorine removal is crucial for reaching high lithium carbonate purities fluorine. Figure 20 shows the diffractogram of the heat-treated black mass and the C-and Li-filter cakes (T3 2.A.). X-ray diffraction was performed at room temperature using a STADI P (STOE Darmstadt) powder diffractometer using an IPPSD detector and monochromatic Cu-Kα1 radiation (λ = 1.54059 Å; flat sample; $1.5 \leq 2\theta \leq 116°$ step rate 0.015° in 2θ) with a measuring time of 2 h.

LiF was still present in the C-filter cake; hence, the solid/liquid ratio was optimized. In order to prove influencing factors on the lithium yield by an adjusted solid/liquid ratio, the parameter 1:15 (solubility of 13.3 g/L lithium carbonate at 20 °C) and 1:30 (solubility of 7.2 g/L lithium carbonate at 100 °C) were tested. Moreover, to prove the mechanism of autoclave carbonation, two parameters were examined additionally: autoclave carbonation by Ar-excess pressure (3.A) and direct and dry autoclave carbonation by CO_2-excess pressure (4.A). Again, since the Ar-pyrolyzed black mass has shown the highest yields in terms of neutral leaching and in terms of autoclave carbonation, only Ar-pyrolyzed black mass is chosen for the parameter improvements.

Table 5 sums up the parameters for the second autoclave carbonation with solid/liquid ratios of 1:15 and 1:30.

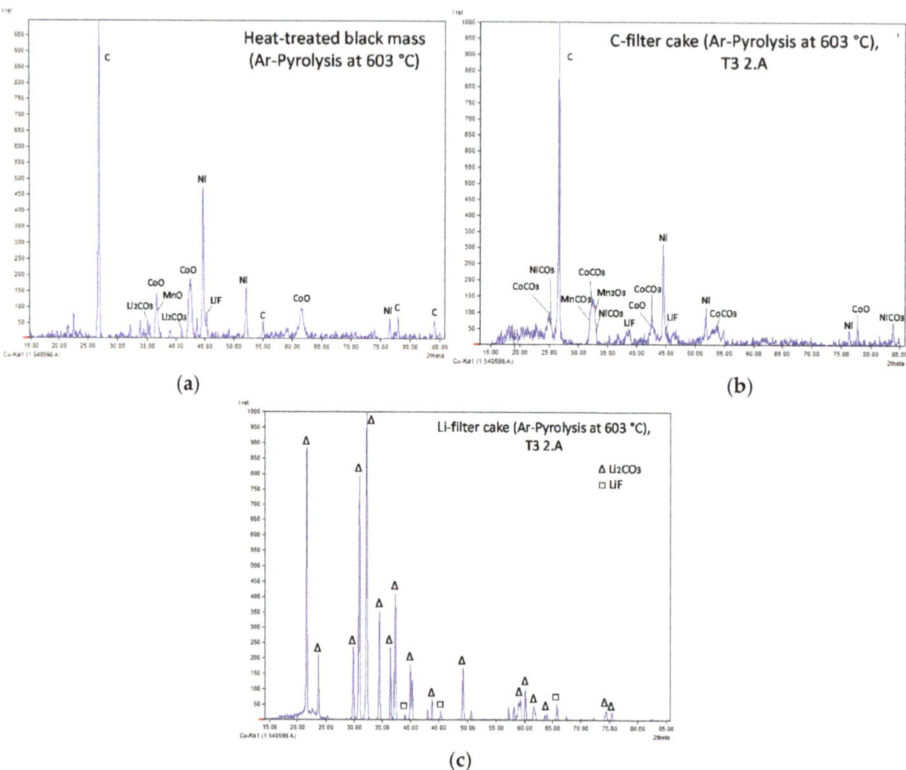

Figure 20. Graphic pattern of the Ar-pyrolyzed black mass at 603 °C (**a**), and the corresponding C-filter cake (**b**) and Li-filter cake (**c**) from trial T3 2.A. The XRD-evaluation was performed using the "match!" Software and the COD Inorganics database.

Table 5. Detailed list of parameters examined for autoclave trials T10–T13 (3.A), T14–T16, T18, T19, T21 and T22 (2.A), and T17 and T20 (4.A) with a solid/liquid ratio of 1:15 and 1:30. In T17/20 s/l ratio refers to leaching after autoclave treatment.

	Solid/Liquid Ratio (s/l) (g/mL)	H$_2$O in Autoclave	Autoclave Gas	Washing C-Filter Cake with H$_2$O
T10	1:15	yes	Ar	no
T11	1:15	yes	Ar	no
T12	1:15	yes	Ar	no
T13	1:15	yes	Ar	no
T14	1:15	yes	CO$_2$	no
T15	1:15	yes	CO$_2$	yes
T16	1:15	yes	CO$_2$	no
T17	1:30	no	CO$_2$	no
T18	1:30	yes	CO$_2$	yes
T19	1:30	yes	CO$_2$	yes
T20	1:30	no	CO$_2$	yes
T21	1:15	yes	CO$_2$	no
T22	1:30	yes	CO$_2$	no

The following illustration (see Figure 21) shows the results of autoclave carbonation with solid/liquid ratios of 1:15 and 1:30:

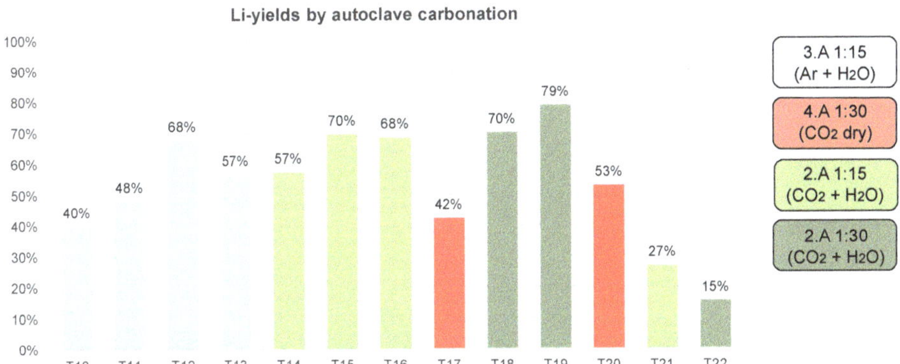

Figure 21. Lithium yields obtained by autoclave carbonation with a solid/liquid ratio of 1:15 (T10–T16, and T21) or 1:30 (T17–T20, and T22). The C-filter-cakes of T15, T18 and T19 were washed, T17 was leached for 5 min and T20 for 90 min. T21 and T22 were filtrated instead of fully boiled.

Thus, the only process window leading to satisfying yields of 79% is an s/l ratio of 1:30 in combination with CO_2 carbonation in an aquatic medium. The underlying mechanism seems to be indirect carbonation.

This can be supported by the detected pH value of all trials. Whereas 1.A.1—1.A.40, T10–T13, and T20 (without CO_2 purging in the liquid) showed a pH value of 11–12 after charging heat-treated black mass in H_2O, the trials T0–T9. T14–T16, T18/T19 and T21/T22 (with CO_2 purging in the liquid) showed a pH value of 7–8 after charging heat-treated black mass in H_2O. Hence, CO_2 was dissolved in the liquid. The generally higher yields in T10–T20 can also be attributed to the avoidance of a second filtration step for recovering lithium. Instead, the solution was boiled until reaching a slurry-state and then was dried in a beaker. Hereby, lithium losses in the residual filtrate are avoided. In addition, when comparing T17 to T20, the advantage of a longer leaching time and washing of the carbon filter cake with deionized H_2O is shown. The washing of the carbon filter cake generally leads to higher yields since dissolved lithium remaining in the filter cakes in the solution can leave the system just by washing. However, comparing T15 to T14 and T16 in terms of C-filter cake washing reveals a rather small impact on the lithium yields (max. 2%). When comparing T21 to T14 and to T16, it can be seen that filtrating of the lithium solution is not expedient. This can be explained by the residual lithium dissolution in the filtrate. The comparison between T22 and T18–T19 confirms this relation. T21 and T22 show very low yields. Although the lithium filter cake was filtrated instead of full boiling and the C-filter cake was not washed, their lithium yield shows disproportionally low yields, which can only be explained by heterogeneity in the heat-treated black mass.

4. Discussion

This study proves the concept of indirect carbonation for treating lithium-ion battery heat-treated black mass with supercritical CO_2. The involvement of supercritical CO_2 in terms of lithium carbonate generation is supported by yields comparing Ar-excess pressure and CO_2-excess pressure. Moreover, indirect carbonation is shown by comparing a dry autoclave process to a liquid-based autoclave process. The lower pH-value of pH = 7–8 when applying CO_2 in comparison to H_2O-leaching (pH = 11–12) can lead to the following statements:

1. When leaching heat-treated black mass in H_2O, the solution is basic. This can be attributed to the dissolution of basic phases in the liquid. $\rightarrow$ LiF and Li_2CO_3 could be detected in the heat-treated black mass by XRD; both phases are slightly soluble and therefore are responsible for the elevated pH-value. Although LiOH and Li_2O could not be detected via XRD-analysis in the heat-treated black mass, they may be present

in small amounts since the SEI-layers consist of Li_2CO_3, LiF, LiOH and Li_2O [93]. However, it was shown that LiF decomposes to HF and LiOH in aqueous solutions, which indicates $Li^+ + OH^-$ in the solution.

2. When leaching heat-treated black mass in H_2O and adding CO_2-gas, the pH value of the solution decreases to 7–8. Mechanisms are in place, which can be attributed to CO_2 and which are leading to a higher lithium leaching efficiency. In the following, hypotheses for the underlying mechanisms are stated:

   a. The formation of carbonic acid and thus the formation of CO_3^{2-} and HCO_3^- as acidic leaching agents. CO_2 is added to a basic solution; it reacts acidic by the release of H^+ ions. This pH-value decrease can be responsible for a higher leaching efficiency by creating quasi-acidic leaching conditions similar to conventional hydrometallurgy.

   b. Recombination of Li^+, stemming from non-lithium carbonate phases like LiF, with present CO_3^{2-} or HCO_3^-. This would entail the following suggested equations (see Equations (19) and (20), schematically shown in Figure 22):

$$Li^+ + CO_3^{2-} \rightarrow Li_2CO_3 \tag{19}$$

$$Li^+ + HCO_3^- \rightarrow LiHCO_3 \tag{20}$$

   c. A combination of both mentioned mechanisms. In this way, the dissolution of lithium phases in the heat-treated black mass is promoted by CO_2, more lithium ions can be formed to Li_2CO_3, and this effect is also promoted by the increased operating temperatures and arising excess pressure.

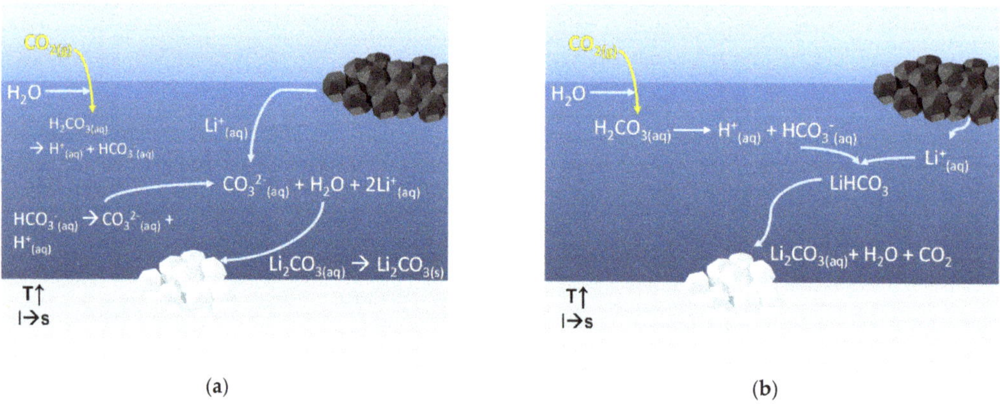

(a) (b)

Figure 22. Schematic process visualization of indirect carbonation promoted by supercritical CO_2 based on CO_3^{2-} (**a**) and HCO_3^- (**b**) in terms of leaching lithium-ion battery heat-treated black mass in deionized water. When increasing the solution's temperature, lithium carbonate is precipitated as a solid lithium salt.

5. Conclusions

The presented "ESLR" process, consisting of thermal treatment, mechanical comminution and a sorting step, followed by a subsequent carbonation process, results in the following scientific findings:

Carbonation by supercritical CO_2 shows an increased lithium yield of around 15%. This value stems from the difference between a maximum lithium yield in neutral leaching of 64% and a maximum lithium yield in autoclave carbonation of 79%. When expressing the yield as leaching efficiency, 88% were reached. The different pyrolysis atmospheres and temperatures show a direct influence on the lithium yield. Further key influencing factors for both H_2O-leaching with and without CO_2 are solid/liquid ratio, filter cake washing and the lithium extraction method (filtration vs. full boiling). It can be concluded that the "ESLR" process shows benefits in comparison to simple H_2O-leaching and that the mechanism for indirect carbonation is beneficial. Moreover, the "ESLR" process is a separate step to ease Ni/Co/Mn recovery and to enhance the degree of lithium mobilization. Hence, the resulting lithium-reduced filter cake (C-filter cake) can be integrated into existing hydro- or pyrometallurgical steps.

The process of technology relevance is shown by the following specific benefits in contrast to the state-of-the-art:

- Conventional lithium carbonation, e.g., by Na_2CO_3, is avoided, and no further chemicals are required, making lithium recovery more environmentally friendly;
- Subsequent treating the C-filter cake hydrometallurgically for metal extraction (Ni, Co, …) requires fewer leaching agents because the input mass is reduced, and hence, fewer additives for pH-adjustments are needed;
- Moreover, in comparison to conventional hydrometallurgical lithium recovery, the liquid volume can be fully evaporated (filtration vs. full boiling). Hereby, no lithium remains in the solution. This is possible since no enhancement of salinity is caused in "ESLR";
- Lithium losses in various byproducts of chemical solution purification and metal winning steps are avoided;
- Costly lithium extraction from a pyrometallurgy treatment and hydrometallurgical purifying of slags is also avoided.

In contrast to other studies, the sequence of thermal and mechanical treatment is inverted. In this study, battery cells are first thermally treated and then shredded to extract heat-treated black mass. This procedure is safer due to the avoidance of ignition during shredding.

Comparing lithium yields by H_2O-leaching in this study to literature, the following statements can be given: In [74], 66% of lithium from NMC black mass are obtained by shredding, then thermal treating and H_2O-leaching. Here, the authors rather focus on LMO-cells and report on one trial, only reaching 66% [74]. However, in this paper, 64% could be recovered by thermal treatment with subsequent shredding and H_2O-leaching at a thermal treatment by 100 °C lower than Xiao et al. and without costly vacuum operations. In comparison to [46,47], where 40% of lithium could be recovered by shredding, thermal treatment and H_2O-leaching of cathode black mass, the yield in this study are up to 24% higher.

A comparison of lithium yields by H_2O-leaching in combination with CO_2 (indirect carbonation) is not straightforward since there is no study in place using the whole black mass from NMC-cells for this process. However, by using cathode black mass with lignite, 85% is reached, whereas, in this study, 79% are reached. This difference might be attributable to the neglection of anode material and/or the use of lignite instead in [46,47]. In comparison to [90], the yields in this study are 19% higher (60% vs. 79%), but yield and matching parameters are given based on a lithium ore treatment. Only the transferability to black mass is mentioned. However, this study also uses heat-treated black mass in contrast to [90]. In comparison to [91], and avoidance of electrodialysis in a Li_2SO_4-solution and of carbonation reagents could be reached.

A comparison of lithium yields by a thermal CO_2-treatment with subsequent H_2O-leaching (direct carbonation) is hardly possible since the autoclave process in this study worked at $T_{max} = 230\,°C$, whereas literature focuses on elevated temperatures (~650–800 °C [73]) with CO_2 as purging instead of excess pressure; moreover, no yield calculation is given [73].

In this study, a proof-of-concept regarding the indirect carbonation using supercritical CO_2 in an autoclave could be shown.

The most important follow-up research comprises a further enhancement of lithium yields to a value of >90%, which is necessary to make the "ESLR" a competitive process option. Then, CO_2-driven carbonation without supercritical CO_2, but by CO_2-gas purging instead. This is crucial because the combination of thermal pretreatment and an autoclave treatment comprise high energy requirements. However, as reported in Section 1.1.1, thermal conditioning is also beneficial for hydrometallurgical treatment. Hence, the connected energy demands cannot particularly and only be counted for the "ESLR" process. First trials with CO_2-gas instead of $SCCO_2$ have shown lithium yields around 70%. Hereby, insights into the role of excess pressure (73.8 bar) and high temperatures (150 °C) are possible. Moreover, this setup would imply economic benefits due to the avoidance of high-pressure operations. This will be one topic of "Early-Stage Recovery of Lithium from Tailored Thermal Conditioned Black Mass Part II: Mobilizing Lithium via gaseous CO_2-Carbonation". Moreover, a refining of the C-filter cake by flotation or acidic leaching should be tested. Upscaling is planned for future research to test possible scale effects due to losses on equipment surfaces, for example, on beakers after boiling the lithium filtrate. In addition, a suitable development for removing fluorine from the heat-treated black mass, filtrates and filter cakes would be an important tool for hazardous-free processing, which would not harm the used equipment by developing HF-gas. Moreover, experimental series 3.O/3.C and 4.O/4.C are to be performed. Moreover, the heat-treated black mass-producing pyrolysis was conducted with a holding time of 60 min. This may be optimized as well to find the perfect match in terms of temperature and holding time.

Author Contributions: Conceptualization, L.S. and B.F.; methodology, L.S.; validation, L.S., T.S. and B.F.; formal analysis, L.S. and T.S.; investigation, L.S. and T.S.; resources, L.S. and T.S.; data curation, L.S. and T.S.; writing—original draft preparation, L.S.; writing—review and editing, L.S.; visualization, L.S.; supervision, B.F. All authors have read and agreed to the published version of the manuscript.

Funding: This research was partly funded by the Research Council of Norway in the frame of the LIBRES project (project number 282328).

Institutional Review Board Statement: Not applicable.

Data Availability Statement: The data presented in this study are available on request from the corresponding author. The data are not publicly available since being part of active research.

Acknowledgments: In cooperation with Access e.V. EDS-analyses were enabled, and in cooperation with the Institute of Inorganic Chemistry (IAC) at RWTH Aachen, XRD-measurements were possible. We are grateful for the support of both partners.

Conflicts of Interest: The authors declare no conflict of interest.

References

1. Wang, X.; Gaustad, G.; Babbitt, C.W.; Richa, K. Economies of scale for future lithium-ion battery recycling infrastructure. *Resour. Conserv. Recycl.* **2014**, *83*, 53–62. [CrossRef]
2. Zheng, X.; Zhu, Z.; Lin, X.; Zhang, Y.; He, Y.; Cao, H.; Sun, Z. A Mini-Review on Metal Recycling from Spent Lithium Ion Batteries. *Engineering* **2018**, *4*, 361–370. [CrossRef]
3. Pinegar, H.; Smith, Y.R. Recycling of End-of-Life Lithium Ion Batteries, Part I: Commercial Processes. *J. Sustain. Metall.* **2019**, *5*, 402–416. [CrossRef]
4. Boyden, A.; Soo, V.K.; Doolan, M. The Environmental Impacts of Recycling Portable Lithium-Ion Batteries. *Procedia CIRP* **2016**, *48*, 188–193. [CrossRef]
5. Zhao, S.; He, W.; Li, G. Recycling Technology and Principle of Spent Lithium-Ion Battery. In *Recycling of Spent Lithium-Ion Batteries*; An, L., Ed.; Springer International Publishing: Cham, Switzerland, 2019; pp. 1–26, ISBN 978-3-030-31833-8.
6. Becker, J.; Beverungen, D.; Winter, M.; Menne, S. *Umwidmung und Weiterverwendung von Traktionsbatterien*; Springer Fachmedien Wiesbaden: Wiesbaden, Germany, 2019; ISBN 978-3-658-21020-5.
7. Werner, D.; Peuker, U.A.; Mütze, T. Recycling Chain for Spent Lithium-Ion Batteries. *Metals* **2020**, *10*, 316. [CrossRef]
8. Harper, G.; Sommerville, R.; Kendrick, E.; Driscoll, L.; Slater, P.; Stolkin, R.; Walton, A.; Christensen, P.; Heidrich, O.; Lambert, S.; et al. Recycling lithium-ion batteries from electric vehicles. *Nature* **2019**, *575*, 75–86. [CrossRef]
9. Pinegar, H.; Smith, Y.R. Recycling of End-of-Life Lithium-Ion Batteries, Part II: Laboratory-Scale Research Developments in Mechanical, Thermal, and Leaching Treatments. *J. Sustain. Metall.* **2020**, *6*, 142–160. [CrossRef]
10. Bai, Y.; Muralidharan, N.; Sun, Y.-K.; Passerini, S.; Stanley Whittingham, M.; Belharouak, I. Energy and environmental aspects in recycling lithium-ion batteries: Concept of Battery Identity Global Passport. *Mater. Today* **2020**, *41*, 304–315. [CrossRef]
11. Peters, J.F.; Simon, B.; Rodriguez-Garcia, G.; Weil, M. Building a common base for LCA benchmarking of Li-Ion batteries. In Proceedings of the SETAC 26th Annual Meeting, Nantes, France, 22–26 May 2016.

12. Dunn, J.B.; Gaines, L.; Kelly, J.C.; Gallagher, K.G. Life Cycle Analysis Summary for Automotive Lithiumion Battery Production and Recycling. In *Rewas 2016: Towards Materials Resource Sustainability*; John Wiley & Sons, Ltd.: Hoboken, NJ, USA, 2016; pp. 73–79.

13. Martens, H.; Goldmann, D. *Recyclingtechnik*; Fachbuch für Lehre und Praxis, 2. Aufl; Springer Fachmedien Wiesbaden: Wiesbaden, Germany, 2016; ISBN 978-3-658-02786-5.

14. Diekmann, J.; Sander, S.; Sellin, G.; Petermann, M.; Kwade, A. Crushing of Battery Modules and Cells. In *Recycling of Lithium-Ion Batteries*; Kwade, A., Diekmann, J., Eds.; Springer International Publishing: Cham, Switzerland, 2018; pp. 127–138, ISBN 978-3-319-70571-2.

15. Sojka, R.T. Sichere Aufbereitung von Lithium-basierten Batterien durch thermische Konditionierung. Safe treatment of Lithium-based Batteries through Thermal Conditioning. In *Recycling und Sekundärrohstoffe, Band 13*; Thomé-Kozmiensky, E., Holm, O., Friedrich, B., Goldmann, D., Eds.; Thomé-Kozmiensky Verlag GmbH: Nietwerder, Germany, 2020; pp. 506–523, ISBN 978-3-944310-51-0.

16. Zhang, G.; Du, Z.; He, Y.; Wang, H.; Xie, W.; Zhang, T. A Sustainable Process for the Recovery of Anode and Cathode Materials Derived from Spent Lithium-Ion Batteries. *Sustainability* **2019**, *11*, 2363. [CrossRef]

17. Jiang, H.; Emmett, R.K.; Roberts, M.E. Thermally induced deactivation of lithium-ion batteries using temperature-responsive interfaces. *Ionics* **2019**, *25*, 2453–2457. [CrossRef]

18. Meshram, P.; Pandey, B.D.; Mankhand, T.R. Recovery of valuable metals from cathodic active material of spent lithium ion batteries: Leaching and kinetic aspects. *Waste Manag.* **2015**, *45*, 306–313. [CrossRef] [PubMed]

19. Lombardo, G.; Ebin, B.; St. J. Foreman, M.R.; Steenari, B.-M.; Petranikova, M. Chemical Transformations in Li-Ion Battery Electrode Materials by Carbothermic Reduction. *ACS Sustain. Chem. Eng.* **2019**, *7*, 13668–13679. [CrossRef]

20. Ekberg, C.; Petranikova, M. Lithium Batteries Recycling. In *Lithium Process Chemistry. Resources, Extraction, Batteries, and Recycling*; Chagnes, A., Światowska, J., Eds.; Elsevier: Amsterdam, The Netherlands, 2015; pp. 233–267, ISBN 9780128014172.

21. Sun, L.; Qiu, K. Vacuum pyrolysis and hydrometallurgical process for the recovery of valuable metals from spent lithium-ion batteries. *J. Hazard. Mater.* **2011**, *194*, 378–384. [CrossRef]

22. Träger, T.; Friedrich, B.; Weyhe, R. Recovery Concept of Value Metals from Automotive Lithium-Ion Batteries. *Chem. Ing. Tech.* **2015**, *87*, 1550–1557. [CrossRef]

23. Stallmeister, C.; Schwich, L.; Friedrich, B. Early-Stage Li-Removal-Vermeidung von Lithiumverlusten im Zuge der Thermischen und Chemischen Recyclingrouten von Batterien. In *Recycling und Sekundärrohstoffe, Band 13*; Thomé-Kozmiensky, E., Holm, O., Friedrich, B., Goldmann, D., Eds.; Thomé-Kozmiensky Verlag GmbH: Nietwerder, Germany, 2020; pp. 545–557, ISBN 978-3-944310-51-0.

24. Vest, M. Weiterentwicklung des Pyrometallurgischen IME Recyclingverfahrens für Li-Ionen Batterien von Elektrofahrzeugen. Ph.D. Thesis, Shaker Verlag GmbH, Aachen, Germany, 2016.

25. Wang, H.; Friedrich, B. Development of a Highly Efficient Hydrometallurgical Recycling Process for Automotive Li–Ion Batteries. *J. Sustain. Metall.* **2015**, *1*, 168–178. [CrossRef]

26. Chen, Y.; Liu, N.; Jie, Y.; Hu, F.; Li, Y.; Wilson, B.P.; Xi, Y.; Lai, Y.; Yang, S. Toxicity Identification and Evolution Mechanism of Thermolysis-Driven Gas Emissions from Cathodes of Spent Lithium-Ion Batteries. *ACS Sustain. Chem. Eng.* **2019**, *7*, 18228–18235. [CrossRef]

27. Shin, S.M.; Kim, N.H.; Sohn, J.S.; Yang, D.H.; Kim, Y.H. Development of a metal recovery process from Li-ion battery wastes. *Hydrometallurgy* **2005**, *79*, 172–181. [CrossRef]

28. Yang, Y.; Song, S.; Lei, S.; Sun, W.; Hou, H.; Jiang, F.; Ji, X.; Zhao, W.; Hu, Y. A process for combination of recycling lithium and regenerating graphite from spent lithium-ion battery. *Waste Manag.* **2019**, *85*, 529–537. [CrossRef]

29. Or, T.; Gourley, S.W.D.; Kaliyappan, K.; Yu, A.; Chen, Z. Recycling of mixed cathode lithium-ion batteries for electric vehicles: Current status and future outlook. *Carbon Energy* **2020**, *61*, 1. [CrossRef]

30. Essay, P.; Foxworth, C.; Strauss, C.; Torres, A. Corby Anderson. Hydrometallurgical Recovery of Materials from Lithium-ion Batteries. 2012. Available online: https://www.researchgate.net/publication/302285317_Hydrometallurgical_Recovery_of_Materials_from_Lithium-ion_Batteries (accessed on 19 January 2021).

31. Liu, P.; Xiao, L.; Chen, Y.; Tang, Y.; Wu, J.; Chen, H. Recovering valuable metals from LiNixCoyMn1-x-yO2 cathode materials of spent lithium ion batteries via a combination of reduction roasting and stepwise leaching. *J. Alloys Compd.* **2019**, *783*, 743–752. [CrossRef]

32. Li, L.; Zhang, X.; Li, M.; Chen, R.; Wu, F.; Amine, K.; Lu, J. The Recycling of Spent Lithium-Ion Batteries: A Review of Current Processes and Technologies. *Electrochem. Energy Rev.* **2018**, *1*, 461–482. [CrossRef]

33. Huang, B.; Pan, Z.; Su, X.; An, L. Recycling of lithium-ion batteries: Recent advances and perspectives. *J. Power Sources* **2018**, *399*, 274–286. [CrossRef]

34. Meshram, P.; Pandey, B.D.; Mankhand, T.R. Extraction of lithium from primary and secondary sources by pre-treatment, leaching and separation: A comprehensive review. *Hydrometallurgy* **2014**, *150*, 192–208. [CrossRef]

35. Li, L.; Bian, Y.; Zhang, X.; Xue, Q.; Fan, E.; Wu, F.; Chen, R. Economical recycling process for spent lithium-ion batteries and macro- and micro-scale mechanistic study. *J. Power Sources* **2018**, *377*, 70–79. [CrossRef]

36. Barik, S.P.; Prabaharan, G.; Kumar, L. Leaching and separation of Co and Mn from electrode materials of spent lithium-ion batteries using hydrochloric acid: Laboratory and pilot scale study. *J. Clean. Prod.* **2017**, *147*, 37–43. [CrossRef]

37. Chen, X.; Zhou, T. Hydrometallurgical process for the recovery of metal values from spent lithium-ion batteries in citric acid media. *Waste Manag. Res.* **2014**, *32*, 1083–1093. [CrossRef]
38. Wang, R.-C.; Lin, Y.-C.; Wu, S.-H. A novel recovery process of metal values from the cathode active materials of the lithium-ion secondary batteries. *Hydrometallurgy* **2009**, *99*, 194–201. [CrossRef]
39. Nayl, A.A.; Elkhashab, R.A.; Badawy, S.M.; El-Khateeb, M.A. Acid leaching of mixed spent Li-ion batteries. *Arab. J. Chem.* **2017**, *10*, S3632–S3639. [CrossRef]
40. Sattar, R.; Ilyas, S.; Bhatti, H.N.; Ghaffar, A. Resource recovery of critically-rare metals by hydrometallurgical recycling of spent lithium ion batteries. *Sep. Purif. Technol.* **2019**, *209*, 725–733. [CrossRef]
41. Gao, W.; Zhang, X.; Zheng, X.; Lin, X.; Cao, H.; Zhang, Y.; Sun, Z. Lithium Carbonate Recovery from Cathode Scrap of Spent Lithium-Ion Battery: A Closed-Loop Process. *Environ. Sci. Technol.* **2017**, *51*, 1662–1669. [CrossRef]
42. Stopic, S.; Dertmann, C.; Modolo, G.; Kegler, P.; Neumeier, S.; Kremer, D.; Wotruba, H.; Etzold, S.; Telle, R.; Rosani, D.; et al. Synthesis of Magnesium Carbonate via Carbonation under High Pressure in an Autoclave. *Metals* **2018**, *8*, 993. [CrossRef]
43. Rahmani, O.; Highfield, J.; Junin, R.; Tyrer, M.; Pour, A.B. Experimental Investigation and Simplistic Geochemical Modeling of CO_2 Mineral Carbonation Using the Mount Tawai Peridotite. *Molecules* **2016**, *21*, 353. [CrossRef] [PubMed]
44. Daval, D.; Martinez, I.; Guigner, J.-M.; Hellmann, R.; Corvisier, J.; Findling, N.; Dominici, C.; Goffe, B.; Guyot, F. Mechanism of wollastonite carbonation deduced from micro- to nanometer length scale observations. *Am. Mineral.* **2009**, *94*, 1707–1726. [CrossRef]
45. Haug, T.A. Dissolution and Carbonation of Mechanically Activated Olivine. Investigating CO_2 Sequestration Possibilities. Ph.D. Thesis, Norwegian University of Science and Technology, Trondheim, Norway, 2010.
46. Hu, J.; Zhang, J.; Li, H.; Chen, Y.; Wang, C. A promising approach for the recovery of high value-added metals from spent lithium-ion batteries. *J. Power Sources* **2017**, *351*, 192–199. [CrossRef]
47. Zhang, J.; Hu, J.; Zhang, W.; Chen, Y.; Wang, C. Efficient and economical recovery of lithium, cobalt, nickel, manganese from cathode scrap of spent lithium-ion batteries. *J. Clean. Prod.* **2018**, *204*, 437–446. [CrossRef]
48. Kunzler, C.; Alves, N.; Pereira, E.; Nienczewski, J.; Ligabue, R.; Einloft, S.; Dullius, J. CO_2 storage with indirect carbonation using industrial waste. *Energy Procedia* **2011**, *4*, 1010–1017. [CrossRef]
49. Jandová, J.; Dvorák, P.; Kondás, J.; Havlák, L. Recovery of Lithium from Waste Materials. *Ceram.-Silik.* **2012**, *56*, 50–54.
50. Hönisch, B.; Ridgwell, A.; Schmidt, D.N.; Thomas, E.; Gibbs, S.J.; Sluijs, A.; Zeebe, R.; Kump, L.; Martindale, R.C.; Greene, S.E.; et al. The geological record of ocean acidification. *Science* **2012**, *335*, 1058–1063. [CrossRef]
51. Haynes, W.M.; Lide, D.R.; Bruno, T.J. CRC handbook of chemistry and physics. In *A Ready-Reference Book of Chemical and Physical Data*; CRC Press: Boca Raton, FL, USA; London, UK; New York, NY, USA, 2017; ISBN 978-1-4987-5429-3.
52. Yi, W.-T.; Yan, C.-Y.; Ma, P.-H. Kinetic study on carbonation of crude Li2CO3 with CO2-water solutions in a slurry bubble column reactor. *Korean J. Chem. Eng.* **2011**, *28*, 703–709. [CrossRef]
53. Yi, W.-T.; Yan, C.-Y.; Ma, P.-H. Crystallization kinetics of Li_2CO_3 from $LiHCO_3$ solutions. *J. Cryst. Growth* **2010**, *312*, 2345–2350. [CrossRef]
54. Trogdon-Stout, D. Chemistry. In *Essential Practice for Key Science Topics*; The 100+ Series; Carson Dellosa Publishing Group: Greensboro, NC, USA, 2015; ISBN 1483817091.
55. European Chemicals Agency ECHA. Lithium Carbonate (EC number: 209-062-5 | CAS Number: 554-13-2). Hydrolysis. European Union. 2010. Available online: https://echa.europa.eu/de/registration-dossier/-/registered-dossier/15034/5/2/3#sLinkToRelevantStudyRecord (accessed on 13 June 2020).
56. Wawra, E.; Dolznig, H.; Müllner, E. Chemie verstehen. In *Allgemeine Chemie für Mediziner und Naturwissenschaften, 5. Aufl*; UTB: Medizin, Naturwissenschaften, 2009; ISBN 9783825282059.
57. House, J.E. *Inorganic Chemistry*, 2nd ed.; Elsevier: San Diego, CA, USA, 2020; ISBN 978-0-12-814369-8.
58. Perry, D.L. *Handbook of Inorganic Compounds*; CRC Press: Hoboken, NJ, USA, 2011; ISBN 978-1-4398-1461-1.
59. Groult, H.; Nakajima, T. *Fluorinated Materials for Energy Conversion, 1. Aufl.*; Elsevier: Amsterdam, The Netherlands; San Diego, CA, USA; Oxford, UK, 2005; ISBN 9780080444727.
60. Scholz, F.; Kahlert, H. *Chemische Gleichgewichte in der Analytischen Chemie*; Springer: Berlin/Heidelberg, Germany, 2020; ISBN 978-3-662-61106-7.
61. Binnewies, M.; Finze, M.; Jäckel, M.; Schmidt, P.; Willner, H.; Rayner-Canham, G. *Allgemeine und Anorganische Chemie*; Springer: Berlin/Heidelberg, Germany, 2016; ISBN 978-3-662-45066-6.
62. Lew, K. Acids and bases. In *Essential Chemistry*; Chelsea House: New York, NY, USA, 2009; ISBN 9781438116617.
63. Sigma-Aldrich Co. Safety Data Sheet according to Regulation (EC) No. 1907/2006. Lithium Fluoride (CAS-No.: 7789-24-4). 2018. Available online: https://www.sigmaaldrich.com/MSDS/MSDS/DisplayMSDSPage.do?country=DE&language=EN-generic&productNumber=449903&brand=ALDRICH&PageToGoToURL=https%3A%2F%2Fwww.sigmaaldrich.com%2Fcatalog%2Fsearch%3Fterm%3DLiF%26interface%3DAll_DE%26N%3D0%26mode%3Dmatch%26lang%3Dde%26region%3DDE%26focus%3Dproduct (accessed on 29 December 2020).
64. Shimonishi, Y.; Zhang, T.; Imanishi, N.; Im, D.; Lee, D.J.; Hirano, A.; Takeda, Y.; Yamamoto, O.; Sammes, N. A study on lithium/air secondary batteries—Stability of the NASICON-type lithium ion conducting solid electrolyte in alkaline aqueous solutions. *J. Power Sources* **2011**, *196*, 5128–5132. [CrossRef]

55. Bi, X.; Wang, R.; Lu, J. Li-Air Batteries: Discharge products. In *Metal-Air Batteries. Fundamentals and Applications*; Zhang, X.-B., Ed.; Wiley-VCH: Weinheim, Germany, 2018; ISBN 3527807667.

56. Ham, B.M.; MaHam, A. Analytical chemistry. In *A Chemist and Laboratory Technician's Toolkit*; John Wiley & Sons, Inc.: Hoboken, NJ, USA, 2016; ISBN 9781119069690.

57. Zilberman, P. The CO_2 Absorber Based on LiOH. *Acta Medica Marisiensis* **2015**, *61*, 4–6. [CrossRef]

58. Johnson, G.K.; Grow, R.T.; Hubbard, W.N. The enthalpy of formation of lithium oxide (Li_2O). *J. Chem. Thermodyn.* **1975**, *7*, 781–786. [CrossRef]

59. Li, J.; Wang, G.; Xu, Z. Environmentally-friendly oxygen-free roasting/wet magnetic separation technology for in situ recycling cobalt, lithium carbonate and graphite from spent LiCoO2/graphite lithium batteries. *J. Hazard. Mater.* **2016**, *302*, 97–104. [CrossRef] [PubMed]

60. Vishvakarma, S.; Dhawan, N. Recovery of Cobalt and Lithium Values from Discarded Li-Ion Batteries. *J. Sustain. Metall.* **2019**, *5*, 204–209. [CrossRef]

61. Xiao, J.; Li, J.; Xu, Z. Recycling metals from lithium ion battery by mechanical separation and vacuum metallurgy. *J. Hazard. Mater.* **2017**, *338*, 124–131. [CrossRef]

62. Kuzuhara, S.; Ota, M.; Tsugita, F.; Kasuya, R. Recovering Lithium from the Cathode Active Material in Lithium-Ion Batteries via Thermal Decomposition. *Metals* **2020**, *10*, 433. [CrossRef]

63. Wang, J.-P.; Pyo, J.-J.; Ahn, S.-H.; Choi, D.-H.; Lee, B.-W.; Lee, D.-W. A Study on the Recovery of Li_2CO_3 from Cathode Active Material NCM(LiNiCoMnO2) of Spent Lithium Ion Batteries. *J. Korean Powder Metall. Inst.* **2018**, *25*, 296–301. [CrossRef]

64. Xiao, J.; Li, J.; Xu, Z. Novel Approach for in Situ Recovery of Lithium Carbonate from Spent Lithium Ion Batteries Using Vacuum Metallurgy. *Environ. Sci. Technol.* **2017**, *51*, 11960–11966. [CrossRef]

65. Yang, Y.; Huang, G.; Xu, S.; He, Y.; Liu, X. Thermal treatment process for the recovery of valuable metals from spent lithium-ion batteries. *Hydrometallurgy* **2016**, *165*, 390–396. [CrossRef]

66. Lombardo, G.; Ebin, B.; St J Foreman, M.R.; Steenari, B.-M.; Petranikova, M. Incineration of EV Lithium-ion batteries as a pretreatment for recycling—Determination of the potential formation of hazardous by-products and effects on metal compounds. *J. Hazard. Mater.* **2020**, *393*, 122372. [CrossRef] [PubMed]

67. Peters, L.; Friedrich, B. *Proven Methods for Recovery of Lithium from Spent Batteries*; DERA Workshop Lithium: Berlin, Germany, 2017.

68. Petranikova, M. Spracovanie Použitých Prenosných Lítiových Akumulátorov. Ph.D. Thesis, Technical University of Košice, Košice, Slovakia, 2012.

69. Mosqueda, H.A.; Vazquez, C.; Bosch, P.; Pfeiffer, H. Chemical Sorption of Carbon Dioxide (CO_2) on Lithium Oxide (Li_2O). *Chem. Mater.* **2006**, *18*, 2307–2310. [CrossRef]

70. Ueda, S.; Inoue, R.; Sasaki, K.; Wakuta, K.; Ariyama, T. CO_2 Absorption and Desorption Abilities of Li_2O–TiO_2 Compounds. *ISIJ Int.* **2011**, *51*, 530–537. [CrossRef]

71. Sloop, S.E.; Parker, R. System and Method for Processing an End-of-Life or Reduced Performance Energy Storage and/or Conversion Device Using a Supercritical Fluid (US 8,067,107 B2). 2011. Available online: https://patentimages.storage.googleapis.com/1f/5d/1b/69963cd0c16466/US8067107.pdf (accessed on 1 April 2020).

72. Nowak, S.; Winter, M. The Role of Sub- and Supercritical CO_2 as "Processing Solvent" for the Recycling and Sample Preparation of Lithium Ion Battery Electrolytes. *Molecules* **2017**, *22*, 403. [CrossRef]

73. Chen, Y.; Tian, Q.; Chen, B.; Shi, X.; Liao, T. Preparation of lithium carbonate from spodumene by a sodium carbonate autoclave process. *Hydrometallurgy* **2011**, *109*, 43–46. [CrossRef]

74. Bertau, M.; Voigt, W.; Schneider, A.; Martin, G. Lithiumgewinnung aus anspruchsvollen Lagerstätten: Zinnwaldit und magnesiumreiche Salzseen. *Chem. Ing. Tech.* **2017**, *89*, 64–81. [CrossRef]

75. Liu, Y. Analysis on Extraction Behaviour of Lithium-ion Battery Electrolyte Solvents in Supercritical CO_2 by Gas Chromatography. *Int. J. Electrochem. Sci.* **2016**, *11*, 7594–7604. [CrossRef]

76. Grützke, M.; Mönnighoff, X.; Horsthemke, F.; Kraft, V.; Winter, M.; Nowak, S. Extraction of lithium-ion battery electrolytes with liquid and supercritical carbon dioxide and additional solvents. *RSC Adv.* **2015**, *5*, 43209–43217. [CrossRef]

77. Rothermel, S.; Grützke, M.; Mönnighoff, X.; Winter, M.; Nowak, S. Electrolyte Extraction—Sub and Supercritical CO_2. In *Recycling of Lithium-Ion Batteries*; Kwade, A., Diekmann, J., Eds.; Springer International Publishing: Cham, Switzerland, 2018; pp. 177–185, ISBN 978-3-319-70571-2.

78. Sloop, S.; Crandon, L.; Allen, M.; Koetje, K.; Reed, L.; Gaines, L.; Sirisaksoontorn, W.; Lerner, M. A direct recycling case study from a lithium-ion battery recall. *Sustain. Mater. Technol.* **2020**, *25*, e00152. [CrossRef]

79. Rothermel, S.; Evertz, M.; Kasnatscheew, J.; Qi, X.; Grützke, M.; Winter, M.; Nowak, S. Graphite Recycling from Spent Lithium-Ion Batteries. *ChemSusChem* **2016**, *9*, 3473–3484. [CrossRef] [PubMed]

80. Bertau, M.; Martin, G. Integrated Direct Carbonation Process for Lithium Recovery from Primary and Secondary Resources. *Mater. Sci. Forum* **2019**, *959*, 69–73. [CrossRef]

81. Bertau, M.; Martin, G.; Pätzold, C. Verfahren zur Gewinnung von Lithiumcarbonat aus lithiumhaltigen Batterierückständen mittels CO2-Behandlung. Deutsche Patentanmeldung (DE102016208407A1), 23 November 2017.

92. Stopic, S.; Dertmann, C.; Koiwa, I.; Kremer, D.; Wotruba, H.; Etzold, S.; Telle, R.; Knops, P.; Friedrich, B. Synthesis of Nanosilica via Olivine Mineral Carbonation under High Pressure in an Autoclave. *Metals* **2019**, *9*, 708. [CrossRef]
93. Peled, E.; Golodnitsky, D.; Penciner, J. The Anode/Electrolyte Interface. In *Handbook of Battery Materials, 2. Aufl.*; Daniel, C., Besenhard, J.O., Eds.; Wiley-VCH: Weinheim, Germany, 2012; pp. 479–524, ISBN 978-3-527-32695-2.

Article

Speciation of Manganese in a Synthetic Recycling Slag Relevant for Lithium Recycling from Lithium-Ion Batteries

Alena Wittkowski [1], Thomas Schirmer [2], Hao Qiu [3], Daniel Goldmann [3] and Ursula E. A. Fittschen [1,*]

[1] Institute of Inorganic and Analytical Chemistry, Clausthal University of Technology, Arnold-Sommerfeld Str. 4, 38678 Clausthal-Zellerfeld, Germany; alena.wittkowski@tu-clausthal.de

[2] Department of Mineralogy, Geochemistry, Salt Deposits, Institute of Disposal Research, Clausthal University of Technology, Adolph-Roemer-Str. 2A, 38678 Clausthal-Zellerfeld, Germany; thomas.schirmer@tu-clausthal.de

[3] Department of Mineral and Waste Processing, Institute of Mineral and Waste Processing, Waste Disposal and Geomechanics, Clausthal University of Technology, Walther-Nernst-Str. 9, 38678 Clausthal-Zellerfeld, Germany; hao.qiu@tu-clausthal.de (H.Q.); daniel.goldmann@tu-clausthal.de (D.G.)

* Correspondence: ursula.fittschen@tu-clausthal.de; Tel.: +49-5323-722205

Abstract: Lithium aluminum oxide has previously been identified to be a suitable compound to recover lithium (Li) from Li-ion battery recycling slags. Its formation is hampered in the presence of high concentrations of manganese (9 wt.% MnO_2). In this study, mock-up slags of the system Li_2O-CaO-SiO_2-Al_2O_3-MgO-MnO_x with up to 17 mol% MnO_2-content were prepared. The manganese (Mn)-bearing phases were characterized with inductively coupled plasma optical emission spectrometry (ICP-OES), X-ray diffraction (XRD), electron probe microanalysis (EPMA), and X-ray absorption near edge structure analysis (XANES). The XRD results confirm the decrease of $LiAlO_2$ phases from Mn-poor slags (7 mol% MnO_2) to Mn-rich slags (17 mol% MnO_2). The Mn-rich grains are predominantly present as idiomorphic and relatively large (>50 μm) crystals. XRD, EPMA and XANES suggest that manganese is present in the form of a spinel solid solution. The absence of light elements besides Li and O allowed to estimate the Li content in the Mn-rich grain, and to determine a generic stoichiometry of the spinel solid solution, i.e., $(Li_{(2x)}Mn^{2+}{}_{(1-x)})_{1+x}(Al_{(2-z)},Mn^{3+}{}_z)O_4$. The coefficients x and z were determined at several locations of the grain. It is shown that the aluminum concentration decreases, while the manganese concentration increases from the start (x: 0.27; z: 0.54) to the end (x: 0.34; z: 1.55) of the crystallization.

Keywords: lithium; engineered artificial minerals (EnAM); X-ray absorption near edge structure (XANES); powder X-ray diffraction (PXRD); electron probe microanalysis (EPMA); melt experiments

Citation: Wittkowski, A.; Schirmer, T.; Qiu, H.; Goldmann, D.; Fittschen, U.E.A. Speciation of Manganese in a Synthetic Recycling Slag Relevant for Lithium Recycling from Lithium-Ion Batteries. *Metals* **2021**, *11*, 188. https://doi.org/10.3390/met11020188

Academic Editor: Mark E. Schlesinger

Received: 30 December 2020

Accepted: 18 January 2021

Published: 21 January 2021

Publisher's Note: MDPI stays neutral with regard to jurisdictional claims in published maps and institutional affiliations.

1. Introduction

Modern technologies, such as renewable energy and e-mobility, demand a new portfolio of technology-critical elements and materials. Limited resources, national policies or monopolies threaten the supply of some technology-critical elements. Hence, the recovery of these elements from waste is crucial. On the one hand, demand for lithium (Li) has increased rapidly due to the popularity and extraordinary performance of Li-ion batteries. On the other hand, Li is produced by mainly two countries, Australia (ca. 55%) and Chile (ca. 23%) [1]. The recycling of some components from Li-ion batteries is already put into practice, e.g., cobalt, in a pyrometallurgical process [2,3]. Pyrometallurgical recycling has the benefit of being adaptable to many waste streams; additionally, the emission of toxic compounds like HF is prevented. Due to its ignoble character, Li is driven into the slag and is usually not recovered. The recovery of Li from pyrometallurgical recycling slags can be accomplished by the targeted formation of "engineered artificial minerals" (EnAM).

The strategy of EnAM formation is to concentrate the elements of interest in a few phases, with a structure and size that allows an efficient separation. Figure 1 shows a scheme of EnAM formation. The target element (yellow triangle) spreads over several different phases (red and green forms) and the matrix (blue) in the unmodified slag (top picture). The target element will be difficult to isolate due to its occurrence in many different phases. The goal in EnAM is to concentrate the target element in a single phase (red pentagon, bottom picture) that differs physically and chemically from the other phases (green hexagon), and allows for efficient separation and further treatment.

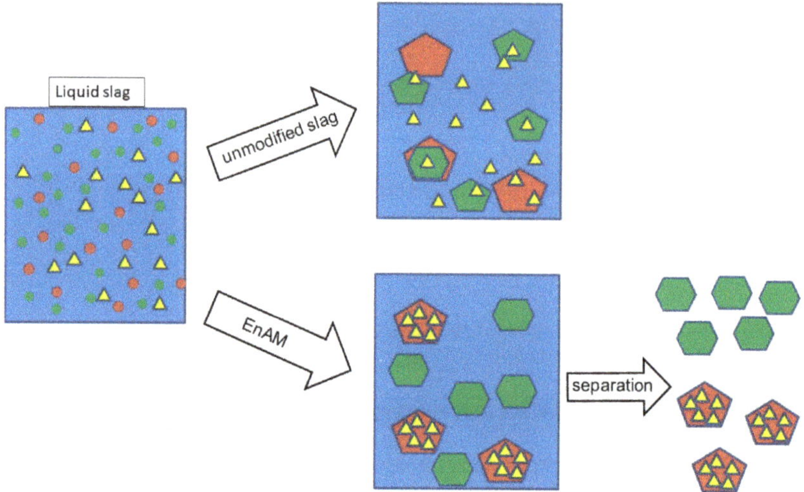

Figure 1. Sketch of engineered artificial minerals (EnAM) formation. On the left, different elements are shown in the liquid slag-matrix: target element (yellow triangle), phase for EnAM (red) and other phases (green). The upper route refers to an unmodified slag–the target element is spread over several different phases and the matrix (blue). The target element is not recovered. The bottom picture illustrates the formation of EnAM. The target element is concentrated in a single phase (red pentagon) that differs physically and chemically from the other phases (green hexagon), and allows for efficient separation and further treatment.

Separation of artificial minerals, enriched in valuable elements from remaining slag components, might be carried out by flotation processes. Here, the composition and structure of phases are crucial. It can be expected that none of the artificially produced minerals will show hydrophobic behavior by nature. Thus, the mineral-collector-interaction has to be studied, and this interaction relies strongly on crystal structure and ions, responsible for the adsorption of the active group of collector molecules.

The recovery of Li from slags has been subject to several studies. Elwert et al. [2] found that Li reacts with aluminum to form predominately large lithium aluminate ($LiAlO_2$) crystals. The aluminum binds Li from the melt uniformly in one phase at an early state of solidifying phase of the molten slag. The idiomorphic to hypidiomorphic lithium aluminate crystals can easily be separated from the matrix by flotation. Hence, lithium aluminate is a promising EnAM candidate. The recovery of Li from $LiAlO_2$ was successfully conducted by Haas et al. [4]. Elwert et al. [5] investigated a hydrometallurgical process to recover Li from slags with low aluminum compared to [2] and enriched in silicone. However, it was found that the formation of $LiAlO_2$ is suppressed in manganese (Mn)-rich slags [2]. In these slags, Li is distributed over several phases of low crystallinity. Elements like Mn seem to have a significant influence on the formation of compounds and grain size during the formation of the slag. Due to the increasing amount of Li-ion batteries with Mn-based cathode materials, i.e., $Li\text{-}Ni_xMn_yCo_zO_2$ (NMC), it must be assumed that the Mn content in battery waste streams will increase soon. Accordingly, it is necessary to understand the role of the redox-active Mn on the genesis of crystalline, and especially amorphous grains from the ionic melts. A careful characterization of the Mn-bearing grains will give insight into these processes. The first survey with mineralogical and thermodynamic methods on the formation of Li-EnAMs was conducted by Schirmer et al. [6] on a similar system, but without Mn. Their study shows that formation of spinel solid solutions is a favorable reaction with a thermodynamically proved potential to scavenge Li from the slag in an early stage of the solidification. Adding Mn to this system would result in more complex solidification reactions with Mn-containing spinel solid solutions in particular, as numerous spinel-like oxides of Li and Mn are already described [7].

Therefore, in this study, Li- and Mn-bearing phases from a synthetic slag, in the following termed mock-up slag (MUS), of the system Li_2O-CaO-SiO_2-Al_2O_3-MgO-MnO_x, with up to 17 mol% MnO_2-content, are studied. The crystalline components are determined by powder X-ray diffraction (PXRD). The main Mn-containing phase could be identified as an oxide solid solution of spinel-type using electron probe microanalysis (EPMA). From a combination of the virtual compounds $LiMnO_2$, $Mn_{0.5}AlO_2$ (1/2 galaxite spinel) and $Mn_{0.5}MnO_2$ (1/2 hausmannite), the formula $(Li_{(2x)}Mn^{2+}_{(1-x)})_{1+x}(Al_{(2-z)},Mn^{3+}_z)O_4$ was calculated. This matches with the PXRD results, as all main reflections of the oxides are located between those of $LiAl_5O_8$ and $MnAl_2O_4$. The EPMA and PXRD analysis showed that a high amount of grains are amorphous or of low crystallinity. The PXRD falls short of giving insight into the Mn species in amorphous phases. EPMA allows recognizing individual crystalline and amorphous phases. For amorphous phases, however, stoichiometric information can usually not be extracted from the data. Hence, laboratory-based X-ray absorption near edge structure (XANES) is applied here to determine the bulk Mn species of the crystalline and the amorphous components of the slag. The findings from this independent method confirmed the Mn oxidation state as being between +2 and +3, as well as the presence of Mn spinel structures.

2. Background

The following section gives a short introduction to the relevant binary subsystems of the Al_2O_3-MnO_x-Li_2O ternary system, which itself is not yet published. This will allow putting the presented results into the mineralogical context. In addition, the methods used to investigate the oxidation state of Mn are briefly described.

2.1. The System Li_2O-MnO_x with Focus on Spinel Structures

The stoichiometric spinel $LiMn_2O_4$ and several other spinel phases of Li-Mn-O are known in the Li_2O-MnO_x system. Paulsen and Dahn [7] created a binary phase diagram of Li-Mn-O in air. They described a spinel with the formula $Li_{(1+x)}Mn_{(2-x)}O_4$ to be stable between 400 and 880 °C. With the rising temperature, x decreases from 1/3 at 400 °C to 0 at 880 °C. Below 400 °C, the only stable spinel phases are $Li_4Mn_5O_{12}$ and $LiMn_{1.75}O_4$. Raising the temperature above 880 °C leads to the replacement of Li by Mn and tetragonal spinel $[Li_{(1-x)}Mn_x]MnO_4$ [7].

2.2. The System Li_2O-Al_2O_3

The binary system Li_2O-Al_2O_3 exhibits several phases. Konar et al. [8] described $LiAl_5O_8$ spinel, $LiAl_{11}O_{17}$, $LiAlO_2$ and Li_5AlO_4.

$LiAl_5O_8$ is described as an inverse spinel structure with Al^{3+} on tetrahedral sites and Li^+ and the remaining Al^{3+} on octahedral sites. A polymorphic transition to this spinel structure occurs from the stoichiometric phase. $LiAl_{11}O_{17}$ is a high-temperature phase, only stable above 1537 °C. For $LiAlO_2$, a polymorphic transition from α- to γ-phase occurs.

Li_5AlO_4 crystallizes below 50 mol% Al_2O_3, in a mixture either with $LiAlO_2$ or with LiO_2 [8].

2.3. The System Al_2O_3-MnO_x

Solid solutions of spinel oxide type (XY_2O_4) with a cubic and tetragonal crystal system are described by Chatterjee et al. [9] to form in the Al_2O_3-MnO-Mn_2O_3 system. Besides

the cubic $MnAl_2O_4$ (galaxite), additionally, $MnMn_2O_4$ (hausmannite) with a tetragon
and cubic crystal system (transformation tetragonal -> cubic: 1172 °C in the air) can occu
While the tetragonal form of hausmannite incorporates only a small amount of galaxit
the cubic variety forms a solid solution.

2.4. Speciation of Oxidation States in Spinel Systems

Recycling slags are usually complex elemental mixtures comprising various comp
nents and phases. Due to uncontrolled and–in geological terms–fast cooling condition
only a few components in these slags are crystalline. Crystalline phases are easily accessib
by established methods like PXRD.

The amorphous components, however, elude the identification by PXRD. Wet chemic
methods like inductively coupled plasma atomic emission spectroscopy (ICP-OES) yield i
formation on the stoichiometric composition, but fall short for the speciation. EPMA allow
studying crystalline and amorphous phases. Stoichiometric information can be obtaine
for crystalline phases with reasonable certainty. Structure predictions on non-crystallir
phases, however, are merely estimations. XANES is a convenient method to analyze th
species of 3d-elements in crystalline, as well as amorphous materials. Asaoka et al. dete
mined the oxidation state of Mn in granulated coal ash via XANES [10], while Kim et a
used XANES for Mn speciation in steel-making slags [11]. In general, the oxidation state, th
coordination sphere, and in some cases the actual compound can be determined. Usuall
this method is exclusively available at synchrotron sources (Sy). Here, a laboratory-base
XANES spectrometer was used, which allows everyday access to measure several routin
samples or improve sample preparation. The set-up is described by Seidler et al. [12,13
Laboratory-based XANES has been used successfully in catalysis research, as well as fc
the determination of vanadium oxidation state in catalysts and vanadium redox flov
batteries [14–16].

3. Materials and Methods

3.1. Synthesis of the Mock-Up Slag

The chemicals used for producing the mock-up slags are lithium carbonate (Merc
KGaA purum, Darmstadt, Germany), calcium carbonate (Sigma-Aldrich, reagent grade, S
Louis, MO, USA), silicon dioxide (Sigma-Aldrich, purum p.a., St. Louis, MO, USA), alu
minum oxide (Merck KGaA, Darmstadt, Germany), magnesium oxide (98%, Roth, Karlsruh
Germany) and manganese dioxide (Merck KGaA, reagent grade, Darmstadt, Germany).

The concentrations of the reactants of the slag synthesis are the following: For V-
32 mol% Al_2O_3 were mixed with 16 mol% CaO, 21 mol% Li_2O, 3 mol% MgO, 7.4 mol%
MnO_2, and 22 mol% SiO_2. For V-2, 29 mol% Al_2O_3 were mixed with 14 mol% CaC
19 mol% Li_2O, 3 mol% MgO, 13 mol% MnO_2, and 21 mol% SiO_2. For V-3, 28 mol%
Al_2O_3 were mixed with 14 mol% CaO, 18 mol% Li_2O, 3 mol% MgO, 17 mol% MnO_2, and
20 mol% SiO_2. The chemicals were manually mixed in a mortar and ground in a dis
mill for 5 min. Each sample was placed in a Pt-Rh crucible and heated in a chambe
furnace (Nabertherm HT16/17, Nabertherm GmbH, Lilienthal, Germany) in an ambient ai
atmosphere. The temperature program is shown in Figure 2. A heating rate of 2.89 °C/mir
was initially employed to reach 720 °C, which is the melting temperature of $Li_2CO_
Subsequently, a heating rate of 1.54 °C/min was used to decompose Li_2CO_3 and reacl
the target temperature of 1600 °C. Finally, the obtained MUS were kept at 1600 °C for two
hours. After that, the samples were cooled to 500 °C at a cooling rate of 0.38 °C/min, anc
quenched in water. Three MUS V-1-3 were obtained. For PXRD and XANES measurement:
parts of the obtained slag were ground for five minutes in a disc mill (Siebtechnik GmbF
Mühlheim an der Ruhr, Germany).

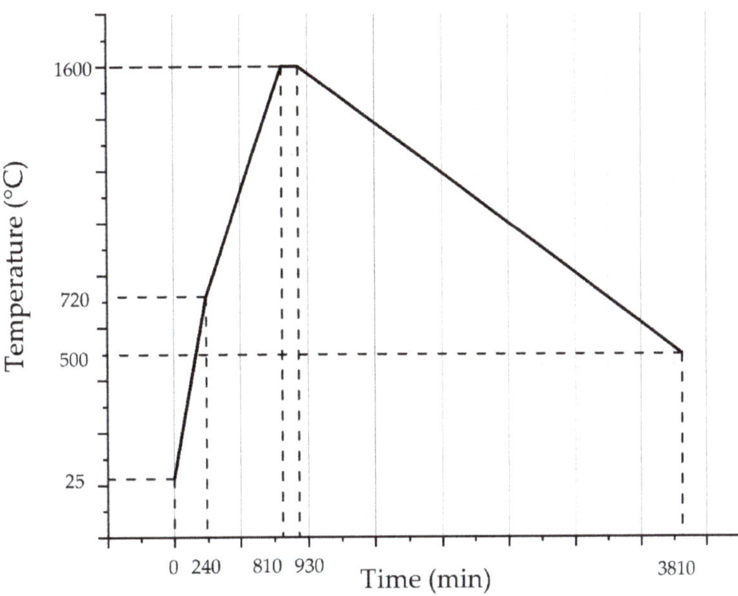

Figure 2. Temperature program of the chamber furnace. A heating rate of 2.89 °C/min was employed to reach the melting temperature of Li_2CO_3 at 720 °C, followed by a heating rate of 1.54 °C to the target temperature of 1600 °C. The temperature was held for 120 min, and afterward, a cooling rate of 0.38 °C/min was employed to reach a temperature of 500 °C.

The elemental content was determined by ICP-OES (ICP-OES 5100, Agilent, Agilent Technologies Germany GmbH & Co. KG, Waldbronn, Germany). The samples were fused with sodium tetraborate in a platinum crucible at 1050 °C, and then leached with diluted hydrochloric acid to determine Al, Ca, Li, Mg, Ti and Si. To determine other elements, the samples were suspended in nitric acid and digested at 250 °C and under a pressure of 80 bar in an autoclave (TurboWAVE, MLS, Leutkirch im Allgäu, Germany).

3.2. Synthesis of Galaxite via Solution Combustion Synthesis

Artificial galaxite was prepared via solution combustion synthesis. Aluminum nitrate nonahydrate (VWR chemicals, Darmstadt, Germany, analytical reagent, min. 98%) and manganese (II) nitrate tetrahydrate (Merck, KGaA, Darmstadt, Germany, pro analysi, min 98.5%) were used as oxidizers and mixed in stoichiometric ratio 1:2. Aluminum nitrate nonahydrate (7.5 g) was mixed with manganese (II) nitrate tetrahydrate (2.5 g). As a fuel, urea (Merck KGaA, Darmstadt, Germany, pro analysi, min. 99.5%) was added in excess (5 g). All components were dissolved in water. The solution was heated with a Bunsen burner until near dryness. The mixture was ignited with a second Bunsen burner. Purity was verified by PXRD (STOE STADI P, STOE & Cie GmbH, Hilpertstraße 10, 64295 Darmstadt, Germany).

3.3. Powder X-ray Diffraction

The analysis of the bulk mineralogical composition was provided by PXRD, using a PANalytical X-Pert Pro diffractometer, equipped with a Co-X-ray tube (Malvern Panalytical GmbH, Nürnberger Str. 113, 34123 Kassel, Germany). For identification of the compounds, the PDF-2 ICDD XRD database [17], the American Mineralogist Crystal Structure Database [18] and the RRUFF-Structure database [19] were evaluated.

3.4. Electron Probe Microanalysis

The analysis of single crystals and grains was performed with EPMA. EPMA is a standard method to characterize the chemical composition in terms of single spot analysis or element distribution patterns, accompanied by electron backscattered Z (ordinal number) contrast (BSE(Z)) or secondary electron (SE) micrographs. The EPMA measurements were performed on samples, which were embedded in epoxy resin, polished and coated with carbon. They were characterized using a Cameca SX^{FIVE} FE (Field Emission) electron probe, equipped with five wavelength dispersive (WDX) spectrometers (CAMECA SAS, 29, quai des Grésillons, 92230 Gennevielliers, Cedex, France). To calibrate the wavelength dispersive X-ray fluorescence spectrometers (WDXRF), an appropriate suite of standards and analyzing crystals was used. The reference materials were provided by P&H Developments (The Shire 85A Simmondley village, Glossop, Derbyshire SK13 9LS, UK and Astimex Standards Ltd., 72 Milicent St, Toronto, ON, M6H 1W4, Canada). The beam size was set to zero, leading to a beam diameter of substantially below 1 μm (100–600 nm with field emitters of Schottky-type, e.g., [20]). To evaluate the measured intensities, the X-PHI-Model was applied [21].

Li, one of the key elements in this study, cannot be directly analyzed, since EPMA uses X-ray emission to detect the elements in the sample, and the extremely low fluorescence yield and long wavelength of the Li Kα render the direct determination of this element merely impossible. With the reasonable assumption that other refractory light elements like Be and B are not present in the investigated material and volatile elements and compounds like F, H_2O, CO_2 or NO_3^- are effectively eliminated during the melt experiment, Li can be calculated using virtual compounds. Where necessary, the balanced Li concentration was included in the matrix correction calculation. To access the analytical accuracy with respect to the determination of Li, the international reference material spodumene (Astimex Standards Ltd., Toronto, ON, Canada) and the in-house standard $LiAlO_2$ were analyzed (Table 1).

Table 1. Recovery of Li-compounds. Spod: spodumene, % StdDev: Relative standard deviation in percent, Repeats: N = 5, R: Recovery, LiAl: $LiAlO_2$.

Wt.%	Average Spod.	%StDev., Spod.	Ref. Spod.	R (%)	Average LiAl	%StDev., LiAl	Ref. LiAl	R (%)
Al	15.04	0.35	14.4	104	41.24	0.22	40.9	101
Mg	0.00	n.a.	0.0	n.a.	0.01	n.a.	0.0	n.a.
Ti	0.00	n.a.	0.0	n.a.	0.00	n.a.	0.0	n.a.
Mn	0.05	n.a.	0.0	n.a.	0.00	n.a.	0.0	n.a.
Fe	0.02	n.a.	0.0	n.a.	0.03	n.a.	0.0	n.a.
Ca	0.01	n.a.	0.0	n.a.	0.01	n.a.	0.0	n.a.
K	0.00	n.a.	0.0	n.a.	0.00	n.a.	0.0	n.a.
Si	28.71	0.56	30.0	96	0.01	n.a.	0.0	n.a.
Na	0.10	2.83	0.09	112	0.00	n.a.	0.0	n.a.

3.5. X-ray Absorption near Edge Structure

The Mn speciation was achieved with XANES. Other than usual, XANES was not conducted at a synchrotron facility but using a laboratory-based device, the easyXES100-extended (short: easyXES100; easyXAFS LLC, Renton, WA, USA). To enable these measurements with high energy resolution comparable to that obtained at synchrotron facilities, a Rowland circle Johann-type monochromator is used, see Figure 3. The instrument comprises an X-ray tube (100 W, air-cooled tube with W/Pd anode, 35 kV maximum accelerating potential), a spherically bent crystal analyzer (SBCA, Si 110) and an SDD detector (AXAS-M1, KETEK, Munich, Germany). The components are positioned on scissor drives.

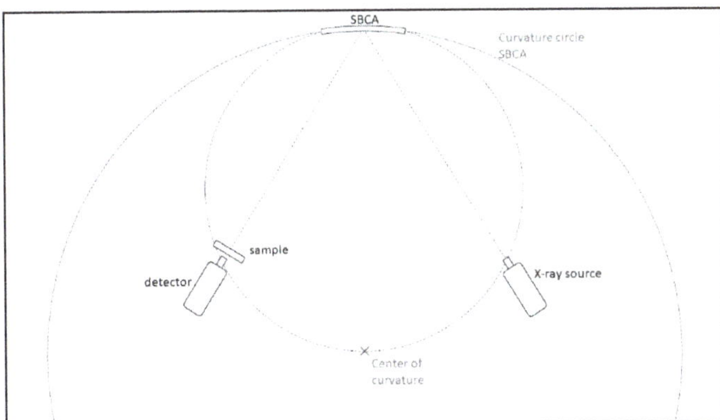

Figure 3. Scheme of the Rowland circle monochromator. The outer circle represents the curvature of the SBCA. The inner circle has a diameter matching the radius of the curvature circle. On this inner circle, the X-ray source and the detector are positioned. The sample is positioned right in front of the detector. The components are set on a scissor drive, allowing fine energy scanning.

This enables energy scanning by synchronously and symmetrically moving the detector and the source. The SBCA sits on a passively driven slider in the Rowland circle, which is coupled to the X-ray source stage. This way, the proper positioning of all three components is elegantly achieved. A helium-filled box is installed in the beam path, to lower background absorption and scattering by air. This set-up allows energy scanning with a resolution of about 1 eV. Further information about the set-up is published by Seidler et al. and Jahrman et al. [12,13,22]. At every energy step, an energy-dispersive X-ray spectrum (EDX) is acquired. The area of interest is automatically integrated by the software based on labVIEW2017 [23]. This way, a file with the relevant information on the energy and counts per lifetime is created and used together with the I_0 measurement to obtain the XANES.

Spectra of manganese dioxide (Merck AG, Darmstadt, Germany, for synthesis) were recorded, validated with spectra from the literature, i.e., spectra from Hokkaido University, Japan [24,25] and subsequently used as reference for energy calibration. The scans in the θ-angle space are converted to the energy space (see Appendix A).

The XANES of references and samples were recorded and processed as follows: The Mn K-edge was scanned from 6482 eV to 6800 eV (SBCA, Si 110, $n = 4$; see Appendix A), with a step width of 0.25 eV. At each step, an EDX spectrum with a 10 s live time was obtained, resulting in a total measurement time of *approx.* 3.5 h per spectrum. Three samples of MUS V-3 were prepared in parallel by mixing the finely ground slag with Vaseline (ISANA; Dirk Rossmann GmbH, Burgwedel, Germany) in a weight ratio of 2/3 Vaseline and 1/3 MUS powder. The mixtures were placed in washers with a height of 0.1 mm to adjust the thickness of the samples. The washer was sealed from both sides with adhesive tape (tesapack, tesa SE, Norderstedt, Germany) to hold the sample inside, and protect it from environmental influences. Each replicate was analyzed three times. The following Mn species were used as references: manganese (II) oxide (*oxidation state of manganese:* +2; alfa aesar, Thermo fisher (Kandel) GmbH, Kandel, Germany, 99%); synthetic galaxite ($Mn^{2+}Al_2^{3+}O_4$; *oxidation state of manganese:* +2), manganese (II, III) oxide (*average oxidation state of manganese:* +2.67; Sigma-Aldrich Chemie GmbH, Steinheim, Germany, 97%); braunite ($Mn^{2+}Mn^{3+}_6[O_8 \mid SiO_4]$; *formal oxidation state of manganese:* +2.85; Friedrichroda, Thuringia, Germany; obtained from Geo collection, Clausthal University of Technology); bixbyite (Mn_2O_3; *oxidation state of manganese:* +3; Paling Mine, Republic of South Africa; obtained from Geo collection, Clausthal university of technology); and manganese (IV) oxide (*oxidation state of manganese: +4*; Merck AG, Darmstadt, Germany, for synthesis). They were prepared in the same way as the actual sample.

An additional I_0 spectrum of an empty washer sealed with adhesive tape was separately acquired for every measurement to calculate the absorption coefficient $\mu(E)$ according to the equation: $\mu(E) = -\ln I \cdot I_0^{-1}$. The average post-edge absorption $\mu(E)$ (post edge line) in each spectrum was normalized to unity using ATHENA software [26]. Spectra from the same Mn species were merged in ATHENA and used for further analysis.

4. Results

The results of the characterization of the MUS V-1-3, which are supposed to match Li-ion battery recycling slags, are presented in this section. The composition was chosen to be similar to recycling slags characterized previously by Elwert et al. [2]. The methodology applied to study the MUS chemistry comprises ICP-OES, PXRD, EPMA and XANES.

Initially, the composition of the reactants and the products was determined by ICP-OES. The suppression of the formation of $LiAlO_2$ EnAM in MUS with increasing Mn content was studied by PXRD. The microstructure and microscopic elemental composition of the MUS with the highest Mn-content (MUS V-3; MnO_2: 17 mol%) were studied by BSE(Z) micrographs, as well as detailed spatially resolved quantitative point measurements and element distribution profiles, recorded with EPMA. From these results, a hypothesis for the genesis of the Mn-rich grains was established. This hypothesis is discussed in detail in Section 4.3. Finally, it was evaluated by studying the Mn species using XANES.

4.1. Bulk Chemistry

The elemental composition of the reactants and the product slags were determined by digestion, followed by ICP-OES. The Mn-content increases from MUS V-1 to V-3, resulting in a concentration of MnO_2 of 7 mol%, 13 mol%, and 17 mol% in the final products. The results are given in Table 2. A loss of about 5% of Li and 0–18% Mn occurred during the melting and cooling of the material. The MUS are therefore close in composition to the actual recycling slags studied by Elwert et al. [2].

Table 2. Comparison of the bulk chemical composition of the three melt experiments, given in mole percent.

	Raw Mix (Mole Fraction)			Product (Mole Fraction)			Recovery %		
	V-1	V-2	V-3	V-1	V-2	V-3	V-1	V-2	V-3
Al_2O_3	32	29	28	33	31	29	103	107	104
CaO	16	14	14	15	15	14	94	107	100
Li_2O	21	19	18	20	18	17	95	95	94
MgO	3	3	3	3	3	3	100	100	100
MnO_2	7.4	13	17	6.1	12	17	82	92	100
SiO_2	22	21	20	23	22	21	105	105	105

4.2. Powder X-ray Diffraction

The crystalline composition of the MUS V-1-3 was studied by PXRD and compared to an Mn-free material with comparable $LiAlO_2$-content. The compounds gehlenite, spinel and $LiAlO_2$ were present in all three slags (Figure 4a). The Li-Al-Oxide reflections are best explained with the diffraction pattern of $LiAlO_2$ (ICDD PDF2 no. 00-038-1464 [17]). A comparison of the reflection height of three mixtures with increasing Mn-concentration shows a negative correlation with the intensity of the $LiAlO_2$ main reflection. A comparison with an Mn-free material with comparable Li_2O content (Figure 4c) indicates a strong negative influence of the Mn-concentration on the formation of $LiAlO_2$. The low intensities of the reflection and the comparable high background imply a high amount of amorphous material being present.

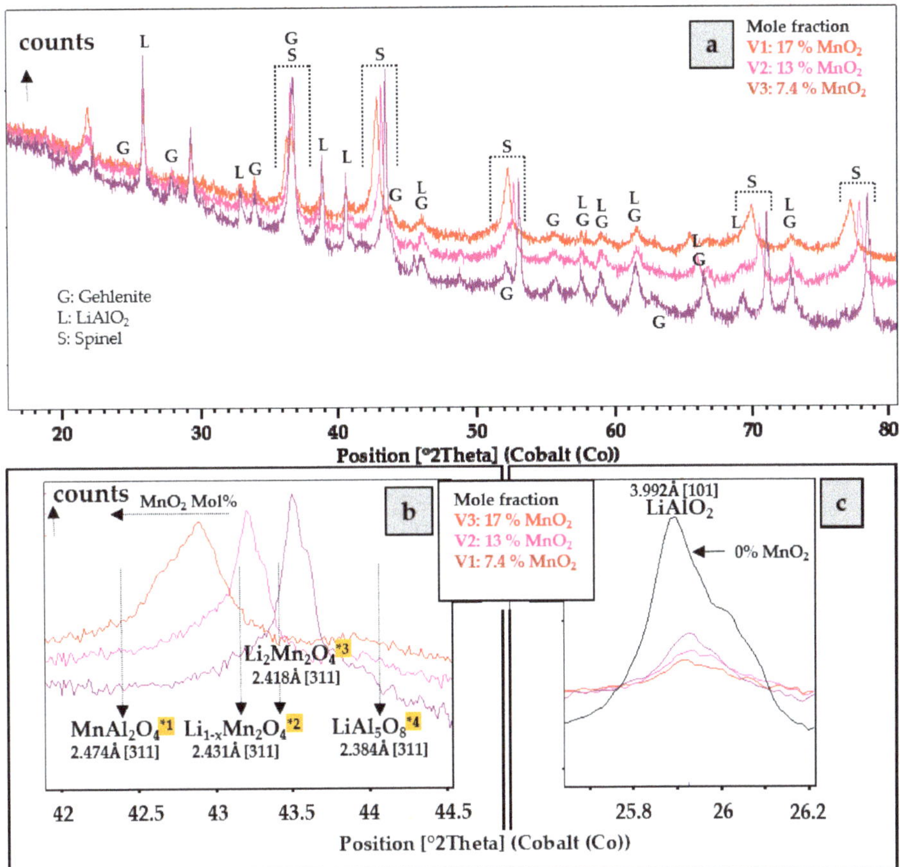

Figure 4. (**a**) Powder X-ray diffraction (PXRD) of the solidified melt. G: Gehlenite, S: Spinel, L: LiAlO₂. (**b**): Enlarged section of the main spinel peak. * 1: the position of the main peak of $MnAl_2O_4$ from the ICDD-PDF2 no. 00-029-0880 [17], * 2: the position of the main peak of the $Li_{(1-x)}Mn_2O_4$ spinel from the ICDD-PDF2 no. 00-038-07891 [17], * 3: the position of the main peak of the $Li_2Mn_2O_4$ spinel from the ICDD-PDF2 no. 01-084-1524 [17], * 4: the position of the main peak of the $LiAl_5O_8$ spinel from the ICDD-PDF2 no. 00-038-1425 [17], (**c**): Enlarged sections of the first two main LiAlO₂ peaks. In (**c**): for comparison, the LiAlO₂ main reflection of an Mn-free solidified melt with comparable Li_2O content is presented.

The enlarged section of the two-theta region of the main spinel reflections gives an indication of the changing composition of the spinel with the change of the Mn-content (Figure 4b). The main spinel reflections (311) of all three MUS lie between those of galaxite spinel $MnAl_2O_4$ and $LiAl_5O_8$. In this range, two reflections of Li/Mn-spinel are located ($Li_{1-x}Mn_2O_4$ and $Li_2Mn_2O_4$) [17]. This indicates the presence of a complex solid solution of a spinel-like oxide with the general formula $(Li_{(2x)}Mn^{2+}_{(1-x)})_{1+x}(Al_{(2-z)},Mn^{3+}_z)O_4$, which was derived by EPMA, as discussed below.

In conclusion, the PXRD results show the decrease of crystalline LiAlO₂ content with an increasing Mn-concentration. They also indicate the presence of a Li/Al/Mn spinel-like solid solution. The findings are consistent with the results of the microscopic EPMA analysis, which are presented in the following section.

4.3. Electron Probe Microanalysis

The MUS with the highest MnO_2-content (MUS V-3, 17 mol%) was subjected to EPMA. From the PXRD results, it was expected that the microstructure of this sample would be most conclusive on the processes, resulting in a $LiAlO_2$-depleted material. The main compounds of the melt experiment V-3, determined with EPMA, were: a spinel phase $(Li_{(2x)}Mn^{2+}_{(1-x)})_{1+x}(Al_{(2-z)},Mn^{3+}_{z})O_4$; lithium aluminate (LiAl) with the stoichiometric formula $Li_{1-x}(Al_{(1-x)}Si_x)O_2$; a gehlenite-like calcium-alumosilicate (GCAS) with the stoichiometric formula $Ca_2Al_2SiO_7$ and with minute amounts of Mg and Na (max. ~0.2 wt.%); and amorphous phases (APh) with various amounts of Al, Si, Ca, Mn, small amounts of Na (<0.3 wt.%), and sometimes with unusual elements like Ba and K (contaminants enriched in the eutectic residual melt).

The LiAl and the GCAS have already been described by Schirmer et al. [6] in a MUS not containing Mn but Li, Ca, Si, Al and Mg. There is strong evidence that the presence of Mn in the slag has an influence on the formation of Li and Al compounds. In particular, the formation of spinel solid solutions suggested by the PXRD is expected to influence the formation of $LiAlO_2$. Accordingly, the EPMA focused on elucidating the genesis of the Mn-rich grains. Overall, it was found that besides negligible amounts in amorphous phases, the Mn is almost exclusively incorporated into pure oxide (spinel) structures. For this reason, a detailed examination of a representative grain of Mn-containing oxide is presented in the following. Early crystallites of this type are predominantly found throughout the whole sample.

The BSE(Z) micrograph (Figure 5) shows a large grain of predominantly idiomorphic Mn-enriched oxide (spinel) surrounded by idiomorphic/hypidiomorphic grains of LiAl in a matrix of GCAS and accompanied by a grain of an APh enriched in unusual elements (contaminants), e.g., Ti and K probably representing the last eutectic melt composition.

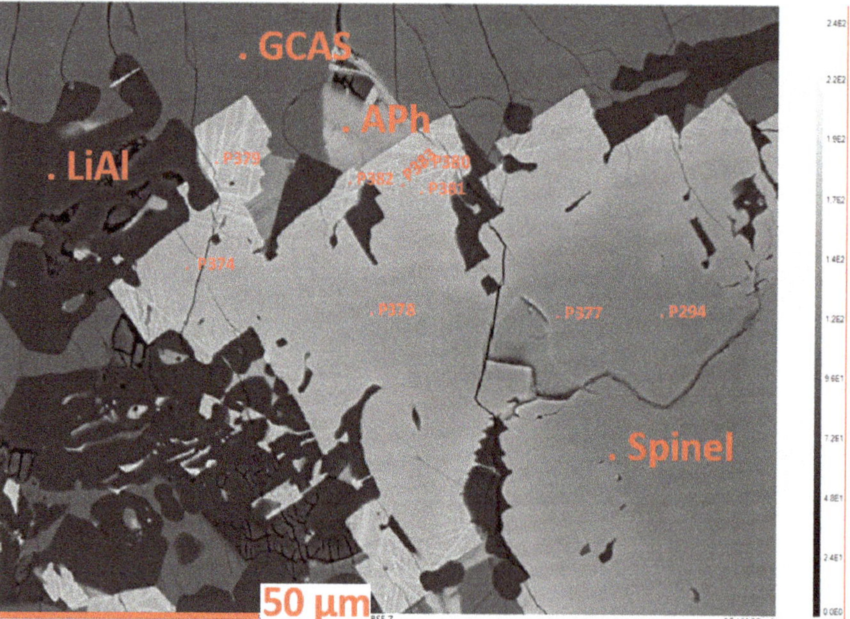

Figure 5. Electron micrograph (BSE(Z)) of the solidified melt. Light grey grain: spinel; dark gray sections, and dendrites: LiAl, surrounded by Ca-alumosilicate (GCAS, light grey sections); amorphous phases (APh): amorphous grain with unusual elements (K, Ba, Ti): contamination; black: pores or preparation damage. The chemical analysis of the points marked in red (P294–P383) is presented in Table 3.

Table 3. Elemental concentrations (wt.%) at the locations depicted in Figure 5, sorted in ascending order according to the Al concentration. For a close-up of the lamellae region, see Appendix C. The distance from the lamellae region points to the nearest rim is given. Regarding the point scans, a virtual line is drawn through all points to the left rim. The distances are given from each point to the intersection of this line with the rim.

Location	Lamellae Region						Point Scan		
No.	379	382	383	380	381	374	378	377	294
Distance from rim in µm	2.6	2.2	6.9	5.6	10.4	10.1	46.5	81.8	101.2
Al	0.7	0.7	1.5	1.5	3.1	6.0	11.9	16.7	22.3
Mg	0.1	0.1	0.1	0.1	0.1	0.1	0.2	0.2	0.5
Ti	0.4	0.4	0.3	0.2	0.1	0.0	0.0	0.0	0.0
Mn	69.7	64.4	62.9	64.0	62.2	60.0	52.3	46.9	39.0
Fe	0.3	0.4	0.5	0.5	0.5	0.4	0.3	0.3	0.3
Ca	0.2	0.1	0.1	0.2	0.1	0.2	0.0	0.0	0.0
Si	0.2	0.2	0.2	0.2	0.1	0.1	0.1	0.0	0.0

The gradient of the grey shade of the Mn-oxide grain indicates an increase of the mean atomic number from the center to the rim. At the outer edge, segregation lamellae can be observed. The brighter grey shade of these lamellae indicates a higher mean atomic number than in the surrounding grain. To investigate the changing composition from the inner part of the grain to the rim, several points were analyzed. Due to the relatively small features (<1–2 µm) of the segregation lamellae, the emphasis was on the precise determination of the whiskers. Therefore, the recording of a line scan was omitted. Table 3 contains the original data.

The concentrations found for Al and Mn were used to calculate normalized contents of Al and Mn^{2+}, Mn^{3+} as well as the fractions of $LiMnO_2$, $Mn_{0.5}AlO_2$ (1/2 galaxite spinel), and $Mn_{0.5}MnO_2$ (1/2 hausmannite) fitting the three spinel components to the elemental amounts (Table 4). The Li-concentration was subsequently obtained from the calculated amount of $LiMnO_2$. The general formula obtained from the fittings is $(Li_{(2x)}Mn^{2+}_{(1-x)})_{1+x}(Al_{(2-z)},Mn^{3+}_{z})O_4$. The minute concentrations of the other elements are omitted in this calculation. The results show that the Al content in the grain decreases during crystallization. The result of this calculation gives an indication that a solid solution of Li-Al-Mn spinel is formed.

Table 4. Spinel formulas, calculated with the Al and Mn concentrations of Table 3, sorted in ascending order according to the Al concentration. The Mn^{2+}/Mn^{3+}-ratio was calculated using the total measured Mn concentration in Table 3. The Li concentration is derived from the calculated fraction of $LiMnO_2$. In the first section of the table, the concentrations of the elements are given in weight percent. In the second section, the calculated fractions of the different virtual components in percent are presented. In the last section, the stoichiometric ratio of the formula $(Li_{(2x)}Mn^{2+}_{(1-x)})_{1+x}(Al_{(2-z)},Mn^{3+}_{z})O_4$ is presented. The distance from the lamellae region points to the nearest rim is given. Regarding the point scans, a virtual line is drawn through all points to the left rim. The distances are given from each point to the intersection of this line with the rim.

Location	Lamellae Region						Point Scan		
No.	379	382	383	380	381	374	378	377	294
Distance from rim in µm	2.6	2.2	6.9	5.6	10.4	10.1	46.5	81.8	101.2
Concentrations (wt.%)									
Al	0.7	0.7	1.5	1.5	3.1	6.0	11.9	16.7	22.3
Mn^{2+}	21.5	12.2	11.5	13.6	14.5	17.9	19.3	21.8	22.7
Mn^{3+}	48.2	52.2	51.4	50.4	47.7	42.1	33.0	25.1	16.7
Li	0.8	3.7	4.0	3.3	3.2	2.4	2.4	2.0	2.1
Calculated fractions (%)									
$Mn_{0.5}AlO_2$	2.1	2.2	4.7	4.9	10.0	19.2	38.2	53.5	71.5
$LiMnO_2$	11.0	49.9	53.6	45.1	42.7	31.8	31.9	26.7	28.5
$Mn_{0.5}MnO_2$	86.9	47.9	41.7	50.0	47.3	48.9	29.9	19.9	0.0
Stoichiometric factors $(Li_{2x}Mn^{2+}_{(1-x)})_{(1+x)}(Al_{(2-z)},Mn^{3+}_{z})O_4$									
x	0.13	0.54	0.58	0.49	0.46	0.34	0.33	0.26	0.27
z	1.95	1.95	1.89	1.88	1.76	1.55	1.15	0.85	0.54

4.4. X-ray Absorption near Edge Structure

To verify the structures suggested from EPMA analysis, XANES was conducted on the sample MUS V-3. For comparison, the spectra of known Mn oxidation states were recorded as well. In Figure 6, Mn K-edge XANES spectra of compounds representing oxidation states from Mn^{2+} to Mn^{4+} are displayed for comparison with the spectrum of the MUS.

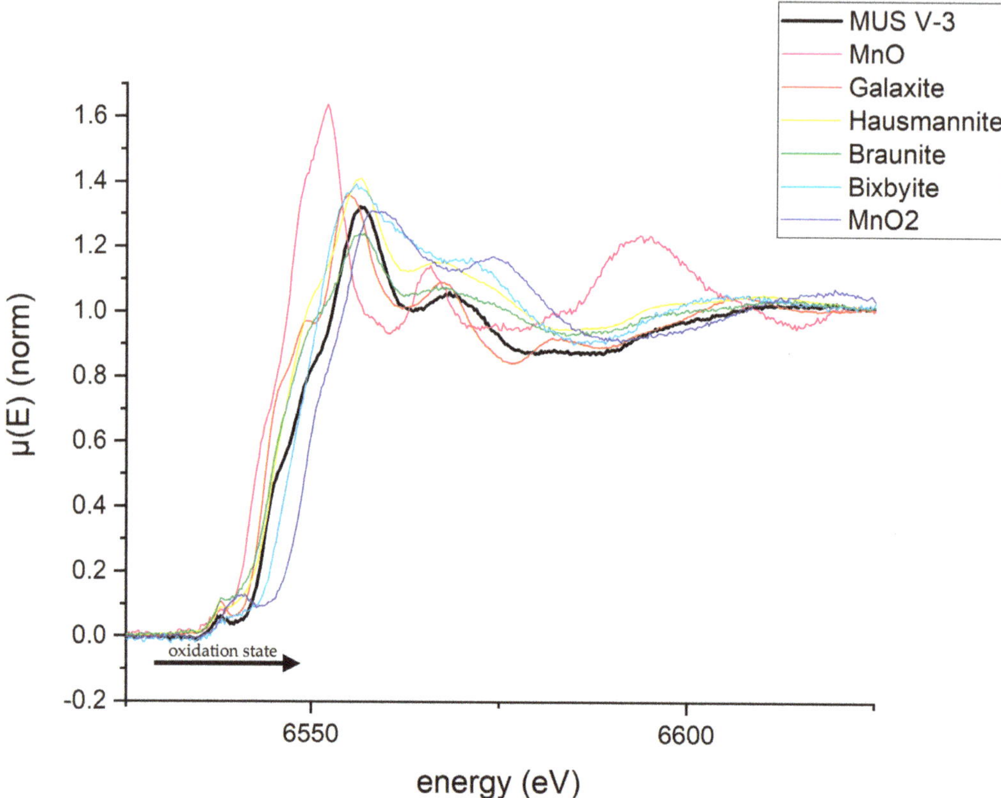

Figure 6. Spectra of Mn samples of different oxidation states. The spectra are compared to the spectrum of the mock-up slag (MUS). A shift of the edge accompanies the increase in the oxidation state which is indicated by the arrow. The mean oxidation states of Mn in the compounds are as follows: MnO: +2; galaxite: +2, hausmannite: +2.67; braunite: +2.85; bixbyite: +3; and MnO_2: +4.

A shift of the edge to higher energies with increasing oxidation state is observed (Figure 6); this is a well-known effect [27]. From the edge shift and the shape of the curve it can be concluded that the oxidation state of Mn in the MUS is a mixture of +2 and +3. A mixture of +4 and +2 is unlikely. A combination of both would result in a relatively flat curve, which is not observed in the MUS spectrum. For a better overview of the correlation with oxidation states ranging from +2 to +3, see Appendix B.

According to the results of the Mn K-edge XANES of the reference samples, the average oxidation state of Mn has to be between +2 and +3. Additionally, the analysis via EPMA and PXRD strongly supports the presence of spinel structures involving Mn und Al. The XANES spectra show that Mn is not present in the pure form of any of the analyzed oxides. The same is concluded from the EPMA, which in addition has shown that the Al-percentage in these Mn phases is lower than in pure galaxite.

Accordingly, linear combinations of the spinels galaxite and hausmannite, as well as of galaxite and bixbyite, were calculated. From these linear combinations, the model XANES spectra in Figures 7 and 8 were obtained. These mixtures present a lower Al-content than galaxite, but still have a spinel structure. The results obtained from these linear combinations can be seen in Figures 7 and 8. A linear combination of hausmannite with bixbyite was also calculated. These results are shown in Appendix D.

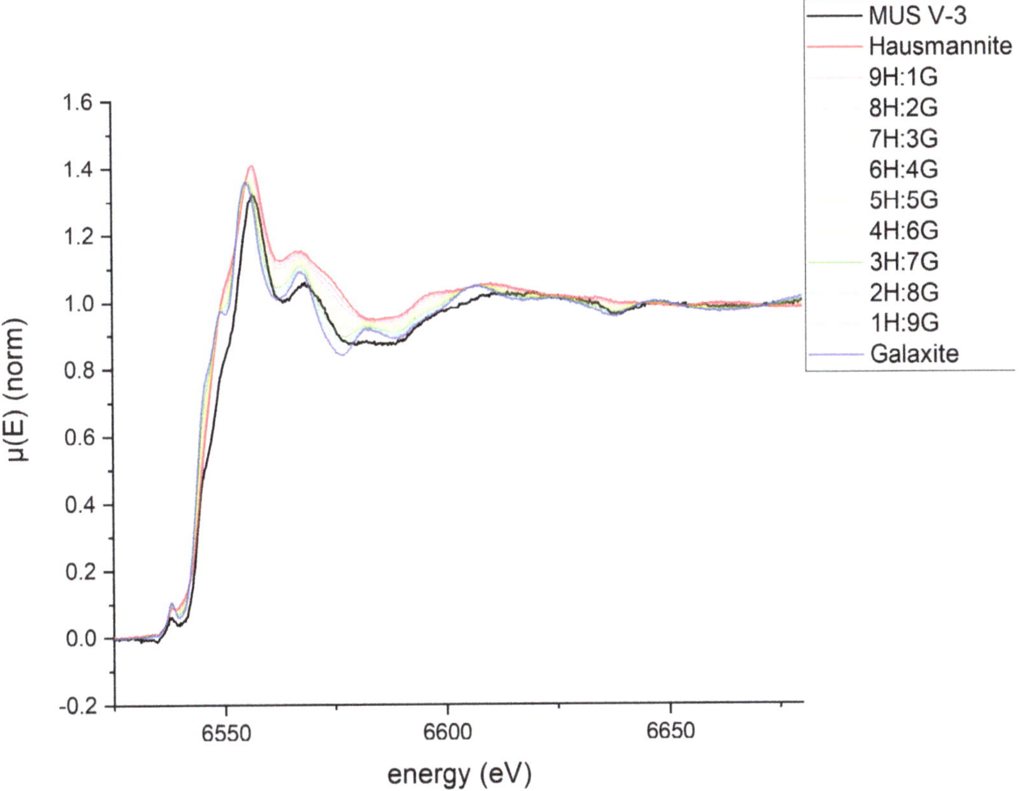

Figure 7. Linear combination of hausmannite (H) and galaxite (G). At the edge jump, the MUS spectrum fits the spectrum of hausmannite, whereas after the jump, the spectrum is more similar to galaxite.

The linear combinations (Figures 7 and 8) of a 50:50 (mass) mixture of both galaxite and bixbyite, as well as galaxite and hausmannite, are quite similar to the measured spectrum of the MUS V-3. A combination of bixbyite and hausmannite is excluded from further investigation, as the shape is significantly different (see Appendix D). Accordingly, the experimental XANES data was obtained from a 50:50 (mass) mixture of galaxite and bixbyite, as well as galaxite and hausmannite. The obtained spectra are shown in Figures 9 and 10. Figure 10 shows a close-up of the edge-jump.

The spectra displayed in Figures 9 and 10 show that the pre-edge, the edge jump and the region after the edge of the MUS and both references are similar. The results suggest that the slag contains a mixture of Mn^{2+} and Mn^{3+}, confirming the EPMA analysis. The combinations of galaxite and bixbyite with an average Mn oxidation state of 2.69—as well as of galaxite and hausmannite, with an average oxidation state of 2.46—mostly match the MUS spectrum. In conclusion, galaxite is present in the MUS.

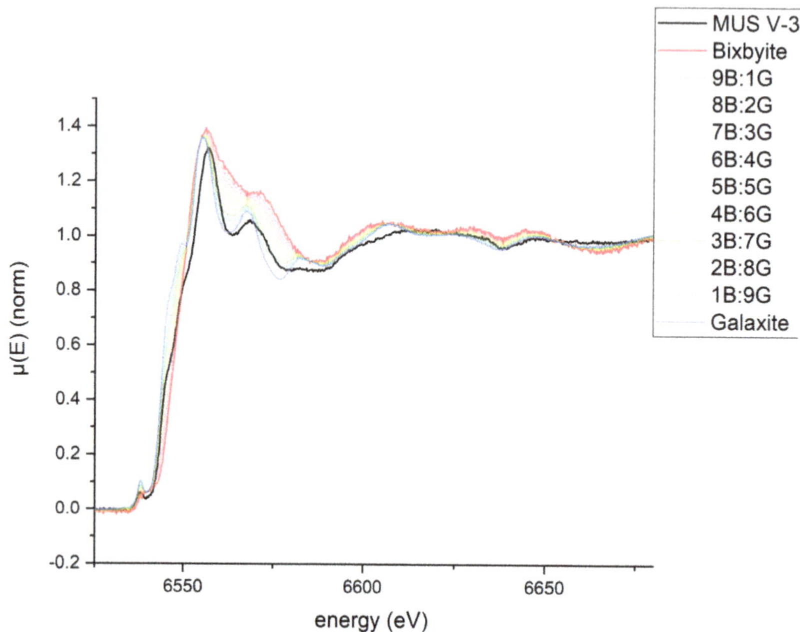

Figure 8. Linear combination of bixbyite (B) and galaxite (G). The spectrum of the MUS is similar to the linear combination. At the edge jump, the MUS spectrum fits the middle of the linear combination, whereas in the region after the edge, the spectrum is more similar to that of galaxite.

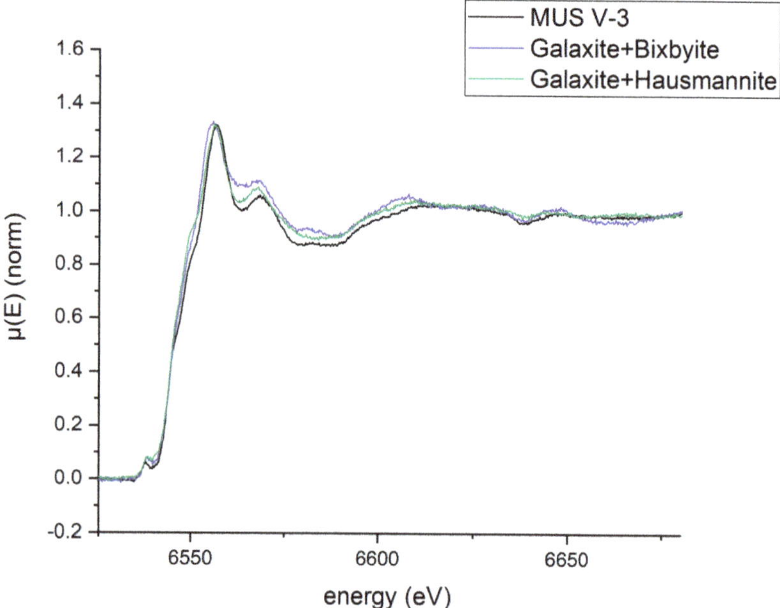

Figure 9. Experimentally derived X-ray absorption near edge structure analysis (XANES) spectra of a 50 wt.% mixture of galaxite and bixbyite, respective of galaxite and hausmannite compared to the spectrum obtained from the MUS V-3. The pre-edge peak, edge jump and course of the spectrum are very similar for all three spectra.

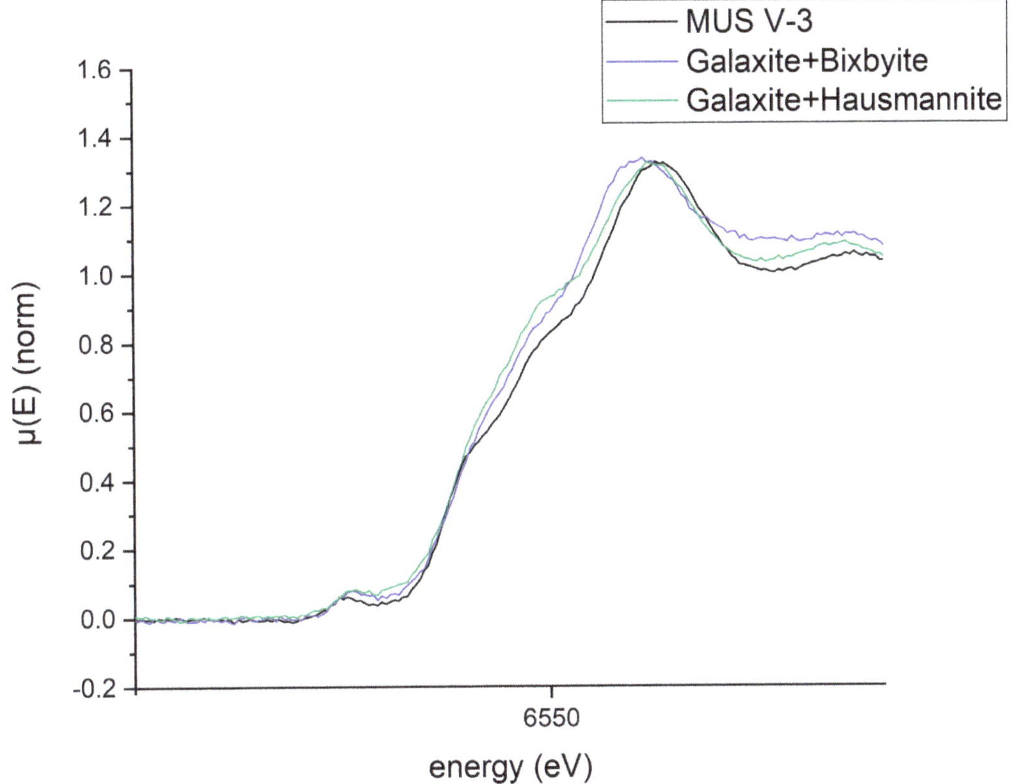

Figure 10. Close-up to the edge jump from Figure 9.

5. Discussion

The experimental investigation of the influence of Mn on the solidification, and especially on the formation of the EnAM $LiAlO_2$ in slags of the six-component oxide system (Li, Mg, Al, Si, Ca and Mn) is crucial to understand. This is also indispensable for the phase relations, as well as the reactions in this complex system. It will also help to predict the slag composition and improve thermodynamic modeling. Slags, unlike most geological features, are formed on a short timescale and with high cooling rates. Hence, non-equilibrium thermodynamic modeling will have to be consulted to develop a route to create the desired EnAM.

In contrast to the other elements in this system, Mn is redox-sensitive, occurring in several oxidation states ranging from +2 to +7. Due to the moderate to high oxygen fugacity in the slag, the expected oxidation numbers are +2, +3 and +4, and mixtures thereof. The purpose of this research was to study the suppression of $LiAlO_2$ formation in Mn-rich Li-ion battery recycling slags. The determination of the Mn-species, including the oxidation state formed in slags, is key to understanding this phenomenon.

Investigations with PXRD and EPMA on Mn-rich MUS reveal that besides LiAl and GCAS, the melt contains large grains of Al/Mn-rich oxides. The PXRD results show that these oxides can be best described as spinel-like compounds. The diffractograms exhibit reflections in the range of the main (311) diffraction line of the spinel-structures $MnAl_2O_4$ (galaxite), $Li_{1-x}Mn_2O_4$, $Li_2Mn_2O_4$ and $LiAl_5O_8$ (Figure 4). Due to the non-direct matching of these diffraction lines, the best explanation is a spinel solid solution with the elements Li, Al and Mn. With increasing Mn concentration within the melt experiments MUS V-1 to

V-3, there is a shift of the diffraction reflection towards galaxite, indicating that the galaxite component is increasing.

The amount of LiAlO$_2$ seems to be suppressed compared to an Mn-free melt with similar Li-concentration (Figure 4c, black line). Due to the high peak to background ratio, a comparable high amount of amorphous phase can be assumed.

The BSE(Z) micrograph observations show large idiomorphous Mn-rich grains (example see: Figure 5), suggesting an early and complex crystallization scenario. EPMA point scan analyses (Table 3) show a distinct decrease in the aluminum concentration from the center to the rim of the predominant Mn-rich crystals. At the edge, the aluminum concentration drops nearly to zero. Additionally, there is a split into two components, one relatively Mn-enriched and one relatively Mn-depleted. If the composition of all measurements is calculated as fractions of the virtual compounds Mn$_{0.5}$AlO$_2$ ($\frac{1}{2}$ galaxite), LiMnO$_2$ and Mn$_{0.5}$MnO$_2$ ($\frac{1}{2}$ hausmannite) a general formula (Li$_{(2x)}$Mn^{2+}$_{(1-x)}$)$_{1+x}$(Al$_{(2-z)}$,Mn^{3+}$_z$)O$_4$ can be calculated. From this calculation, a Li-content is derived and used to assess a gradient of the Li-bearing compounds. In accordance with the elemental gradients, a constant decrease of the galaxite fraction from the center to the rim is observed. In contrast, the hausmannite fraction is increasing. The Li-Mn compound fraction stays more or less constant except for a steep increase at the last ~10 µm from the rim. Directly at the rim, a split into a "normal" and a Mn$_{0.5}$MnO$_2$-dominated region can be observed. The increase and decrease of the individual species over the point scans are shown in Figure 11.

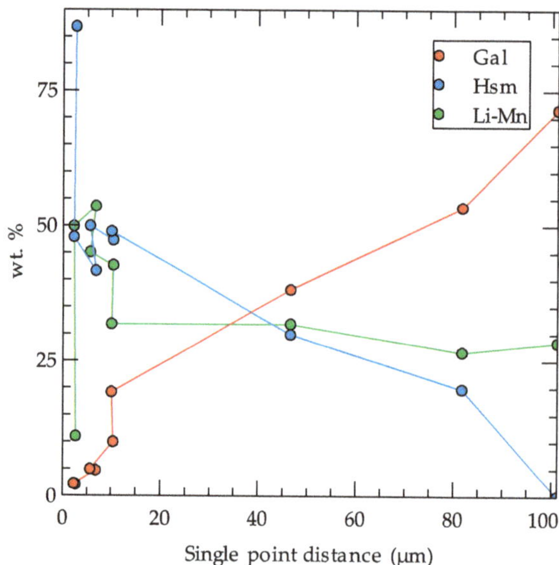

Figure 11. Fractions of the virtual compounds Mn$_{0.5}$AlO$_2$ ($\frac{1}{2}$ galaxite, Gal), LiMnO$_2$ (Li-Mn), and Mn$_{0.5}$MnO$_2$ ($\frac{1}{2}$ hausmannite, Hsm) in the grain presented in Figure 5.

This observation indicates that from the beginning to the end of the crystallization, Li is incorporated into the spinel structure. The spinel composition itself changes from a galaxite dominated to hausmannite-dominated chemistry. Directly at the rim, the oversaturation of the melt with Mn is such that the spinel solid solution segregates (most probably during cooling down to room temperature) to form two different (most probably spinel like) oxides.

In this respect, it is interesting that at lower temperatures, the hausmannite converts to the tetragonal crystal system with low solubility of the spinel compound galaxite as reported by Chatterjee et al. [9]. This could indicate an exsolution of the hausmannite component due to crystal lattice incompatibility. The hypothesis is backed by the results from the Mn K-edge analysis, which suggests a mixture of galaxite and Mn^{2+}, Mn^{3+} oxide spinels. The virtual Li compound would mix into the cubic galaxite-like spinel phase.

By combining the above results, a scenario of the large crystal genesis is established. The crystallization starts with a high aluminum galaxite-like composition that is subsequently enriched in Mn during the crystal growth. At the end of the crystallization, the solid solution becomes unstable, indicated by exsolution whiskers with a higher mean atomic number, surrounded by the massive crystal.

6. Conclusions

In this study, Mn-rich grains in mock-up slags (system: Li_2O-CaO-SiO_2-Al_2O_3-MgO-MnO_x) were characterized to understand the suppression of the $LiAlO_2$ formation in Mn-rich Li-ion battery recycling slags. The PXRD, EPMA and XANES data suggest that Mn-rich grains crystallize early on as a spinel solid solution. A generic stoichiometry, i.e., $(Li_{(2x)}Mn^{2+}{}_{(1-x)})_{1+x}(Al_{(2-z)},Mn^{3+}{}_z)O_4$ of the solid solution was determined assuming a combination of the virtual components $Mn_{0.5}AlO_2$, $LiMnO_2$ and $Mn_{0.5}MnO_2$. From the spatially resolved data, it was concluded that the solid solution is relatively Al-rich at the beginning of the crystallization and becomes depleted during the process. The formation of spinel solid solution with Mn and Al seems to scavenge Li from the melt before the $LiAlO_2$ crystallization can begin.

In conclusion, the experimental evaluation of mock-up slags has provided valuable insights into the Li, Mg, Al, Si, Ca, Mn and O system, and emphasized the benefit to study model melts and slags. In the future, however, an approach allowing for a faster synthesis of variable composition would be desirable. This approach will help to design suitable EnAM. The extraction of the EnAM from the slag and further processing will be part of subsequent studies.

In addition, it is not clear how these early crystals form on a molecular level. The solid solution could be a product of a solid phase process. It could also be driven by the ion-pair formation in the melt. In this respect, it is crucial to evaluate the primary crystallization fields in the system Li_2O-Al_2O_3-MnO_x in the presence of the other slag compounds Mg, Si and Ca. Despite this, the influence of the viscosity and the oxygen concentration on the early formation of Mn-rich compounds needs to be studied. The impact of viscosity changes, and pair formation in the ionic melt could be accessed by molecular dynamic modeling.

Author Contributions: A.W. conceived the paper. A.W., U.E.A.F. and T.S. conducted the literature review. All melting experiments were designed and performed by H.Q. and D.G. The chemical bulk analysis was executed by the analysis laboratory of the Institute of Mineral and Waste Processing. The phase analysis (PXRD) and the mineralogical investigation (EPMA) were conducted by T.S. The speciation analysis with XANES and interpretation of the spectra was conducted by A.W. and U.E.A.F. Interpretation, discussion and conceptualization were conducted by all authors. All authors have read and agreed to the published version of the manuscript.

Funding: This research was funded by the Clausthal University of Technology in the course of a joint research project, "Engineering and Processing of Artificial Minerals for an Advanced Circular Economy Approach for Finely Dispersed Critical Elements" (EnAM).

Acknowledgments: We acknowledge support by Open Access Publishing Fund of Clausthal University of Technology. We thank Jörg Wittrock from the Institute of Inorganic and Analytical Chemistry for the idea and implementation of the galaxite synthesis. We thank Joanna Kolny-Olesiak for discussions and proofreading.

Conflicts of Interest: The authors declare no conflict of interest. The funders had no role in the design of the study, in the collection, analyses, or interpretation of data, in the writing of the manuscript, or in the decision to publish the results.

Appendix A

The conversion from θ-angle to energy space is done by following Bragg's law, with the order of diffraction (n), Planck constant (h), speed of light (c), interplanar distance (d).

$$E = \frac{n \cdot h \cdot c}{2 \cdot d \cdot \sin(\theta)} \tag{A1}$$

For a cubic system, d is defined as: $d = \dfrac{a_0}{\sqrt{h^2 + k^2 + l^2}}$, with lattice spacing ($a_0$) and Miller indices ($h$, k, l). Therefore, the term $\dfrac{n \cdot h \cdot c}{2d}$ is dependent on the chosen crystal and the order of diffraction.

Appendix B

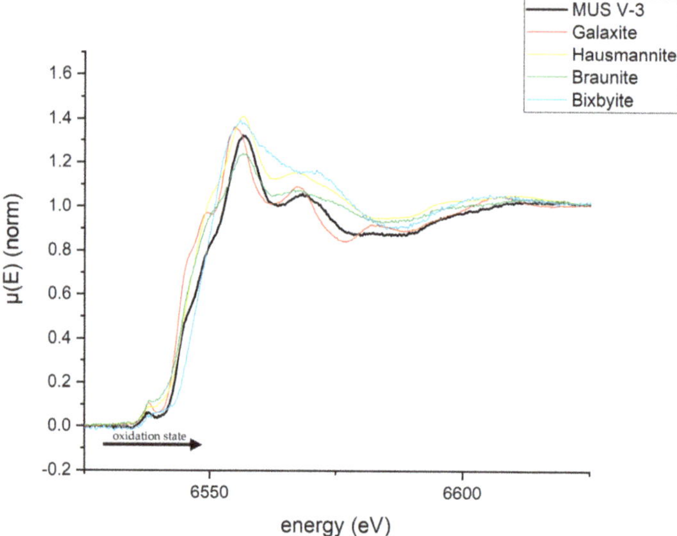

Figure A1. Spectra of Mn samples of different oxidation states. The spectra are compared to the measurement of the MUS. Oxidation states ranging for a better overview in contrast to Figure 6 from +2 to +3.

Appendix C

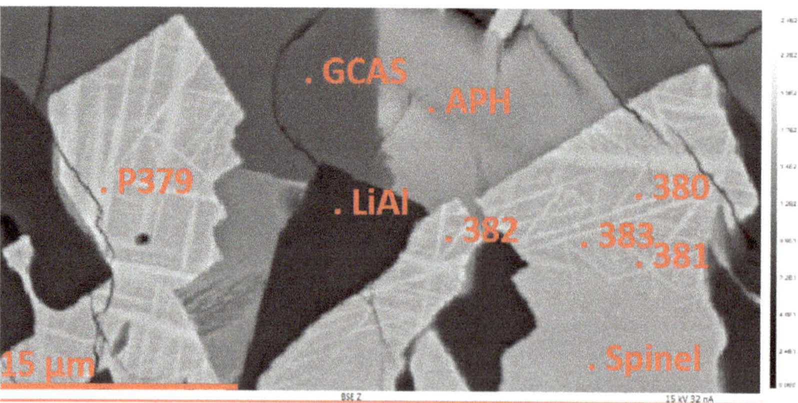

Figure A2. Close-up of Figure 4 showing the lamellae region.

Appendix D

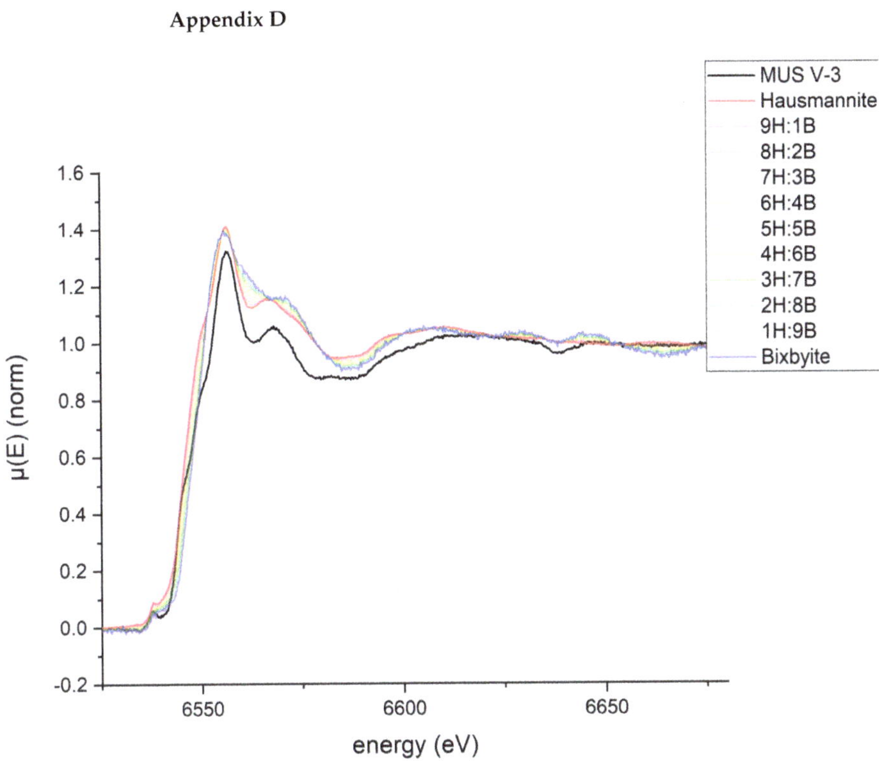

Figure A3. Linear combination of hausmannite (H) and bixbyite (B).

References

1. Jaskula, B.W. Lithium. Available online: https://pubs.usgs.gov/periodicals/mcs2020/mcs2020-lithium.pdf (accessed on 15 December 2020).
2. Elwert, T.; Strauss, K.; Schirmer, T.; Goldmann, D. Phase composition of high lithium slags from the recycling of lithium ion batteries. *World Metall. -Erzmetall* **2012**, *65*, 163–171.
3. Velázquez-Martínez, O.; Valio, J.; Santasalo-Aarnio, A.; Reuter, M.; Serna-Guerrero, R. A Critical Review of Lithium-Ion Battery Recycling Processes from a Circular Economy Perspective. *Batteries* **2019**, *5*, 68. [CrossRef]
4. Haas, A.; Elwert, T.; Goldmann, D.; Schirmer, T. Challenges and research needs in flotation of synthetic metal phases: EMPRC 06/25-06/26/2018. In Proceedings of the European Mineral Processing & Recycling Congress, Essen, Germany, 25–26 June 2018; EMPRC, 2018. GDMB Verlag GmbH: Clausthal-Zellerfeld, Germany, 2018. ISBN 978-3-940276-84-1.
5. Elwert, T.; Goldmann, D.; Schirmer, T.; Strauss, K. *Recycling von Li-Ionen-Traktionsbatterien*; Recycling und Rohstoffe: Neuruppin, Germany, 2012; pp. 679–690.
6. Schirmer, T.; Qiu, H.; Li, H.; Goldmann, D.; Fischlschweiger, M. Li-Distribution in Compounds of the Li$_2$O-MgO-Al$_2$O$_3$-SiO$_2$-CaO System—A First Survey. *Metals* **2020**, *10*, 1633. [CrossRef]
7. Paulsen, J.M.; Dahn, J.R. Phase Diagram of Li–Mn–O Spinel in Air. *Chem. Mater.* **1999**, *11*, 3065–3079. [CrossRef]
8. Konar, B.; van Ende, M.-A.; Jung, I.-H. Critical Evaluation and Thermodynamic Optimization of the Li$_2$O-Al$_2$O$_3$ and Li$_2$O-MgO-Al$_2$O$_3$ Systems. *Met. Mater. Trans. B* **2018**, *49*, 2917–2944. [CrossRef]
9. Chatterjee, S.; Jung, I.-H. Critical evaluation and thermodynamic modeling of the Al–Mn–O (Al2O3–MnO–Mn2O3) system. *J. Eur. Ceram. Soc.* **2014**, *34*, 1611–1621. [CrossRef]
10. Asaoka, S.; Okamura, H.; Akita, Y.; Nakano, K.; Nakamoto, K.; Hino, K.; Saito, T.; Hayakawa, S.; Katayama, M.; Inada, Y. Regeneration of manganese oxide as adsorption sites for hydrogen sulfide on granulated coal ash. *Chem. Eng. J.* **2014**, *254*, 531–537. [CrossRef]
11. Kim, K.; Asaoka, S.; Yamamoto, T.; Hayakawa, S.; Takeda, K.; Katayama, M.; Onoue, T. Mechanisms of hydrogen sulfide removal with steel making slag. *Environ. Sci. Technol.* **2012**, *46*, 10169–10174. [CrossRef]

12. Seidler, G.T.; Mortensen, D.R.; Remesnik, A.J.; Pacold, J.I.; Ball, N.A.; Barry, N.; Styczinski, M.; Hoidn, O.R. A laboratory-based hard X-ray monochromator for high-resolution X-ray emission spectroscopy and X-ray absorption near edge structure measurements. *Rev. Sci. Instrum.* **2014**, *85*, 113906. [CrossRef]
13. Seidler, G.T.; Mortensen, D.R.; Ditter, A.S.; Ball, N.A.; Remesnik, A.J. A Modern Laboratory XAFS Cookbook. *J. Phys. Conf. Ser.* **2016**, *712*, 12015. [CrossRef]
14. Le, H.V.; Parishan, S.; Sagaltchik, A.; Göbel, C.; Schlesiger, C.; Malzer, W.; Trunschke, A.; Schomäcker, R.; Thomas, A. Solid-State Ion-Exchanged Cu/Mordenite Catalysts for the Direct Conversion of Methane to Methanol. *ACS Catal.* **2017**, *7*, 1403–1412. [CrossRef]
15. Dimitrakopoulou, M.; Huang, X.; Kröhnert, J.; Teschner, D.; Praetz, S.; Schlesiger, C.; Malzer, W.; Janke, C.; Schwab, E.; Rosowski, F.; et al. Insights into structure and dynamics of (Mn, Fe)Ox-promoted Rh nanoparticles. *Faraday Discuss.* **2018**, *208*, 207–225. [CrossRef]
16. Lutz, C.; Fittschen, U.E.A. Laboratory XANES to study vanadium species in vanadium redox flow batteries. *Powder Diffr.* **2020**, 1–5. [CrossRef]
17. Gates-Rector, S.; Blanton, T. The Powder Diffraction File: A quality materials characterization database. *Powder Diffr.* **2019**, *34*, 352–360. [CrossRef]
18. Downs, R.T.; Hall-Wallace, M. The American Mineralogist crystal structure database. *Am. Mineral.* **2003**, *88*, 247–250.
19. Lafuente, B.; Downs, R.T.; Yang, H.; Stone, N. The power of databases: The RRUFF project. In *Highlights in Mineralogical Crystallography*; Walter de Gruyter GmbH: Berlin, Germany, 2015; pp. 1–30. [CrossRef]
20. Jercinovic, M.J.; Williams, M.L.; Allaz, J.; Donovan, J.J. Trace analysis in EPMA. *IOP Conf. Ser. Mater. Sci. Eng.* **2012**, *32*, 12012. [CrossRef]
21. Merlet, C. Quantitative Electron Probe Microanalysis: New Accurate Φ (ρz) Description. *Mikrochim. Acta* **1992**, *12*, 107–115. [CrossRef]
22. Jahrman, E.P.; Holden, W.M.; Ditter, A.S.; Mortensen, D.R.; Seidler, G.T.; Fister, T.T.; Kozimor, S.A.; Piper, L.F.J.; Rana, J.; Hyatt, N.C.; et al. An improved laboratory-based X-ray absorption fine structure and X-ray emission spectrometer for analytical applications in materials chemistry research. *Rev. Sci. Instrum.* **2019**, *90*, 24106. [CrossRef]
23. easyXAFS, LLC. *easyXES100-Extended Executable Progra, Object Files, Documentation, and Source Code*; easyXAFS, LLC: Seattle, WA, USA, 2018.
24. Asakura, K.; Abe, H.; Kimura, M. The challenge of constructing an international XAFS database. *J. Synchrotron Radiat.* **2018**, *25*, 967–971. [CrossRef]
25. Itoh, T. K_MnO2_Si111_20110423. Available online: https://www.cat.hokudai.ac.jp/catdb/index.php?action=xafs_dbinpbrowsedetail&opnid=2&resid=22&r=32 (accessed on 30 November 2020).
26. Ravel, B.; Newville, M. ATHENA, ARTEMIS, HEPHAESTUS: Data analysis for X-ray absorption spectroscopy using IFEFFIT. *J. Synchrotron Radiat.* **2005**, *12*, 537–541. [CrossRef]
27. De Vries, A.H.; Hozoi, L.; Broer, R. Origin of the chemical shift in X-ray absorption near-edge spectroscopy at the MnK-edge in manganese oxide compounds. *Int. J. Quantum Chem.* **2003**, *91*, 57–61. [CrossRef]

 *metals*

Article

The COOL-Process—A Selective Approach for Recycling Lithium Batteries

Sandra Pavón, Doreen Kaiser, Robert Mende and Martin Bertau *

Institute of Chemical Technology, TU Bergakademie Freiberg, Leipziger Straße 29, 09599 Freiberg, Germany; sandra.pavon-regana@chemie.tu-freiberg.de (S.P.); doreen.kaiser@chemie.tu-freiberg.de (D.K.); robert.mende@chemie.tu-freiberg.de (R.M.)
* Correspondence: martin.bertau@chemie.tu-freiberg.de; Tel.: +49-3731-392384

Abstract: The global market of lithium-ion batteries (LIB) has been growing in recent years, mainly owed to electromobility. The global LIB market is forecasted to amount to \$129.3 billion in 2027. Considering the global reserves needed to produce these batteries and their limited lifetime, efficient recycling processes for secondary sources are mandatory. A selective process for Li recycling from LIB black mass is described. Depending on the process parameters Li was recovered almost quantitatively by the COOL-Process making use of the selective leaching properties of supercritical CO_2/water. Optimization of this direct carbonization process was carried out by a design of experiments (DOE) using a 3^3 Box-Behnken design. Optimal reaction conditions were 230 °C, 4 h, and a water:black mass ratio of 90 mL/g, yielding 98.6 ± 0.19 wt.% Li. Almost quantitative yield (99.05 ± 0.64 wt.%), yet at the expense of higher energy consumption, was obtained with 230 °C, 4 h, and a water:black mass ratio of 120 mL/g. Mainly Li and Al were mobilized, which allows for selectively precipitating Li_2CO_3 in battery grade-quality (>99.8 wt.%) without the need for further refining. Valuable metals, such as Co, Cu, Fe, Ni, and Mn, remained in the solid residue (97.7 wt.%), from where they are recovered by established processes. Housing materials were separated mechanically, thus recycling LIB without residues. This holistic zero waste-approach allows for recovering the critical raw material Li from both primary and secondary sources.

Keywords: lithium recycling; circular economy; lithium batteries; supercritical CO_2; black mass

Citation: Pavón, S.; Kaiser, D.; Mende, R.; Bertau, M. The COOL-Process—A Selective Approach for Recycling Lithium Batteries. *Metals* **2021**, *11*, 259. https://doi.org/10.3390/met11020259

Academic Editor: Bernd Friedrich
Received: 15 December 2020
Accepted: 30 January 2021
Published: 3 February 2021

Publisher's Note: MDPI stays neutral with regard to jurisdictional claims in published maps and institutional affiliations.

1. Introduction

Since the market launch in 1991, the global market of lithium-ion batteries (LIBs) has been growing steadily. The global LIB market was valued at \$36.7 billion in 2019 and is expected to reach \$129.3 billion by 2027 [1]. One reason for this strong growth is the rising market for electric mobility. In 2018, 5.12 million electric passenger cars were registered worldwide, which corresponds to an increase of 63% compared to the previous year [2]. Furthermore, rechargeable LIBs are used extensively in the growing market of cableless electronic devices and applied in electric tools and grid storage applications [3]. Since the global reserves required to produce LIBs, as well as the lifetime of LIBs, are limited, efficient recycling approaches are necessary. The chemistry and technology of LIBs are still in development, resulting in a wide variety of different battery types, which in turn makes recycling more sophisticated. Battery recycling is also supported by the directive 2006/66/EC of the European Union, which requires a recycling rate of spent batteries of at least 50 wt.% of whole spent battery [4].

Despite structural diversity, the basic structure of all LIBs is mostly the same [5]. Usually, the cathode is an aluminum foil with an intercalated Li compound, and the anode a copper foil with a graphite coating. The anode and cathode compartments are separated by a porous polyolefin and the electrolyte is a mixture of an organic solvent and a lithium salt. These cells are enclosed by a sealed container made of aluminum, steel, special plastics, or highly refined aluminum composite foils [6]. Depending on the used cathode materials current commercial LIBs can be categorized into five types [7]:

(1) $LiCoO_2$ (LCO),
(2) $LiNi_xCo_yMn_zO_2$ (NCM, $x + y + z = 1$),
(3) $LiMn_2O_4$ (LMO),
(4) $LiNi_xCo_yAl_zO_2$ (NCA, $x + y + z = 1$), and
(5) $LiFePO_4$ (LFP) series.

To simplify the battery recycling the process should be independent of the type of the spent LIBs and should be applicable for mixtures of different LIBs.

Most of the already developed recycling processes are pyrometallurgical and/or hydrometallurgical approaches. Pyrometallurgical processes are associated with high energy consumption, high capital costs, and potential hazardous gas emission, as well as complex extraction procedures [8,9]. Furthermore, the selective recovery of lithium is very difficult [8]. Moreover, recycling of plastics and electrolyte is not possible. As both components make up 40–50 wt.% of the spent battery, it is difficult to meet the required recycling rate of 50 wt.% [9]. Hydrometallurgical approaches allow for recycling lithium, as well as cobalt and nickel with high purity [7]. Leaching procedures with inorganic or organic acids followed by precipitation and/or solvent extraction obtain the desired products with high recycling efficiencies [10–12]. However, the high recycling rates can only be achieved by using high quantities of acid which in turn not only produce high amounts of wastewater [7]. Already the costs of the chemicals for acidic digestion and subsequent neutralization exceed the intrinsic metal value by far. Furthermore, the low leaching selectivity, especially in the case of inorganic acids, necessitates extensive purification steps, which render the entire process complex and costly. A promising alternative is the COOL-Process (CO_2-leching), the core step of which is leaching with supercritical CO_2 (*sc* CO_2).

Supercritical fluids are interesting alternatives to conventional solvents for metal extraction. There are more than 100 plants worldwide that extract using supercritical solvents, thus creating a broad field of application for these processes [13]. Probably the best-known process is the decaffeination of coffee [14]. A study by Rentsch et al. was able to show that the higher investment costs compared to conventional processes are already compensated by low operating costs after about two years. The low operating costs are due to a low chemical requirement and less complex wastewater treatment [15].

In the field of battery recycling, *sc* CO_2 currently plays only a minor role, but this is incomprehensible. Only for the recycling of the electrolyte of the LIBs have several studies [14,16–19] been published. The application of *sc* CO_2 for the extraction of metals has only been published in one paper on cobalt extraction [20]. The recovery of the electrolyte is a challenging task, especially regarding the different compositions of the LIBs. Several studies have shown that extraction with *sc* CO_2 is an efficient way to recycle the electrolyte [14,16–19], but this requires LIBs with the same composition, which is associated with a high sorting effort and therefore does not appear economical. Supercritical CO_2 has also been employed for metal extraction from several materials, like ores, resins, and foils. For instance, Bertuol et al. developed a process that allows the recovery of cobalt from LIBs using *sc* CO_2 and H_2O_2 (4% *v/v*) as co-solvent. This process allows the extraction of more than 95 wt.% cobalt in a very short time (5 min) [20]. Other metals, such as nickel, manganese, and lithium, are not considered in this study. Research on the recovery of lithium from LIB by means of *sc* CO_2 is not published yet.

Originally, the COOL-Process was developed for the production of Li_2CO_3 from lithium containing ores, like zinnwaldite and spodumene (Figure 1) [21].

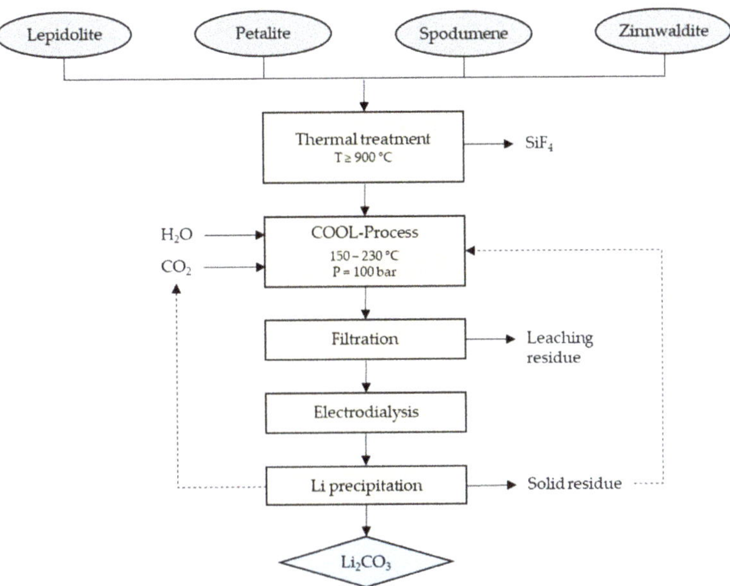

Figure 1. Flowsheet for the production of Li_2CO_3 from Lepidolite, Petalite, Spodumene, and Zinnwaldite minerals by direct carbonation [21].

Considering that this direct-carbonation process promises a selective leaching of Li with subsequent precipitation and separation of Li_2CO_3 without the addition of further chemicals [21,22], the COOL-Process has been applied to recover Li from black mass in the current work as depicted (Figure 2).

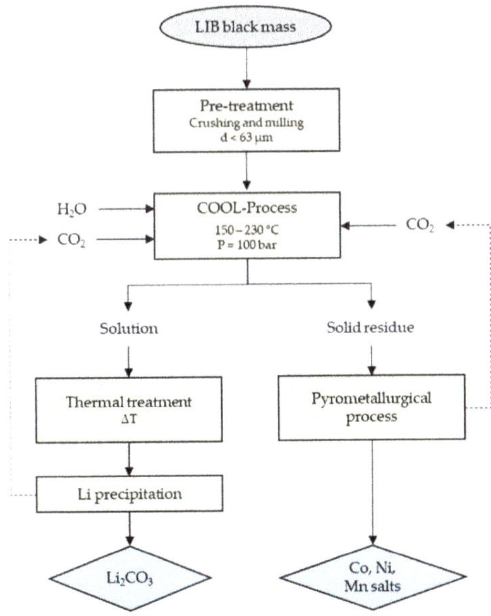

Figure 2. Recycling process scheme of lithium-ion batteries (LIB) black mass.

2. Materials and Methods

2.1. LIB Black Mass Pre-Treatment and Characterization

LIB black mass sample, type battery Li-NMC, was kindly supplied by the Institute of Mechanical Process Engineering and Mineral Processing from TU Bergakademie Freiberg and pre-treated before carrying out the optimization process by multi-stage crushing using a planetary ball mill PM 100 (Retsch GmbH, Haan, Germany) and subsequent milling by a vibrating cup mill (AS 200, Retsch GmbH, Haan, Germany) for grinding the black mass sample to a particle size of <63 μm (d_{90}: 61.18 μm).

The elemental composition of the LIB black mass was determined by atomic emission spectrometry with inductively coupled plasma (ICP-OES, Optima 4300 DV, Perkin Elmer, Waltham, MA, USA) and atomic absorption spectroscopy (AAS, ContrAA 700, Analytik Jena, Jena, Germany). The LIB black mass sample was treated with aqua regia in a liquid:solid ratio (L:S) of 100 at 180 ± 2 °C for 30 min using a Microwave MARS 6 (CEM Corporation, London, UK). The procedure was repeated three times.

Carbon measurement was carried out with the vario EL MICRO cube system made by Elementar Analysensysteme GmbH, Langenselbold, Germany, based on a combustion method.

Thermogravimetric analysis (TGA) and differential thermal analysis (DTA) were conducted by using TGA/DSC (Differential scanning calorimetry) 1 with a DSC sensor and mass-flow controller GC 200 (Mettler Toledo, Gießen, Germany) to examine the thermal behavior of the LIB black mass. TGA/DTA was carried out placing 22.36 mg of the black mass in a 150 μL alumina crucible heated from 25 up to 1000 °C with a heating rate of 10 K/min under pure oxygen or nitrogen flow of 40 mL/min.

2.2. Optimization

The aim was to determine the reaction conditions at which the highest yield of the target value can be obtained. However, processes reported in the literature are usually conducted using the one-factor-at-a-time method. The influence of different factors is evaluated by varying one after another, keeping the other ones constant. This method often fails to determine the global optimum because the correlation between different factors is not in consideration. Hence, the obtained optimum is a local instead of a global one, and the process efficiency from an economic and environmental point of view is not properly evaluated. Furthermore, the optimization cannot be considered accurate, because the influence of some factors (binary correlations) on the target yield is often significant yet not determined. Therefore, the current work employs a statistical experimental design by considering both binary correlations and squared effect in order to determine the global optimum.

2.2.1. 3^3 Box-Behnken Design

To optimize Li recovery from the LIB black mass sample, a 3^3 Box-Behnken design was used to determine the global optimum by consideration of all the factor combinations. This design of experiment (DOE) requires tests on every half of the edges and in the center, which was conducted threefold to determine the experimental error. The factors investigated were: temperature T [°C], residence time t [h], and water:black mass ratio *L:S ratio* [mL/g] in a range comprising three levels (Table 1).

Table 1. Factors and levels in the 3^3 Box-Behnken experimental design.

Factors	Factor Levels		
	−1	**0**	**+1**
Temperature T [°C]	150	190	230
Residence time t [h]	2	3	4
water:black mass ratio *L:S ratio* [mL/g]	30	60	90

Statgraphics v.18 (Statpoint Technologies Inc., Warrenton, VA, USA) was used as the evaluation statistical software to determine the global optimum, as well as the model equation which describes how Li_{yield} depends on each nine effects (linear, squared, and binary correlations). The model equation was obtained using Equation (1) via multi-linear regression.

$$y = b_0 + \sum_{i=1}^{N} b_i x_i + \sum_{1 \leq i \leq j}^{N} b_{ij} x_i x_j + \sum_{i=1}^{N} b_{ii} x_i^2 \,, \tag{1}$$

where:

y: Target value: Li_{yield} [wt.%];
x_i: Factors: T [°C], t [h], L:S ratio [mL/g];
N: Number of factors (3);
b_0: Ordinate section; and
b_i, b_{ij}, b_{ii}: Regression parameters of linear, squared and cross effects.

2.2.2. Experimental Procedure

Digestion experiments of the LIB black mass, using the conditions in random order, were performed under elevated pressure (100 bar) using the autoclave Hastelloy C4 (Berghof Products + Instruments GmbH, Eningen unter Achalm, Germany) and are depicted in Figure 3.

Figure 3. Hastelloy C4 autoclave used to carry out the COOL-Process.

Digestion experiments by the COOL-process, which uses CO_2 as a reagent, were carried out adding the LIB black mass sample (d_{50} = 9.1 μm) in the BR-300 autoclave to a volume of distilled water of 400 mL in accordance with the operating conditions depicted in Table 2. The suspension was heated to a range of temperature of 150–230 °C at a heating rate of 5 K/min and 500 rpm. CO_2 was added and a pressure of 100 bar was set after reaching the target temperature. The digestion time varied between 2–4 h. Afterward, the reaction mixture was cooled down to T < 30 °C under pressure and subsequently decompressed to normal pressure. The suspension was filled up with distilled water to 1 L and the residue was separated by vacuum filtration using an ash-free paper filter MN 640 dd (Macherey-Nagel, Düren, Germany). The leachates were analyzed by ICP-OES to determine the Al, Cu, Co, Fe, Ni, and Mn content and AAS for the Li content.

Table 2. Composition of LIB black mass in wt.% analyzed by ICP-OES, AAS, and combustion.

	Composition [wt.%]							
	Al	Co	Cu	Fe	Li	Mn	Ni	C
Mean	1.89	2.37	2.21	0.29	3.18	23.89	8.31	26.04
Std. Dev.	0.29	0.10	0.06	0.09	0.02	0.69	0.34	0.01

2.3. Li₂CO₃ Precipitation

The leachate (100 mL) obtained at T = 230 °C, t = 3 h, and L:S ratio = 30 mL/g was heated T = 100 °C. Li_2CO_3 precipitation was complete at V = 2.5 mL. The solid product was separated by filtration and washed with deionized water (5 mL). The liquid fractions were combined and recirculated for the next run in order to not loose residual Li. Product purity was determined as 99.8% by ICP-OES and AAS after dissolving with HNO_2 1 vol.%.

3. Results and Discussion

3.1. LIB Black Mass Characterization

After discharging the LIB, a black mass was obtained by mechanical treatment involving crushing and magnetic separation. Housing material, polyethylene, aluminum, and copper were separated as initial products. The resulting black mass is a powder consisting of anode and cathode material, coating material, electrode foils, and small parts of aluminum, copper, and polyethylene from the separator foil. During mechanical comminution, the release of highly volatile compounds of the electrolyte (dimethyl carbonate, diethyl carbonate) also occurs. Table 2 shows the mean value of black mass analysis and the standard deviation of the sample set (Std. Dev.).

The Li content was 3.18 wt.% and the main metals were Co (2.37 wt.%), Cu (2.21 wt.%), Mn (23.89%), and Ni (8.31 wt.%). The high Mn content compared to Co and Ni is noticeable, from which was evident that the LIB was a ($LiNi_xMn_yCo_zO_2$, x + y + z = 1) type battery (Li-NMC). Iron was introduced into the sample through the mechanical processing of the LIB. However, due to the low concentration (0.29 wt.%), no negative impact on the process was to be expected.

To be able to exclude the possibility of toxic fluorine compounds formed during the digestion, a thermogravimetric analysis (TGA) was carried out. TGA under nitrogen and oxygen showed a mass loss of 20.42 wt.% and 33.34 wt.% (Figure 4). The mean value of both measurements (26.88 wt.%) was in accordance with the carbon content measured by the elemental analysis (26.04 wt.%). Under oxygen atmosphere differential thermal analysis (DTA) showed, as expected, an exothermic peak in the range between 450 and 750 °C, which correlates with the combustion of the contained carbon. However, heat formation occurring here is only of minor importance for the process, which works at $T \leq 230$ °C. Only at $T > 800$ °C a slight release of fluorine compounds was observed. Therefore, it is not expected that volatile fluorine compounds will form during digestion, which means that no additional safety measures or special materials are required to operate the COOL-Process safely.

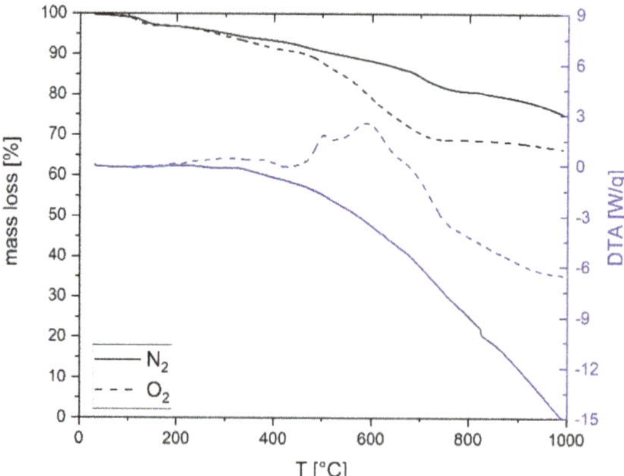

Figure 4. Thermogravimetric (TG) curve (black) and differential thermal analysis (DTA) curve (blue) of black mass heating under O_2 (dashed line) and N_2 (continuous line) atmosphere from 25 to 1000 °C at 10 K/min.

3.2. Optimization

3.2.1. Significant Influences on Lithium Yield

The Li leaching efficiency of each experiment obtained experimentally following the DOE was determined and listed in Table 3 as Li yield. To evaluate which of the nine effects: linear (A, B, C), squared (AA, BB, CC), and binary correlation (AB, BC, AC) contribute significantly to Li yield, an analysis of Variance called ANOVA was conducted. The results are depicted in Table 4. Terms that are insignificant for the target value were removed by the stepwise method.

Table 3. Li yield in each experiment obtained experimentally following the 3^3 Box-Behnken experimental design (*A: T* [°C], *B: t* [h], *C: L:S ratio* [mL/g]). The experiments shaded in grey correspond to the replicated central point.

	Factors			Li yield [wt.%]
	A: T [°C]	*B: t* [h]	*C: L:S ratio* [mL/g]	
1	190	3	60	75.2
2	190	2	30	66.8
3	150	2	60	52.1
4	230	4	60	91.5
5	190	4	30	66.3
6	190	4	90	80.2
7	150	3	30	52.7
8	190	3	60	72.5
9	230	3	90	93.1
10	190	2	90	74.6
11	230	2	60	87.0
12	150	3	90	58.2
13	150	4	60	57.8
14	230	3	30	71.3
15	190	3	60	72.8

Table 4. ANOVA results for Li yield from the 3^3 Box-Behnken design with three central points.

Source	Sum of Squares	DF *	Mean Square	F-Value	p-Value
A: Temperature	1264.5	1	1264.5	158.87	0.0001
B: Residence time	13.2914	1	13.2914	1.67	0.2528
C: Ratio	242.857	1	242.857	30.51	0.0027
AA	18.9075	1	18.9075	2.38	0.1839
AB	0.403225	1	0.403225	0.05	0.8308
AC	66.1782	1	66.1782	8.31	0.0344
BB	2.71234	1	2.71234	0.34	0.5847
BC	9.09023	1	9.09023	1.14	0.3341
CC	21.0541	1	21.0541	2.65	0.1648
Total error	39.7955	5	7.9591	-	-
Total (corr.)	2349.67	14	-	-	-
R^2	98.3063	-	-	-	-
R^2 adjusted for DF	95.2577	-	-	-	-

* Degrees of freedom.

According to the Pareto diagram depicted in Figure 5, the temperature in terms of a linear effect takes the most pronounced effect on Li yield. The water:black mass ratio (*L:S ratio*) showed the second highest effect, whereas the influence of residence time was rather poor. Consequently, leaching is almost fully completed after 2 h, as Li yield was affected only minorly within the residence time range studied (Table 3). This observation indicates that supercritical CO_2 is an efficient leaching agent for Li. The fact that residence time has only a slight impact on leaching black mass for Li mobilization has been observed in other studies, too: For inorganic (e.g., HCl, H_2SO_4, H_2O_2), as well as organic (e.g., oxalic acid, tartaric acid, citric acid) leaching agents, residence times between 30 and 240 min were reported [10]. Hence, Li is only weakly bound in the matrix of the black mass and it can therefore be easily extracted. This is also supported by the excellent extraction properties of supercritical (*sc*) CO_2. In previous studies, it was shown that the optimal residence time for Li mobilization from zinnwaldite through leaching with *sc*-CO_2/H_2O is 3 h [21], whereas leaching with HCl takes 7 h [22].

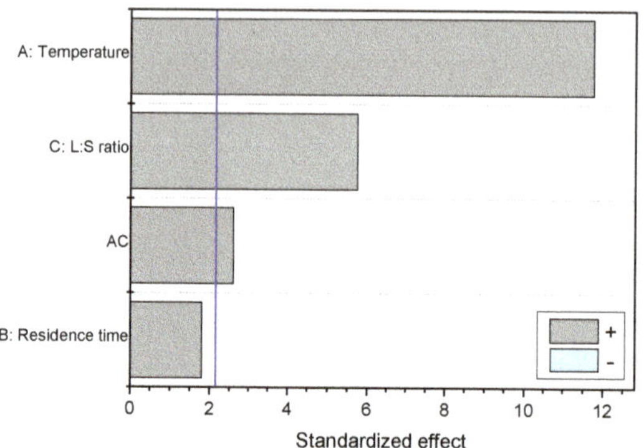

Figure 5. Pareto diagram with the significant effects on Li yield.

Furthermore, the AC correlation contributes to increasing the Li yield, too. It represents an interaction between temperature and water:black mass ratio. After Urbańska et al. had observed the same trend when leaching black mass with H_2SO_4 and H_2O_2 [23], this interaction was not unexpected.

The three squared correlations (AA, BB, and CC), as well as AB and BC as binary interactions were removed from Equation (1) by the stepwise method because of their insignificant effect to the Li yield optimization.

3.2.2. Model Equation and Optimum

In accordance with the experimental obtained results from the DOE, a mathematical model equation was determined considering all significant effects on Li yield. This equation allows for predicting the Li yield at any desired point within the investigated factors levels range. Equation (2) predicts that Li yield reaches its all-time maximum with 98.8 wt.% using the following reaction conditions: $T = 230\,°C$, $t = 4$ h and *L:S ratio* = 90 mL/g.

$$Li_{yield}(\text{wt.\%}) = 19.6125 + 0.178 \cdot A + 1.9275 \cdot B - 0.43977 \cdot C + 0.00339 \cdot A \cdot C \qquad (2)$$

where:

A: Temperature (°C);
B: Residence time (h); and
C: L:S ratio (mL_{water}/$g_{black\ mass}$).

To validate the mathematical model, a twofold experiment involving these optimized parameters was carried out. Both experiments provided a Li yield of 94.5 ± 0.33 wt.%. With a difference of <5 wt.%, one may consider the calculated model by employing a 3^3 Box-Behnken design in accordance with the experimental data. However, bearing in mind that only three of the nine effects studied took effect on Li yield, together with the optimum being in a corner of the DOE, an evaluation of the statistical design will be carried out in follow-up work. For instance, by employing a full factorial design, an improvement of the statistical experimental design can be achieved, thus obtaining a better mathematical model that describes how Li yield depends on the three chosen factors: temperature, residence time, and water:black mass ratio.

The model in terms of surface response is shown in Figure 6, where the bold points correspond to the experimental data and the star to the two replicates which were carried out using the optimal reaction conditions (*T* = 230 °C, *t* = 4 h and *L:S ratio* = 90 mL/g).

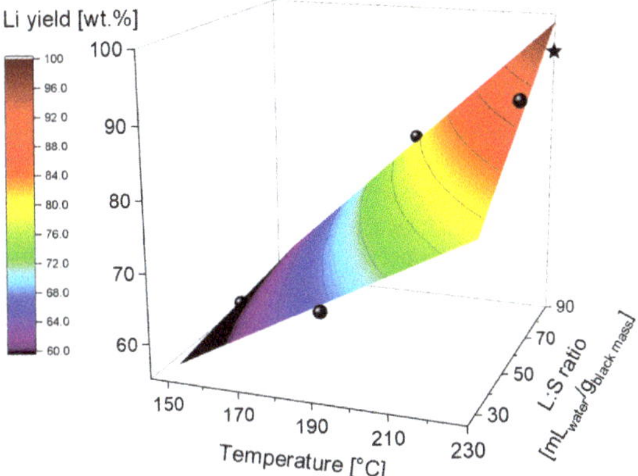

Figure 6. Li yield determined by the mathematical model equation varying the temperature and L:S ratio, maintaining the residence time constant at 4 h. Bold points correspond to the experimental data and the star point to the optimum obtained experimentally.

The Li yield (94.5 ± 0.33 wt.%) obtained under the optimal reaction conditions is roughly in line with several studies using inorganic and organic acids. For instance, Takacova et al. achieved a quantitative Li mobilization from the black mass with 2 M HCl (60–80 °C, 90 min, L:S ratio = 50) with simultaneous quantitative cobalt mobilization [24]. Similar results were described by Urbańska et al. using 1.5 M H_2SO_4 and 30% H_2O_2 (90 °C, 120 min, L:S ratio = 10). In their study up to 99.91 wt.% Li was leached and 87.85 wt.% cobalt and 91.46 wt.% nickel were co-extracted [23]. A high Li (<90 wt.%) and <30 wt.% Mn yield were obtained by Li et al. with 1 M oxalic acid (95 °C, 12 h, L:S ratio = 10) [25]. In other words, Co, Ni, and Mn were co-mobilized. All published results have in common, though, that their Li selectivity is poor, thus requiring additional purification and consumption of chemicals, not to speak of process costs. In contrast, the current work exhibits not only a high Li selectivity but also a high degree of Li mobilization. Under optimal reaction conditions, only Al was co-extracted (Table 5), but the presence of this metal has no influence on the further process of the Li_2CO_3 precipitation. The reason is for the insensitivity of the COOL-Process towards Al is the inability of Al to form neither carbonates nor hydrogen carbonates under this condition. Precipitation of Al salts with CO_2 only occurs from pH 9 and not in the acetic range [26]. With CO_2/water being a weak acid only, any other acidic leachate reagent will consequently co-mobilize Co, Ni, and/or Mn, the interaction of which with Al^{3+} and Fe^{3+} inevitably requires tedious and complex separation of these metals prior to Li_2CO_3 precipitation. The latter step is gaining further complexity through the mutually interacting chemistry of these metal cations with hydroxide, thus severely taking effect on the process economy of these approaches. A further advantage of the COOL-Process is the effect that, in contrast to other carbonates, the solubility of Li_2CO_3 decreases with increasing temperature. For this reason, the digestion solution is heated to 90 to 95 °C and the target product is precipitated in the process. The precipitation behavior of Li_2CO_3 in such digestion solutions has been extensively investigated in previous studies, so it was omitted here [15,21,22,27].

Table 5. Co-mobilization of selected elements at 230 °C, 4 h, and L:S ratio = 90 mL/g.

Element	Mobilization [%]
Al	52.34
Co	0.52
Cu	0.08
Fe	2.27
Mn	0.66
Ni	0.66

Direct carbonation (COOL-Process) of black mass has the advantage of leaving other valuable metals, such as Co, Mn, and Ni, in the leaching residue from where they can be recycled with ease according to established techniques. There exist pyrometallurgical processes for this purpose, so that the COOL-Process can be understood in terms of an enabling technology, which gives way to isolate lithium prior to known pyrometallurgy in a preliminary stage. The CO_2 released during the pyrometallurgical recovery of Co, Ni, and Mn can, in turn, be used for carbonization, this way contributing to both a zero-waste approach and circular economy.

Another advantage of the COOL-Process is its efficiency in terms of Li recovery regardless of the composition of the raw material. Particularly in the field of LIB recycling, a broad and robust feedstock variability is a prerequisite to operate the process economically, which in turn is mandatory in terms of successfully establishing a circular economy. Each battery manufacturer uses different compositions, for which reason there are a plethora of different battery types on the market. Most processes for recycling LIBs are specialized in certain compositions, which entails complex sorting processes. This is usually only possible by hand, which renders these processes highly cost intensive. The flammability of damaged LIB is susceptible to danger, what is an issue when hand-sorting. In the COOL-Process, LIB can be processed regardless of their composition. Moreover, previous studies have already shown that this process is suitable for extracting Li from ores, like zinnwaldite, too. The optimum reaction conditions determined in these studies were 230 °C, 3 h, and a L:S ratio of 30 and are thus comparable to the conditions determined in the current work [21]. Therefore, it can be concluded that the COOL-Process probably allows for recovering Li from both primary and secondary sources. It is a textbook example of circular resources chemistry, which comprises origin-independent processes for the production of chemical raw materials that do not differentiate between primary and secondary raw materials [28].

However, the maximum is placed in a corner, which raises the question of whether another factor (e.g., pressure) needs to be explored to ensure that the optimization covers all effective factors. From the viewpoint of process engineering, temperature increase appears as the factor of choice to check for higher Li mobilization. This is not possible, though, which is one of the limitations of the materials in contemporary LIB. Conventional sealing material (polytetrafluoroethylene, PTFE) is only stable up to 230 °C, so special materials, such as perfluoro rubber (FFKM), would be necessary. Particularly, on an industrial scale, these special sealing materials, together with the energy input required to reach temperatures >230 °C, are associated with considerable additional costs. Since a high Li mobilization was already achieved at 230 °C, only a small yield increase can be expected from a further temperature increase, which, however, is not justified in terms of additional energy and raw material demand which is in sharp contrast to the plus of Li to be expected. For these reasons, this factor remained unaltered and the maximum level was set to 230 °C. Since the second highest effect on the Li yield was provided by water:black mass ratio, the highest factor level was increased to 120 mL/g to evaluate how much the target value can be increased (Figure 7). 98.6 ± 0.19 wt.% Li was recovered by the COOL-Process at 230 °C, 4 h, and 120 mL/g. Considering the increase of 4.6 wt.% on the target value using 120 instead of 90 mL/g, the DOE could be improved by redefining the factor levels considering this enhancement on the Li yield. However, the small increase in yield represents a 22 wt.% reduced Li concentration in the digestion solution, which in turn will necessitate a higher energy input for its concentration prior to precipitating the target product Li_2CO_3. Again, this additional energy costs, in combination with resulting CO_2 emissions for energy generation, may hardly be compensated for the rather small plus in lithium yield.

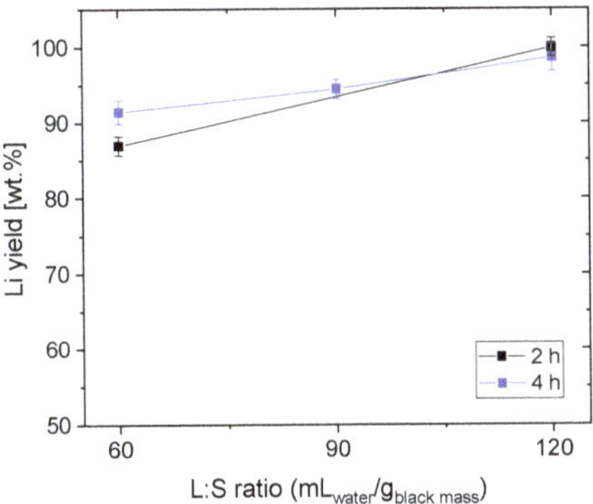

Figure 7. Effect of L:S ratio on the Li yield carrying out the COOL-Process at 230 °C for 2 or 4 h as residence time.

With this in mind, the 4.6 wt.% higher Li yield obtained from varying the water:black mass ratio cannot compensate for the lower energy efficiency. This latter issue can be equalized by reducing residence time *t*. Although the theoretical optimum is placed in a corner of the 3^3 box, the obtained information is sufficient to recognize the potential that lies in reducing *t* without investing in further optimization work. If *sc*-CO_2-leaching is done at 230 °C, 120 mL/g, and 2 h instead of 4 h, 99.05 ± 0.64 wt.% Li was recovered. As can be seen in Figure 7, the differences between the Li yields (~1%) when the COOL-Process was carried out at 230 °C for 2 or 4 h are not significant when using 120 mL/g. Hence, almost quantitative lithium recovery was reached by simply increasing the L:S ratio from 90 to 120 mL/g and conducting the leaching for 2 h. As pointed out before, the economic impact on the entire Li recycling process is an essential factor to be considered. Process performance depends not only on maximizing target values but also on economic efficiency.

It appears evident from these considerations that in general quantitative metal recovery from whatever feedstock may technologically be feasible, yet the bill is paid in terms of higher energy consumption, higher CO_2 emissions, and lacking economy. Applied to the circular economy, where the intrinsic metal value of the secondary raw material should exceed process costs, it is obvious that real-world processes always will constitute a compromise between what is desirable, what is feasible, and what is realizable. A way out of this situation is integrated processes, where the (secondary) raw material is converted into marketable products to the most possible extent. This is given here, since Co, Ni, and Mn, as well as housing material, are products, too, and CO_2 is re-circulated. A follow-up economical assessment will be conducted to provide the essential information to which extent additional efforts towards quantitative recovery are justified.

3.3. Li₂CO₃ as a Final Product

Black mass leaching under the conditions identified optimal in the optimisation study (Section 3.2) to reach the highest Li concentration (T = 230 °C, t = 3 h, L:S ratio = 30 mL/g) yielded an aqueous solution of $LiHCO_3$. The final product, Li_2CO_3, was obtained making use of its solubility anomaly. The carbonate's solubility in water is 13.3 g/L at T = 20 °C, while it is 7.2 g/L at T = 100 °C. Heating the solution to T = 100 °C not only decomposes $LiHCO_3$ to give Li_2CO_3, it also serves to reduce the solution volume in order to obtain best possible precipitation results. Filtration of the solid product gave pure X which was washed with deionised water with twice the amount of the volume of the residual precipitation solution (Figure 8). After drying, Li$_2$CO$_3$ was dissolved in HNO$_3$ 1 vol.%. Product purity was 99.8% as determined by ICP-OES and AAS.

Figure 8. Li$_2$CO$_3$ as a product after precipitation of LiHCO$_3$ by heating up at 100 °C.

The concentration of all other cations, such as Al, Mn, Fe, Co, Ni, and Cu, in sum accounted to <0.17 wt.%. With Li$_2$CO$_3$ purity >99.8 wt.% it was shown that the COOL Process is capable of producing battery grade Li$_2$CO$_3$ as crude product, which needs no further purification.

The remaining liquid fractions from product precipitation and washing were combined and recirculated. They serve as aqueous phase for the next run. Although this way no lithium is lost, Li$_2$CO$_3$ precipitation remains an issue, since precipitation efficiency in our experiments ranged widely between 43 and 85 wt.%. Optimizing product precipitation is therefore matter of follow-up studies. The same applies for Al, which under the given conditions is not susceptible for precipitation from carbonatic solutions. Upon recirculating the aqueous solutions, Al will accumulate and may interfere with the process. Exploratory experiments showed that Al can be eliminated as oxalate. If, however, aluminium oxalate precipitation interferes with Li leaching beyond what is tolerable, purging the solution is an option.

3.4. Industrial Application Feasibility

The COOL-Process has been successfully tested on a lab-scale and the high Li yield obtained using LIB as secondary sources demonstrates the efficiency of the direct-carbonation process.

The purpose of realizing holistic research approaches with practical relevance that can be carried out by using different raw materials is challenging. According to the obtained results, it can be affirmed that the goal has been reached because COOL-Process has been used for recovering Li from primary [21], as well as secondary raw material. This success essentially contributes to safeguarding the raw material base of the European industry for LIB production.

Furthermore, considering the lack of Li recycling from secondary sources due to the uneconomic methods, this new approach offers selective leaching where Li can be recovered and subsequently precipitated to obtain Li$_2$CO$_3$. Present pyro- and hydrometallurgical processes developed for LIB recycling are focused on the recovery from other valuable metals, such as Co, Mn, and Ni, among others. Li remains in the solid residue and its recovery is cost intensive (if feasible at all). The current alternative allows for efficiently recovering Li and offers the possibility of recycling other metals, such as Co, Cu, Mn, and Ni since they are not affected by the COOL-Process. Their selective separation can be carried out using different techniques, such as solvent extraction, membrane technologies, and precipitation [10,11,29,30]. Therefore, the developed process shown in Figure 2 contributes to a zero-waste concept, as well as the development of sustainable recycling processes.

4. Conclusions

The current work shows a selective process to mobilize Li from LIB black mass by leaching with supercritical CO_2. Process parameter optimization was done by using a 3^3 Box-Behnken design as DOE. The maximum Li yield of 94.5 wt.% was reached at 230 °C, 4 h, and a water:black mass ratio of 90 mL/g. With a water:black mass ratio of 120 mL/g Li yield was almost quantitative (99.05 ± 0.64 wt.%), yet requiring higher energy input. In contrast to all other studies, only Li and Al were mobilized, which allows for selectively precipitating Li_2CO_3 in high purity without much effort, yielding battery grade-quality (>99.5 wt.%) as the crude product. There is no further refining required. Other valuable metals, such as Co, Cu, Ni, and Mn, remained in the solid residue, which can be separated selectively and recovered by established processes. The CO_2 released in these processes can be fed back to the COOL-Process. Therefore, this holistic approach for LIB recycling comes very close to the goals of zero-waste. Last but not least, this approach allows for simultaneously treating primary and secondary raw materials for Li recycling.

Author Contributions: Conceptualization, S.P., D.K., and M.B.; methodology, R.M.; validation, R.M.; formal analysis, R.M.; investigation, R.M., S.P. and D.K.; resources, D.K. and M.B.; writing—original draft preparation, S.P., D.K., and R.M.; writing—review and editing, M.B., D.K., S.P.; supervision, S.P. and D.K.; project administration, M.B. and D.K.; funding acquisition, M.B. All authors have read and agreed to the published version of the manuscript.

Funding: Financial support by the German Federal Ministry of Education and Research (Grant nr. 033RC020A) is gratefully acknowledged.

Institutional Review Board Statement: Not applicable.

Informed Consent Statement: Not applicable.

Data Availability Statement: It was followed according to MDPI Research Data Policies.

Acknowledgments: Further thanks are owed to Sebastian Hippmann, Andrea Schneider and Jan Walter for conducting TGA/DTA, ICP-OES and AAS analyses and Resource technology & Metal processing Freiberg GmbH (RMF) for providing LIB black mass sample.

Conflicts of Interest: The authors declare no conflict of interest. The funders had no role in the design of the study; in the collection, analyses, or interpretation of data; in the writing of the manuscript, or in the decision to publish the results.

References

1. Applied Market Research. Global Lithium-ion Battery Market. Opportunities and Forecast 2020–2027. Available online: https://www.alliedmarketresearch.com/lithium-ion-battery-market (accessed on 20 October 2020).
2. IEA. Global EV Outlook 2019–Analysis-IEA. Available online: https://www.iea.org/reports/global-ev-outlook-2019 (accessed on 20 October 2020).
3. U.S. Geological Survey. *Mineral Commodity Summaries 2020*; U.S. Geological Survey: Reston, VA, USA.
4. Directive 2006/99/EC of the European parliament and of the council of 6 September 2006 on batteries and accumulators and waste batteries and accumulators and repealing Directive 91/157/EEC. 2006. Available online: https://www.legislation.gov.uk/eudr/2006/99/contents (accessed on 20 October 2020).
5. Velázquez-Martínez, O.; Valio, J.; Santasalo-Aarnio, A.; Reuter, M.; Serna-Guerrero, R. A Critical Review of Lithium-Ion Battery Recycling Processes from a Circular Economy Perspective. *Batteries* **2019**, *5*, 68. [CrossRef]
6. Werner, D.; Peuker, U.A.; Mütze, T. Recycling Chain for Spent Lithium-Ion Batteries. *Metals* **2020**, *10*, 316. [CrossRef]
7. Fan, E.; Li, L.; Wang, Z.; Lin, J.; Huang, Y.; Yao, Y.; Chen, R.; Wu, F. Sustainable Recycling Technology for Li-Ion Batteries and Beyond: Challenges and Future Prospects. *Chem. Rev.* **2020**, *120*, 7020–7063. [CrossRef] [PubMed]
8. Gao, W.; Zhang, X.; Zheng, X.; Lin, X.; Cao, H.; Zhang, Y.; Sun, Z. Lithium Carbonate Recovery from Cathode Scrap of Spent Lithium-Ion Battery: A Closed-Loop Process. *Environ. Sci. Technol.* **2017**, *51*, 1662–1669. [CrossRef] [PubMed]
9. Harper, G.; Sommerville, R.; Kendrick, E.; Driscoll, L.; Slater, P.; Stolkin, R.; Walton, A.; Christensen, P.; Heidrich, O.; Lambert, S.; et al. Recycling lithium-ion batteries from electric vehicles. *Nature* **2019**, *575*, 75–86. [CrossRef] [PubMed]
10. Lv, W.; Wang, Z.; Cao, H.; Sun, Y.; Zhang, Y.; Sun, Z. A Critical Review and Analysis on the Recycling of Spent Lithium-Ion Batteries. *ACS Sustain. Chem. Eng.* **2018**, *6*, 1504–1521. [CrossRef]
11. Chagnes, A.; Pospiech, B. A brief review on hydrometallurgical technologies for recycling spent lithium-ion batteries. *J. Chem. Technol. Biotechnol.* **2013**, *88*, 1191–1199. [CrossRef]

12. He, L.-P.; Sun, S.-Y.; Mu, Y.-Y.; Song, X.-F.; Yu, J.-G. Recovery of Lithium, Nickel, Cobalt, and Manganese from Spent Lithium-Ion Batteries Using l-Tartaric Acid as a Leachant. *ACS Sustain. Chem. Eng.* **2017**, *5*, 714–721. [CrossRef]

13. Lin, F.; Liu, D.; Maiti Das, S.; Prempeh, N.; Hua, Y.; Lu, J. Recent Progress in Heavy Metal Extraction by Supercritical CO 2 Fluids. *Ind. Eng. Chem. Res.* **2014**, *53*, 1866–1877. [CrossRef]

14. Grützke, M.; Mönnighoff, X.; Horsthemke, F.; Kraft, V.; Winter, M.; Nowak, S. Extraction of lithium-ion battery electrolytes with liquid and supercritical carbon dioxide and additional solvents. *RSC Adv.* **2015**, *5*, 43209–43217. [CrossRef]

15. Rentsch, L.; Martin, G.; Bertau, M.; Höck, M. Lithium Extracting from Zinnwaldite: Economical Comparison of an Adapted Spodumene and a Direct-Carbonation Process. *Chem. Eng. Technol.* **2018**, *41*, 975–982. [CrossRef]

16. Rothermel, S.; Evertz, M.; Kasnatscheew, J.; Qi, X.; Grützke, M.; Winter, M.; Nowak, S. Graphite Recycling from Spent Lithium-Ion Batteries. *ChemSusChem* **2016**, 3473–3484. [CrossRef] [PubMed]

17. Liu, Y.; Mu, D.; Zheng, R.; Dai, C. Supercritical CO 2 extraction of organic carbonate-based electrolytes of lithium-ion batteries. *RSC Adv* **2014**, *4*, 54525–54531. [CrossRef]

18. Mönnighoff, X.; Friesen, A.; Konersmann, B.; Horsthemke, F.; Grützke, M.; Winter, M.; Nowak, S. Supercritical carbon dioxide extraction of electrolyte from spent lithium ion batteries and its characterization by gas chromatography with chemical ionization. *J. Power Source* **2017**, *352*, 56–63. [CrossRef]

19. Nowak, S.; Winter, M. The Role of Sub- and Supercritical CO2 as "Processing Solvent" for the Recycling and Sample Preparation of Lithium Ion Battery Electrolytes. *Molecules* **2017**, *22*, 403. [CrossRef]

20. Bertuol, D.A.; Machado, C.M.; Silva, M.L.; Calgaro, C.O.; Dotto, G.L.; Tanabe, E.H. Recovery of cobalt from spent lithium-ion batteries using supercritical carbon dioxide extraction. *Waste Manag.* **2016**, *51*, 245–251. [CrossRef]

21. Bertau, M.; Martin, G. Integrated Direct Carbonation Process for Lithium Recovery from Primary and Secondary Resources. *MSF* **2019**, *959*, 69–73. [CrossRef]

22. Martin, G.; Pätzold, C.; Bertau, M. Integrated process for lithium recovery from zinnwaldite. *Int. J. Miner. Process.* **2017**, *160*, 8–15. [CrossRef]

23. Urbańska, W. Recovery of Co, Li, and Ni from Spent Li-Ion Batteries by the Inorganic and/or Organic Reducer Assisted Leaching Method. *Minerals* **2020**, *10*, 555. [CrossRef]

24. Takacova, Z.; Havlik, T.; Kukurugya, F.; Orac, D. Cobalt and lithium recovery from active mass of spent Li-ion batteries: Theoretical and experimental approach. *Hydrometallurgy* **2016**, *163*, 9–17. [CrossRef]

25. Li, Q.; Fung, K.Y.; Xu, L.; Wibowo, C.; Ng, K.M. Process Synthesis: Selective Recovery of Lithium from Lithium-Ion Battery Cathode Materials. *Ind. Eng. Chem. Res.* **2019**, *58*, 3118–3130. [CrossRef]

26. Aghazadeh, V.; Shayanfar, S.; Hassanpour, P. Aluminum hydroxide crystallization from aluminate solution using carbon dioxide gas: Effect of pH and seeding. *Miner. Proc. Extr. Metall.* **2019**, *1*, 1–7. [CrossRef]

27. Martin, G.; Schneider, A.; Bertau, M. Lithiumgewinnung aus heimischen Rohstoffen. *Chem. Unserer Zeit* **2018**, *52*, 298–312. [CrossRef]

28. Bertau, M.; Eschment, J.; Fröhlich, P. Wertstoffchemie: Die Rohstoffbasis sichern. *Nachr. Chem.* **2017**, *65*, 1206–1209. [CrossRef]

29. Nayl, A.A.; Elkhashab, R.A.; Badawy, S.M.; El-Khateeb, M.A. Acid leaching of mixed spent Li-ion batteries. *Arabian J. Chem.* **2017**, *10*, S3632–S3639. [CrossRef]

30. Or, T.; Gourley, S.W.D.; Kaliyappan, K.; Yu, A.; Chen, Z. Recycling of mixed cathode lithium-ion batteries for electric vehicles: Current status and future outlook. *Carbon Energy* **2020**, *2*, 6–43. [CrossRef]

 *metals*

Review

Challenges in Ecofriendly Battery Recycling and Closed Material Cycles: A Perspective on Future Lithium Battery Generations

Stefan Doose [1,2,*], Julian K. Mayer [1,2], Peter Michalowski [1,2] and Arno Kwade [1,2]

1   Institute of Particle Technology, TU Braunschweig, 38104 Braunschweig, Germany; j.mayer@tu-braunschweig.de (J.K.M.); p.michalowski@tu-braunschweig.de (P.M.); a.kwade@tu-braunschweig.de (A.K.)
2   Battery LabFactory Braunschweig, TU Braunschweig, 38106 Braunschweig, Germany
*   Correspondence: s.doose@tu-braunschweig.de; Tel.: +49-531-391-94668

Abstract: The global use of lithium-ion batteries of all types has been increasing at a rapid pace for many years. In order to achieve the goal of an economical and sustainable battery industry, the recycling and recirculation of materials is a central element on this path. As the achievement of high 95% recovery rates demanded by the European Union for some metals from today's lithium ion batteries is already very challenging, the question arises of how the process chains and safety of battery recycling as well as the achievement of closed material cycles are affected by the new lithium battery generations, which are supposed to enter the market in the next 5 to 10 years. Based on a survey of the potential development of battery technology in the next years, where a diversification between high-performance and cost-efficient batteries is expected, and today's knowledge on recycling, the challenges and chances of the new battery generations regarding the development of recycling processes, hazards in battery dismantling and recycling, as well as establishing a circular economy are discussed. It becomes clear that the diversification and new developments demand a proper separation of battery types before recycling, for example by a transnational network of dismantling and sorting locations, and flexible and high sophisticated recycling processes with case-wise higher safety standards than today. Moreover, for the low-cost batteries, recycling of the batteries becomes economically unattractive, so legal stipulations become important. However, in general, it must be still secured that closing the material cycle for all battery types with suitable processes is achieved to secure the supply of raw materials and also to further advance new developments.

Keywords: lithium-ion battery; LIB; battery recycling; mechanical recycling processes; hydrometallurgy; pyrometallurgy; battery generation; circular economy; solid state batteries

Citation: Doose, S.; Mayer, J.K.; Michalowski, P.; Kwade, A. Challenges in Ecofriendly Battery Recycling and Closed Material Cycles: A Perspective on Future Lithium Battery Generations. *Metals* **2021**, *11*, 291. https://doi.org/10.3390/met11020291

Academic Editor: Yoshikazu Ito
Received: 31 December 2020
Accepted: 3 February 2021
Published: 8 February 2021

Publisher's Note: MDPI stays neutral with regard to jurisdictional claims in published maps and institutional affiliations.

1. Introduction

The goal of economical and sustainable battery cell production remains a key element on the way to establishing electromobility as a green technology of the future [1]. Sustainable process management and development also includes the economic recycling and recirculation of materials used in cell production with a simultaneously low energy input, which leads to a reduction of the ecological CO_2 footprint in battery cell production [2–5]. Therefore, the establishment and sustainable further development of an internationally leading, competitive battery cell production must go hand in hand with the development of appropriate recycling technologies [6–9]. The recycling technologies must be flexible and adaptable to future production technologies and especially materials that are processed in the future with regard to new battery generations [10].

Moreover, closing the material cycles for batteries on the basis of scalable production and recycling technologies is a central component for a CO_2-reduced or CO_2-neutral battery cell production and thus for electromobility (e.g., achieving the "Green Deal" goal of the

European Union (EU)). Only closed material cycles in batteries can enable a conversion from carbon-based energy sources to sustainably produced electrical energy in ecological, economic, and social terms [7,11,12]. To achieve closed loop material cycles, appropriate recycling technologies must be developed. For example, according to the new battery directive proposal of the EU Nr. 2019/1020, 95% of cobalt (Co), copper (Cu), and nickel (Ni) as well as 70% of lithium (Li) have to be recovered from spent lithium-ion traction batteries by 1 January 2030 [13]. Moreover, according to the report on the circular economy of traction batteries published by the Circular Economy Initiative Deutschland of acatech (National Academy of Science and Engineering in Germany) 90% of Co, Ni, and Cu as well as 85% of Li should be recycled from spent lithium battery systems by 2030. In addition, a recycling rate of 70% of the entire battery should be aimed for [14]. In a worldwide comparison, the EU sets very high requirements with the Battery Directive. In the USA there are no generally applicable requirements for the return of lithium ion batteries (LIBs). However, voluntary consortia (e.g., the End of Live Vehicle (ELV) Solutions consortium) work together here to close material cycles. China follows a similar approach to the EU. Producers are encouraged to take back batteries that have been put into the market and to return them to the materials cycle [15]. Other countries in Asia (South Korea, Japan) are pursuing similar goals as the EU (South Korean RoHS/ELV/WEEE Act, 2007 and Japan's End-of-Life Vehicle Recycling Law).

For the recycling of lithium-ion batteries today, usually at first, a deactivation of the battery system is realized, which is followed often (but not compulsory) by a dismantling of the battery system down to the cell modules (or more seldom individual cells). The deactivation can be achieved by full electrical discharge and subsequent short circuiting, by treatment in a saline solution, or by pyrolysis (high heat treatment) of the battery systems at temperatures of more than 200 °C [16,17]. Afterwards, in general three types of process technologies are used, which are combined in a different manner: mechanical, hydrometallurgical, and pyrometallurgical processes [7,17–22]. Figure 1 gives an overview of different possibilities to combine the different process types. The different process steps can be applied in different sequences and above all in different processing depths [19,23]. One lean way is deactivation by pyrolysis, pyrometallurgical processing and slagging. Pyrometallurgical processes are well established for processing primary materials and can achieve high recovery yields concerning the metals cobalt, nickel, and copper but they show challenges regarding the recovery of lithium. Therefore, to recover lithium and manganese, pyrometallurgical processes have to be combined with hydrometallurgical processes for processing the slag. Overall, a relatively small overall recovery of the batteries material can be expected in this case due to graphite, polymers, and electrolyte being burned, although a very high recovery of Ni, Co, and Cu is possible [24]. A more elaborate way puts together another combination of processes. Here, the battery cells or modules are discharged in the first step, for example, before they are mechanically crushed [25]. Subsequently, the black mass or the shredded battery material is pyrometallurgically processed before it goes into a final hydrometallurgical step. Here, for example, the degree of mechanical processing can be varied [18,24,26]. The amount of recovered materials increases, as polymer components as well as aluminum can also be recovered. Furthermore, after deactivation and mechanical processing, there is also the possibility of proceeding directly to hydrometallurgical processing (as it was proposed for the LithoRec process [20,27,28]). This route enables early recovery of the polymer battery components as well as Cu and Al. In addition, hydrometallurgical processing can also recover Mn that cannot be recovered by pyrometallurgical processes. In addition to the recovery of the individual substances such as Ni and Co, also, the direct reconditioning of the cathode material by hydrometallurgical treatment is carried out on an industrial scale. Avoiding pyrometallurgical processing during battery preparation theoretically reduces the energy requirement and thus improves the ecological footprint [29]. Furthermore, other process routes are theoretically possible, but they shall not further be discussed here; they are discussed in more detail in relevant literature.

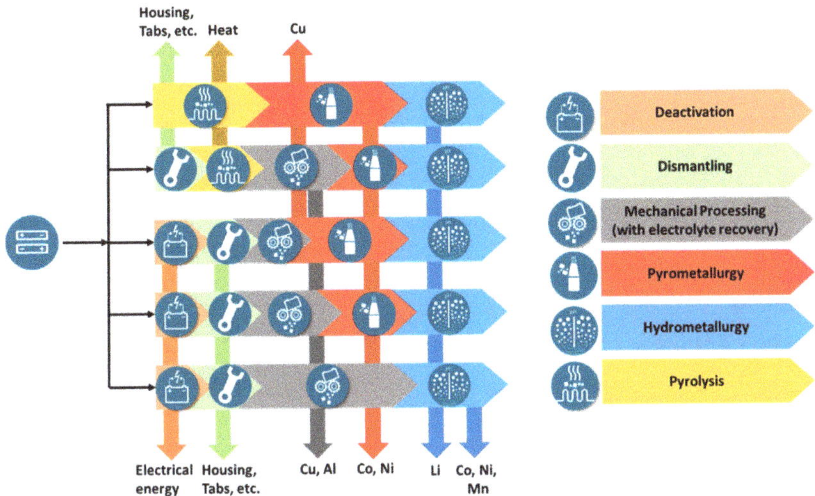

Figure 1. General overview of some potential recycling process chains in different combinations.

Overall, the overview of the recycling processes shows that many different recycling process routes are possible and in industrial use or under development at pilot scale at least [17,28–30]. As a requirement for a future process chain, besides achieving high recovery rates of more than 90% or even 95% [13,14] and in parallel sufficient material purities for further usage as battery material, it is therefore to be set that it should be highly flexible in order to achieve the most energy-efficient multi-material recovery possible. To reach this goal, mechanical, thermal, and chemical process steps are to be used and combined in different ways. Future recycling processes for Li-ion batteries must not only be able to process new materials but should also replace energy-intensive processes currently used for classic Li-ion batteries with low-energy, environmentally friendly process steps.

2. Lithium Battery Development

Today, lithium-ion batteries with liquid electrolyte dominate the market for electromobility. They are also used in portable devices as well as in the field of stationary energy storage. On the cathode's side, particles of lithium transition metal oxides such as lithium nickel manganese cobalt oxide (NMC, $LiNi_xMn_yCo_zO_2$) and lithium nickel cobalt aluminum oxide (NCA, $LiNi_xCo_yAl_zO_2$) or also lithium metal phosphates, especially lithium iron phosphate (LFP), are mainly used. In the case of the NMC and NCA materials, the nickel content increases steadily, and the cobalt content decreases continuously. Polyvinylidene fluoride (PVDF) is currently used as a standard binder on the cathode side. On the anode's side, graphite is usually used as the anode material; in rare cases, lithium titanium oxide (LTO) is used. Moreover, the first cells with the addition of very small amounts of silicon to the graphite are offered by the cell manufactures. A mixture of carboxymethyl cellulose (CMC) and styrene butadiene rubber (SBR) usually acts as a binder within the anodes. As shown in Figure 2 within the near and medium-term future, the following developments in next-generation lithium battery technology are expected:

1. **High-performance lithium-ion batteries** with liquid electrolyte, but anodes with higher content of nanosized silicon and cathodes with minimal or no cobalt content [31,32].

2. **Cost-efficient lithium-ion batteries** with liquid electrolyte, graphite anode, and cathode material based mainly on iron and/or manganese and only small amounts of nickel and eventually cobalt. In addition to lithium-ion based batteries, also sodium-ion based batteries are under development, which could replace at least partly the named cost-efficient lithium-ion batteries [33,34].

3. **Solid-state lithium batteries** with <u>lithium or lithium-free anode structure</u> (eventually graphite anode as intermediate stage) and solid-state electrolytes on the cathode side and as separator [35].
4. **Lithium sulfur batteries** with <u>lithium anode</u> and a cathode made out of sulfur–carbon composites [36].

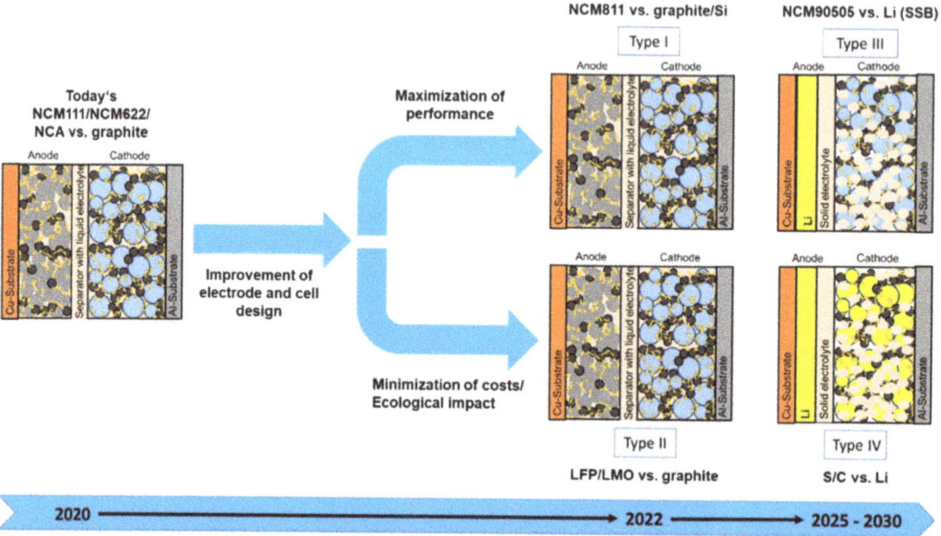

Figure 2. Diversification of lithium battery technologies.

An overview of the materials and their potential contents in the different battery types is given in Table 1. According to Table 1, it can be concluded that with the use of the upcoming battery generations, the composition in terms of recyclables will also vary. While the Ni content will increase significantly in type I, the presence of Fe or Mn is expected in type II. In both types, the liquid electrolyte including Li conducting salt is also expected to have high potential for recovery. However, for type II, also sodium instead of Li has a significant potential. In contrast, for type III and eventually also type IV, solid electrolytes are employed for the separator and as electrolyte within the composite cathode. The solid electrolytes can have an oxide, sulfide, or polymer nature, whereby the compositions and properties can be highly variable (Table 2).

Table 1. Material contents of different battery generations (with a focus on lithium-based batteries).

Amounts of Materials in Each Battery Type (%)	Type I (High Ni-NMC) [37,38] (%)	Type II (e.g., LFP) [37,38] (%)	Type III (SSB) [39] (%)	Type IV (LiS) (%)
Housing	22 (cylindrical)	27 (cylindrical)	27 (pouch)	34 (pouch)
Cathode current collector	7 (Al)	6 (Al)	4 (Al)	5 (Al)
Cathode active material	26 (high Ni content, i.e., NMC 811)	25 (e.g., Fe content, i.e., LFP)	42 (high Ni content, i.e., NMC 90505)	21 (S-C composite)
Anode current collector	17 (Cu)	10 (Cu)	10 (stainless steel)	14 (Cu)
Anode active material	15 (C/Si)	13 (C)	3 (Ag-C composite)	7 (Li)
Electrolyte	10 (liquid)	16 (liquid)	13 (solid, Li_6PS_5Cl)	19 (solid, Li_6PS_5Cl)
Separator	3	3		

Table 2. Typical components and properties of separator technologies.

	Classic Electrolyte	Oxides	Sulfides	Polymeres
Aggregate state	liquid	solid	solid	solid
Conducting salt	1M LiPF$_6$			LiTFSI
Liquid organic solvents	EC, EMC, DMC, PC			
Solid electrolyte compounds		LLZO, LATP	e.g., Li$_3$PS$_4$, Li$_6$PS$_5$Cl, Li$_{10}$SnP$_2$S$_{12}$ [40]	e.g., PEO, PC, PS, and variations [41]

From these points, it is clear that next-generation technologies will include much fewer critical components, such as cobalt, but also new materials such as silicon, or even germanium. In the medium to long term, metallic lithium is expected on the anode side in case of type III and especially type IV. The use of lithium anodes will require the use of polymeric, sulfidic, and/or oxidic solid electrolytes as separator and as electrolyte on the cathode's side. However, more and more also lithium-free anode structures are shown in the literature [42,43]. In this case, the lithium from the cathode active material moves to the anode side and is deposited on the lithium-free anode layer. In order to minimize the formation of dangerous lithium dendrites, also lanthanum, titanium, zirconium or phosphorus in the solid electrolyte area are presented. Lithium is still indispensable for the time being, but the commercialization of sodium ion or other batteries in the near future is also conceivable. Overall, a further increase in the complexity of the battery cells developed and produced and the use of other economically strategic raw materials can be expected. The recovery of these should be tackled urgently in order to establish sustainable battery production with respect to the increasing sales figures.

In view of the rapidly evolving battery technologies, the best possible recycling routes for the newly developed battery cells should be determined and their recyclability assessed at an early stage. In addition to battery cells with a higher proportion of silicon, pure lithium, and/or solid-state electrolytes, the use of novel binders and fibrous additives or the increase in the adhesive strength of the electrodes can also significantly influence the recycling processes. In addition, active material mixtures are increasingly being used on the cathode's side, so that iron, among other substances, enters the metallurgical material preparation process. Moreover, a direct reconditioning of the active material gets very difficult to impossible. Sustainable recycling concepts must be evaluated in advance by process-based economic and ecological models for the entire life cycle.

Of great interest is the assessment of the coming battery generations with regard to the change of a future circular production of batteries, the recycling process itself, and the hazard potential within the recycling process. Accordingly, in the first step, the three criteria for today's LIB are briefly presented, and based on this, the potential challenges posed by the introduction of the different cell generations mentioned above are assessed.

3. State of the Art Recycling Processes for Lithium-Ion Batteries

As shown in Figure 1, different process chains and routes are developed for the recycling of today's lithium-ion batteries. Today, the most common ones are a combination of mechanical and hydrometallurgical processes as well as a combination of dismantling and pyrometallurgical processing. Casing and connection materials of the battery pack and module are removed in advance or even after comminution in the processes presented and fed, in the normal case, to the conventional recycling methods for aluminum, iron, polymers, and others.

3.1. Mechanical–Hydrometallurgical Recycling Technology and Challenges

A common recycling route to be found combines mechanical processing with direct hydrometallurgical processing of the batteries. The mechanical processing can be fulfilled in a dry or a wet mode. Moreover, before mechanical treatment, the battery system has to be deactivated by complete discharge and usually short circuiting and dismantled down to at least the module level. However, it is also reported that the dismantling goes down to the electrode level in order to separate anode and cathode materials already before the mechanical treatment.

The economically and ecologically attractive mechanical processing of the battery cells is carried out down to the level of active and inactive materials, and it can be combined with an evaporation of electrolyte components [20,27,28,44]. Discharged battery cells and modules are comminuted under inert atmosphere in a shredder process or under water [45] Then, the components are separated by classification and sieving processes. To achieve a high separation quality of the materials, several steps can be run consecutively with different process settings [20,28]. Furthermore, some alternative methods also include electrolyte recovery steps. The pre-dried shredded material is divided in the separation process into metallic components (casing material), current collector foils (Al and Cu) black mass (Li, Co, Ni, and Mn oxides, graphite, PVDF, carbon black, and impurities) and organic components (separator) [19,25]. However, today, the mechanical processes are not able to achieve practical recycling rates of more than 95% regarding cathode active materials and the necessary purities. Subsequently, the black mass is processed without any intermediate steps in the hydrometallurgy.

A particular challenge without pyrolysis or pyrometallurgical processes is the handling of the remaining fluorine components, which can be deviated on the routes including pyrolysis or pyrometallurgical processes due to high temperatures. Hydrometallurgy is a process at low temperatures in aqueous phases and can be performed in three major steps Leaching is the first step and describes the dissolution of metals via the usage of acids or bases. Typically, at first, the black mass is dissolved in NaOH and afterwards leached using H_2SO_4, with initial impurities such as iron, aluminum, and copper precipitated by small amounts of NaOH and sieving [46,47]. The purification is the second step in the process chain where the metals are separated and purified via e.g., selective chemical reactions. The last step is the final recovery of the metals or the salts. This can be done by means of e.g., crystallization, ionic precipitation, electrolytic deposition, or further methods For example, the individual metals can be further precipitated as sulfate salts by adding NaOH or other basic agents and increasing the pH. Selectively, this can be controlled by considering the different solubilities of the metal salts [46,47]. Hydrometallurgical process steps are capable of producing high product purities. However, plants for counter-flow are larger than those used in pyrometallurgy and require a larger financial investment volume for their construction [21,48,49]. In general, important challenges are to achieve the demanded high recovery rates and, at the same time, high material purities.

An example of the mechanical–hydrometallurgical process chain is the LithoRec process [19,20,28], which is commercialized in similar form by Duesenfeld and Redux (Figure 3). The developed process has a high potential to close the loop of the circular economy in the battery production as well to reduce the CO_2 footprint in the battery production and utilization phase due to the high recovery rates of the recyclables and the low energy consumption. Maximizing the recycling rate was a core objective of the projects Among other things, a process was developed in which, depending on the active material 85 to potentially over 95% of the lithium can be recovered by mechanical (includes drying step) and hydrometallurgical means (leaching and subsequent precipitation of the lithium)

Figure 3. Developed process chain of mechanical/thermal processing of the LithoRec process prior to hydrometallurgical steps [20,28].

Another example is the process used by Accurec GmbH® (Krefeld, Germany). Here, the batteries are discharged prior to mechanical processing and subjected to vacuum thermal pretreatment (pyrolysis) [44,50]. In this process step, volatile organic (electrolyte) as well as polymeric components (separator) and halogenic compounds can be pyrolyzed at 200–400 °C and low pressure. In the further course, the pyrolyzed mass is subjected to mechanical processing, pyrometallurgical processing, and subsequently hydrometallurgical processing to recover also the lithium. The developed process also has a noticeable potential to close the loop of the circular economy. In addition, the company Erlos GmbH represents another interesting approach in the recycling of lithium-ion batteries, in which the battery cells are separated by type (anode, separator, cathode, casing). Subsequently, in the example, the entire cathode coating is separated from the aluminum substrate foil. Then, the cathode coating is separated into its components—active material, binder, and carbon—by wet chemical processes (leaching), and the active material is reconditioned. This step allows the active materials on the cathode and also anode side to be reused after the treatment without having to perform a resynthesis based on the purified substances, i.e., metals. However, the electrolyte, binder, and conductive additives are lost for reuse. Nevertheless, the active material chemistry is retained and cannot be adapted to new requirements [47,51].

3.2. Pyrometallurgical Recycling Technology and Challenges

Another approach for the recycling of lithium ion batteries is the pyrometallurgical (mechanical–pyrometallurgical) way. In this process, the battery cells or modules can be placed in a direct pyrometallurgical step or they can be mechanically processed in a first step (analogous to the mechanical–hydrometallurgical), and the obtained valuable black mass is fed into the pyrometallurgy. Pyrometallurgical processing involves high-temperature processes such as melting and roasting to produce battery slag [52]. A pyrometallurgical recycling process of LIBs starts with an initial heating in the temperature range of 150–500 °C, during which electrolyte components and organic solvents are removed. Subsequently, a high-temperature process with temperatures up to 1450 °C using reducing agents (graphite, coke, NaHSO$_4$, CaCl$_2$, or NH$_4$Cl) is carried out in the furnace to obtain battery alloy and slag (Li$_2$O as well as Li$_2$CO$_3$) as products [17,53,54]. Due to the different meting points, the metals Ni, Co, and Cu can be individually recovered. However, lithium, manganese, and aluminum get into the slag or the kiln dust. The battery alloy produced in this way contains the valuable materials (such as Co, Cu, and Ni) and can then be processed hydrometallurgically. Lithium and other battery components can only be recovered at great expense or not at all by this process. The advantage of this process technology is the comparatively high robustness against changing feed material and a comparatively small plant size with given throughput. The disadvantage is the comparatively lower overall quantity of recyclable materials. However, the critical metals are recycled at a high recovery rate. The processes generate only intermediates that have to be purified by further steps to enable reuse (e.g., further processing by hydrometallurgical steps). In addition, they show low economic efficiency when low concentrations of recyclable materials (e.g., Co, Cu, and Ni) are present [24,52,55].

The process from Umicore Valéas™ (Bruxelles, Belgium) can be cited as an example of the pyrometallurgical processing route. As a great advantage in advance can be mentioned the great robustness of this way to various types of batteries [17]. Prior to thermal processing, the batteries are dismantled, whereby plastic and metal housing parts in particular are removed, and the cells are exposed [56]. Then, these are pyrolyzed in the shaft furnace at three different process temperatures (400–1450 °C). The alloy obtained contains valuable materials such as Cu, Co, and Ni. In turn, the slag contains Li. Both are subsequently hydrometallurgically processed in a leaching process to recover the valuable materials contained for reuse [17].

3.3. Potential Hazards of Lithium-Ion Batteries in Recycling Processes

Battery systems of electrically powered vehicles (e.g., EV, PHEV, HEV) contain chemically stored energy. The systems contain energy quantities of up to 20–100 kWh (TESLA Model S) and reach system voltages of 300–800 V [2,57]. The associated hazards are particularly relevant when the battery is handled separately from its application, such as in the dismantling and the recycling process [28]. In recycling, due to mechanical process steps in particular, this should be of major importance. Today's and future battery generations combine built-in active materials with high energy densities and partly highly inflammable electrolytes. In normal use, particularly external factors such as short circuits (internal and external), high temperatures or mechanical deformation can trigger critical events and lead to the thermal runaway [58]. Thus, the hazards that can be caused by these faults from the battery in recycling can be divided into three main areas: (1) electrical, (2) thermal, and (3) chemical hazards. However, in the real case, they never occur alone but are usually a combination of hazards. Therefore, in general, a recycling process should be designed and engineered in such a way that the risks can be avoided as far as possible.

- Direct electrical shock is one of the main types of hazard when handling batteries. A direct electric shock can cause severe skin burns at the point of entry and exit, depending on the current, voltage and type of current (AC/DC). In addition, the paralysis of muscles and, in the worst case, electrolysis of the blood may occur.
- The thermal hazards of a battery cell are mainly due to the electrolyte components used. The main components of the current electrolytes are a mixture of organic solvents (e.g., ethyl carbonate, EC; ethyl methyl carbonate, EMC; and others) and a conducting salt (lithium hexaflourophosphate, $LiPF_6$). The carbonates used are highly flammable hydrocarbons. The reaction of $LiPF_6$ with water can result in the highly toxic and corrosive hydrogen fluoride (HF) [59,60]. Partially high evaporation rates of electrolyte components in moderate temperature ranges and partially closed process rooms can lead to explosive mixtures in combination with an oxygen-containing atmosphere [61]. In addition, if higher temperatures have occurred, the reaction products of reactions of the different components of a battery cell can also lead to fire and explosions in the processes (hydrogen, methane, carbon monoxide).
- Chemical hazards of battery cells are mainly determined by the ingredients but also by accessible reaction products in case of failure. The materials and products have irritating, human-toxic, carcinogenic, respiratory, environmentally harmful, and water-damaging effects. Particularly noteworthy in this case is the active material of the cathode. The cathode active materials consist mainly of lithium transition metal oxides such as NMC and NCA or also lithium metal phosphates, especially lithium iron phosphate (LFP). Especially the significant amounts of the heavy metals nickel and cobalt are both known to be carcinogenic and toxic for mammals. In addition, the small particle size (10–15 μm) of these can increase the exposure through the human respiratory system.

Uncontrolled temperature rises up to the so-called thermal runaway of the battery cells can be caused by external as well as internal short circuits if a critical temperature is exceeded, which can be generated by handling in recycling. These effects can be triggered during the process by mechanical penetration of foreign bodies, internal cell defects, or external arrester/electrode contacts [62–65]. In addition, overloading or high ambient temperatures can cause the thermal runaway to occur. This type of hazard mainly occurs during the storage and discharge of spent batteries and not during the recycling process. The thermal runaway involves the reaction of cathode, anode, and electrolyte (Figure 4). In the first step, the process runaway begins with the decomposition of the solid electrolyte interface (SEI) on the anode. At a critical temperature of about 90–120 °C in a battery, the chemical decomposition processes start. Under an exothermic reaction process, the formed SEI is decomposed and leads to different gaseous reaction products (e.g., carbon dioxide, ethane, ethane) [66]. The energy released from the first exothermic reactions leads to further heating of the battery cell and can dissolve the subsequent chain reaction processes. Then, the intercalated lithium begins to react with the electrolyte.

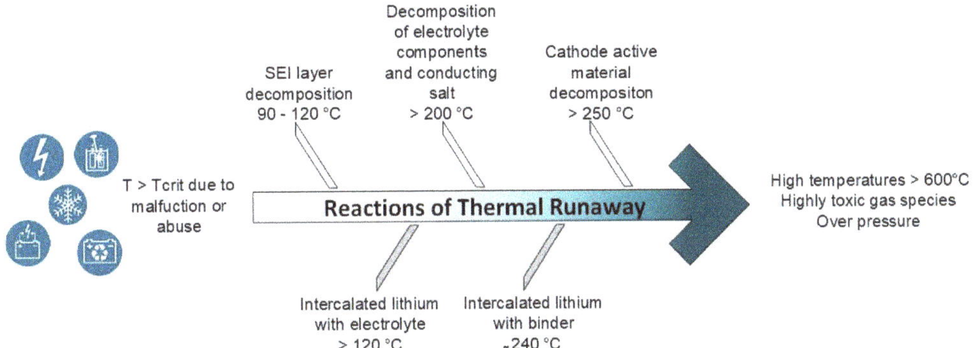

Figure 4. Schematic sequence of thermal runaway between different battery cell components after a critical cell temperature has been exceeded.

Electrolyte decomposition starts at approximately 200 °C with the formation of CO_2, hydrogen fluoride, ethene, and other hydrocarbons containing fluoride [67,68]. The exothermic reactions of the embedded lithium with the binder and the decomposition of the cathode active material start at 240 °C and 250 °C, respectively [69]. In actual recycling processes, the steps that still involve electrolytes are to be regarded as particularly critical. To reduce hazards related to the electrolyte, it is advisable to remove the electrolyte as early as possible in the recycling process chain and, in an ideal case, to purify and return it back into the material cycle. However, how valuable a repeated use of the electrolyte is regarding costs and environmental protection is an open question. If a primarily mechanical process strategy is used, comminution can be carried out under an inert gas atmosphere (e.g., nitrogen), followed by evaporation of the electrolyte at moderate temperature increases and/or at negative pressures [20,28] or even by further extraction methods [70,71]. Then, the evaporated electrolyte components can be condensed out as in the Lithorec process. In the case of upstream pyrolysis, the electrolyte can already be removed in advance at temperatures of approximately 200–400 °C in specialized ovens (Accurec GmbH®). The comminution takes place downstream. The pyrometallurgical process route can handle the electrolyte removal even more easily. Due to the high temperatures during the processing of the battery materials (up to 1450 °C) and the addition of reducing agents, the electrolyte can be safely removed from the cell materials [17]. Moreover, at temperatures above about 650 °C also, the binder PVDF can be decomposed and removed under the challenge of handling hydrofluorocarbons. In summary, an improperly handled battery cell represents some potential hazards. However, with appropriate and orderly process control, these risks can be reduced to a minimum, and safe process control should be ensured.

4. Circular Economy in the Context of Battery Production

Establishing closed material cycles for batteries on the basis of scalable production and recycling technologies is a central component for CO_2-reduced or CO_2-neutral battery cell production and thus for electromobility, the provision of energy in household and handicraft appliances, as well as for the stationary storage of renewable energies [7,72,73]. As a matter of fact, closed material cycles in batteries are the only way to convert carbon-based energy sources into sustainably generated electrical energy in ecological, economic, and social contexts. Consequently, a circular economy for traction batteries, i.e., lithium-ion battery systems is demanded by the European Union as written in the Battery Directive 2006/66/EC and [13] as well as the Circular Economy Initiative Deutschland (CEID) of the National Academy of Science and Engineering in Germany [14]. In addition, many research studies are being done on this topic while highlighting the challenges that still exist and need to be overcome [17,21,52,72]. Such closed circles improve the sustainability of lithium-ion batteries; they especially decrease the carbon footprint, decrease the material costs, and ensure secondary material resources for Europe [73,74]. However, in general when closing material cycles, it should be noted that the energy used is in proportion to the products. With regard to recycling, mechanical treatments are energy-wise recommended as they are less energy-intensive than metallurgical processes. Examples of specific consumption parameters of energy are electrical energy and wastewater. Pyrometallurgical processing requires 4.68 MJ of electrical energy per kg of battery; hydrometallurgical processing requires 0.125 MJ of electrical energy per kg of battery [75,76]. In addition, approximately 3.76 L of wastewater are produced per kg of battery during hydrometallurgy. However, as described above, a combined process strategy is necessary for good product properties and qualities [29].

A possible implementation of such a closed-loop production is shown in Figure 5. After the utilization phase, batteries are mechanically disassembled, whereby the safety of the processes is an important aspect of development. Depending on the level of detail of the mechanical process chain, components such as copper and aluminum foil can already be separated. Pyro- or hydrometallurgical processes must be applied for further purification at the latest after the active materials, the so-called black mass, have been exposed. Depending on the process, graphite and binder can also be separated. Finally, metal salts are to be precipitated and re-synthesized to produce new active materials [48,52]. Alternatively, the anode and cathode active materials can be reconditioned by means of purification processes, lithium enhancement, and functionalization [47,51,77–79]. Direct reconditioning has the advantage of being a fast and less energy-consuming process, but it cannot directly adapt the cathode chemistry to new developments. For short life cycle phases or rejects from production (e.g., losses of battery cells during formation), direct reconditioning can be very attractive [80].

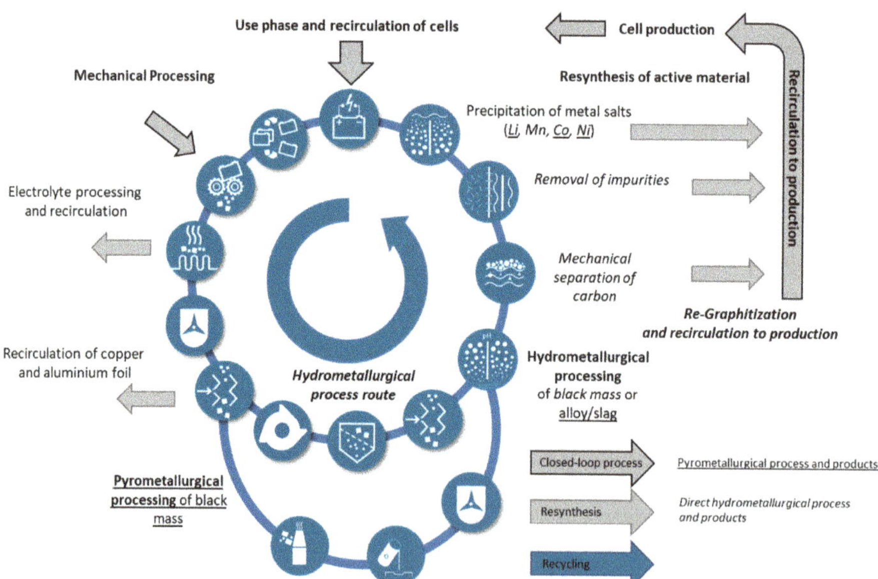

Figure 5. Exemplary approach for the cycle of circular battery production with the presentation of both established processing routes, direct hydrometallurgical and upstream pyrometallurgical processing after previous mechanical processing.

In order to realize a circular battery cell production, a number of technological and organizational prerequisites must be created:

- Almost 100% of end-of-life batteries must be collected and recycled at the latest after any second-life application [73]. The expected lifetime of batteries in the automotive sector is at least 8 years, and, thus, in the mean, it tends to be more than 10 years. In the future, the lifetime will probably increase further. However, this value is highly dependent on the loads (fast charging, temperatures, etc.) [81]. Before recycling, it is important to check whether a second-life application can be reasonable.

- The condition of the batteries, especially the material composition, must be documented for the subsequent recycling process [82]. Alternatively, a uniform interface for reading out specific battery data could be implemented in the systems.
- With regard to the material composition of batteries, robust recycling processes must be developed and industrially implemented, especially with regard to future battery generations.
- The re-synthesis and eventual reconditioning of the active materials, such as Si-containing anode materials and cathode materials from lithium mixed oxides, has to function on a large scale without any loss of performance of the later battery. The synthesis processes should be as robust as possible against material contamination [83,84].
- The design of the battery cells should not only be based on requirements such as performance, cost, and safety, but also on sustainability and thus recyclability [85].
- The production of the battery cells themselves must be ecologically and economically sustainable [2].
- For objective evaluation of the individual technologies, new software tools should be developed for an "as objective as possible" cost and environmental life cycle assessment of different battery cells and process technologies.

In order to get an overview of the technological difficulties, an example is given below to obtain a recycling rate of 95% from a recyclable material in five process steps. If each of these process steps has a yield of 99%, a total yield of 95.09% can just be obtained with all five steps. Thus, it is very important to use few very good process steps in the reprocessing to generate the highest possible yield.

5. Perspective on Recycling and Circular Economy of Future Battery Generations

The focus in future developments is a highly energy-efficient multi-material recovery with recovery rates of at least more than 90%, most probably more than 95%, which should be adaptable to variations of respective input streams and current output demand. This results in the requirements for future recycling processes, which must either have a high degree of flexibility or which are focused on certain battery types, which in this case have to be sorted efficiently. In addition, the process routes to be developed must be specialized with regard to the individual components to be purified, e.g., electrolyte, electrode components, and active materials. To achieve this, combinations of mechanical, thermal, and chemical process steps have to be used and interconnected. These processes have to be designed also with regard to novel battery materials and chemistry (e.g., solid-state batteries). They should over a long term replace the current energy-intensive processes for classical LIB. This will render future recycling processes significantly more environmentally friendly and less energy-intensive.

Various challenges and requirements for recycling processes, closed material cycles (circular economy), and safety aspects arise from the large-scale use of the four battery types expected to dominate the future as mentioned in Section 1.

Battery type I (Si-containing anodes, Ni-rich cathodes, liquid electrolyte): Closed-loop production is economically attractive because of the high nickel content, and regeneration of the anode materials should be very feasible both ecologically and economically. The recycling processes known today, using hydrometallurgy alone or a combination of pyro- and hydrometallurgy, should be very well able to process these battery cells, which will be available on the market in the near future. However, the recycling process chains must ensure high recovery rates of 95% and higher for critical metals and most possible lithium. For active materials with high nickel and low cobalt content and already high specific capacities such as NMC 811, also a reconditioning of the active material can be attractive because a further increase in specific capacity due to new material developments is probably relatively small. These reconditioned active materials could also be used for the production of battery type II in the near future, especially if the performance is slightly reduced due to the multiple usage. Regarding safety, battery type I cells will be more sensitive to mechanical damage and external short circuits due to the high nickel content and the presence of nanoparticulate silicon. This increases the risk of thermal runaway and resulting fires.

Battery type II (Graphite anodes, Mn- and Fe-rich cathodes, liquid electrolyte): This type of battery, which is mainly used in stationary applications and low-cost mobility concepts, can be well recycled with the existing processes such as the previous one. The preferential use of Mn and Fe leads to a low-cost batteries and thus, economically less attractive recycling processes. Therefore, regulations have to be set up to ensure that these batteries are recycled at the EoL. Due to the high risk of contamination, cells with active materials containing iron must most probably be separated from batteries of other types even before recycling. Otherwise, complex purification processes must be used to separate the transition metals (Ni, Co) from the iron especially in case of hydrometallurgical process routes. For such batteries, the reconditioning of the active materials is probably also a cost-efficient and sustainable option for material recovery.

Battery type III (Lithium or lithium-free anodes, Ni-rich cathodes, solid electrolyte): The solid-state electrolytes used in this type of battery will lead to obstacles in establishing a closed-loop economy [35]. It makes it difficult to recover the individual materials used in the battery in high degrees of purity. On the anode's side, thin lithium foils or, in the future, metallic lithium deposited on 3D structures (depending on the transfer of current research results to industry) or even lithium-free 3D-structures are used. These anodes require that the mechanical treatment of the battery cells be performed in an inert gas atmosphere. Furthermore, this results in additional equipment and work safety requirements during processing. This is also due to additionally required hydrometallurgical steps of solving and recovering lithium in the form of salt (LiOH), which leads to the formation of gaseous H_2. When considering the possible recycling processes for cathodes and separators, a distinction must be made between the solid-state electrolytes used:

1. The use of polymer electrolytes both in the cathode and in the separator results in a complex task of separation of the individual materials used. As it stands now, there are two options: On the one hand, the polymer electrolyte can be burned using thermal processes, and the materials exposed can be further processed in a similar way to classical LIB. However, the polymer-type solid electrolyte is lost. On the other hand, complex wet chemical processes can be chosen. Here, the electrolyte is dissolved in a suitable solvent, and the polymer can be recovered in the process, but it is not known today if this can be fulfilled with a sufficient quality or purity, respectively. However, the wet chemical route is not expected to be economically or environmentally viable, despite an associated increase in recycling yield.

2. When using sulfidic solid electrolytes, the formation of toxic hydrogen sulfide compounds must be avoided during recycling. A mechanical separation of the solid sulfidic electrolyte from the active material is very difficult and probably not possible with the required purity or separation efficiency, respectively. Therefore, a reconditioning of the solid electrolyte and the active materials seems to be not possible from the today´s experience. In addition, the frequent use of other elements, such as germanium, makes it more difficult to recycle these substances in a pure form by hydrometallurgical processes. Therefore, complex hydrometallurgical processes are probably required to recover the different materials.

3. If oxidic solid electrolytes are used in the separator and/or cathode, the electrolyte particles will be firmly sintered together. Thus, mechanical separation is associated with significantly higher costs, so that pyrometallurgical treatment of entire cells or at least larger cell fragments probably becomes more attractive compared to a mechanical/hydrometallurgical process.

Battery type IV (Lithium anodes, S/C-containing cathodes, liquid or solid electrolyte): Apart from lithium, copper, and aluminum, no further valuable materials are used in lithium–sulfur batteries. This results in the task of efficiently separating these three materials from the mechanically treated battery mixture. Presumably, the recovery of sulfur and carbon for direct use in a battery is not practical, since there are other inexpensive and reliable sources of sulfur and carbon that promise less effort and the needed higher purity. Accordingly, a closed-loop production of this cell type does not necessarily make sense economically and probably also ecologically according to the current state of knowledge. If solid instead of liquid electrolytes are used, the challenges described for battery type III apply additionally.

In conclusion, with the exception of battery type II, all other battery types that potentially lead to a higher energy density in Wh/L (battery types I and III) or an increased specific energy in Wh/kg (Li-sulfur battery) cause additional challenges with regard to circular economy and closed material cycles, recycling processes as well as safety in the handling and recycling of batteries. From the current state of research and industrial implementation, already established or partially established processes can be adapted and expanded to meet the challenges of the coming battery generations. Nevertheless, this can be ambitious for some battery types (i.e., battery type III/IV) with regard to the required high recycling rates of the total battery of 70% and higher or 95% of individual metals and make the development of new process steps necessary. Today's established recycling processes allow a recycling rate for individual materials of up to more than 90% in some cases, depending on the combination of the processes. However, if for example the recovery of the transition metals such as nickel and cobalt is maximized by a mainly pyrometallurgical process, a recovery of graphite as an anode material is not practical; i.e., the graphite serves as an energy source for the heating. Therefore, depending on the regulations, a certain process route can be worthwhile or not useful at all. Processes routes based on only one process type (e.g., only mechanical) enable at least lower overall recovery values. Thus, with the position today, it can be assumed that almost all battery materials can be recycled and reused within new batteries. However, an open question is how far the substances have to be purified so that there is no effect on the electrochemical performance, and if also a reconditioning of the active materials, especially also the cathode materials, is possible on large scale, as it has been shown on a lab scale. This question is the subject of actual research and can hopefully be quantitatively answered in the near future. However, it is expected that the original performance will be achieved in any case when the current active material is reprocessed to the original material purity via complex processes similar to the ones applied for primary materials [65,66,68]. For types I and III, this is also expected for the cathode materials, as the reprocessing processes will be very similar. For type II, a good performance retention result can also be expected for the cathode, since these battery chemistries and materials are already well researched. For type IV, reuse cannot be reasonably implemented from today's economic perspectives. In addition, direct reconditioning of the cathode material can be used to recycle type II and, in some cases, type I materials. These materials are subject to a longer period of use and application than new and even higher-energy materials in example for type III. With regard to the graphite-containing anode in particular (types I and III), recycling is not yet an option from an economic point of view. However, technically, there are approaches that show a reuse with the same performance [86]. Since types II and IV contain a lithium anode, recycling is an economical option for reuse here.

It becomes clear that the diversification and new developments demand a proper separation of battery types before recycling, for example by a transnational network of dismantling and sorting locations, and flexible and high sophisticated recycling processes with case-wise higher safety standards than today. Moreover, for the low-cost batteries, recycling of the batteries becomes economically unattractive, so that legal stipulations become important. However, in general, it must be still secured that closing the material cycle for all battery types with suitable processes is achieved to secure the supply of raw materials and also to further advance new developments. Last but not least, it can be stated that in absolute percentage values, a relatively small step has to be realized until we meet the targets, but it will be enormously demanding to achieve the last percentage points in the quotas especially if more than only the transition metals shall be recovered.

Author Contributions: Writing, S.D., J.K.M. and P.M., with support from A.K.; project supervision, A.K with support from P.M. All authors have read and agreed to the published version of the manuscript.

Funding: We have received financial support for the preparation of this manuscript from the German Research Foundation and the Open Access Publication Funds of Technische Universität Braunschweig.

Data Availability Statement: Not applicable.

Acknowledgments: We acknowledge support by the German Research Foundation and the Open Access Publication Funds of Technische Universität Braunschweig.

Conflicts of Interest: The authors declare no conflict of interest.

References

1. *The European Dimension of Germany's Energy Transition. Opportunities and Conflicts*; Gawel, E.; Strunz, S.; Lehmann, P.; Purkus, A (Eds.) Springer International Publishing: Cham, Switzerland, 2019; ISBN 978-3-030-03374-3.
2. Kwade, A.; Haselrieder, W.; Leithoff, R.; Modlinger, A.; Dietrich, F.; Droeder, K. Current status and challenges for automotive battery production technologies. *Nat. Energy* **2018**, *3*, 290–300. [CrossRef]
3. Peters, J.F.; Baumann, M.; Zimmermann, B.; Braun, J.; Weil, M. The environmental impact of Li-Ion batteries and the role of key parameters—A review. *Renew. Sustain. Energy Rev.* **2017**, *67*, 491–506. [CrossRef]
4. Habib, K.; Hansdóttir, S.T.; Habib, H. Critical metals for electromobility: Global demand scenarios for passenger vehicles 2015–2050. *Resour. Conserv. Recycl.* **2020**, *154*, 104603. [CrossRef]
5. *Behaviour of Lithium-Ion Batteries in Electric Vehicles. Battery Health, Performance, Safety, and Cost*; Pistoia, G.; Liaw, B. (Eds.) Springer Cham, Switzerland, 2018; ISBN 978-3-319-69950-9.
6. Mohr, M.; Peters, J.F.; Baumann, M.; Weil, M. Toward a cell-chemistry specific life cycle assessment of lithium-ion battery recycling processes. *J. Ind. Ecol.* **2020**, *24*, 1310–1322. [CrossRef]
7. Velázquez-Martínez, O.; Valio, J.; Santasalo-Aarnio, A.; Reuter, M.; Serna-Guerrero, R. A Critical Review of Lithium-Ion Battery Recycling Processes from a Circular Economy Perspective. *Batteries* **2019**, *5*, 68. [CrossRef]
8. Dai, Q.; Kelly, J.C.; Gaines, L.; Wang, M. Life Cycle Analysis of Lithium-Ion Batteries for Automotive Applications. *Batteries* **2019** *5*, 48. [CrossRef]
9. Chen, M.; Zheng, Z.; Wang, Q.; Zhang, Y.; Ma, X.; Shen, C.; Xu, D.; Liu, J.; Liu, Y.; Gionet, P.; et al. Closed Loop Recycling of Electric Vehicle Batteries to Enable Ultra-high Quality Cathode Powder. *Sci. Rep.* **2019**, *9*, 173. [CrossRef] [PubMed]
10. Fan, E.; Li, L.; Wang, Z.; Lin, J.; Huang, Y.; Yao, Y.; Chen, R.; Wu, F. Sustainable Recycling Technology for Li-Ion Batteries and Beyond: Challenges and Future Prospects. *Chem. Rev.* **2020**, *120*, 7020–7063. [CrossRef]
11. Zeng, X.; Li, J.; Singh, N. Recycling of Spent Lithium-Ion Battery: A Critical Review. *Crit. Rev. Environ. Sci. Technol.* **2014**, *44*, 1129–1165 [CrossRef]
12. Xu, C.; Dai, Q.; Gaines, L.; Hu, M.; Tukker, A.; Steubing, B. Future material demand for automotive lithium-based batteries *Commun. Mater.* **2020**, *1*, 437. [CrossRef]
13. Proposal for a Regulation of the European Parliament and of the Council Concerning Batteries and Waste Batteries, Repealing Di rective 2006/66/EC and Amending Regulation (EU) No 2019/1020. 2020. Available online: https://ec.europa.eu/environment/ waste/batteries/pdf/Proposal_for_a_Regulation_on_batteries_and_waste_batteries.pdf (accessed on 22 December 2020).
14. Kwade, A.; Hagelüken, C.; Kohl, H.; Buchert, M.; Herrmann, C.; Vahle, T.; von Wittken, R.; Carrara, M.; Daelemans, S.; Ehrenberg, H. et al. *Ressourcenschonende Batteriekreisläufe—mit Circular Economy die Elektromobilität Antreiben*; Acatech: München, Germany London, UK, 2020; Available online: https://www.acatech.de/publikation/ressourcenschonende-batteriekreislaeufe/ (accessed on 22 December 2020).
15. Siqi, Z.; Guangming, L.; Wenzhi, H.; Juwen, H.; Haochen, Z. Recovery methods and regulation status of waste lithium-ion batteries in China: A mini review. *Waste Manag. Res.* **2019**, *37*, 1142–1152. [CrossRef] [PubMed]
16. Wang, W.; Wu, Y. An overview of recycling and treatment of spent LiFePO 4 batteries in China. *Resour. Conserv. Recycl.* **2017** *127*, 233–243. [CrossRef]
17. Georgi-Maschler, T.; Friedrich, B.; Weyhe, R.; Heegn, H.; Rutz, M. Development of a recycling process for Li-ion batteries *J. Power Sources* **2012**, *207*, 173–182. [CrossRef]
18. Sommerfeld, M.; Vonderstein, C.; Dertmann, C.; Klimko, J.; Oráč, D.; Miškufová, A.; Havlík, T.; Friedrich, B. A Combined Pyro- and Hydrometallurgical Approach to Recycle Pyrolyzed Lithium-Ion Battery Black Mass Part 1: Production of Lithium Concentrates in an Electric Arc Furnace. *Metals* **2020**, *10*, 1069. [CrossRef]
19. Hanisch, C.; Loellhoeffel, T.; Diekmann, J.; Markley, K.J.; Haselrieder, W.; Kwade, A. Recycling of lithium-ion batteries: A novel method to separate coating and foil of electrodes. *J. Clean. Prod.* **2015**, *108*, 301–311. [CrossRef]
20. Diekmann, J.; Hanisch, C.; Froböse, L.; Schälicke, G.; Loellhoeffel, T.; Fölster, A.-S.; Kwade, A. Ecological Recycling of Lithium-Ion Batteries from Electric Vehicles with Focus on Mechanical Processes. *J. Electrochem. Soc.* **2017**, *164*, A6184–A6191. [CrossRef]

21. Chagnes, A.; Pospiech, B. A brief review on hydrometallurgical technologies for recycling spent lithium-ion batteries. *J. Chem. Technol. Biotechnol.* **2013**, *88*, 1191–1199. [CrossRef]

22. Sommerville, R.; Shaw-Stewart, J.; Goodship, V.; Rowson, N.; Kendrick, E. A review of physical processes used in the safe recycling of lithium ion batteries. *Sustain. Mater. Technol.* **2020**, *25*, e00197. [CrossRef]

23. Sojka, R.T. Safe Treatment of Lithium-based Batteries through Thermal Conditioning. In *Recycling und Sekundärrohstoffe, Band 13*; Thomé-Kozmiensky, E., Holm, O., Friedrich, B., Goldmann, D., Eds.; Thomé-Kozmiensky Verlag GmbH: Nietwerder, Germany, 2020; pp. 506–523. ISBN 3944310519.

24. Hiskey, B. Metallurgy, Survey. In *Encyclopedia of Chemical Technology*; Kirk, R.E., Othmer, D.F., Eds.; Wiley: New York, NY, USA, 2003; ISBN 0471238961.

25. Zhang, G.; Yuan, X.; He, Y.; Wang, H.; Zhang, T.; Xie, W. Recent advances in pretreating technology for recycling valuable metals from spent lithium-ion batteries. *J. Hazard. Mater.* **2020**, 124332. [CrossRef]

26. Träger, T.; Friedrich, B.; Weyhe, R. Recovery Concept of Value Metals from Automotive Lithium-Ion Batteries. *Chem. Ing. Tech.* **2015**, *87*, 1550–1557. [CrossRef]

27. Diekmann, J.; Hanisch, C.; Loellhoeffel, T.; Schalicke, G.; Kwade, A. (Invited) Ecologically Friendly Recycling of Lithium-Ion Batteries—The LithoRec Process. *ECS Trans.* **2016**, *73*, 1–9. [CrossRef]

28. Kwade, A.; Diekmann, J. *Recycling of Lithium-Ion Batteries*; Springer International Publishing: Cham, Germany, 2018; ISBN 978-3-319-70571-2.

29. Dai, Q.; Spangenberger, J.; Ahmed, S.; Gaines, L.; Kelly, J.C.; Wang, M. *EverBatt. A Closed-loop Battery Recycling Cost and Environmental Impacts Model*; Argonne National Laboratory: Lemont, IL, USA, 2019. Available online: https://publications.anl.gov/anlpubs/2019/0-7/153050.pdf (accessed on 3 February 2021).

30. Ciez, R.E.; Whitacre, J.F. Examining different recycling processes for lithium-ion batteries. *Nat. Sustain.* **2019**, *2*, 148–156. [CrossRef]

31. Zhao, X.; Lehto, V.-P. Challenges and prospects of nanosized silicon anodes in lithium-ion batteries. *Nanotechnology* **2021**, *32*, 42002. [CrossRef]

32. Heck, C.A.; Horstig, M.-W.; von Huttner, F.; Mayer, J.K.; Haselrieder, W.; Kwade, A. Review—Knowledge-Based Process Design for High Quality Production of NCM811 Cathodes. *J. Electrochem. Soc.* **2020**, *167*, 160521. [CrossRef]

33. Cheruvally, G. *Lithium Iron Phosphate. A Promising Cathode-Active Material for Lithium Secondary Batteries*; Trans Tech Publishers: Zurich, Switzerland, 2008; ISBN 978-0-87849-477-4.

34. Omar, N.; Monem, M.A.; Firouz, Y.; Salminen, J.; Smekens, J.; Hegazy, O.; Gaulous, H.; Mulder, G.; van den Bossche, P.; Coosemans, T.; et al. Lithium iron phosphate based battery—Assessment of the aging parameters and development of cycle life model. *Appl. Energy* **2014**, *113*, 1575–1585. [CrossRef]

35. Manthiram, A.; Yu, X.; Wang, S. Lithium battery chemistries enabled by solid-state electrolytes. *Nat. Rev. Mater.* **2017**, *2*, 294. [CrossRef]

36. Wild, M.; O'Neill, L.; Zhang, T.; Purkayastha, R.; Minton, G.; Marinescu, M.; Offer, G.J. Lithium sulfur batteries, a mechanistic review. *Energy Environ. Sci.* **2015**, *8*, 3477–3494. [CrossRef]

37. Golubkov, A.W.; Fuchs, D.; Wagner, J.; Wiltsche, H.; Stangl, C.; Fauler, G.; Voitic, G.; Thaler, A.; Hacker, V. Thermal-runaway experiments on consumer Li-ion batteries with metal-oxide and olivin-type cathodes. *RSC Adv.* **2014**, *4*, 3633–3642. [CrossRef]

38. Olivetti, E.A.; Ceder, G.; Gaustad, G.G.; Fu, X. Lithium-Ion Battery Supply Chain Considerations: Analysis of Potential Bottlenecks in Critical Metals. *Joule* **2017**, *1*, 229–243. [CrossRef]

39. Lee, Y.-G.; Fujiki, S.; Jung, C.; Suzuki, N.; Yashiro, N.; Omoda, R.; Ko, D.-S.; Shiratsuchi, T.; Sugimoto, T.; Ryu, S.; et al. High-energy long-cycling all-solid-state lithium metal batteries enabled by silver–carbon composite anodes. *Nat. Energy* **2020**, *5*, 299–308. [CrossRef]

40. Kato, Y.; Hori, S.; Saito, T.; Suzuki, K.; Hirayama, M.; Mitsui, A.; Yonemura, M.; Iba, H.; Kanno, R. High-power all-solid-state batteries using sulfide superionic conductors. *Nat. Energy* **2016**, *1*, 652. [CrossRef]

41. Wang, H.; Sheng, L.; Yasin, G.; Wang, L.; Xu, H.; He, X. Reviewing the current status and development of polymer electrolytes for solid-state lithium batteries. *Energy Storage Mater.* **2020**, *33*, 188–215. [CrossRef]

42. Son, I.H.; Hwan Park, J.; Kwon, S.; Park, S.; Rümmeli, M.H.; Bachmatiuk, A.; Song, H.J.; Ku, J.; Choi, J.W.; Choi, J.-M.; et al. Silicon carbide-free graphene growth on silicon for lithium-ion battery with high volumetric energy density. *Nat. Commun.* **2015**, *6*, 7393. [CrossRef] [PubMed]

43. Song, T.; Cheng, H.; Choi, H.; Lee, J.-H.; Han, H.; Lee, D.H.; Yoo, D.S.; Kwon, M.-S.; Choi, J.-M.; Doo, S.G.; et al. Si/Ge double-layered nanotube array as a lithium ion battery anode. *ACS Nano* **2012**, *6*, 303–309. [CrossRef]

44. Li, L.; Zhang, X.; Li, M.; Chen, R.; Wu, F.; Amine, K.; Lu, J. The Recycling of Spent Lithium-Ion Batteries: A Review of Current Processes and Technologies. *Electrochem. Energ. Rev.* **2018**, *1*, 461–482. [CrossRef]

45. Zhang, T.; He, Y.; Ge, L.; Fu, R.; Zhang, X.; Huang, Y. Characteristics of wet and dry crushing methods in the recycling process of spent lithium-ion batteries. *J. Power Sources* **2013**, *240*, 766–771. [CrossRef]

46. Zou, H.; Gratz, E.; Apelian, D.; Wang, Y. A novel method to recycle mixed cathode materials for lithium ion batteries. *Green Chem.* **2013**, *15*, 1183. [CrossRef]

47. Larouche, F.; Tedjar, F.; Amouzegar, K.; Houlachi, G.; Bouchard, P.; Demopoulos, G.P.; Zaghib, K. Progress and Status of Hydrometallurgical and Direct Recycling of Li-Ion Batteries and Beyond. *Materials* **2020**, *13*, 801. [CrossRef]

48. Yao, Y.; Zhu, M.; Zhao, Z.; Tong, B.; Fan, Y.; Hua, Z. Hydrometallurgical Processes for Recycling Spent Lithium-Ion Batteries: A Critical Review. *ACS Sustain. Chem. Eng.* **2018**, *6*, 13611–13627. [CrossRef]

49. Zheng, Z.; Chen, M.; Wang, Q.; Zhang, Y.; Ma, X.; Shen, C.; Xu, D.; Liu, J.; Liu, Y.; Gionet, P.; et al. High Performance Cathode Recovery from Different Electric Vehicle Recycling Streams. *ACS Sustain. Chem. Eng.* **2018**, *6*, 13977–13982. [CrossRef]

50. Pinegar, H.; Smith, Y.R. Recycling of End-of-Life Lithium Ion Batteries, Part I: Commercial Processes. *J. Sustain. Metall.* **2019**, *5*, 402–416. [CrossRef]

51. Sloop, S.; Crandon, L.; Allen, M.; Koetje, K.; Reed, L.; Gaines, L.; Sirisaksoontorn, W.; Lerner, M. A direct recycling case study from a lithium-ion battery recall. *Sustain. Mater. Technol.* **2020**, *25*, e00152. [CrossRef]

52. Brückner, L.; Frank, J.; Elwert, T. Industrial Recycling of Lithium-Ion Batteries—A Critical Review of Metallurgical Process Routes. *Metals* **2020**, *10*, 1107. [CrossRef]

53. Xiao, S.; Ren, G.; Xie, M.; Pan, B.; Fan, Y.; Wang, F.; Xia, X. Recovery of Valuable Metals from Spent Lithium-Ion Batteries by Smelting Reduction Process Based on MnO–SiO_2–Al_2O_3 Slag System. *J. Sustain. Metall.* **2017**, *3*, 703–710. [CrossRef]

54. Assefi, M.; Maroufi, S.; Yamauchi, Y.; Sahajwalla, V. Pyrometallurgical recycling of Li-ion, Ni–Cd and Ni–MH batteries: A minireview. *Curr. Opin. Green Sustain. Chem.* **2020**, *24*, 26–31. [CrossRef]

55. *Encyclopedia of Chemical Technology*; Kirk, R.E.; Othmer, D.F. (Eds.) Wiley: New York, NY, USA, 2003; ISBN 0471238961.

56. Werner, D.; Peuker, U.A.; Mütze, T. Recycling Chain for Spent Lithium-Ion Batteries. *Metals* **2020**, *10*, 316. [CrossRef]

57. Iclodean, C.; Varga, B.; Burnete, N.; Cimerdean, D.; Jurchiş, B. Comparison of Different Battery Types for Electric Vehicles. *Iop Conf. Ser. Mater. Sci. Eng.* **2017**, *252*, 12058. [CrossRef]

58. Ruiz, V.; Pfrang, A.; Kriston, A.; Omar, N.; van den Bossche, P.; Boon-Brett, L. A review of international abuse testing standards and regulations for lithium ion batteries in electric and hybrid electric vehicles. *Renew. Sustain. Energy Rev.* **2018**, *81*, 1427–1452. [CrossRef]

59. Tebbe, J.L.; Fuerst, T.F.; Musgrave, C.B. Mechanism of hydrofluoric acid formation in ethylene carbonate electrolytes with fluorine salt additives. *J. Power Sources* **2015**, *297*, 427–435. [CrossRef]

60. Yang, H.; Zhuang, G.V.; Ross, P.N. Thermal stability of LiPF6 salt and Li-ion battery electrolytes containing LiPF6. *J. Power Sources* **2006**, *161*, 573–579. [CrossRef]

61. Nedjalkov, A.; Meyer, J.; Köhring, M.; Doering, A.; Angelmahr, M.; Dahle, S.; Sander, A.; Fischer, A.; Schade, W. Toxic Gas Emissions from Damaged Lithium Ion Batteries—Analysis and Safety Enhancement Solution. *Batteries* **2016**, *2*, 5. [CrossRef]

62. Lamb, J.; Orendorff, C.J. Evaluation of mechanical abuse techniques in lithium ion batteries. *J. Power Sources* **2014**, *247*, 189–196. [CrossRef]

63. Feng, X.; Sun, J.; Ouyang, M.; Wang, F.; He, X.; Lu, L.; Peng, H. Characterization of penetration induced thermal runaway propagation process within a large format lithium ion battery module. *J. Power Sources* **2015**, *275*, 261–273. [CrossRef]

64. Diekmann, J.; Doose, S.; Weber, S.; Münch, S.; Haselrieder, W.; Kwade, A. Development of a New Procedure for Nail Penetration of Lithium-Ion Cells to Obtain Meaningful and Reproducible Results. *J. Electrochem. Soc.* **2020**, *167*, 90504. [CrossRef]

65. Doose, S.; Haselrieder, W.; Kwade, A. Effects of the Nail Geometry and Humidity on the Nail Penetration of High-Energy Density Lithium Ion Batteries. *Batteries* **2021**, *7*, 6. [CrossRef]

66. Yoon, T.; Milien, M.S.; Parimalam, B.S.; Lucht, B.L. Thermal Decomposition of the Solid Electrolyte Interphase (SEI) on Silicon Electrodes for Lithium Ion Batteries. *Chem. Mater.* **2017**, *29*, 3237–3245. [CrossRef]

67. Campion, C.L.; Li, W.; Lucht, B.L. Thermal Decomposition of $LiPF_6$-Based Electrolytes for Lithium-Ion Batteries. *J. Electrochem. Soc.* **2005**, *152*, A2327. [CrossRef]

68. Diaz, F.; Wang, Y.; Weyhe, R.; Friedrich, B. Gas generation measurement and evaluation during mechanical processing and thermal treatment of spent Li-ion batteries. *Waste Manag.* **2019**, *84*, 102–111. [CrossRef] [PubMed]

69. Wang, Q.; Ping, P.; Zhao, X.; Chu, G.; Sun, J.; Chen, C. Thermal runaway caused fire and explosion of lithium ion battery. *J. Power Sources* **2012**, *208*, 210–224. [CrossRef]

70. Grützke, M.; Mönnighoff, X.; Horsthemke, F.; Kraft, V.; Winter, M.; Nowak, S. Extraction of lithium-ion battery electrolytes with liquid and supercritical carbon dioxide and additional solvents. *RSC Adv.* **2015**, *5*, 43209–43217. [CrossRef]

71. Nowak, S.; Winter, M. The Role of Sub- and Supercritical CO_2 as "Processing Solvent" for the Recycling and Sample Preparation of Lithium Ion Battery Electrolytes. *Molecules* **2017**, *22*, 403. [CrossRef] [PubMed]

72. Baars, J.; Domenech, T.; Bleischwitz, R.; Melin, H.E.; Heidrich, O. Circular economy strategies for electric vehicle batteries reduce reliance on raw materials. *Nat. Sustain.* **2020**, *69*, 37. [CrossRef]

73. Pagliaro, M.; Meneguzzo, F. Lithium battery reusing and recycling: A circular economy insight. *Heliyon* **2019**, *5*, e01866. [CrossRef] [PubMed]

74. Richa, K.; Babbitt, C.W.; Gaustad, G. Eco-Efficiency Analysis of a Lithium-Ion Battery Waste Hierarchy Inspired by Circular Economy. *J. Ind. Ecol.* **2017**, *21*, 715–730. [CrossRef]

75. Xie, Y.H.; Yu, H.J.; Ou, Y.N.; Li, C.D. Environmental impact assessment of recycling waste traction battery. *Inorg. Chem. Ind.* **2015**, *47*, 43–46.

76. Heulens, J.; Van Horebeek, D.; Quix, M.; Brouwe, S. Process for smelting lithium-ion batteries. EP2015/067809, 8 March 2015.

77. Xu, P.; Dai, Q.; Gao, H.; Liu, H.; Zhang, M.; Li, M.; Chen, Y.; An, K.; Meng, Y.S.; Liu, P.; et al. Efficient Direct Recycling of Lithium-Ion Battery Cathodes by Targeted Healing. *Joule* **2020**, *4*, 2609–2626. [CrossRef]

78. Gao, H.; Yan, Q.; Xu, P.; Liu, H.; Li, M.; Liu, P.; Luo, J.; Chen, Z. Efficient Direct Recycling of Degraded $LiMn_2O_4$ Cathodes by One-Step Hydrothermal Relithiation. *ACS Appl. Mater. Interfaces* **2020**, *12*, 51546–51554. [CrossRef]

79. Wang, T.; Luo, H.; Bai, Y.; Li, J.; Belharouak, I.; Dai, S. Direct Recycling of Spent NCM Cathodes through Ionothermal Lithiation. *Adv. Energy Mater.* **2020**, *10*, 2001204. [CrossRef]

80. Hanisch, C.; Schunemann, J.-H.; Diekmann, J.; Westphal, B.; Loellhoeffel, T.; Prziwara, P.F.; Haselrieder, W.; Kwade, A. In-Production Recycling of Active Materials from Lithium-Ion Battery Scraps. *ECS Trans.* **2015**, *64*, 131–145. [CrossRef]

81. Tomaszewska, A.; Chu, Z.; Feng, X.; O'Kane, S.; Liu, X.; Chen, J.; Ji, C.; Endler, E.; Li, R.; Liu, L.; et al. Lithium-ion battery fast charging: A review. *eTransportation* **2019**, *1*, 100011. [CrossRef]

82. Bai, Y.; Muralidharan, N.; Sun, Y.-K.; Passerini, S.; Stanley Whittingham, M.; Belharouak, I. Energy and environmental aspects in recycling lithium-ion batteries: Concept of Battery Identity Global Passport. *Mater. Today* **2020**, *41*, 304–315. [CrossRef]

83. Krüger, S.; Hanisch, C.; Kwade, A.; Winter, M.; Nowak, S. Effect of impurities caused by a recycling process on the electrochemical performance of Li[Ni$_{0.33}$Co$_{0.33}$Mn$_{0.33}$]O$_2$. *J. Electroanal. Chem.* **2014**, *726*, 91–96. [CrossRef]

84. Zhang, R.; Zheng, Y.; Yao, Z.; Vanaphuti, P.; Ma, X.; Bong, S.; Chen, M.; Liu, Y.; Cheng, F.; Yang, Z.; et al. Systematic Study of Al Impurity for NCM622 Cathode Materials. *ACS Sustain. Chem. Eng.* **2020**, *8*, 9875–9884. [CrossRef]

85. Thompson, D.L.; Hartley, J.M.; Lambert, S.M.; Shiref, M.; Harper, G.D.J.; Kendrick, E.; Anderson, P.; Ryder, K.S.; Gaines, L.; Abbott, A.P. The importance of design in lithium ion battery recycling—A critical review. *Green Chem.* **2020**, *22*, 7585–7603. [CrossRef]

86. Rothermel, S.; Evertz, M.; Kasnatscheew, J.; Qi, X.; Grützke, M.; Winter, M.; Nowak, S. Graphite Recycling from Spent Lithium-Ion Batteries. *ChemSusChem* **2016**, *9*, 3473–3484. [CrossRef] [PubMed]

Article

Task Planner for Robotic Disassembly of Electric Vehicle Battery Pack

Martin Choux *,†, **Eduard Marti Bigorra** †,‡ **and Ilya Tyapin**

Department of Engineering Sciences, University of Agder, 4604 Kristiansand, Norway;
edumabi@gmail.com (E.M.B.); Ilya.tyapin@uia.no (I.T.)
* Correspondence: martin.choux@uia.no
† These authors contributed equally to this work.
‡ Current address: Northvolt AB, 112 47 Stockholm, Sweden.

Abstract: The rapidly growing deployment of Electric Vehicles (EV) put strong demands on the development of Lithium-Ion Batteries (LIBs) but also into its dismantling process, a necessary step for circular economy. The aim of this study is therefore to develop an autonomous task planner for the dismantling of EV Lithium-Ion Battery pack to a module level through the design and implementation of a computer vision system. This research contributes to moving closer towards fully automated EV battery robotic dismantling, an inevitable step for a sustainable world transition to an electric economy. For the proposed task planner the main functions consist in identifying LIB components and their locations, in creating a feasible dismantling plan, and lastly in moving the robot to the detected dismantling positions. Results show that the proposed method has measurement errors lower than 5 mm. In addition, the system is able to perform all the steps in the order and with a total average time of 34 s. The computer vision, robotics and battery disassembly have been successfully unified, resulting in a designed and tested task planner well suited for product with large variations and uncertainties.

Keywords: robotic disassembly; electric vehicle battery; task planner

Citation: Choux, M.; Marti Bigorra, E.; Tyapin, I. Task Planner for Robotic Disassembly of Electric Vehicle Battery Pack. *Metals* **2021**, *11*, 387.
https://doi.org/10.3390/met11030387

Academic Editor: Alexandre Chagnes

Received: 25 January 2021
Accepted: 21 February 2021
Published: 26 February 2021

1. Introduction

As the adoption rate for electric vehicles (EV) is now accelerating worldwide, EV Lithium-Ion Batteries (EVBs) repurposing or recycling volumes are expected to be larger in 5–10 years (120 GWh/year available by 2030 [1]) and legislation will likely demand higher collection and recycling rates as for example in the newly proposed regulation of the European parliament and of the council concerning batteries and waste in December 2020. Today, the automotive Lithium-Ion Batteries (LIBs) dismantling process is mainly carried out manually and the use of robotics in this process is limited to simple tasks or human assistance [2]. These manual processes are time consuming and must be done by highly skilled personnel. As a direct consequence, the manual total disassembly of Li-ion EVBs might not be profitable and would be stopped at an optimal level, i.e., partial disassembly, that achieves maximum profit while decreasing the environmental impact [3]. In comparison, automated systems are more robust, have a lower-cost, reduce injuries and/or sickness, make the workplace more attractive for those hard-to-recruit-and-retain skilled workers, and are best suited for up-scaling to high-volumes. Therefore, fully automated disassembly of EVBs is inevitable. The main challenges for the success of the automated systems in dismantling are the variations and uncertainties in used products [4]. These challenges in the robotic disassembly of Electrical and Electronic components in electrical vehicles have been presented in article [5] where the need for cognitive systems is identified to enhance the effectiveness of automated disassembly operations. In the case of automated disassembly of EV batteries, advances in Computer Vision (CV) and cognitive robotics offer promising tools but this topic remains an open research challenge [6]. The disassemblability of industrial batteries, as described in articles [7,8], can be improved either by modifying their design and increasing its standardisation, for example redesigning a battery module to make it remanufacturable [9], or by developing new technologies to ease and eventually remove some of the challenges, as for example, making the recognition of fasteners an easy task. This second route, i.e., making the disassembly process smarter and more efficient through better cognitive capabilities is the one chosen in this research, motivated by the fact that many EV battery innovations are emerging making the design of modules and packs prone to rapid changes.

Over the last 15 years, research has been conducted on the recycling of Lithium-Ion Batteries (LIB) cells, mostly focusing on the mechanical and metallurgical recycling processes [10]. However none of the described recycling methods is integrating robotic disassembly in their pre-processing of EVBs, i.e., processes which does not alter the structure of the LIB cells, and the mechanical or pyrometallurgical processes start with EVB cell modules as input. However, a large portion of metals to be remanufactured or recycled comes from housings (pack and module), electrical wire and connectors. For a Nissan Leaf first generation, the weight of the cells alone represents only 60% of weight of the total battery pack [6]. The disassembly process has been extensively studied in the literature, as shown in the survey [11] where disassembly processes of Waste of Electrical and Electronic Equipment (WEEE) are also present [5,12]. In article [13] authors presented an automatic mechanical separation methodology for End-of-Life (EOL) pouch LIBs with Z-folded electrode-separator compounds (ESC). Customised handling tools were designed, manufactured, and assembled into an automatic disassembly system prototype. While this aspect is still an active research field, the focus is now shifting towards automated solutions to support the whole recycling chain. Industrial solutions for the automated disassembly of battery-operated devices have been implemented, but they are limited to specific and often small-sized products. Apple has implemented an automated disassembly line for Iphone6 [14], however the process is not flexible nor adaptive and can only disassemble one model phone in perfect conditions. In article [15], cooperative control techniques are developed and demonstrated on the robotic disassembly of PC. In article [16], a vision system to identify components for extraction and simple robotic processes are used to disassemble printed circuit boards. Using visual information to automate the disassembly process is further developed with the concept of cognitive robotic systems [17] and is applied for disassembly of Liquid Crystal Display (LCD) screens [18].

Building upon the existing work in this area, this paper aims to improve the design of the task planner responsible for automatically generating the disassembly plan and sequences without precise a priori knowledge of the product to dismantle. The developed functions are presented and the results are validated through experiments conducted on a Hybrid Audi battery pack.

2. Materials and Methods

2.1. Task Planner Design

The experimental setup is composed of IRB4400 robot (ABB, Zürich, Switzerland), IRBT4004 track (ABB, Zürich, Switzerland), and Zivid One 3D camera (Zivid, Oslo, Norway) mounted on the robot arm, all connected to a PC with Ubuntu and running the Robot Operating System (ROS), and a A3 Sportback e-tron hybrid Li-ion battery pack (Audi, Ingolstadt, Germany). More information about the connection setup and use of the ROS as a middleware can be found in article [19], whereas a complete description of the Audi battery pack and its disassembly sequences has been presented in article [3]. The task planner proposed in this paper and shown in Figure 1 analyses the information provided by the vision system based on Zivid camera, makes decisions regarding dismantling actions and sends a predefined path to the industrial robot's controller.

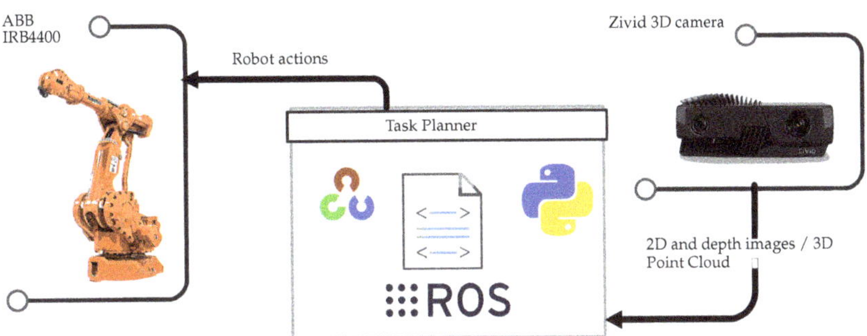

Figure 1. Task planner concept.

The main loop of the task planner is organised as the following: takes 2D and 3D images, detects and identifies components, finds the component's positions in the world reference frame, defines an order of operations, removes the components, and repeats this actions until the goal state is reached (refer to steps A–E respectively in Figure 2).

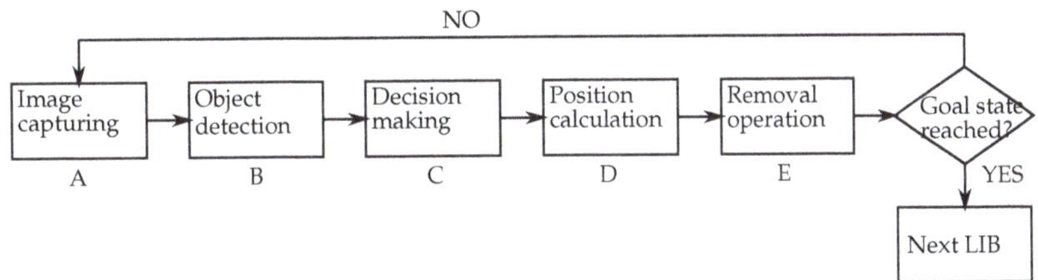

Figure 2. Task planner main loop.

2.1.1. Image Capturing (A)

The task planner moves the robot into several predefined poses to ensure that the system is able to observe different parts of the LIB pack. In order to reach better accuracy, up to eight pictures are required, especially when screws placed in the lateral sides of the battery are present.

2.1.2. Object Detection (2D Image) (B)

Different components in the image such as screws, battery modules, connecting plates, and Battery Management System (BMS) are detected using the YOLO (You Only Look Once) algorithm [20] which also provides the bounding box positions of all constituting components. The output of the object detection procedure as shown in Figure 3 is a file containing all detected objects, labels and corresponding coordinates.

The positions of the detected objects in different images are merged using a weighted mean of each positions.

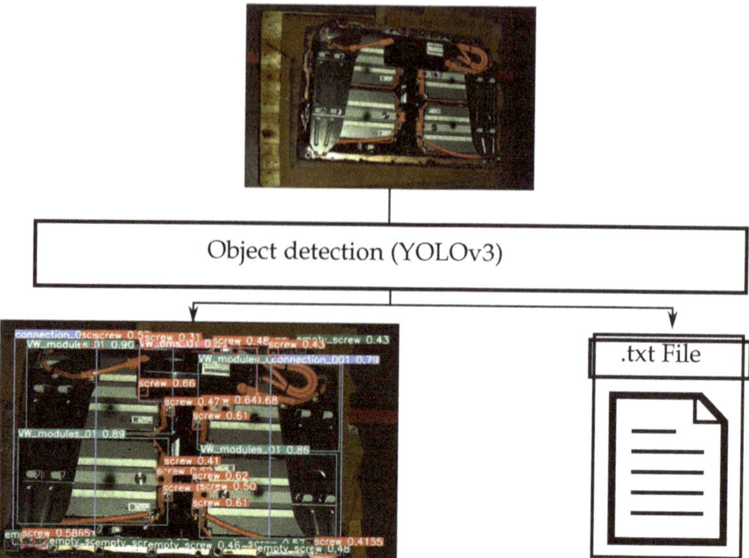

Figure 3. The YOLOv3 algorithm takes as input the 2D image of the EV Lithium-Ion Battery (EVB) pack, detects the components, finds the bounding box coordinates and class probabilities, and stores the information in a text file. The labels shown on the picture have one color for each class.

2.1.3. Decision Making (C)

In order to set a sequence of the removal operations, only one image is used, which is taken with lower camera angle with respect to a horizontal plane. The images with a higher inclination are used for object detection only. A list of the component positions in the removal order is created by adding the detected screw positions and following by a computer vision analysis of each specific component. Based on the probability of being over the other components the remaining parts are added in the list in a correct disassembly order.

2.1.4. Position Calculation (D)

At this point the system defines the positions of the objects in 2D camera coordinates (pixels). In order to move the robot to those positions, they must be converted into 3D points. The 2D object coordinates in camera frame are transformed into 3D coordinates using the depth information of the 3D vision system. Then, the captured positions are known, so that the world reference position of the objects can be obtained. This action is done for all images. Once all the positions from the different points of view are found the nearby points (representing the same object or component) are merged.

2.1.5. Robot Communication and Removal Operation (E)

Once the order of the operations and all the positions are known, the last step is to be proceed in dismantling. The task planner calls the removal operation of every single component. Details on the design of the removal operation are not presented in this paper since a scope is limited to moving the robot arm onto the calculated positions, where the selected removal operation based on the classification of the component is achieved.

2.2. Main Functions

The main functions and scripts used by the task planner connect the computed information by the mean of tests and loops as shown in the complete flow chart in Figure 4.

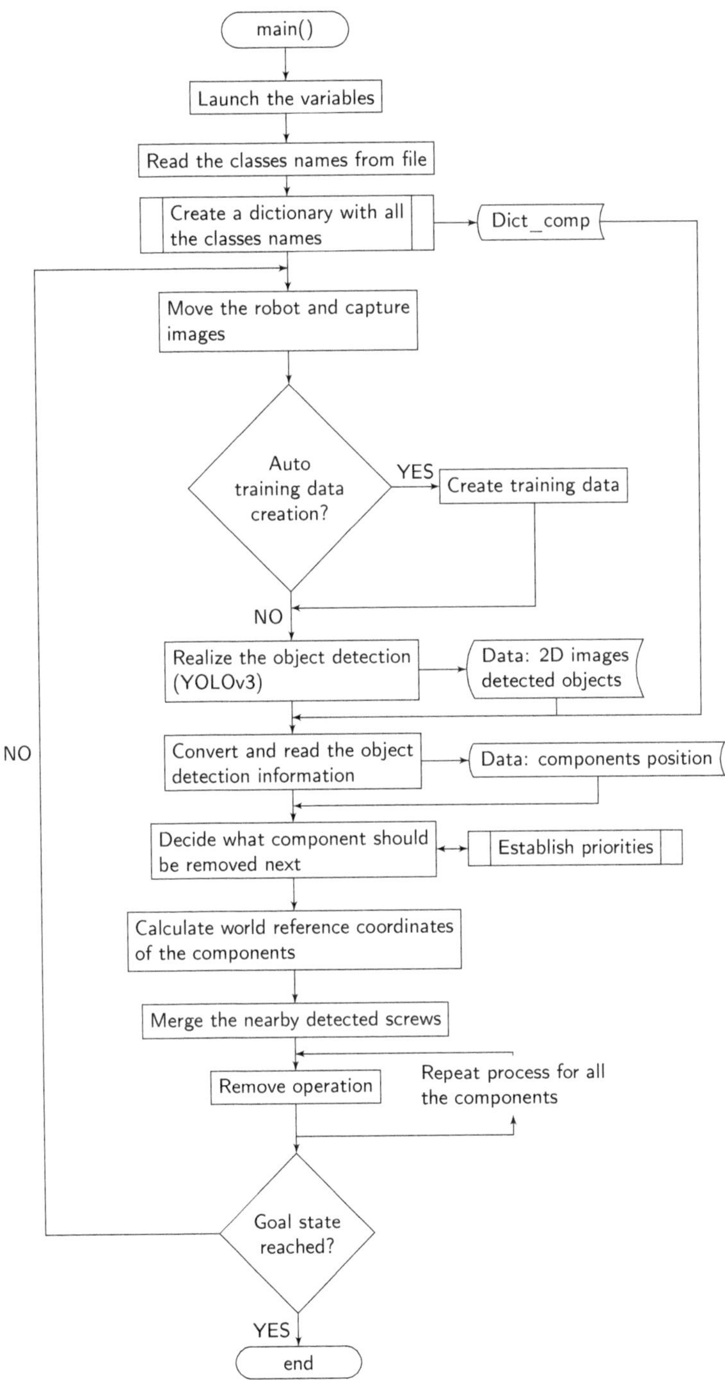

Figure 4. Task planner flow chart.

2.2.1. Function: *main()*

Fist of all, the main function *main()* whose flow chart is shown in Figure 4, declares and initialises all the variables used and transferred in and between the subsequent functions, as for example the screws, connecting plate, Battery Management System (BMS), or module positions and number. Then, it reads the file containing the classes names (classes.names), and saves them into a dictionary *(dict_comp)*. The creation of this dictionary aims to allow the system to have access to the nomenclature in order to relate the detected classes with their name and characteristics necessary for further analysis.

Once the dictionary is created, the function starts its principal loop. Note that this loop keeps running until the dismantling is completed. The first action of the loop is to move the robot to predefined positions and take 3D images, i.e., XYZ + color (RGB) + quality (Q) for each pixel, of the battery pack. Next, once the algorithm has been trained, it runs the object detection to detect the components placed in these images, using the YOLOv3 algorithm.

After the detection, classification and pose estimation of the objects in the image frame are done. The function analyses each taken image, finds different components and their characteristics from Dict_comp, and converts the positions of the objects from the image base frame to the camera reference frame. Following this step, a sub-process establishes priorities to enable the decision-making operation on what component should be removed next. The respective functions are named *what_component* and *has_comp_over* and are further described in the next section. Thus, when the detected positions have been converted into the camera frame, they are further transformed into the world reference frame coordinates. At this stage, since several 3D pictures of the same components have been captured, the resulting multiple positions of the same components are merged, which increases the position accuracy. The output of this function is the positions of the different components in world reference frame. The components are then placed in order of removal preference and finally, the task planner's main function runs the removal operations for all the detected component before starting the main loop again if the goal state is not reached.

2.2.2. Cognitive Functions: *what_component()* and *has_comp_over()*

The function *what_component()* flow chart is shown in Figure 5 and the function is responsible for the decision making. The function analyses the detected objects, and decides the order of the removal operations using the computer vision based sub-function *has_comp_over()* . The sub-function *has_comp_over()* flow chart is shown in Figure 6, and the function establishes different probabilities of a specific component to have a component over.

The inputs for the function *what_component()* are:

- *screw_pos* (array): An array containing coordinates of detected screws, referred to the image base frame.
- *connect_pos* (array): An array containing coordinates of detected connective components, referred to the image base frame.
- *compon_pos* (array): An array containing coordinates of detected general components, referred to the image base frame.
- *empty_screw_pos* (array): An array containing coordinates of detected empty screw holes, referred to the image base frame.
- *co_tot* (array): An array containing coordinates of all detected components, referred to the image base frame.

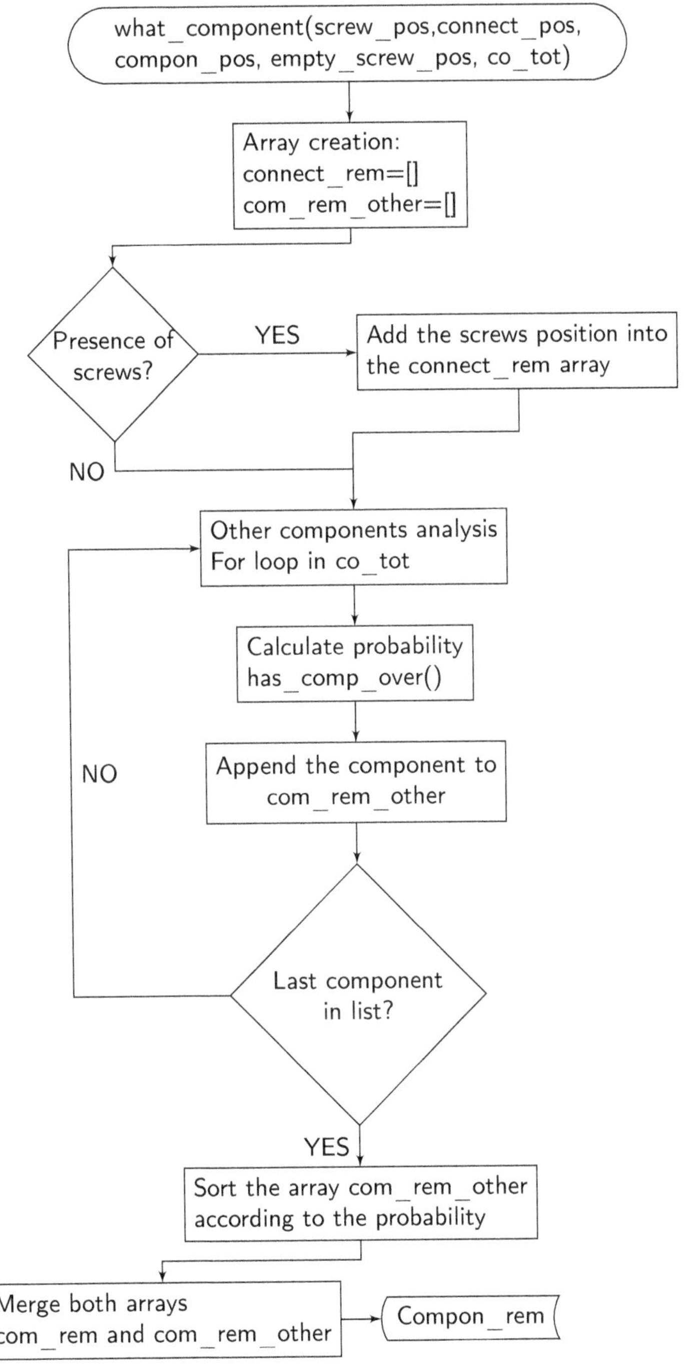

Figure 5. Function *what_component* flow chart.

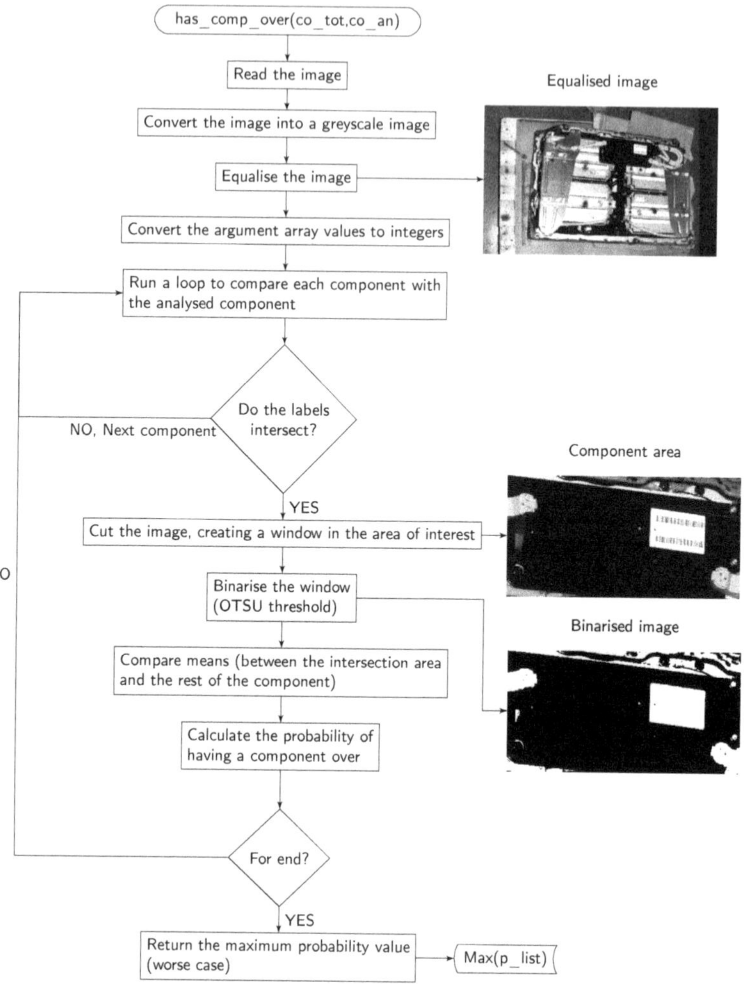

Figure 6. Function *has_comp_over* flow chart.

First of all, the function creates the variable *compon_rem* (array). The aim of this array is to contain the list of the components positions in the removal order. Given the nature of the disassembly process, the first components to be removed are screws, thus, their positions are the first to be added to the *compon_rem* array.

To add the rest of the components and decide the order of the operations, the function *what_component* runs over the list co_tot analysing each component using the function *has_comp_over*.

Essentially, *has_comp_over* realize a computer vision analysis of each specific component, and returns the probability of being over the other components. Thus, the components are ordered from more probability to be over to less probability.

The inputs for the sub-function *has_comp_over* are:

- *co_tot* (array)
- *co_an* (array): Array containing the image coordinates of a specific component (the component being analysed).

The function *has_comp_over* runs a nested for loop over the list containing all the components, analyzing the ones that are intersecting the component that is being inspected in the outer for loop.

To realise this comparison, the full image is converted into grey-scale and equalised. The output image contains a better distribution of the intensities maintaining the relevant image information [21].

After obtaining the equalised image, the function crops the image into the area of interest, in this case, the area of the analysed component. Then, segmentation is applied to the window, binarising it using an Otsu threshold. In the Otsu method, the threshold is determined by minimizing intra-class intensity variance [22].

When the window has been binarized, the mean of the pixel intensities in the intersection area is compared to the means of the intersecting components areas excluding the intersection. The component for which these two means (intersection alone and component area excluding intersection) are the most similar is the most probable to be on the top layer, i.e., over all the others, and hence to be removed first. An example with two overlapping components is shown in Figure 7. In this example, component 2 is over component 1 because:

$$||f(I) - f(A_2 - I)|| < ||f(I) - f(A_1 - I)|| \tag{1}$$

where $f(A)$ is a function calculating the mean of pixel intensity in area A.

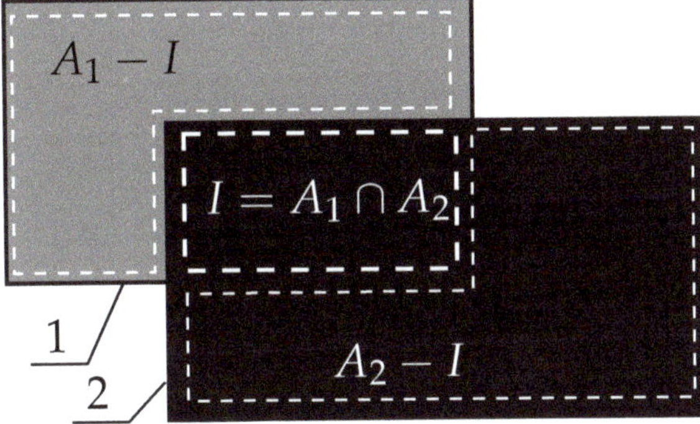

Figure 7. Component 2 with total area A_2 is overlapping component 1 with total area A_1. The intersection area is I.

After repeating this procedure for all the components, the function returns the maximum value, max(p_list), of the calculated probabilities of having another component overlapping the analyzed one.

2.2.3. Merging the Pictures: *merge_detection()*

The function *merge_detection* aims to merge positions of the components (for this version only the screws) commonly detected in one or more images. The function flow chart is shown in Figure 8.

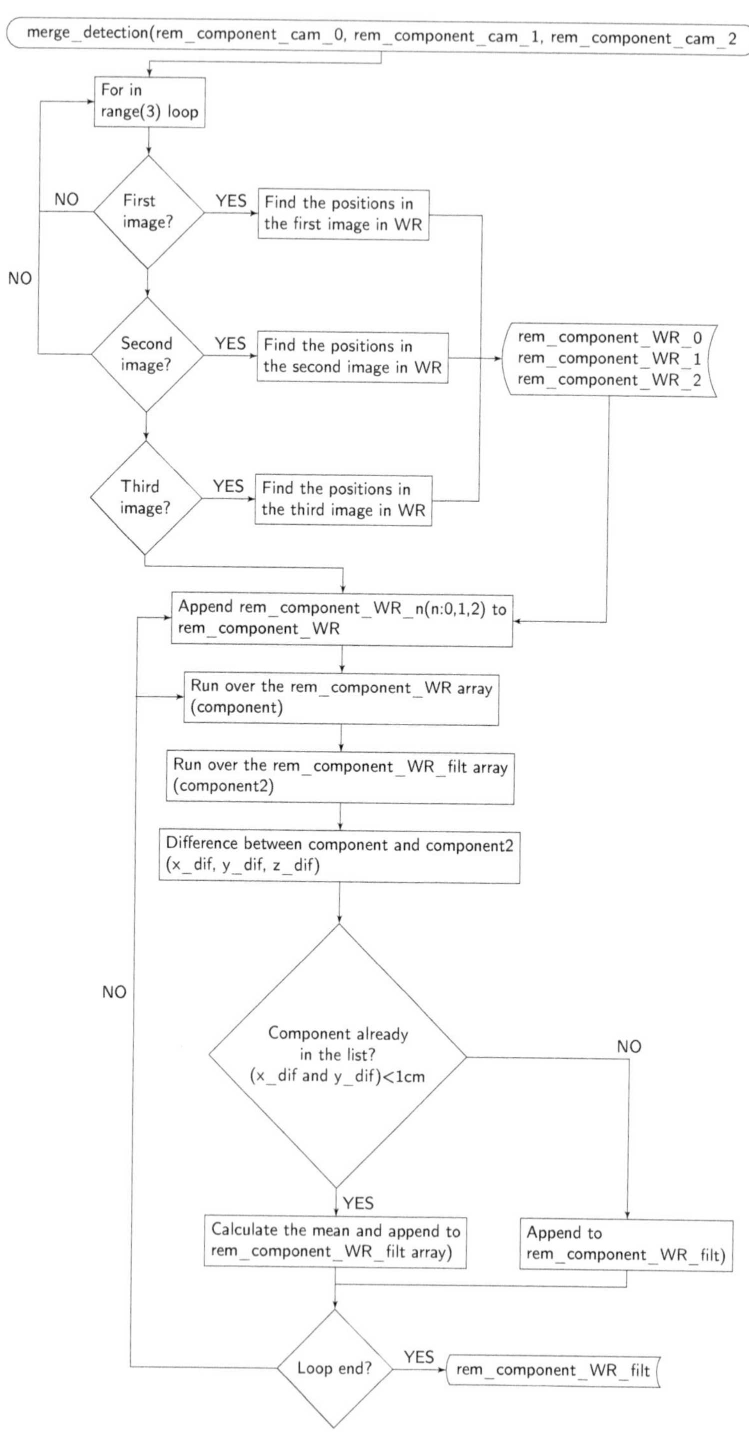

Figure 8. Function *merge_detection flow chart*.

The inputs of the function *merge_detection* are:

- *rem_component_cam_0* (Array): Array containing positions of all screws detected in the first image in camera reference frame.
- *rem_component_cam_1* (Array): Array containing positions of all components detected in the second image in camera reference frame. The components are placed in the removal order list, because the decision-making has been previously done in this image (given its inclinations and consequently its light conditions).
- *rem_component_cam_2* (Array): Array containing positions of all screws detected in the last image in camera reference frame.

The *merge_detection()* function runs a "for in range" loop to find positions of components in the world reference frame. It calls the *WR_pos* function, where the positions are stored in the variable *rem_component_WR* including repeated positions.

In the next step the function merges the repeated positions to find the output list (*rem_component_WR_filt*). In order to filter the points, the function runs over the *rem_component_WR* array adding not-repeated components to the filtered list. The function considers two components as one, when the x and y distances are lower than 1cm, and defines the final position as the mean of both points.

3. Results

The aim of this section is to validate the order suggested by the task planner and to characterise its performance regarding time and accuracy. The Audi A3 Sportback e-tron Hybrid Li-ion Battery Pack serves as the case study. The description of the EVB pack and its components as well as the disassembly process of the battery are detailed in article [3] whereas Table 1 presents the composition, i.e., relative weight of each components and materials. For safety reason when testing the concept, all modules have been manually discharged separately. However, when integrated in the pilot plant, the EVB packs will be discharged prior complete disassembly at a certain state of charge depending if the modules are to be repaired, re-purposed, re-manufactured, or recycled. In addition, damaged EVBs that represent high risk for thermal runaway or gas emission will be sorted out. These last two steps are outside the scope of the present work.

Table 1. Audi A3 Sportback e-tron battery pack constructive components and materials.

Component	%*w*/*w*	Material
Upper housing shell	2.8%	Composite
Upper and lower insulator	0.4%	Expanded polyethylene
BMS (Battery Junction Box and Battery Management Controller)	2.2%	Plastic, electronics
Connecting plates (Top transverse covers) (2)	0.7%	Al
Modules (8)	75.3%	Li, Co, Mg, Ni, Cu, Al, Graphite, plastic
Cooling system	0.4%	Ruber, plastic, Al
Lower housing shell	16.6%	Al
High-voltage cables and connectors	1.2%	Cu
Screws	0.4%	Fe

3.1. Object Detection Results

Figure 9 shows the results of the YOLOv3 algorithm implementation, where the red filled boxes indicate the position of the connective components, the blue-filled boxes indicate the screws positions, the pink-filled boxes indicate the position of the BMS and the black-filled boxes indicate the position of the battery modules.

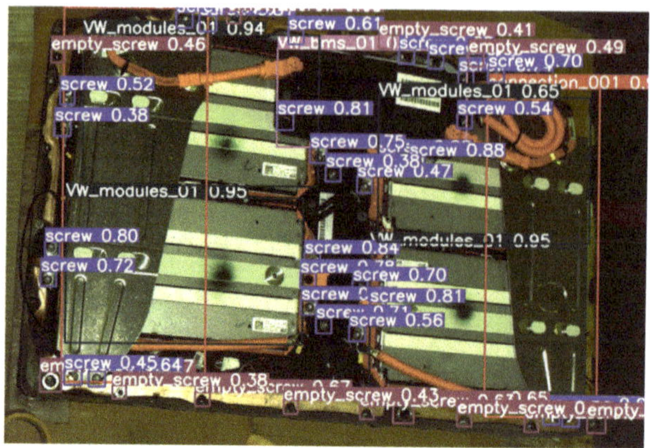

Figure 9. YOLOv3 output results: all screws, modules, transverse covers, and BMS are detected classified, and localised.

3.2. Time Analysis

The timings of the operations realized by the task planner, i.e., image capture, object detection, decision making, and motion to estimated pose, have been recorded on 20 repetitions with the physical setup, resulting in the mean times summarized in Figure 10. During experiments, the speed of the industrial robot has been reduced to 25% speed for safety reasons. The expected timing at production speed (100% speed) are shown in black and red in Figure 10.

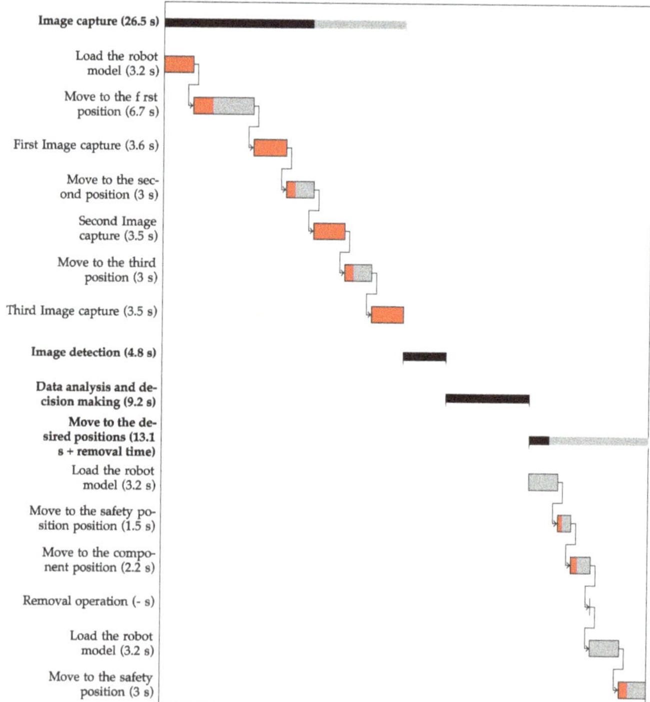

Figure 10. Timing summary at 25% speed in grey and expected timing at full speed in black and red.

3.2.1. Image Capture (Mean Time: 29.1 s)

In this case, the image capture process refers to the robot movement into the image taking positions and image capturing. In order to implement the capturing, the process has been divided into seven different actions. The first action refers to the robot model loading, and the rest refer to the robot movements and image captures, see Figure 11.

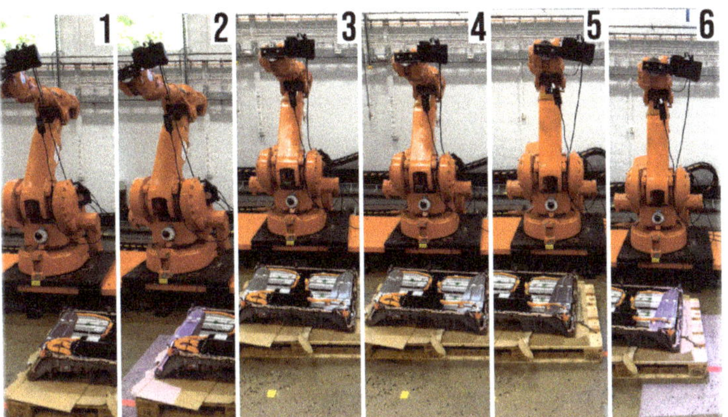

Figure 11. Image capturing actions.

1. Move to the first position. Mean time: 6.7 s.
2. First image capture. Mean time: 3.6 s.
3. Move to the second position. Mean time: 3.0 s.
4. Second image capture. Mean time: 3.5 s.
5. Move to the third position. Mean time: 6.0 s.
6. Third image capture. Mean time: 3.6 s.

3.2.2. Object Identification (Mean Time: 4.8 s)

In this stage the YOLO algorithm is applied to three taken images to detect and identify different components on 2D images. Mean time: 4.8 s. In this study 3 images are analysed but up to 8 images are required to increase an accuracy.

3.2.3. Data Analysis and Decision Making (Mean Time 9.2 s)

Data analysis and decision-making refers to calculating the object positions in the world coordinate frame and define the optimal path for the operations. Mean time: 9.2 s.

3.2.4. Move to the Desired Positions (Mean Time: 13.1 s)

The robot approaches the components to perform the removal operation. First, the robot rapidly moves to a safety position displaced thirty centimetres in the Z-axis above the object and then moves to the desired location. After that, the robot moves back to the safety position. The timings for this operation have been divided into six sub-processes.

- Load the robot model (MoveIt!). Mean time: 3.2 s.
- Move to the safety position. Mean time: 1.8 s.
- Move to the component position. Mean time: 2.2 s.
- Removal operation. Mean time: not applicable, since it depends on the removal operations, which is not considered in this project.
- Load the robot model (MoveIt!). Mean time: 3.2 s.
- Move to the safety position. Mean time: 2.6 s.

3.3. Decision-Making: Optimal Path

The decision-making of the system (the order of the removal operations) affects directly onto the dismantling time. For this reason, the aim of this section is to analyse the order suggested by the task planner for the Audi A3 Sportback e-tron Hybrid Battery Pack.

3.3.1. Optimal Dismantling Plans

After manually dismantling and analysing the battery pack, the optimal dismantling plan is to first remove the screws; the second step is to remove the two connective components and the battery management system (the order of the removal operations for these three elements is not critical); and finally to remove the four battery modules in a arbitrary order.

3.3.2. Dismantling Plans Proposed by the System

A set of tests have been carried out under different conditions (i.e., different orientations, different ambient lights conditions, etc.), the system has given a good response.

It has been observed, within the proposed dismantling plans, that the system follows the guidelines defined in Section 3.3.1. Because the BMS and the two connective components (left and right) are not overlapping, the system is proposing two different plans that are equivalent. These are referred as the (A) and (B) plans and are illustrated in Figure 12.

In the (A) plan, the system begins removing screws. The screws are always the first components to be removed. Afterwards, the system removes one connective component (in some tests the left one in other tests the right one), the BMS, and the other connective component in that order. Finally, it removes the battery modules. See Figure 12. In the (B) plan, screws are still the first components to be removed. Then, the system proposes to remove the BMS and the connective components in that order. As in the previous plan, the battery modules are removed the last. Thus, the main difference observed between these two plans is that in the plan (B) the BMS is removed after the screws instead of connective plates. This has no impact on the final disassembly process.

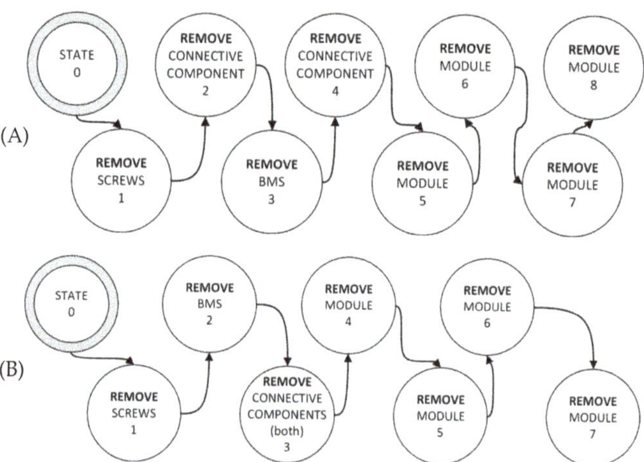

Figure 12. Dismantling plans (**A**) or (**B**) proposed by the task planner.

3.4. Accuracy

To analyse the system's accuracy, a 3D printed pointer has been used as Tool Center Point (TCP). The tool consists of a thin 25 cm long bar with a sharp end. With the tool mounted, the task planner has been run in debug mode. For safety reasons, the robot TCP has been moved 3 cm above in the Z-axis (word base frame) in order to avoid collisions with the battery pack in case of failure. Some of the tests are shown in Figure 13. In the majority of the cases, the system has an accuracy of (<5 mm). The accuracy has been measured by the mean of a laser beam attached to the red bar and pointing towards the target position, whereas the distance of the laser pointer to the target position is measured with a caliper.

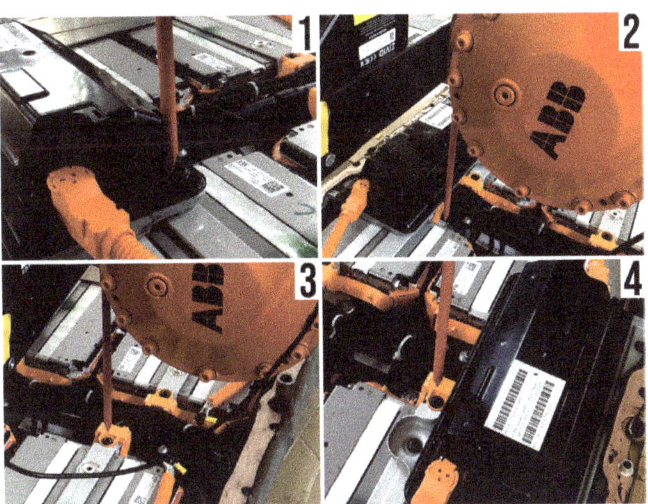

Figure 13. Accuracy tests showing the red pointer 3 cm above one detected screw. Four different views of the same position.

4. Discussion

In this paper the proposed objectives have been achieved. Different research areas in computer vision, robotics, circular economy and electr(on)ic components (battery) disassembly have been successfully unified.

The assigned main hardware elements such as an industrial robot, 3D camera and PC have been interconnected to carry out the principal system tasks, such as object detection, pose estimation, decision-making, and robot displacement. Therefore, the system is able to recognise the dismantling object main components, to find their position, and to move the robot to the defined positions in a specific order. Lab tests have been used to validate the designed task planner. In this case, experiments were limited to Audi LIB pack, however, a similar procedure might be applied to any EV battery pack. Compared to other disassembly processes as reviewed by Zhou et al. [11], the proposed task planner relies on state-of-the-art 3D camera system with high accuracy and does not require Computer Aided Design (CAD) models of the battery pack and its components. This presents a great advantage since EoL products are often different than their original CAD models, due to possible maintenance, deformations, or corrosion. Recognising the model and date of production of the EVB to be disassembled will help to determine a first disassembly sequence based on a self-updating database, but the system must also be flexible and robust enough to handle the above-mentioned variations or in the case of new or unrecognised model. Therefore, combined with reinforcement learning and machine reasoning algorithms the proposed disassembly framework will be able in future developments to learn how to disassemble new battery pack models, if not by itself, with only limited information from the human operator. The concept of cognitive robotics in disassembly has been developed and validated on End-of-Life treatment of LCD screen monitors [17]. However, some challenges remained as (1) the too high processing time making the process economically infeasible, (2) the remaining need for human assistance, and (3) the too high inaccuracy of the vision system leading to low success rate. This paper demonstrates that the recent advances in 3D vision system, fast object detection and localisation algorithms, as well as task planner design place EVBs with inherent uncertainties and large variations in design as a good candidate for achieving eventually autonomous and complete disassembly through the cognitive robotic concept.

The proposed task planner for disassembly of EVB pack into modules can also be extended in future work to a deeper level of disassembly, i.e., to battery cell level or even to the cell components (cell casing, electrodes, electrolyte, separator) which will increase the concentration of active materials in the subsequent steps for battery recycling and hence reduce the complexity and energy consumption of the pyrometallurgical and hydrometallurgical processes. Removing manual operations in the pre-processing stages will move the optimum disassembly level determined in the article [3] deeper toward complete disassembly when still considering techno-economic and environmental constraints.

The algorithm You Only Look Once (YOLO) is implemented to detect and find the components placed in the dismantling arena. The results show that the algorithm performs well, giving expected results and detects main components. For example the presence of screws can be distinguished from the presence of screw holes where the screw has been removed. The developed vision system can hence also be used to validate removal operations. The information extracted from the object detection was used in a pose estimation to find coordinates of components, where 2D images and the YOLO results have been matched with the 3D data sets. In future versions of the task planner, object detection and pose estimation might be realised directly in 3 dimensions based on the point cloud data with techniques such as complex-Yolo [23] or DeepGCNs [24]. However, higher processing time or computing resources are to be expected.

When validating the task planner and measuring the timings, the robot model has been loaded every time that the robot had to move. Thus, in future stages the robot model should be loaded just once, at the beginning making the disassembly sequence at least 9.6 s faster. Moreover, the ROS main has been run in manual mode at 25% of the maximum speed of the robot. In automatic mode, i.e., at 100% speed, the total time of the disassembly sequence, excluding the removal operations time, is expected to decrease from 53.6 s to 34 s

Using an eye-in-hand configuration performs well, and has some advantages (i.e., only one camera is needed), but it presents some drawbacks too. In an industrial application, the continuous moves and removal operations could have negative consequences like unexpected collisions of the camera with the environment or causing miscalibrations.

The efforts in future stages of the research should be focused on instrumentation and tool design for the dismantling system. It is also essential to detect flexible bodies such as high-voltage wires, and wires transmitting data. Thus, the direction of the research on the object detection and pose estimation part should concern how to find a feasible solution for such the objects and remove them.

5. Conclusions

A new task planner has been designed for the disassembly of electric vehicle Li-ion battery packs, with as main objective to increase the flexibility and robustness of the system. Lab tests have been used to validate the designed task planner based on a Audi A3 Sportback e-tron hybrid Li-ion battery pack. The results obtained in the tests demonstrate that the obtained solution is able to recognise which component to remove first and the complete disassembly plan without a priori knowledge of the disassembly strategy and battery CAD models. This method is therefore well suited for product with large variations and hence increases the disassemblability. The achieved performances measured in term of accuracy, time to generate the disassembly plan and success rate validated the task planner concept and its ability to make autonomous decision. Further testing on a larger set of EV battery packs with other geometries and connections and addition of learning capabilities will be needed to further increase the robustness of the proposed method and the technological readiness level. However, the results already cast a new light on the use of automation in the EV LIB batteries disassembly process by bringing the technology one step closer to eventually fully automated operations and hence redefine the optimum level of disassembly for the batteries to enter the subsequent stages of recycling and metal recovery.

The experience in this field could also be adapted to be used for other dismantling processes and opens new doors and research challenges to other fields directly related to robotics.

Author Contributions: Conceptualization, E.M.B., M.C. and I.T.; methodology, E.M.B., M.C., and I.T.; software, E.M.B.; validation, E.M.B., M.C., and I.T.; formal analysis, M.C., I.T., and E.M.B.; resources, M.C. and I.T.; writing—original draft preparation, M.C., E.M.B., and I.T.; writing—review and editing, M.C., E.M.B., and I.T.; visualization, E.M.B., M.C. and I.T.; supervision, M.C., I.T.; project administration, E.M.B., M.C., and I.T.; funding acquisition, M.C. All authors have read and agreed to the published version of the manuscript.

Funding: This research was funded by The Research Council of Norway grant number 282328.

Institutional Review Board Statement: Not Applicable.

Informed Consent Statement: Not Applicable.

Data Availability Statement: Not Applicable.

Acknowledgments: Special thanks go to BatteriRetur AS for the battery pack used for experiments. This work contributes to the research performed at TRCM (Priority Research Center Mechatronics at the University of Agder, Norway).

Conflicts of Interest: The authors declare no conflict of interest.

References

1. Awan, A. Batteries: An essential technology to electrify road transport. In *Global EV Outlook 2020*; International Energy Agency: Paris, France, 2020; pp. 185–219.
2. Wegener, K.; Andrew, S.; Raatz, A.; Dröder, K.; Herrmann, C. Disassembly of Electric Vehicle Batteries Using the Example of the Audi Q5 Hybrid System. *Procedia CIRP* **2014**, *23*. [CrossRef]
3. Alfaro-Algaba, M.; Ramirez, F.J. Techno-economic and environmental disassembly planning of lithium-ion electric vehicle battery packs for remanufacturing. *Resour. Conserv. Recycl.* **2020**, *154*, 104461. [CrossRef]
4. Pistoia, G.; Liaw, B. *Behaviour of Lithium-Ion Batteries in Electric Vehicles: Battery Health, Performance, Safety, and Cost*; Springer: Berlin/Heidelberg, Germany, 2018.
5. Li, J.; Barwood, M.; Rahimifard, S. Robotic disassembly for increased recovery of strategically important materials from electrical vehicles. *Robot. Comput. Integr. Manuf.* **2018**, *50*, 203–212. [CrossRef]
6. Harper, G.; Sommerville, R.; Kendrick, E.; Driscoll, L.; Slater, P.; Stolkin, R.; Walton, A.; Christensen, P.; Heidrich, O.; Lambert, S.; et al. Recycling lithium-ion batteries from electric vehicles. *Nature* **2019**, *575*, 75–86. [CrossRef] [PubMed]
7. Kroll, E.; Beardsley, B.; Parulian, A. A Methodology to Evaluate Ease of Disassembly for Product Recycling. *IIE Trans.* **1996**, *28*, 837–846. [CrossRef]
8. Mok, H.; Kim, H.; Moon, K. Disassemblability of mechanical parts in automobile for recycling. *Comput. Ind. Eng.* **1997**, *33*, 621–624.
9. Schäfer, J.; Singer, R.; Hofmann, J.; Fleischer, J. Challenges and Solutions of Automated Disassembly and Condition-Based Remanufacturing of Lithium-Ion Battery Modules for a Circular Economy. *Procedia Manuf.* **2020**, *43*, 614–619. [CrossRef]
10. Velázquez-Martínez, O.; Valio, J.; Santasalo-Aarnio, A.; Reuter, M.; Serna-Guerrero, R. A Critical Review of Lithium-Ion Battery Recycling Processes from a Circular Economy Perspective. *Batteries* **2019**, *5*, 68. [CrossRef]
11. Zhou, Z.; Liu, J.; Pham, D.T.; Xu, W.; Ramirez, F.J.; Ji, C.; Liu, Q. Disassembly sequence planning: Recent developments and future trends. *Proc. Inst. Mech. Eng. Part B J. Eng. Manuf.* **2019**. [CrossRef]
12. Xia, K.; Gao, L.; Chao, K.; Wang, L. A Cloud-Based Disassembly Planning Approach towards Sustainable Management of WEEE. In Proceedings of the 2015 IEEE 12th International Conference on e-Business Engineering, Beijing, China, 23–25 October 2015; pp. 203–208. [CrossRef]
13. Li, L.; Zheng, P.; Yang, T. Disassembly Automation for Recycling End-of-Life Lithium-Ion Pouch Cells. *J. Miner. Met. Mater. Soc.* **2019**, *71*, 4457–4464. [CrossRef]
14. Rujanavech, C.; Lessard, J.; Chandler, S.; Shannon, S.; Dahmus, J.; Guzzo, R. *Liam-An Innovation Story*; Apple Inc.: Cupertino, CA, USA, 2016.
15. Torres, F.; Puente, S.; Díaz, C. Automatic cooperative disassembly robotic system: Task planner to distribute tasks among robots. *Control Eng. Pract.* **2009**, *17*, 112–121. [CrossRef]

16. Knoth, R.; Hoffmann, M.; Kopacek, B.; Kopacek, P. Intelligent disassembly of electr(on)ic equipment. In Proceedings of the Second IEEE International Symposium on Environmentally Conscious Design and Inverse Manufacturing, Tokyo, Japan, 11–15 December 2001; pp. 557–561. [CrossRef]
17. Vongbunyong, S.; Chen, W.H. *Cognitive Robotics*; Springer: Cham, Switzerland, 2015; pp. 95–128. [CrossRef]
18. Vongbunyong, S.; Kara, S.; Pagnucco, M. Application of cognitive robotics in disassembly of products. *CIRP Ann.* **2013**, *62*, 31–34. [CrossRef]
19. Aalerud, A.; Dybedal, J.; Ujkani, E.; Hovland, G. Industrial Environment Mapping Using Distributed Static 3D Sensor Nodes. In Proceedings of the 2018 14th IEEE/ASME International Conference on Mechatronic and Embedded Systems and Applications, MESA 2018, Oulu, Finland, 2–4 July 2018; [CrossRef]
20. Redmon, J.; Farhadi, A. Yolo9000: Better faster stronger. In Proceedings of the IEEE Conference on Computer Vision and Pattern Recognition (CVPR), Honolulu, HI, USA, 21–26 July 2017; pp. 6517–6525.
21. Nixon, M.; Aguado, A. *Feature Extraction and Image Processing for Computer Vision*; Academic Press: Cambridge, MA, USA, 2019.
22. Otsu, N. A Threshold Selection Method from Gray-Level Histograms. *IEEE Trans. Syst. Man Cybern.* **1979**, *9*, 62–66. [CrossRef]
23. Simony, M.; Milzy, S.; Amendey, K.; Gross, H.M. Complex-YOLO: An Euler-Region-Proposal for Real-time 3D Object Detection on Point Clouds. In Proceedings of the European Conference on Computer Vision (ECCV) Workshops, Munich, Germany, 8–14 September 2018.
24. Li, G.; Muller, M.; Thabet, A.; Ghanem, B. DeepGCNs: Can GCNs go as deep as CNNs? In Proceedings of the IEEE International Conference on Computer Vision, Seoul, Korea, 27 October–2 November 2019; pp. 9266–9275. [CrossRef]

MDPI
St. Alban-Anlage 66
4052 Basel
Switzerland
Tel. +41 61 683 77 34
Fax +41 61 302 89 18
www.mdpi.com

Metals Editorial Office
E-mail: metals@mdpi.com
www.mdpi.com/journal/metals